Cancer Atlas of the United Kingdom and Ireland 1991–2000

Studies on Medical and Population Subjects No. 68

Editors:

Mike Quinn
Helen Wood
Nicola Cooper
Steve Rowan

palgrave
macmillan

First published 2005 by
PALGRAVE MACMILLAN
Houndmills, Basingstoke, Hampshire RG21 6XS and
175 Fifth Avenue, New York, NY 10010
Companies and representatives throughout the world.

PALGRAVE MACMILLAN is the global academic imprint of the Palgrave Macmillan division of St. Martin's Press, LLC and of Palgrave Macmillan Ltd. Macmillan® is a registered trademark in the United States, United Kingdom and other countries. Palgrave is a registered trademark in the European Union and other countries.

ISBN 1-4039-9645-8

This book is printed on paper suitable for recycling and made from fully managed and sustained forest sources.

A catalogue record for this book is available from the British Library.

10	9	8	7	6	5	4	3	2	1
14	13	12	11	10	09	08	07	06	05

Printed and bound in Great Britain by
Ashford Colour Press Ltd, Gosport.

A National Statistics publication

National Statistics are produced to high professional standards as set out in the National Statistics Code of Practice. They are produced free from political influence.

About the Office for National Statistics

The Office for National Statistics (ONS) is the government agency responsible for compiling, analysing and disseminating economic, social and demographic statistics about the United Kingdom. It also administers the statutory registration of births, marriages and deaths in England and Wales.

The Director of ONS is also the National Statistician and the Registrar General for England and Wales.

For enquiries about this publication, contact:
National Cancer Intelligence Centre
Tel: 01329 813759
E-mail: cancer@ons.gov.uk

For general enquiries, contact the National Statistics Customer Contact Centre.
Tel: **0845 601 3034** (minicom: 01633 812399)
E-mail: info@statistics.gsi.gov.uk
Fax: 01633 652747
Post: Room 1015, Government Buildings,
          Cardiff Road, Newport NP10 8XG

You can also find National Statistics on the internet at
www.statistics.gov.uk

Contents

List of tables

List of figures

List of maps

Maps in Chapters 3-23 and Appendix D illustrate the *ratio* of the directly age-standardised rate in each health authority to the relevant average rate for the UK and Ireland. Maps in Appendix E illustrate the *actual* rates. See Appendix H for further details.

Foreword

Maps – of all types – are fascinating, and are an extremely useful way of enabling the assimilation and understanding of large amounts of often complex geographical information. Many atlases of cancer incidence or mortality have been published around the world, and have led to the identification of areas with previously unsuspected high rates or to advances in knowledge about the causes of cancer.

Two cancer atlases for England and Wales (one of incidence and one of mortality) and one for Scotland (of incidence) have previously been published, and the UK was included in a cancer mortality atlas of the European Economic Community; these atlases, all published in the past 20 years, used data for different periods during the late 1960s to the mid-1980s. This atlas brings the information on geographical patterns of cancer up to date; it includes both incidence and mortality; and it covers not only all the four countries of the UK, but also Ireland.

It is well known that there are wide inequalities within the UK and Ireland in terms of who gets cancer, and what happens to them when they do. People in deprived areas are more likely to get some types of cancer and their survival from most types of cancer is lower. This atlas complements our current knowledge, and will prove invaluable in several ways. It enables rapid visual assessment of the range of variability in cancer incidence and mortality at the health authority level; it shows the locations of groups of areas adjacent to each other that have higher rates within larger areas for which the overall rate is not raised; and it identifies geographical patterns that cross administrative boundaries. The charts and maps also facilitate the assessment of the similarity – or otherwise – of the geographical patterns for diseases with related aetiology, such as those for which smoking tobacco is a major risk factor. The vast majority of cases of lung cancer are avoidable. But the wide differences around the world in the incidence of most of the other major cancers suggest that they too are largely avoidable. The results and analyses in this atlas show that despite all previous efforts to reduce the cancer burden, wide geographical differences in incidence still exist for many cancers in the UK and Ireland. This atlas highlights those cancers and areas where further education, provision of services, or attention to the environment – in the broadest sense, including diet – could markedly reduce the numbers of cancer cases and deaths.

Better recognition and understanding of the geographical patterns in cancer incidence and mortality will assist in ensuring that resources can be appropriately targeted, and that suitable baselines can be set against which the impact of policies and initiatives to tackle the problems can be measured.

This atlas was edited, collated, and in part written by staff at the National Cancer Intelligence Centre at the Office for National Statistics (ONS) with the collaboration of many experts, mostly from the cancer registries of the UK and Ireland. The authors of the 21 chapters on each of the specific types of cancer have taken particular care to discuss those aspects of data collection and data quality that may influence the interpretation of the results. They have also discussed the probable impact of the distribution of known risk factors and of socio-economic deprivation on the observed geographical patterns of cancer incidence and mortality. All of these chapters were peer reviewed by experts in the field of cancer epidemiology.

ONS and its predecessors have for many years published annual statistics on both cancer incidence and mortality (for England and Wales), and a compendium of trends in cancer incidence, mortality and survival was published in 2001. In collaboration with the London School of Hygiene and Tropical Medicine, extensive analyses of cancer survival trends over time by region and socio-economic deprivation have been published. Annual data on incidence and mortality and analyses of survival trends have also long been published for Scotland, and by the cancer registries in Northern Ireland and Ireland that began operation in the early 1990s. In addition, they and the regional cancer registries in England publish a vast amount of detailed information relating to their geographical areas, as well as conducting research which is published in peer reviewed scientific and medical journals.

So much more is known about cancer than for many other diseases because for many years population based – and hence unbiased – data have been collected and collated through the cancer registration system. The cancer registries are essential for monitoring incidence, the effectiveness of screening programmes, and outcomes – particularly survival rates in relation to treatment. The NHS Cancer Plan for England, and similar plans in the other countries of the UK, recognise that these public health benefits depend on the completeness of cancer registration, and on its quality and timeliness. The Government strengthened the cancer registries in England following the review undertaken by Professor Charles Gillis in 2000. And with the current development of large and complex IT systems in the NHS, and the general concerns about the confidentiality of patients' information, the Government is determined to secure the future of the registries. The comprehensive information on cancer mortality presented in this atlas was based on the data collected by the four General Register Offices in the UK and Ireland. The high quality of those data and the validity of the results described in this atlas are due to the expertise and vigilance of all their staff.

I warmly welcome the publication of this atlas, which expertly illustrates the detailed picture of the cancer burden in the UK and Ireland and relates the geographical patterns in all of the major cancers to known risk factors and to levels of socio-economic deprivation.

Professor M A Richards, National Cancer Director (England)

Acknowledgements

The editors in the London team of the National Cancer Intelligence Centre (NCIC) at the Office for National Statistics acknowledge with gratitude the following contributors to this atlas:

The directors and all the staff of the regional cancer registries in England and of the Welsh Cancer Intelligence and Surveillance Unit for their continued co-operation with the NCIC at ONS in the processing of the extremely large numbers of cancer registrations and death records.

All the staff of the NCIC teams in Titchfield and in Southport. The high quality of the national cancer database and the validity of the outputs based on it depend critically on their commitment and attention to detail.

The ONS staff in the General Register Office and in the Social and Vital Statistics Division for their work in collecting and processing the mortality data for England and Wales.

The directors and all the staff of the cancer registries and general register offices in Scotland, Northern Ireland and Ireland who collected the original cancer registration and mortality data for those countries; and those staff who liaised with ONS and supplied the aggregated information for inclusion in this atlas.

The authors of chapters 3-23 – their full details and affiliations are:

Peter J Adamson BSc MSc
Research Fellow, Epidemiology and Genetics Unit, University of York

David H Brewster MD MSc FFPH MRCGP DCH DRCOG
Director, Scottish Cancer Registry, Edinburgh
Honorary Clinical Senior Lecturer, Division of Community Health Sciences, University of Edinburgh

Caroline Brook BA(Hons) MSc
Information Services Manager, Northern and Yorkshire Cancer Registry and Information Service, Leeds

Ray A Cartwright MA PhD FFPH FFOM FRC(Edin)
Emeritus Professor of Cancer Epidemiology, University of Leeds

David Forman BA PhD FFPH
Director of Information and Research, Northern and Yorkshire Cancer Registry and Information Service, Leeds
Professor of Cancer Epidemiology, Centre for Epidemiology and Biostatistics, University of Leeds

Anna T Gavin MB BCH BAO MSc FFPHM
Director, Northern Ireland Cancer Registry, Belfast
Senior Lecturer, Queen's University Belfast

Robert A Haward FFPHM DPH MBChB QHP
Medical Director, Northern and Yorkshire Cancer Registry and Information Service, Leeds
Professor of Cancer Studies, University of Leeds

Richard McNally BSc MSc DIC PhD
Reader in Epidemiology, University of Newcastle upon Tyne

Henrik Møller BA BSc MSc DM FFPH
Director, Thames Cancer Registry
Professor of Cancer Epidemiology, Guy's, King's and St Thomas' School of Medicine, London; and the London School of Hygiene and Tropical Medicine

Paul Silcocks BSc BM BCh MSc FRCPath FFPH CStat
Medical Advisor, Trent Cancer Registry, Sheffield
Clinical Senior Lecturer Trent RDSU (University of Nottingham), Queen's Medical Centre, Nottingham

John Steward MBBCh BA MSc PhD FFPH
Director, Welsh Cancer Intelligence and Surveillance Unit, Velindre NHS Trust, Cardiff
Honorary Senior Lecturer, Epidemiology, Statistics, Public Health, University of Cardiff

Paul M Walsh BSc MSc PhD
Epidemiologist, National Cancer Registry (Ireland), Cork

The experts in the field of cancer epidemiology who reviewed Chapters 3-23.

Colleagues at ONS involved in the publication process:

    James Twist, Andy Leach, Tony Castro – Design
    Paul Hyatt – Publications
    Ali Dent, Pam Blunt, Jeremy Brocklehurst, Deborah Rhodes – ONS Geography
    Phil Hodgson – Editorial
    Diane Bennett, Kim Slatter, Sue Wilde, Dave Pike, Angela Cannell – Desk Top Publishing
    Marged Lloyd and Claudine Munro-Lafon – Web Team

Cindy Robinson and Maya Malagoda – former members of the NCIC team in London who worked on this atlas in its early stages.

Editors:

Mike Quinn BSc MSc PhD CStat, Director, NCIC
Helen E Wood BA MA DPhil, Research Officer, NCIC
Nicola Cooper BSc, Senior Epidemiologist, NCIC
Stephen D Rowan, Research Officer, NCIC

Cover picture: Getty Images

Abbreviations

ALL	acute lymphoblastic leukaemia
AML	acute myeloid leukaemia
CIN	cervical intraepithelial neoplasia
CLL	chronic lymphocytic leukaemia
CML	chronic myeloid leukaemia
CT	computed tomography
DCO	death certificate only
DH	Department of Health
EBV	Epstein-Barr virus
EEC	European Economic Community
FOBt	faecal occult blood test
GOR	government office region
GRO	General Register Office (England and Wales)
GRONI	General Register Office for Northern Ireland
GROS	General Register Office for Scotland
H pylori	helicobacter pylori
HD	Hodgkin's disease
HHV-6	human herpes virus type 6
HIV	human immunodeficiency virus
HPV	human papillomavirus
HRT	hormone replacement therapy
ICD	International Classification of Diseases (ICD9, ninth revision; ICD10, tenth revision)
ICDO	International Classification of Diseases for Oncology (ICDO2, second edition)
LA	local authority
MGUS	monoclonal gammopathy of unknown significance
M:I	mortality-to-incidence ratio
MRI	magnetic resonance imaging
NAW	National Assembly of Wales
NCIC	National Cancer Intelligence Centre (at the Office for National Statistics)
NCRI	National Cancer Registry of Ireland
NHL	non-Hodgkin's lymphoma
NHS	National Health Service
NHSCR	National Health Service Central Register
NICR	Northern Ireland Cancer Registry
NMSC	non-melanoma skin cancer
NSAID	non-steroidal anti-inflammatory drug
ONS	Office for National Statistics
OPCS	Office of Population Censuses and Surveys
PAS	patient administration system
PCT	primary care trust
PSA	prostate-specific antigen (test)
SCC	squamous cell carcinoma
SHA	strategic health authority
SIR	standardised incidence ratio
SMR	standardised mortality ratio
UK	United Kingdom
UKACR	United Kingdom Association of Cancer Registries
USA	United States of America
WAG	Welsh Assembly Government
WCISU	Welsh Cancer Intelligence and Surveillance Unit
WHO	World Health Organisation

Chapter 1
Introduction

Mike Quinn

Aim

The aim of this atlas is to analyse the recent broad geographical patterns – at the country, region (of England) and health authority level – in the incidence of, and mortality from, 21 common cancers in the UK and Ireland in the last ten years of the twentieth century, and to relate these patterns to variations in known aetiological (risk) factors and socio-economic deprivation.

Background

Cancer is a major cause of morbidity and mortality in the UK and Ireland. During the 1990s, there were on average about 270,000 new cases of cancer diagnosed each year (excluding non-melanoma skin cancer – see Appendix G), with almost equal numbers in males and females. There were on average almost 165,000 deaths from cancer each year, with slightly more deaths in males than females. Although mortality from infectious diseases declined rapidly to very low rates during the 1950s, and mortality from heart disease and stroke fell – albeit more gradually – in both males and females, mortality from cancer changed relatively little in both males and females during the second half of the twentieth century. In England and Wales, cancer became the most common cause of death in females in the late 1960s, and in males in the mid-1990s.[1]

Information on the geography, population and economy of each of the five countries covered in this atlas, and on the regions of England, is given in Appendix I; highly detailed information about Scotland was given in the *Atlas of Cancer in Scotland* (published in 1985).[2] Further information on the regions of England can be found in *Regional Trends*.[3]

Disease mapping

Maps enable a rapid visual summary to be made of large amounts of complex geographical information. Disease maps are more powerful than either a table or a bar chart of disease levels for small areas because they intrinsically give the reader additional information on whether or not two (or more) areas are adjacent to each other – and if not, how far apart they are. A comprehensive historical survey of disease mapping has been given by Walter.[4] Disease mapping began in the early nineteenth century. It is used to describe and evaluate geographical patterns, generate hypotheses about risk factors and causes (aetiology),[5] highlight areas at apparently high risk, and to aid policy making and resource allocation.[2,6-8]

One of the most famous examples of disease mapping is the 'spot' map produced by John Snow in the 1850s,[9] with which he demonstrated the spread of cholera through contaminated water from the Broad Street pump in London. Spot maps, however, show only the location of cases and do not take into account the size and characteristics of the underlying populations. There was great interest in infectious diseases towards the end of the nineteenth century, but Havilland[10] produced maps of chronic diseases, including cancer, in England and Wales using mortality data for 1851-60. He was among the first to take population denominators into account, calculating crude mortality rates using population data from a census. Another early example was the map of cancer mortality in Switzerland in the period 1901-10.[11]

Swerdlow and dos Santos Silva have given a detailed history of the mapping of cancer in England and Wales.[6] The first maps for specific cancer sites were produced by Greenwood in 1925. These showed mortality from cancers of the breast and uterus by county over the period 1911-20. A major series of cancer maps was subsequently published by Stocks, including mortality from all cancers combined by county[12,13] and 28 cancer-specific maps by county,[14-16] in which adjustment was made for differences in age and sex between areas. Maps of mortality for 13 major causes of death, including some cancers, using local authority boundaries, were published in 1963[17] and updated in 1970.[18] Maps of mortality from 16 cancers over the period 1969-73 at the regional level were published by the Office of Population Censuses and Surveys (OPCS) in a decennial supplement.[19]

Over the past 20 years, publications of cancer atlases covering all or part of the UK have included:

- *Atlas of Cancer Mortality in England and Wales 1968-1978*, by Gardner et al;[20] this presented maps of mortality from all cancers combined and 13 cancer sites at the local authority level (at that time there were almost 1,400 such areas, compared with the current 370 or so), and mortality maps for the same 13 cancers and 23 others at the county level.

- *Atlas of Cancer in Scotland 1975-1980. Incidence and Epidemiological Perspective*, edited by Kemp et al;[2] this was the first atlas of cancer incidence for anywhere in the UK, with maps for about 40 different cancer sites, for 56 administrative areas.

- *Atlas of Cancer Incidence in England and Wales 1962-85*, by Swerdlow and dos Santos Silva;[6] maps, again for about 40 different cancers, were at the county level.

The book *Mortality and Geography - a Review in the mid-1980s*, edited by Britton,[21] covered England and Wales and included stomach cancer in 1921-30, 1950-53 and 1979-83 at the county level, and lung cancer in 1946-49 and 1979-83 in London boroughs. *Geographic Variations in Health*, edited by Griffiths and Fitzpatrick,[22] covered cancer incidence in 1991-93 for Great Britain (Chapter 9) and cancer mortality in 1991-97 for the UK (Chapter 10) for the three most common cancers in each sex (breast, colorectal, lung and prostate); maps were at the local authority level.

In addition, the *Atlas of Cancer Mortality in the European Economic Community*, edited by Smans et al[7] included the UK along with the (then eight) other members of the EEC with data for 23 types of cancer for varying periods in the 1970s, at various small-area levels – the UK data were for 1976-80 at the county level. An atlas of cancer mortality that will include all 25 current (2005) members of the European Union with data at small-area levels (broadly equivalent to health or local authority in the different countries) for 1993-97 is in preparation.[23]

Many cancer atlases for other countries were published during the 1980s. Smans et al[7] gave details of about 30 such atlases (incidence or mortality), and Walter and Birnie compared the characteristics of many of these (and others).[24] They found that there were wide differences between the atlases in almost all methodological and presentational aspects, including: the types and grouping of cancers; information on case numerators or population denominators; the size of the 'small' areas (many were so small that random variation dominated the rates); criteria for selection of cancers; the measure of cancer risk being mapped; the method of presenting the results on the maps; the use of smoothing of small-area rates (few did); age standardisation (many used the indirect method); the use of colour (half were black and white), and of the colour maps there were differences in the colours used to indicate high, average and low risk; and assessment of time trends and spatial patterns. Few of the atlases attempted any substantive aetiological interpretation or provided supplementary data or maps on important factors such as smoking or socio-economic status.

Several cancer atlases have been published since the early 1990s.[25-37] Three of these incorporated innovative or unusual features that had particular influence on the methodology and presentation used in this atlas. The atlas of cancer mortality (and other causes) in Spain[30] presented maps for about 40 cancer sites at the level of the 52 provinces. It includes separate maps of age-standardised rates and of time trends and an

examination of the spatial patterns using the rank adjacency statistic pioneered in the *Atlas of Cancer in Scotland*.[2] For the USA, owing to the limited availability of cancer incidence data,[38] cancer mortality data (along with mortality from other diseases) were mapped at the level of health service areas aggregated to a minimum size of 250 square miles.[31] Maps showed age-standardised death rates relative to the USA average, and a separate (quarter size) map identifying the health service areas with statistically significantly high rates, and within these the 80 health service areas with the highest and lowest values. Smoothed maps (again quarter size) were produced for death rates at particular ages. The *Cancer Atlas of Northern Europe*[33] included both incidence and mortality where possible. Data for quite small areas were smoothed using a weighting function (the weight declined with distance to 0.5 at 25 km and to zero at 150 km). Maps were produced with both a relative scale and an absolute scale. Rates for cities with populations of over 350,000 were shown as disks with a diameter proportional to the population. As there are many geographically large areas with very sparse populations in Northern Europe, white hatching was used to de-emphasise the colours for these areas that would otherwise have unrealistically dominated the appearance of the maps.

In addition, worldwide figures for national cancer mortality and incidence (estimated for many countries from their mortality data) have been published in GLOBOCAN.[39]

This cancer atlas is the first to cover any part of the UK that has included both incidence and mortality for a large number of cancers (almost 90 per cent of all cancer cases), and the first of any kind to cover the UK and Ireland.

The major potential problems involved in constructing a cancer atlas and analysing the geographical patterns include (numerator) data quality and completeness, estimation of the (denominator) population at risk, and identification of 'true' excess rates. All these problems become more acute the smaller the geographical areas being considered, particularly because the numbers of cases (or deaths) tend to be very small and apparently large differences between areas across a map may simply reflect random variation.[8] In addition to the many methodological problems involved, 'presentational' issues affect not just the appearance but also the utility of disease maps.[7,40-42]

Together with our colleagues in the United Kingdom Association of Cancer Registries (UKACR – see Appendix J), we collated the most up-to-date cancer incidence and mortality data available. We carefully considered all the methodological and presentational issues, and adopted and adapted what we thought were some of the best features from the atlases mentioned above.

Appendix K contains a discussion of the range of methodological issues involved in producing a cancer atlas, and explains our reasoning for the decisions we took about: the time periods covered; the cancers included; data collection and quality (incidence, mortality and populations); measures of incidence and mortality for comparison of areas; the choice of 'small' geographical area; the provision of detailed tables of incidence, mortality and populations by small area; and divergences from the guidelines for disease atlases set out by Walter and Birnie.[24]

The uses of cancer atlases

Cancer atlases often have as a main aim or justification that their real purpose lies in identifying geographical areas or hypotheses that require more detailed epidemiological study.[2,7] But, as Barker noted about 25 years ago, 'maps of disease compel speculation about aetiology but only rarely has such speculation by itself led directly to the discovery of causes'.[43]

There are a few examples of where atlases or other geographical investigations have led to the identification of high-risk areas and/or advances in aetiological knowledge. These include: the highlighting of many previously unsuspected areas of high cancer risk in China following the production of a cancer mortality atlas in the mid-1970s;[44] the uncovering of a number of areas in the USA with high cancer death rates following the publication of cancer mortality atlases,[4,42] which led to some aetiological discoveries;[45] the elevated risks of nasal cancer in workers in the furniture and leather industries in England and Wales;[46] and the identification of the mineral erionite as a carcinogen and a cause of respiratory cancer following investigations of reports of high mortality in a Turkish village.[47] Neutra[48] pointed out in 1990 that erionite was, however, the only one of 35 agents for which the International Agency for Research on Cancer then believed there was sufficient evidence for carcinogenicity in humans that was discovered in this way – the others were all classified on the basis of medical or occupational clusters.[49] The populations affected in these examples were generally small, or the types of cancer fairly rare, and so the overall excesses of cancer cases or deaths were small. Areas of high risk can be just as easily identified from a table of rates as from a map, especially if the rates are ranked in descending order.

Major risk factors are known for many cancers, for example: smoking for lung cancer (and cancers of the bladder, kidney, larynx, lip, mouth and pharynx, oesophagus and stomach); hormonal and reproductive factors for cancers of the breast, ovary and uterus; and diets high in fats and animal proteins and low in fruit, vegetables and fibre for colorectal cancer. Clearly there is now little aetiological insight to be gained from

atlases of cancer incidence or mortality in countries such as the UK and Ireland where such atlases covering earlier years of data have already been published, and there is a long history of cancer registration which has enabled the identification of areas of high risk for cancers or areas for which the incidence trends are rising.

The value of this and similar cancer atlases lies in enabling:

-   the rapid visual confirmation of the range of variability at the small-area level indicated in the charts displaying the average rates with their confidence intervals (Figures x.3 and x.4 in Chapters 3 to 23 – see also Appendix H);

-   assessment of whether there are groups of small areas adjacent to each other which have higher rates within larger regions or countries where the overall rate is not elevated;

-   the identification of any patterns which cross the administrative or other artificial boundaries used;

-   rapid visual assessment of the similarity, or otherwise, of geographical patterns for diseases with related aetiology;

-   the relating of any observed incidence patterns to geographical variation in known risk factors and to socio-economic deprivation;

-   the planning of health education campaigns, the provision of health services,[2] and actions to reduce environmental hazards;[6] and

-   a focus on geographical variations in the potential for reductions in cancer incidence and mortality.

Geographical patterns in cancer incidence and mortality in the UK and Ireland

Each of Chapters 3 to 23 covers a single type of cancer. A brief description is given of the annual average numbers of cases and deaths and the corresponding incidence and mortality rates, of any trends in these over time, and of survival rates. The geographical variations in incidence and mortality at the country, and regional level in England are illustrated in bar charts, and variations at the health authority level are illustrated in charts showing the rate for each area within its country or region, and in maps. Cognitive research on disease maps[50,51] has shown that a 'double-ended' scale best illustrates what, and where, are the patterns in the high and low rates, and whether the patterns are similar for males and females, for incidence and mortality, and for different diseases. The maps in Chapters 3 to 23 indicate where rates are above or below the relevant overall average rate for the UK and Ireland, banded into ratios of 1.1 to 1.33, 1.33 to 1.5, and more than 1.5 above or below the average (see Appendix H for details). The maps are coloured conventionally with a 'red' colour (purple) for high rates and blue for low, the depth of colour indicating the extent of the difference of the rate from the overall average. The risk factors for the particular cancer, and their relationship to the observed geographical patterns, are discussed, and the relationships (if any) between the patterns in the incidence and mortality and socio-economic deprivation are considered. Chapters 3 to 23 were peer reviewed by experts on the epidemiology of the relevant cancer.

In addition to the 'ratio' (or 'relative') maps in each cancer-specific chapter, further maps using the same 'absolute' scale for every cancer are presented in Appendix E; these are coloured from white at the low end to red and violet at the high end. These show, for example, that for colorectal cancer (for which the maps are mostly red) there is some geographical variation in incidence, but the rates are high across the whole of the UK and Ireland compared with a less common cancer such as brain (for which the maps are mostly orange or yellow). See Appendix H for further details of the two different types of maps.

Chapter 2 starts with some background on the trends and patterns in the major causes of death and in the incidence of, and mortality from, the major cancers. There is then a brief description of the patterns in the incidence of, and mortality from, all cancers combined. This is followed by a summary of the results from the 21 cancer-specific chapters. The cancers are considered in six broad groups: those strongly related to smoking and/or drinking alcohol; gastrointestinal cancers; cancers occurring only (or predominantly) in one sex; urological cancers; brain cancer and malignant melanoma of the skin; and lymphomas and leukaemias. The geographical variations in the potential for the prevention of cancer cases and deaths are then considered.

References

1. Quinn MJ, Babb PJ, Brock A, Kirby L, Jones J. *Cancer Trends in England and Wales 1950-1999.* Studies on Medical and Population Subjects No. 66. London: The Stationery Office, 2001.

2. Kemp I, Boyle P, Smans M, Muir C (eds). *Atlas of Cancer in Scotland 1975-1980. Incidence and epidemiological perspective.* IARC Scientific Publications No. 72. Lyon: International Agency for Research on Cancer, 1985.

3. McGinty J, Williams T. *Regional Trends 2001 edition No. 36.* London: The Stationery Office, 2001.

4. Walter SD. Disease mapping: a historical perspective. In: Elliott P, Wakefield JC, Best NG, Briggs DJ (eds). *Spatial Epidemiology - Methods and Applications.* Oxford: Oxford University Press, 2000.

5. Blot WJ, Fraumeni Jr JF, Mason TJ, Hoover RN. Developing clues to environmental cancer: a stepwise approach with the use of cancer mortality data. *Environmental Health Perspectives* 1979; 32: 53-58.

6. Swerdlow AJ, dos Santos Silva I. *Atlas of Cancer Incidence in England and Wales 1962-85.* Oxford: Oxford University Press, 1993.

7. Smans M, Muir CS, Boyle P (eds). *Atlas of Cancer Mortality in the European Economic Community.* IARC Scientific Publications No. 107. Lyon: International Agency for Research on Cancer, 1992.

8. Elliott P, Wakefield JC, Best NG, Briggs DJ. Spatial epidemiology: methods and applications. In: Elliott P, Wakefield JC, Best NG, Briggs DJ (eds). *Spatial Epidemiology - Methods and Applications.* Oxford: Oxford University Press, 2000.

9. Snow J. On the mode of communication of cholera. London: Churchill, 1855. Reproduced in: Buck C, Llopis A, Najera E, Terris M (eds). *The Challenge of Epidemiology - Issues and Selected Readings.* Pan-American Health Organization Scientific Publication No. 505. Washington DC, USA: PAHO and WHO, 1985. Map reproduced in: Jarup L. The role of geographical studies in risk assessment. In: Elliott P, Wakefield JC, Best NG, Briggs DJ (eds). *Spatial Epidemiology - Methods and Applications.* Oxford: Oxford University Press, 2000.

10. Havilland A. *The geographical distribution of heart disease and dropsy, cancer in females and phthisis in females, England and Wales.* London: Smith, Elder & Co., 1875.

11. *Atlas graphique et statistique de la suisse.* Statistique de la Suisse No. 191. Berne: Bureau de statistique de Département federal de l'intérieur, 1914.

12. Stocks P. *On the evidence for a regional distribution of cancer prevalence in England and Wales. Report of the International Conference on Cancer.* London: British Empire Campaign, 1928.

13. Stocks P, Karn MN. The distribution of cancer and tuberculosis mortality in England and Wales. *Annals of Eugenics* 1930-31; 4: 341-361.

14. Stocks P. Distribution in England and Wales of cancer of various organs. *Annual Report.* London: British Empire Campaign, 1936.

15. Stocks P. Distribution in England and Wales of cancer of various organs. *Annual Report.* London: British Empire Campaign, 1937.

16. Stocks P. Distribution in England and Wales of cancer of various organs. *Annual Report.* London: British Empire Campaign, 1939.

17. Howe GM. *National atlas of disease mortality in the United Kingdom.* London: Nelson, 1963.

18. Howe GM. *National atlas of disease mortality in the United Kingdom (second Edition).* London: Nelson, 1970.

19. Office of Population Censuses and Surveys. *Area Mortality Decennial Supplement 1969-73, England and Wales.* OPCS Series DS No. 4. London: HMSO, 1981.

20. Gardner MJ, Winter PD, Taylor CP, Acheson ED. *Atlas of Cancer Mortality in England and Wales, 1968-1978.* Chichester: John Wiley & Sons, 1983.

21. Britton M (ed). *Mortality and Geography - a Review in the mid-1980s.* OPCS Series DS No.9. London: HMSO, 1990.

22. Griffiths C, Fitzpatrick J. *Geographic Variations in Health*. Decennial Supplements No. 16. London: The Stationery Office, 2001.

23. Boyle P and Smans M, in collaboration with Benichou J, Boniol M, Gillis C, La Vecchia C et al. *Cancer Mortality Atlas of European Union and European Economic Area Member States, 1993-1997*: To be published by Oxford University Press, in preparation.

24. Walter SD, Birnie SE. Mapping mortality and morbidity patterns: an international comparison. *International Journal of Epidemiology* 1991; 20: 678-689.

25. Baburin A, Gornoi K, Leinsalu M, Rahu M. *Atlas of Mortality in Estonia*. Tallinn: Institute of Experimental and Clinical Medicine, 1997.

26. Health and Welfare Canada. *Mortality Atlas of Canada*. Ottawa: Government of Canada, 1991.

27. Hisamichi S, Tsuji I, Sauvaget C, Sakka M. *Geographical Distribution of Cancer Mortality in Japan: Standardized Mortality Ratio at all Municipalities from 1980 to 1984*. Sendai: Tohoku Radiological Science Centre, 2000.

28. Le ND, Marrett LD, Robson DL, Semenciw R et al. *Canadian Cancer Incidence Atlas*. Ottawa: Government of Canada, 1996.

29. Marrett LD, Nishri ED, Swift MB, Walter SD et al. *Geographical Distribution of Cancer in Ontario. Vol II: Atlas of Cancer Incidence 1980-91*. Toronto: Ontario Cancer Treatment and Research Foundation, 1995.

30. Ortega GL-A, Santamaría MP, Pujolar AE, Saizar ME et al. *Atlas de Mortalidad per Cáncer y Otras Causes en España 1978-1992*. Madrid: Fundacíon Científica de la Asociación Española contra el Cáncer y Instituto de Salud Carlos III, 2001.

31. Pickle LW, Mungiole M, Jones GK, White AA. *Atlas of United States Mortality*. Hyattsville MD: National Centre for Health Statistics, 1996.

32. Devesa SS, Grauman DG, Blot WJ, Pennello GA et al. *Atlas of Cancer Mortality in the United States: 1950-94*. Washington DC: US Government Print Office, 1999.

33. Pukkala E, Söderman B, Okeanov A, Storm H et al. *Cancer Atlas of Northern Europe*. Cancer Society of Finland Publication No.62. Helsinki: Cancer Society of Finland, 2001.

34. Regional Oncologic Centres. *Atlas of Cancer Incidence in Sweden*. Stockholm: Karolinska Hospital, 1996.

35. World Health Organisation. *Atlas of Mortality in Europe*. WHO Regional Publications, European Series No.75. Geneva: WHO, 1997.

36. Zatonski W, Pukkala E, Didkowska J, Tyczynski J et al. *Atlas of Cancer Mortality in Poland 1986-90*. Warsaw: Cancer Centre, 1993.

37. Zatonski W, Smans M, Tyczynski J, Boyle P (eds). *Atlas of Cancer Mortality in Central Europe*. IARC Scientific Publications No.134. Lyon: International Agency for Research on Cancer, 1996.

38. National Cancer Institute. Surveillance, Epidemiology and End Results (SEER). 2005. Available at *http://seer.cancer.gov/*.

39. Ferlay J, Parkin DM, Pisani P. *GLOBOCAN 1: Cancer Incidence and Mortality Worldwide*. Lyon: IARC Press, 1998.

40. Smans M, Estève J. Practical approaches to disease mapping. In: Elliott P, Cuzick J, English D, Stern R (eds). *Geographical and Environmental Epidemiology: Methods for Small-Area Studies*. Oxford: Oxford University Press, 1992.

41. Semenciw RM, Le ND, Marrett LD, Robson DL et al. Methodological issues in the development of the Canadian Cancer Incidence Atlas. *Statistics in Medicine* 2000; 19: 2437-2449.

42. Pickle LW. Mapping mortality in the United States. In: Elliott P, Wakefield JC, Best N, Briggs DJ (eds). *Spatial Epidemiology - Methods and Applications*. Oxford: Oxford University Press, 2000.

43. Barker DJP. Geographical variations in disease in Britain. *British Medical Journal* 1981; 283: 398-400.

44. The editorial committee for the Atlas of Cancer Mortality in the People's Republic of China. *Atlas of Cancer Mortality in the People's Republic of China*. Beijing: China Map Press, 1979.

45. Anderson L. *Research contributions made possible by the NCI cancer atlases published in the 1970s*. Bethesda MD: National Cancer Institute, Office of Cancer Communications, 1987.

46. Gardner MJ, Winter PD, Acheson ED. Variations in cancer mortality among local authority areas in England and Wales: relations with environmental factors and search for causes. *British Medical Journal* 1982; 284: 784-787.

47. Baris YI, Simonato L, Saracci R, Winkelmann R. The epidemic of respiratory cancer associated with erionite fibres in the Cappadocian region of Turkey. In: Elliott P, Cuzick J, English D, Stern R (eds). *Geographical and Environmental Epidemiology: Methods for Small-Area Studies*. Oxford: Oxford University Press, 1992.

48. Neutra RR. Counterpoint from a cluster buster. *American Journal of Epidemiology* 1990; 132: 1-8.

49. Wakefield JC, Kelsall JE, Morris SE. Clustering, cluster detection, and spatial variation in risk. In: Elliott P, Wakefield JC, Best NG, Briggs DJ (eds). *Spatial Epidemiology - Methods and Applications*. Oxford: Oxford University Press, 2000.

50. Herrmann DJ, Pickle LW. A cognitive subtask model of statistical map reading. *Visual Cognition* 1996; 3: 165-190.

51. Pickle LW, Herrmann DJ (eds). *Cognitive aspects of statistical mapping*. Working Paper Series Report 18. Hyattsville MD: USA: National Center for Health Statistics, 1995.

Chapter 2

Geographical patterns in cancer in the UK and Ireland

Mike Quinn, Helen Wood, Steve Rowan, Nicola Cooper

Summary

- Incidence and mortality for cancers strongly related to smoking and alcohol (larynx; lip, mouth and pharynx; lung) were generally below average in the south and midlands of England and higher in a band across the formerly highly industrialised north of England and across the central belt of Scotland.

- Although there were clear differences in the trends over time in the incidence of cancers of the bladder, kidney, oesophagus and stomach compared with those for lung cancer, their geographical patterns were somewhat similar.

- There was little geographical variation in the incidence of cancers of the breast (in women), ovary and prostate; and virtually none in mortality. The incidence of cancer of the uterus was slightly higher than average in the south and midlands of England and below average in the rural north of England and in much of Scotland, Northern Ireland and Ireland.

- Very wide geographical variations existed in the incidence of cervical cancer, with much higher than average rates in the urban West Midlands, in a band across the north of England, and in parts of Scotland. The geographical patterns were not related to local differences in the uptake or efficiency of the cervical screening programme.

- Incidence and mortality for all cancers combined varied little across the whole of the UK and Ireland, because high rates in some cancers were often balanced by low rates in others.

- For several major cancers which have well-defined risk factors, wide geographical variations existed in incidence and mortality across the UK and Ireland. Reducing the incidence and mortality rates everywhere to those found in the areas with the lowest rates would prevent about 25,600 cases of cancer and 17,500 deaths from cancer each year. Around three quarters of these would be in cancers related to smoking tobacco or drinking alcohol.

Introduction

This chapter is in four main parts. It starts with some background on the trends and patterns in the major causes of death and in the incidence of, and mortality from, the major cancers, and on socio-economic deprivation. This is followed by a brief description of the geographical patterns in all cancers combined. The results from the 21 cancer-specific chapters are then summarised and, where appropriate, compared and contrasted. Finally, the potential for the prevention of cancer cases and deaths, based on the observed geographical variations, is considered.

Background

Over the 50-year period 1950-99, age-standardised mortality from all causes of death in England and Wales fell by around 45 per cent in both males and females;[1] the trends were similar in Scotland.[2] There were large declines in mortality from heart disease, stroke and infectious diseases at different times in males and females for each disease and with varying rates of decline (Figure 2.1). In contrast, age-standardised mortality from cancer in both males and females changed relatively little during the 50-year period. There was a slight increase for males during the 1950s and 1960s to a plateau in the 1970s, followed by a slight decline, but the rate remained mostly between 250 and 280 per 100,000. The rate in females also rose slightly to a plateau – in the late 1980s – followed by a slight decline, but remained mostly in the range 170 to 180 per 100,000 throughout the period. As a result of the large reductions in mortality from the other major causes, the proportion of deaths due to cancer rose over the period from 15 to 27 per cent in males and from 16 to 23 per cent in females. Cancer became the most common cause of death in females in 1969 and in males in 1995. (All the mortality rates have been adjusted for coding and other procedural changes in 1984 and 1992.[3,4,5])

In contrast to the relatively stable mortality trends, the registrations of new cases of all cancers combined (excluding non-melanoma skin cancer – see Appendix G) have risen since the early 1970s by about 20 per cent in males and 30 per cent in females[1] (Figure 2.2). The real increases in the overall incidence of cancer are lower than these figures suggest, however, because the trends are inflated by improvements in diagnostic techniques for many cancer types and increased ascertainment of cases by the cancer registries.

The overall trends for cancer incidence and mortality are, however, case-weighted mixes of the widely varying trends in the different cancer types (Figures 2.3 and 2.4).

Summary

- In the last 50 years of the twentieth century, mortality from heart disease, stroke and infectious diseases fell markedly in both males and females. In contrast, mortality from cancer was fairly stable. Consequently, in England and Wales, cancer became the most common cause of death in females in 1969 and in males in 1995.

- The incidence of all cancers combined in England and Wales rose gradually over the 1970s and 1980s by about 20 per cent in males and 30 per cent in females, then levelled off in the 1990s. The trends were similar in Scotland. The patterns were the result of the combination of very widely different trends for most of the major cancers, improvements in diagnostic techniques, and some increase in case ascertainment by the cancer registries.

- Cancer is predominantly a disease of elderly people. Cancers in children (0-14 years) account for about 0.5 per cent of the total. Incidence rates in women initially rise more steeply with age than in men, but rates in elderly men are double those in elderly women.

Cancers of the lung, breast, prostate and colorectum constitute about 50 per cent of both the total cases of, and deaths from, cancer (Table 2.1). Over the period 1950-99, lung cancer incidence and mortality in England and Wales declined markedly in men following earlier reductions in the prevalence of smoking, but in women rates increased gradually to a plateau in the 1990s. The incidence of breast cancer in women rose steadily, with additional increases in the early 1990s resulting from the bringing forward of diagnosis during the first, 'prevalent' round of the national breast screening programme which covered women aged 50-64. Mortality from breast cancer rose gradually to reach a plateau in the late 1980s, when the rate was among the highest in the world. By the end of the 1990s, mortality had fallen by just over 20 per cent – about one third of this was directly due to the screening programme, the remainder to improvements in chemotherapy, the increasingly widespread use of adjuvant tamoxifen, and to indirect effects of the screening programme.[6] The incidence of colon cancer increased slightly in men but not in women, and rectal cancer rates remained stable in both sexes; mortality for

Figure 2.1a

Selected major causes of death: age-standardised[1] mortality trends
Males, England and Wales 1950-2003

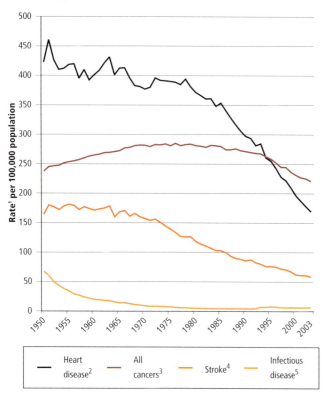

Figure 2.1b

Selected major causes of death: age-standardised[1] mortality trends
Females, England and Wales 1950-2003

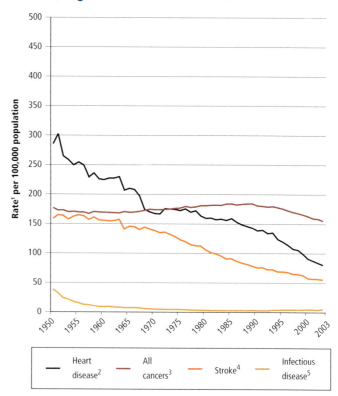

1 Directly age standardised using the European standard population

2 Ischaemic heart disease (ICD10 codes I20-I25)

3 All malignant and benign neoplasms (ICD10 codes C00-D48)

4 Cerebrovascular disease (ICD10 codes I60-I69)

5 Infectious and parasitic diseases (ICD10 codes A00-B99)

1 Directly age standardised using the European standard population

2 Ischaemic heart disease (ICD10 codes I20-I25)

3 All malignant and benign neoplasms (ICD10 codes C00-D48)

4 Cerebrovascular disease (ICD10 codes I60-I69)

5 Infectious and parasitic diseases (ICD10 codes A00-B99)

both colon and rectal cancers in both men and women declined steadily from the 1950s onwards. The incidence of prostate cancer increased slowly up to the early 1990s but over the following ten years rates doubled owing to the increasingly widespread use of prostate-specific antigen (PSA) testing. The PSA test enables invasive prostate cancer to be identified earlier than it might otherwise have been, but it also identifies latent tumours that may never have caused symptoms and been diagnosed during the man's lifetime; mortality increased in parallel with incidence up to the early 1990s and then stabilised.

Both the incidence of, and mortality from, stomach cancer have declined steadily for a very long time in both males and females. The incidence of bladder cancer in males increased steadily in England and Wales and in Scotland during the 1970s

and 1980s and then fell slightly. The sharper downturn in rates after 1999 is largely due to the alignment of coding practices among the cancer registries (see Chapter 3). Mortality from bladder cancer in males declined very gradually. The incidence of ovarian cancer increased very slowly up to the late 1990s, while mortality declined slightly in the 1970s and then remained stable.

Notable trends in the less common cancers include the increase in incidence of melanoma of the skin by a factor of around four since the early 1970s, although due to improving survival rates, mortality has risen less. The incidence of non-Hodgkin's lymphoma (NHL) has increased by a factor of two and a half. The incidence of testicular cancer has doubled, and there were on average about 1,800 cases each year in the UK and Ireland in the 1990s. But with large improvements in survival in the

Figure 2.2a

All cancers[1]: age-standardised[2] incidence and mortality, by sex
England and Wales 1950-2003

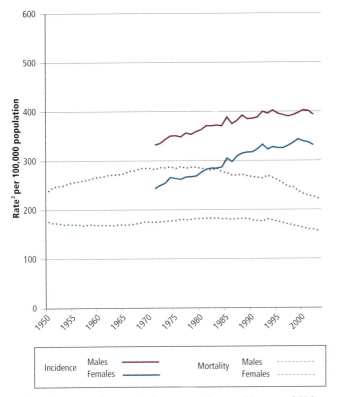

1 All malignant neoplasms excluding non-melanoma skin cancer (ICD9 codes 140-208 excluding 173; ICD10 codes C00-C97 excluding C44)
2 Directly age standardised using the European standard population

Figure 2.2b

All cancers[1]: age-standardised[2] incidence and mortality, by sex
Scotland 1950-2003

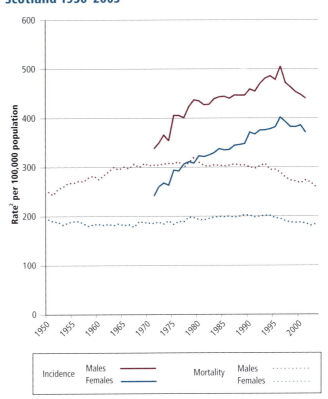

1 All malignant neoplasms excluding non-melanoma skin cancer (ICD9 codes 140-208 excluding 173; ICD10 codes C00-C97 excluding C44)
2 Directly age standardised using the European standard population

1970s resulting from the use of serum markers for its detection and the introduction of treatment with platinum-based drugs which are effective against metastatic disease, there are now only about 100 deaths from testicular cancer each year. The incidence of cancer of the uterus increased slightly. For cervical cancer, incidence was fairly stable up to the late 1980s, but this concealed very large differences in risk by birth cohort. There is no doubt that if the cervical screening programme had not

been re-organised in the late 1980s and coverage increased to the level of around 85 per cent, there would have been an epidemic of cervical cancer owing to the high risk levels in women born from the late 1940s onwards.[7-9] Instead of increasing, the incidence of cervical cancer actually fell dramatically, by around 50 per cent compared with the levels in the 1980s.

Figure 2.3a

Selected major cancers: age-standardised[1] incidence trends
Males, England and Wales 1971-2002

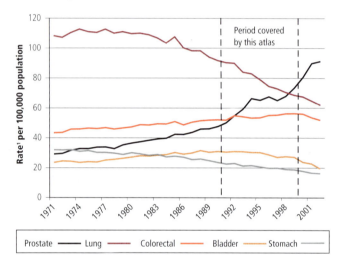

Figure 2.3b

Selected major cancers: age-standardised[1] incidence trends
Females, England and Wales 1971-2002

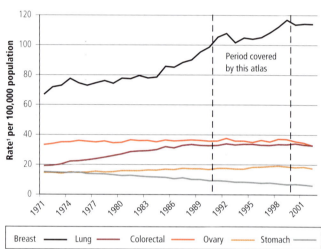

Figure 2.4a

Selected major cancers: age-standardised[1] mortality trends
Males, England and Wales 1971-2003

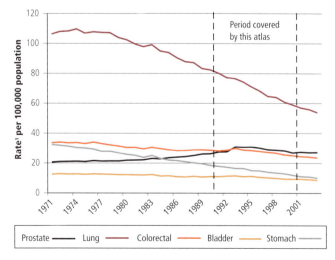

Figure 2.4b

Selected major cancers: age-standardised[1] mortality trends
Females, England and Wales 1971-2003

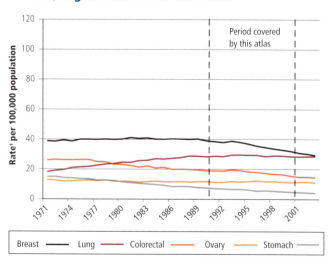

1 Directly age standardised using the European standard population

Cancer is predominantly a disease of elderly people. The age-specific rates rise continuously from around the age of 30 for both males and females, although the rates initially rise more steeply in females than in males (Figure 2.5). The rate in women aged 40-44 is about double that in males, but subsequently the rates rise more rapidly for males and the rate in men aged 80-84 is about double that in women. The age distribution of cases is shown in Figure 2.6. Cancers in children (0-14 years) account for about 0.5 per cent of all cancers, and the proportions occurring in people under 50 are about 8 per cent in males and 14 per cent in females.

The average numbers of cases and deaths for the 21 major cancers included in this atlas are given in Tables 2.1 and 2.2, respectively; the corresponding age-standardised incidence and mortality rates are illustrated in Figures 2.7 and 2.8, respectively (in descending order of frequency). These cancers

constitute just under 90 per cent of the total cancer cases, and about 85 per cent of the total cancer deaths in the UK and Ireland.

Socio-economic deprivation

In the early 1990s, there was a clear north-south divide in Great Britain in socio-economic deprivation measured by the Carstairs index,[10] which is based on four variables from the census – low social class, unemployment, overcrowding and no access to a car (see Appendix F). In 1991, almost 50 per cent of the population of Scotland lived in wards that were in the most deprived fifth of the distribution of the index;[11] the levels of deprivation were as high in the North East of England, and only slightly lower in the North West and in Yorkshire and the Humber (Figure 2.9). Only one tenth of the population in the East, South East and South West regions lived in wards which were in the lowest fifth of the distribution of deprivation. In

Figure **2.5**

All cancers[1]: age-specific incidence by sex and age group, UK and Ireland 1991-99[2]

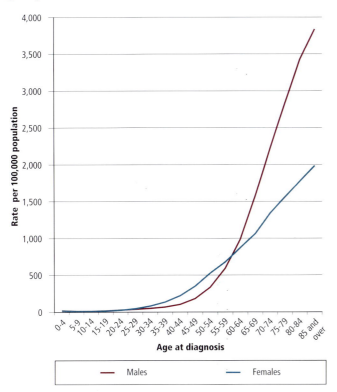

1 *All malignant neoplasms excluding non-melanoma skin cancer (ICD9 codes 140-208 excluding 173; ICD10 codes C00-C97 excluding C44)*
2 *Northern Ireland 1993-99, Ireland 1994-99*

Figure **2.6**

All cancers[1]: frequency distribution of cases by sex and age group, UK and Ireland 1991-99[2]

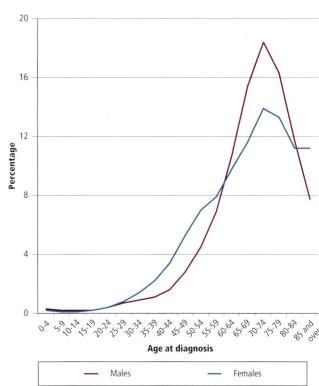

1 *All malignant neoplasms excluding non-melanoma skin cancer (ICD9 codes 140-208 excluding 173; ICD10 codes C00-C97 excluding C44)*
2 *Northern Ireland 1993-99, Ireland 1994-99*

Table 2.1

Incidence of the 21 cancers covered in Chapters 3-23, UK and Ireland, 1991-99[1]

Males

Rank	Site	Number of cases[2]	% of all cancers
1	Lung	27,000	20.2
2	Prostate	22,500	16.8
3	Colorectal	17,600	13.1
Sub-total 1-3		67,100	50.1
4	Bladder	9,700	7.2
5	Stomach	7,000	5.2
6	Non-Hodgkin's lymphoma	4,500	3.4
7	Oesophagus	4,200	3.2
8	Leukaemia	3,800	2.9
9	Pancreas	3,500	2.6
10	Kidney	3,100	2.3
11	Lip, mouth and pharynx	3,100	2.3
12	Brain	2,400	1.8
13	Melanoma of skin	2,400	1.8
14	Larynx	2,000	1.5
15	Testis	1,800	1.4
16	Multiple myeloma	1,800	1.3
17	Hodgkin's disease	820	0.6
Sub-total 4-17		50,300	37.6
	Other	16,500	12.3
	All cancers[3]	133,900	100.0

Females

Rank	Site	Number of cases[2]	% of all cancers
1	Breast	38,900	28.5
2	Colorectal	16,300	11.9
3	Lung	15,000	11.0
Sub-total 1-3		70,100	51.5
4	Ovary	6,700	4.9
5	Uterus	4,900	3.6
6	Stomach	4,200	3.1
7	Non-Hodgkin's lymphoma	4,000	3.0
8	Bladder	3,900	2.9
9	Cervix	3,700	2.7
10	Pancreas	3,700	2.7
11	Melanoma of skin	3,500	2.6
12	Leukaemia	3,000	2.2
13	Oesophagus	2,900	2.1
14	Kidney	1,900	1.4
15	Brain	1,900	1.4
16	Multiple myeloma	1,700	1.2
17	Lip, mouth and pharynx	1,700	1.2
18	Hodgkin's disease	620	0.5
Sub-total 4-18		48,900	35.9
	Other	17,300	12.7
	All cancers[3]	136,300	100.0

Numbers and percentages may not sum exactly to totals shown, due to rounding
1 Northern Ireland 1993-99, Ireland 1994-99
2 Average number per year
3 All malignant neoplasms excluding non-melanoma skin cancer (ICD9 codes 140-208 excluding 173; ICD10 codes C00-C97 excluding C44)

Table 2.2

Mortality from the 21 cancers covered in Chapters 3-23, UK and Ireland, 1991-2000[1]

Males

Rank	Site	Number of deaths[2]	% of all cancers
1	Lung	24,300	28.6
2	Prostate	10,000	11.8
3	Colorectal	9,500	11.1
Sub-total 1-3		43,800	51.5
4	Stomach	5,100	6.0
5	Oesophagus	4,200	5.0
6	Bladder	3,600	4.2
7	Pancreas	3,400	4.0
8	Non-Hodgkin's lymphoma	2,400	2.8
9	Leukaemia	2,300	2.7
10	Brain	1,900	2.2
11	Kidney	1,800	2.1
12	Lip, mouth and pharynx	1,400	1.6
13	Multiple myeloma	1,300	1.5
14	Melanoma of skin	780	0.9
15	Larynx	770	0.9
16	Hodgkin's disease	210	0.2
17	Testis	110	0.1
Sub-total 4-17		29,100	34.2
	Other	12,200	14.3
	All cancers[3]	85,100	100.0

Females

Rank	Site	Number of deaths[2]	% of all cancers
1	Breast	14,600	18.6
2	Lung	13,400	17.1
3	Colorectal	9,100	11.5
Sub-total 1-3		37,000	47.2
4	Ovary	4,600	5.9
5	Pancreas	3,600	4.6
6	Stomach	3,300	4.2
7	Oesophagus	2,700	3.4
8	Non-Hodgkin's lymphoma	2,200	2.8
9	Leukaemia	1,900	2.4
10	Bladder	1,800	2.3
11	Cervix	1,600	2.0
12	Uterus	1,500	1.9
13	Brain	1,400	1.8
14	Multiple myeloma	1,300	1.6
15	Kidney	1,200	1.5
16	Melanoma of skin	790	1.0
17	Lip, mouth and pharynx	740	0.9
18	Hodgkin's disease	160	0.2
Sub-total 4-18		28,900	36.9
	Other	12,500	15.9
	All cancers[3]	78,500	100.0

Numbers and percentages may not sum exactly to totals shown, due to rounding
1 Scotland 1991-99, Ireland 1994-2000
2 Average number per year
3 All malignant neoplasms excluding non-melanoma skin cancer (ICD9 codes 140-208 excluding 173; ICD10 codes C00-C97 excluding C44)

London, however, over 40 per cent of the population lived in wards that were in the lowest fifth. A map of socio-economic deprivation by local authority (with health authority boundaries superimposed) is given in Appendix F. Maps of deprivation at the ward level for all the regions in England, and for Wales and Scotland, are given in Chapter 4 of *Geographic Variations in Health*.[11] When attempting to relate geographic variations in cancer to socio-economic deprivation using such area-based

indices, it must be borne in mind that not everyone living in a deprived ward is themselves socio-economically disadvantaged, nor do all those who are so disadvantaged live in deprived wards. These factors tend to dilute the real relationships at the individual level between the disease of interest and deprivation or any risk factor(s), such as smoking, for which deprivation is a marker.

Figure **2.7a**

Major cancers: age-standardised[1] incidence Males, UK and Ireland 1991-99[2]

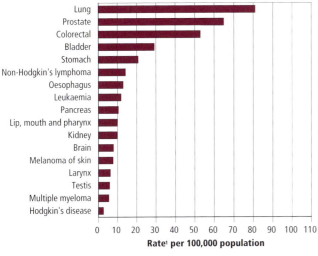

1 Directly age standardised using the European standard population
2 Northern Ireland 1993-99, Ireland 1994-99

Figure **2.7b**

Major cancers: age-standardised[1] incidence Females, UK and Ireland 1991-99[2]

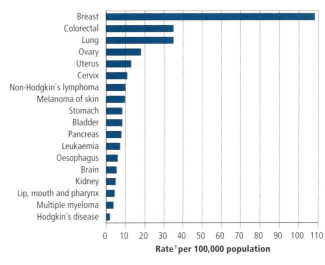

1 Directly age standardised using the European standard population
2 Northern Ireland 1993-99, Ireland 1994-99

Figure **2.8a**

Major cancers: age-standardised[1] mortality Males, UK and Ireland 1991-2000[2]

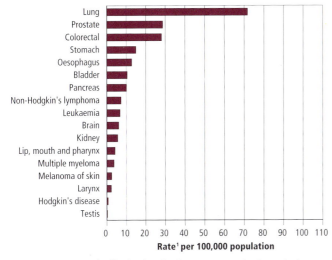

1 Directly age standardised using the European standard population
2 Scotland 1991-99, Ireland 1994-2000

Figure **2.8b**

Major cancers: age-standardised[1] mortality Females, UK and Ireland 1991-2000[2]

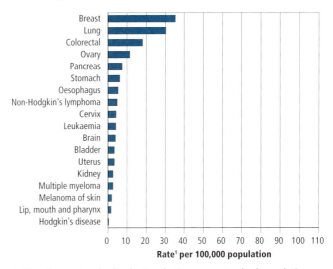

1 Directly age standardised using the European standard population
2 Scotland 1991-99, Ireland 1994-2000

Figure **2.9**

Percentage of the population by socio-economic deprivation category[1], Great Britain 1991

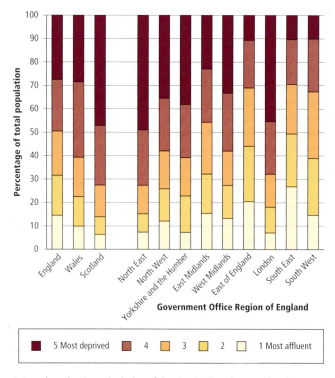

1 *Based on the Carstairs index of deprivation[10] at the ward level (see Appendix F)*

Geographical patterns in the incidence of, and mortality from, all cancers combined

Summary

- For all cancers combined, the geographical variations in incidence were closely similar to those in mortality. This is largely because relatively little geographical variation in survival existed for the major cancers.

- Very little geographical variation existed in overall cancer incidence and mortality across the five countries and the regions of England. Both incidence and mortality in Scotland were about 15 per cent above the UK and Ireland average.

- There was also little variation in the overall cancer incidence and mortality rates across the health authorities, with the differences between the highest and lowest health authority rates within each country, and region of England mostly in the narrow range 10-25 per cent.

The main uses that can be made of this atlas (see Chapter 1) are specific to each type of cancer, or group of cancers. The *total* cancer incidence or mortality, that is from all types of

cancer combined, is therefore of little interest or value for identifying areas for further investigation or public health actions. In addition, false reassurance may be given by an overall cancer rate for an area that is close to the average for the UK and Ireland as a whole, because the rate for one or more cancers could be extremely high, but this is offset by a very low rate in a cancer that is approximately as frequent, or by slightly lower than average rates for several other cancers. However, for completeness, a brief summary of the geographical patterns in overall cancer incidence and mortality is given below. The same types of charts and maps that are presented in Chapters 3 to 23 for the individual cancers are given for all cancers combined in Appendix D.

The geographical patterns in overall cancer incidence and mortality were closely similar to each other for both males and females. This is largely because compared with the wide geographical variations in incidence and mortality for many cancers, there is relatively little geographical variation in survival for most of the major cancers.[12-15]

There was relatively little geographical variation in either overall cancer incidence or mortality across the countries and regions of England. Both incidence and mortality were about 15 per cent above the UK and Ireland average in Scotland (Table 2.3). In Wales, incidence was about 7 per cent higher than average, but mortality was not raised. Within England, in the Eastern region incidence in both males and females was around 10 per cent below the UK and Ireland average and mortality about 5 per cent below. Mortality was slightly above average in the Northern and Yorkshire, and North West regions, and slightly below average in the South East and South West regions. The slight differences between the patterns in incidence and mortality are in large part due to regional differences in the proportions of cancers with different survival rates. For example, lung cancer, which has very low survival, is more common in the north of England than in the south.

There was relatively little variation in the overall rates for cancer incidence and mortality among the health authorities within the countries, or regions of England. In general, the differences between the highest and lowest rates in a country or region were extremely small, most falling in the narrow range 10-25 per cent, as shown in Table 2.4. There were, however, three health authorities which had noticeably high rates for both cancer incidence and mortality in both males and females: Liverpool and Manchester in the North West region of England, and Greater Glasgow in Scotland.

Geographical patterns in the incidence of, and mortality from, specific cancers

This section summarises the geographical patterns for the 21 individual cancers covered in Chapters 3 to 23. The cancers are considered in six broad groups: those strongly related to smoking tobacco and/or drinking alcohol; gastrointestinal cancers; cancers occurring only (or predominantly) in one sex; urological cancers; brain cancer and malignant melanoma of the skin; and lymphomas and leukaemias.

For these 21 cancers, Table 2.5 gives the male-to-female ratio for the age-standardised incidence rates; an indication of the trends during the 1990s in incidence and mortality, by sex; and an indication of the relationships between incidence and mortality and socio-economic deprivation.

Maps 2.1 to 2.24 illustrate the geographical patterns in cancer at the health authority level using the *ratio* of the directly age-standardised rate in each health authority to the relevant average rate for the UK and Ireland. Health authorities with lower than average rates are coloured blue, those with higher than average rates purple. Maps 2.25 to 2.32 illustrate the *actual* age-standardised rates, coloured from white for the 'lowest' rates to dark red for the 'highest'. See Appendices E and H for further details

Cancers strongly related to smoking tobacco and/or alcohol consumption

Larynx; lip, mouth and pharynx; and lung

Summary

- Although smoking tobacco is a major risk factor for all these cancers, the time trends and birth cohort trends for lung cancer were different from those for the other cancers.

- Incidence and mortality for these cancers was generally lower than the UK and Ireland average in the south and midlands of England, and higher in a band across the formerly highly industrialised parts of northern England. Rates were also higher than average across the central belt of Scotland.

- The areas with the highest incidence and mortality rates for these cancers were mostly those with high levels of socio-economic deprivation, there were higher than average rates of cancers of the lip, mouth and pharynx – but not larynx or lung – in Northern Ireland and Ireland.

- The differences between the geographical patterns for lung cancer and those for the other cancers in this group are most likely due to differences in the levels of alcohol consumption.

Table 2.3

Incidence of, and mortality from, all cancers combined[1]: difference (%) from UK and Ireland average by country and region of England

	Incidence		Mortality	
	Males	Females	Males	Females
England				
Northern and Yorkshire	+1.6	-1.3	+7.3	+4.7
Trent	-3.3	-3.9	+0.4	-0.2
West Midlands	+0.4	-2.2	+1.0	-1.2
North West	+4.6	+1.8	+8.7	+6.8
Eastern	-10.3	-5.3	-9.4	-6.4
London	-2.4	-3.2	-1.2	-0.8
South East	-4.2	-0.6	-7.4	-5.8
South West	-3.7	+0.0	-9.9	-8.1
Wales	+7.9	+5.9	+1.1	+1.3
Scotland	+16.4	+13.2	+14.6	+13.2
Northern Ireland	+0.1	+2.6	-3.2	-3.2
Ireland	-2.5	-4.7	-0.1	+1.0

1 All malignant neoplasms excluding non-melanoma skin cancer (ICD9 codes 140-208 excluding 173, ICD10 codes C00-C97 excluding C44)

Table 2.4

Incidence of, and mortality from, all cancers combined[1]: difference (%) between highest and lowest health authority rates by country and region of England

	Incidence		Mortality	
	Males	Females	Males	Females
England				
Northern and Yorkshire	21	19	20[2]	31
Trent	22	18	25	22
West Midlands	24	6	29	18
North West[3]	25	17	25	18
Eastern	10	9	12	9
London	26	12	36	22
South East	21	16	17	16
South West	19	18	9	8
Wales	14	10	14	8
Scotland[4]	16	13	25	22
Northern Ireland	11	6	12	4
Ireland	15	16	8	9

1 All malignant neoplasms excluding non-melanoma skin cancer (ICD9 codes 140-208 excluding 173, ICD10 codes C00-C97 excluding C44)
2 Excluding the very low rate in North Yorkshire
3 Excluding the high rates in Liverpool and Manchester
4 Excluding the high rate in Greater Glasgow and the highly variable rates in Orkney, Shetland and the Western Isles

All of these cancers have very strong relationships with socio-economic deprivation for both incidence and mortality. The rates in patients living in the most deprived areas were generally about three times higher than in the affluent. The exception was for cancers of the lip, mouth and pharynx in females where rates in the most deprived areas were higher by a factor of about two.[1] The time trends and birth cohort trends in England and Wales in incidence for both cancers of the lip, mouth and pharynx, and of the larynx, are, however, very different from those for lung cancer.[1] And the ratio of male to female incidence rates is much higher for cancer of the larynx (over 5:1) than for the other two cancers (just over 2:1).

The incidence of lung cancer was generally lower than average in the south and midlands of England, with the exception of some health authorities in London and the Birmingham conurbation, and in Northern Ireland and Ireland (Map 2.1). There was a band of higher than average rates across the formerly highly industrialised parts of northern England, from Liverpool and Manchester in the west to Hull in the east. Rates were also above average in Teesside and Tyneside. There was another band of higher than average incidence across the central, more densely populated, belt of Scotland from Glasgow in the west to Edinburgh in the east. Although rates of lung cancer are generally much higher in males than in females, the map showing the variations in incidence for females was similar to that for males (Map 13.1b). The maps for mortality from lung cancer (Map 13.2) were closely similar to those for incidence, owing to the very low survival from this cancer – only 5 per cent at five years from diagnosis.[12-15] The maps of the incidence of lung cancer on an absolute scale (Map 2.25 and Appendix E) illustrate that although there are geographical variations, the rates are very high everywhere in comparison to those for most other cancers.

The map for incidence of cancer of the larynx in males shows many similarities with that for lung cancer, with generally lower than average rates in the south and midlands of England and in Ireland (Map 2.2). There are, however, differences in that there are some higher than average rates in Northern Ireland, including the area of Belfast, and the area of Dublin in Ireland; and more health authorities in Scotland had rates that were more than 50 per cent above the UK and Ireland average. The absolute scale maps in Appendix E confirm the generally higher incidence and mortality rates in Scotland and in London and the urban areas of the midlands and north of England, and the lower rates in most of the rest of England.

For cancers of the lip, mouth and pharynx, the map for incidence in males showed many similarities with those for cancers of the lung and larynx, with generally lower than average rates across most of the south and midlands of

England, and above average rates in London and parts of the urban north of England (Map 2.3). Rates were also mostly above average in Wales. The pattern in the rates in Scotland was more similar to that for cancer of the larynx than that for lung cancer. But the major differences were for Northern Ireland and Ireland, where the incidence of cancers of the lip, mouth and pharynx was generally above average. The geographical patterns for females were similar to those for males, but generally less pronounced and rates were generally below average in Ireland (Map 2.4). The similarity between the sexes in the geographical patterns can also be seen in the absolute scale maps in Appendix E. These also illustrate the higher rates in males than in females, and that survival for these cancers as a group is relatively high – the mortality maps are considerably paler in colour than those for incidence.

Gastrointestinal cancers

Colon and rectum combined (colorectal), oesophagus, pancreas and stomach

Summary

- Only small geographical variations existed in incidence and mortality for both colorectal and pancreatic cancer; the patterns were not strongly related to socio-economic deprivation.

- In contrast, the incidence of cancer of the oesophagus was noticeably above average in much of northern England and Scotland.

- The incidence of cancer of the stomach was higher than average in parts of London and the urban West Midlands, in bands across northern England and central Scotland, and in the health authorities containing Belfast and Dublin.

- Stomach cancer had similar geographical patterns to those for lung cancer, despite the clear differences in their time trends and birth cohort trends.

Geographical studies, migrant studies and the time trends in colorectal cancer suggest that 'environmental' factors (in the widest sense), especially diet, play an important role,[16-20] and obesity is a risk factor, although about a third of colorectal cancers might be explained by heritable factors.[21] The age-standardised rate is about 50 per cent higher in males than in females. Hormonal factors may explain at least in part, the lower risk in females. In the early 1990s, there was little or no relationship between either incidence or mortality and socio-economic deprivation in England and Wales.[1] In the last 30 years of the twentieth century, the incidence of colon cancer increased slightly in men, but the incidence rates of colon cancer in females, and of rectal cancer in both males and females, were stable.

Map 2.1

**Lung: incidence* by health authority
Males, UK and Ireland 1991-99**

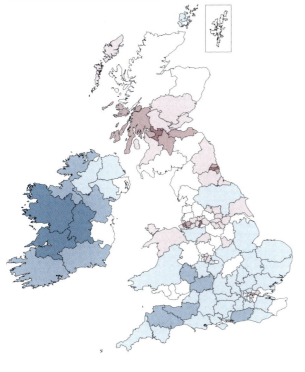

Map 2.2

**Larynx: incidence* by health authority
Males, UK and Ireland 1991-99**

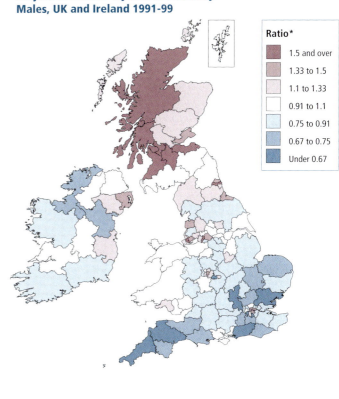

Ratio*

	1.5 and over
	1.33 to 1.5
	1.1 to 1.33
	0.91 to 1.1
	0.75 to 0.91
	0.67 to 0.75
	Under 0.67

Map 2.3

**Lip, mouth and pharynx: incidence* by health authority
Males, UK and Ireland 1991-99**

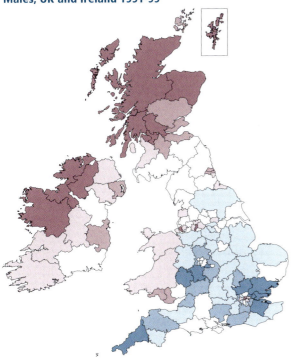

Map 2.4

**Lip, mouth and pharynx: incidence* by health authority
Females, UK and Ireland 1991-99**

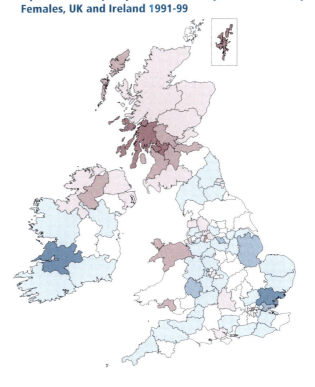

* Ratio of directly age-standardised rate in health authority to UK and Ireland average

Table **2.5**

Characteristics of the 21 cancers covered in Chapters 3 to 23

Cancer group/type	Chapter	Male-to-female ratio[1]	Trends in the 1990s[2]				Relationship to deprivation[3]	
			Incidence		Mortality		Incidence	Mortality
			Males	Females	Males	Females		
Smoking/alcohol								
Larynx	10	5.3	↓	..	↓	..	+++	+++
Lip, mouth and pharynx	12	2.3	↑	↑	0	0	+++	+++
Lung	13	2.3	↓↓↓	0	↓↓↓	0	+++	+++
Gastrointestinal								
Colorectal	7	1.5	(↑)	0	↓	↓	0	0
Oesophagus	13	2.2	↑↑	↑	↑↑	↑	+	+
Pancreas	19	1.3	↓↓	0	↓↓	0	(+)	(+)
Stomach	21	2.5	↓↓	↓↓	↓↓	↓↓	++	++
Single sex								
Breast	5	:	:	↑	:	↓↓↓	–	0
Cervix	6	:	:	↓↓↓	:	↓↓↓	+++	+++
Ovary	18	:	:	↑↑	:	0	–	(–)
Uterus	23	:	:	↑	:	↑	–	0
Prostate	20	:	↑↑↑	:	0	:	–	0
Testis	22	:	↑↑	:	0	:	–	x
Urological								
Bladder	3	3.5	↓	0	↓	0	0	0
Kidney	9	2.0	↑↑	↑↑	↑	↑	0	0
Brain, melanoma of skin								
Brain	4	1.5	0	0	0	0	–	–
Melanoma of skin	14	0.8	↑↑	↑↑	0	0	– – –	–
Lymphomas and leukaemias								
Hodgkin's disease	8	1.4	0	0	↓↓	↓↓	y	y
Non-Hodgkin's lymphoma	16	1.4	↑↑↑	↑↑↑	↑	↑	–	0
Leukaemia	11	1.7	0	0	0	0	–	0
Multiple myeloma	15	1.5	0	0	0	0	0	0

.. Not included   : Not applicable   x Not available (very low numbers of deaths)   y Not available

1 Ratio of the age-standardised rates

2 ↑ Upward trend; 0 No trend; ↓ Downward trend
  (England and Wales; strength of trend indicated by the number of arrows)

3 + Positive relationship with deprivation (higher rates in those living in deprived areas)
  0 No relationship with deprivation
  – Negative relationship with deprivation (that is higher rates in those living in affluent areas)
  (England and Wales; strength of relationship indicated by the number of symbols)

Sources:

Quinn MJ, Babb PJ, Brock A, Kirby L, Jones J. Cancer Trends in England and Wales 1950-1999. Studies on Medical and Population Subjects No.66. London: The Stationery Office, 2001.

ONS. Cancer Statistics Registrations: Registrations of cancer diagnosed in 2000, England. Series MB1 No.31. London: Office for National Statistics, 2003.

At the country, and region of England level, the patterns in incidence and mortality for colorectal cancer were broadly similar in males and females. In England, around two thirds of health authorities had incidence rates that were within 10 per cent of the UK and Ireland average (Maps 2.5, 7.1b). Incidence was slightly above the average in a few health authorities in northern England; and in Scotland, Northern Ireland and Ireland, the rates in all but a few health authorities were more than 10 per cent above average. Rates in most of London and parts of the south east of England were slightly below average. In males, there was slightly more of a north-south divide in mortality from colorectal cancer than in incidence. The maps of incidence and mortality on an absolute scale (Map 2.26 and Appendix E) illustrate, as for lung cancer, that although there are geographical variations, the rates are very high everywhere in comparison to those in most other cancers.

Little is known about the aetiology of cancer of the pancreas. Smoking appears to double a person's risk, but is estimated to account for less than 30 per cent of cases.[22-24] Dietary factors are also likely to be involved.[17] The ratio of male to female incidence (1.3:1) is much lower than that for lung cancer. The time trends and birth cohort trends in the incidence of pancreatic cancer show some similarity with those for lung cancer.[12] There were parallel, but less strong, declines in pancreatic and lung cancer in males, and increases to a plateau in the 1990s in females; and the birth cohorts at highest risk for both lung and pancreatic cancer were those born at the end of the nineteenth century for males and in the early 1920s for females. There is, however, only a weak relationship between the incidence of pancreatic cancer and socio-economic deprivation, with rates in people living in the most deprived areas only about 30 per cent higher than those in affluent areas.[1]

There was very little variation in the incidence of cancer of the pancreas among the countries and regions of England. There were no clear patterns at the health authority level, and little consistency between rates for males and females. Incidence was, however, slightly higher than the UK and Ireland average in parts of London and some of the urban parts of the midlands and north of England, and slightly below average in the more rural areas (Maps 2.6, 19.1b). There was no obvious link between incidence and any known risk factors or any strong association with socio-economic deprivation.

There are two main types of cancer of the oesophagus. The risk factors for squamous cell carcinoma (SCC), which is most common in the upper two thirds of the oesophagus, are smoking and alcohol consumption, which in combination have a greater than additive effect. Adenocarcinomas, for which smoking is also a risk factor, are preceded by a condition called Barrett's oesophagus. Poor diet is also associated with

adenocarcinomas, but the main risk factor is obesity and the increasing prevalence of obesity in the UK and Ireland may have contributed to the increasing incidence of this type of oesophageal cancer.[25] About 70 per cent of deaths from cancer of the oesophagus in England have been attributed to smoking.[24] Smoking and poor diet, together with chronic infection with H pylori, are the main risk factors for cancer of the stomach. About 35 per cent of cases of stomach cancer in men and just over 10 per cent in women have been attributed to smoking.

The time trends and birth cohort trends for cancers of both the oesophagus and stomach are, however, very different from those for lung cancer. The incidence of cancer of the oesophagus has increased by about 50 per cent in both males and females since the early 1970s.[1] The birth cohort risks have increased steadily for males since the 1880s, and for women up to those born in the 1930s, after which the risks declined slightly. In contrast, the incidence of stomach cancer has declined strongly and continuously since the 1970s (when reliable records began in England and Wales) – and the mortality has declined from the 1950s (and earlier) (see Figures 2.3 and 2.4). The birth cohort risks for both men and women were highest for those born towards the end of the nineteenth century, and declined continuously for 70 years, but the risk has levelled off for those born from the 1950s onwards.[1] Further differences are that the incidence of stomach cancer in both males and females in the more deprived areas is about twice that in the affluent, and for cancer of the oesophagus the excess is only about 50 per cent, compared with a factor of about three for lung cancer. The opposite trends in the incidence of cancers of the oesophagus and stomach mean that although there were about three times as many cases of cancer of the stomach than of the oesophagus in the early 1970s, by the end of the 1990s there were only about 50 per cent more.[26]

One interesting feature of the incidence and mortality figures for cancers of the stomach and oesophagus is that although the survival rates are broadly similar[12-15] the ratio of deaths to cases is much higher than would be expected for cancer of the oesophagus, and lower than would be expected for stomach cancer, especially for males. There is obviously some misclassification of stomach cancer deaths as being from cancer of the oesophagus[26] but it is not clear why this occurs more in males than in females. This inconsistency between the incidence and mortality figures occurs in many other countries.[27]

The incidence of cancer of the oesophagus was higher than average in most of the health authorities in Scotland, and more than one third above the average in most of central Scotland

(Maps 2.7, 17.1b). There was also a band of higher than average incidence from North Wales across to the east coast of England. Rates elsewhere in England, and in Northern Ireland and Ireland were slightly below or not markedly different from the average. The maps for males and females were closely similar, although in females the rates in all health authorities on the mainland of Scotland were higher than average,
several of these by more than 50 per cent, and there were few health authorities with below average rates in the north of England. Given the marked differences from lung cancer in the patterns of the time trends and the lack of any strong relationship with socio-economic deprivation for cancer of the oesophagus, it is slightly surprising that there are close similarities between their geographical patterns (Maps 2.1, 13.1b).

The geographical patterns in the incidence of stomach cancer were closely similar in males and females (Maps 2.8, 21.1b). Incidence was higher than the UK and Ireland average in both males and females in most of the central belt in Scotland and in Wales. In England, there was a clear north-south divide, with higher than average rates (some more than 50 per cent above average) in the north; rates were also slightly raised in the urban part of the West Midlands and in a small number of health authorities in London. In the area of Northern Ireland that contains Belfast, the rate in males, but not that in females, was raised. In Ireland, the rates for both males and females were raised in the Eastern area which contains Dublin, but were mostly below average elsewhere. The patterns in incidence can be seen clearly in the absolute scale maps (Map 2.27 and Appendix E) which highlight parts of London and the West Midlands, the bands across the north of England and central Scotland, and the areas around Belfast and Dublin.

Despite the wide differences in the time trends from those for lung cancer, particularly the birth cohort trends for females, the geographical patterns in the incidence of stomach cancer are closely similar to those for lung cancer. The patterns for stomach cancer are also more similar to those for lung cancer than to those for cancer of the oesophagus. As the geographical patterns in both lung and stomach cancer are related to socio-economic deprivation, the differences in the trends suggest that for stomach cancer, deprivation is a marker for a major risk factor other than smoking.

Cancers in only one (or predominantly one) sex

Breast, cervix, ovary and uterus in women; prostate and testis in men

Summary

- There was very little geographical variation across the whole of the UK and Ireland in the incidence of breast cancer in women, and virtually none in mortality.

- Ovarian cancer shares many of the risk factors for breast cancer, and the weak geographical patterns in both incidence and mortality were closely similar to those for breast cancer.

- The geographical patterns in both incidence and mortality for cancer of the uterus were different from those for cancers of the breast and ovary, with slightly above average rates in the south and midlands of England, and below average rates in the rural north of England and in much of Scotland, Northern Ireland and Ireland.

- In complete contrast, very wide geographical variations existed in the incidence of cervical cancer, which is related to patterns in sexual behaviour and infection with human papillomavirus (HPV). There was higher than average incidence in the West Midlands, in a band across the north of England, and in Scotland; and there was a clear link with socio-economic deprivation. The geographical patterns were not related to local differences in the uptake or efficiency of the cervical screening programme.

- Although the overall levels of recorded incidence of prostate cancer have been affected by increasing uptake of prostate-specific antigen (PSA) testing, the geographical pattern was closely similar to that for breast cancer in women – in England, most of the health authorities with above average rates were in the south, and those with below average rates in the north. As with breast cancer, there was virtually no geographical variation in mortality from prostate cancer.

- The incidence of testicular cancer tended to be higher than average in the more affluent areas in the south of England and in Scotland, and below average in London, the north of England, and much of Ireland.

The main risk factors for breast cancer relate to a woman's reproductive history. (Breast cancer in men is extremely rare, with age-standardised incidence of about 1 per 100,000 – about 300 cases each year in the UK and Ireland.) Most of the risk factors for female breast cancer are shared by cancers of the ovary and uterus. Some part of the apparent geographical

Map **2.5**

**Colorectal: incidence* by health authority
Males, UK and Ireland 1991-99**

Map **2.6**

**Pancreas: incidence* by health authority
Males, UK and Ireland 1991-99**

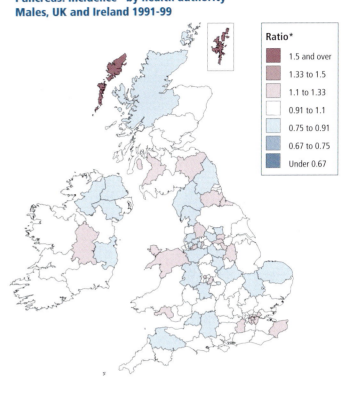

Ratio*

■	1.5 and over
■	1.33 to 1.5
■	1.1 to 1.33
□	0.91 to 1.1
■	0.75 to 0.91
■	0.67 to 0.75
■	Under 0.67

Map **2.7**

**Oesophagus: incidence* by health authority
Males, UK and Ireland 1991-99**

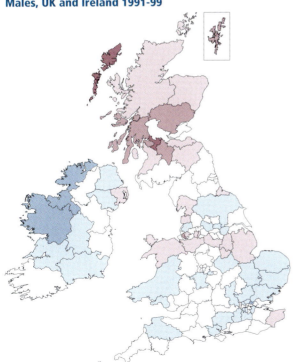

Map **2.8**

**Stomach: incidence* by health authority
Males, UK and Ireland 1991-99**

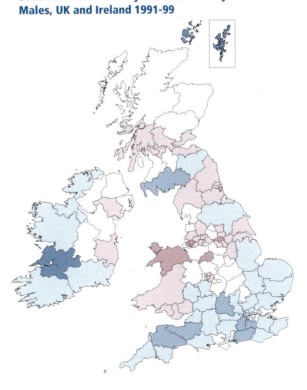

* Ratio of directly age-standardised rate in health authority to UK and Ireland average

21

patterns in the incidence of cancer of the ovary may be due to variations in the registration of 'borderline' malignancies (see Chapter 18). And some part of the apparent patterns in cancer of the uterus may be due to variations in the proportions of cases registered as 'uterus, not specified whether cervix or body of uterus', or to geographical variations in the proportion of women who have had a hysterectomy (see Chapter 23).

There was very little geographical variation in the incidence of breast cancer, with the vast majority of health authorities having rates within 10 per cent of the UK and Ireland average (Map 2.9). There were several health authorities in the south east of England with slightly raised rates, and there were slightly lower than average rates in parts of the rural north of England and in much of Ireland. There was virtually no geographical variation in breast cancer mortality across the whole of the UK and Ireland (Maps 2.10, 5.2). This is consistent with survival from breast cancer being better in the more affluent.[12-15]

The geographical pattern in the incidence of cancer of the ovary was broadly similar to that for breast cancer, with the vast majority of health authorities having rates within 10 per cent of the UK and Ireland average (Map 2.11). There were, however, some health authorities with slightly raised rates for ovarian, but not breast, cancer in Wales. The lack of any wide geographical variations is confirmed in the map of incidence on an absolute scale (Map 2.28 and Appendix E). As with breast cancer, there was even less geographical variation in mortality from ovarian cancer than for incidence (Map 18.2). Again, this is consistent with survival being better in women living in the more affluent areas.

For cancer of the uterus, the map of incidence is somewhat different from those for breast and ovarian cancers. It shows slightly higher than average rates in two bands across England: one across the West Midlands and parts of the Trent region, and the other across the South West and South East regions from Cornwall to Norfolk (Map 2.12). The pattern in mortality was broadly similar, with above average rates in the south and midlands of England and in Wales, and below average rates in the rural north of England, and in much of Scotland, Northern Ireland and Ireland (Map 23.2).

The incidence of cervical cancer is closely related to patterns in sexual behaviour and infection with HPV. In complete contrast to the relatively low levels of geographical variation in cancers of the breast, ovary and uterus, there were very wide variations in the incidence of cervical cancer among the countries and the regions of England. There were higher than average rates in parts of the West Midlands, in a band across the northern part

of England, and in Scotland; and lower than average rates in the south and east of England, and in Northern Ireland and Ireland (Map 2.15). The map of mortality from cervical cancer (Map 6.2) was closely similar to that for incidence. The absolute scale maps (Map 2.29 and Appendix E) confirm that almost all of the health authorities with low incidence rates were in England, south of a line running north-eastwards across the country from Bristol; and that mortality rates were considerably lower than those for incidence. There was a clear link between both incidence and mortality rates and levels of socio-economic deprivation. The geographical patterns in incidence (and mortality) in the UK were not related to differences in the uptake or effectiveness of the revised cervical screening programme: the regional variations were similar in the 1970s and 1980s (although the rates were much higher).

The apparent incidence of prostate cancer in the 1990s has been heavily affected by the increasingly widespread use of PSA testing.[28-30] Nevertheless, the map for prostate cancer shows many similarities to that for breast cancer in women. There were relatively few health authorities with rates more than 10 per cent above or below the UK and Ireland average (although slightly more than for breast cancer) (Map 2.14). In England, most of the health authorities with above average rates were, as with breast cancer, in the south, and those with below average rates in the north. Despite the apparent geographical differences in incidence, there were virtually no geographical variations in mortality from prostate cancer in the UK, although rates were slightly elevated in Ireland. This is consistent with survival being better in the more affluent.[12-15] Despite evidence of wide variations in incidence from international comparisons and studies of migrants, causal factors (such as environmental, life style, diet and occupation) have not been identified conclusively (Chapter 20); the higher incidence rates in London and the south of England suggest an association with affluence, but what underlying risk factor this is a marker for is not known.

The known risk factors for testicular cancer are congenital malformations of the testicles (particularly undescended testicle), low birth weight and growth retardation of the foetus in the uterus, low maternal parity, and sub-fertility;[31] in the USA, rates are five times higher in white males than in black.[32] The incidence of testicular cancer was higher than average in the south of England and in Scotland, and below average in London, the north of England, and in much of Ireland (Map 2.16). Although the areas with higher incidence tended to be the more affluent, this most likely reflects the proportions of white males living in these areas, rather than any causative factor inversely related to socio-economic deprivation.

Urological cancers

Bladder and kidney

Summary

- Geographical patterns in the incidence of bladder cancer were strongly affected by differences in coding practice among the cancer registries. Mortality from bladder cancer was below average in both Northern Ireland and Ireland, and slightly lower in Wales (for men), and in most of the south and midlands of England. The rates in the north of England and central Scotland were higher than average.

- Although the time trends and birth cohort trends for bladder cancer differed noticeably from those for lung cancer, there was some weak similarity in the geographical patterns, particularly the north-south divide in England.

- The geographical patterns in mortality from cancer of the kidney showed some similarity with those for bladder cancer, with higher than average rates in the formerly industrialised north of England and in Scotland, but the patterns were only weakly similar to those for lung cancer.

- The marked differences for cancers of the bladder and kidney from lung cancer in their time trends and birth cohort trends, male-to-female ratio, and relationship with socio-economic deprivation, and the lack of strong similarity in their geographical patterns, all suggest that smoking tobacco is not a dominant risk factor for these cancers.

There is substantial evidence that there is some relationship between smoking and bladder cancer, with relative risks for smokers of two to three fold.[33-38] It has been suggested that possibly 40 per cent of male and 10 per cent of female cases might be attributable to smoking.[33] Smoking is also a risk factor for cancer of the kidney[39-45] along with obesity.[44,46-48] Although the evidence for the effect of smoking is strong, the time trends and birth cohort trends for cancers of both the bladder and kidney are quite different from those for lung cancer; the male-to-female ratio is higher for bladder cancer (3.5:1) and lower for cancer of the kidney (2.0:1); and for both bladder and kidney cancers there is no relationship in either males or females between either incidence or mortality and socio-economic deprivation.

There are differences between the cancer registries in the classification and registration of some bladder tumours which are regarded as malignant by some registries and non-malignant (benign) by others (Chapter 3); but these differences have been consistent over time, and so the overall trends are reliable. The time trends, and the birth cohort trends, for the incidence of bladder cancer are markedly different from those for lung cancer. In particular, in contrast to the steep decline in lung cancer in males, the incidence of bladder cancer rose steeply during the 1970s and 1980s, then declined slightly in the 1990s.[1]

The interpretation of the apparent geographical patterns in the incidence of bladder cancer is difficult because of the differences in registration practice among the cancer registries. This section therefore describes the patterns in mortality from bladder cancer. In Scotland, mortality rates were above average in Grampian and Greater Glasgow and its surrounding areas. In England, there was a band of above average rates from Liverpool and Manchester extending across the Pennines to Hull on the east coast. Both of these patterns were stronger in females than in males (Maps 2.17, 3.2b). Mortality rates were at, or slightly below the average in most of the midlands and south of England, and in almost all of Wales, Northern Ireland and Ireland in both males and females. These patterns can also be clearly seen in the absolute scale maps (Map 2.30 and Appendix E), particularly the higher rates in the north of England and central Scotland, and the lower rates in Wales and the English areas bordering Wales, and in the whole of Northern Ireland and Ireland.

There is some weak similarity between the geographical patterns for mortality from bladder and lung cancers, but for lung cancer in both males and females the north-south divide in England was much more marked, as were the higher than average rates in central Scotland.

As with bladder cancer, the time trends and birth cohort trends for cancer of the kidney are quite different from those for lung cancer: incidence has doubled in both males and females since the 1970s.[1]

At all geographical levels, the patterns in incidence and mortality for cancer of the kidney were broadly similar to each other, in both males and females. Incidence in both Wales and Scotland was above average in both males and females, but there was little variation at the regional level in England except for slightly lower than average rates in the Eastern and London regions. The maps of incidence and mortality (Maps 2.18, 9.1, 9.2b) generally confirm the regional patterns, with most health authorities in Scotland having higher than average rates. There were also higher than average rates in some, but not all, parts of the urban north of England. Rates were at or slightly below the overall average in almost all of the health authorities in the Eastern and London regions for both males and females.

Map **2.9**

**Breast: incidence* by health authority
Females, UK and Ireland 1991-99**

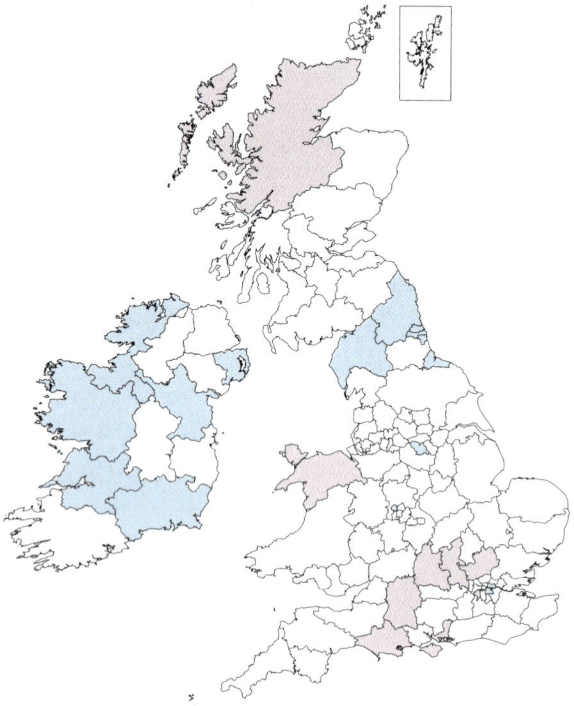

Map **2.10**

**Breast: mortality* by health authority
Females, UK and Ireland 1991-2000**

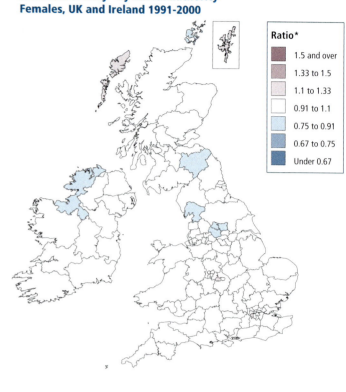

Ratio*	
	1.5 and over
	1.33 to 1.5
	1.1 to 1.33
	0.91 to 1.1
	0.75 to 0.91
	0.67 to 0.75
	Under 0.67

Map **2.11**

**Ovary: incidence* by health authority
UK and Ireland 1991-99**

Map **2.12**

**Uterus: incidence* by health authority
UK and Ireland 1991-99**

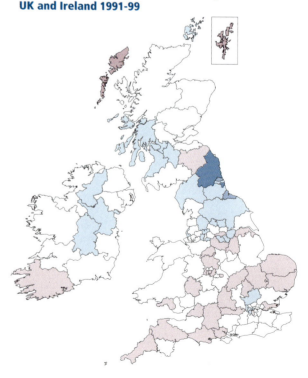

* Ratio of directly age-standardised rate in health authority to UK and Ireland average

Map **2.13**

**Prostate: incidence* by health authority
UK and Ireland 1991-99**

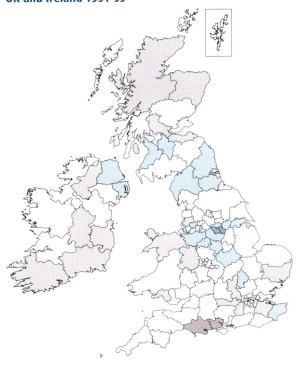

Map **2.14**

**Prostate: mortality* by health authority
UK and Ireland 1991-2000**

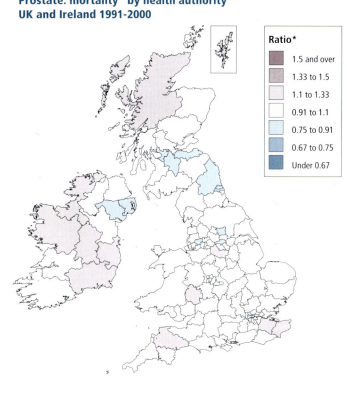

Ratio*

	1.5 and over
	1.33 to 1.5
	1.1 to 1.33
	0.91 to 1.1
	0.75 to 0.91
	0.67 to 0.75
	Under 0.67

Map **2.15**

**Cervix: incidence* by health authority
UK and Ireland 1991-99**

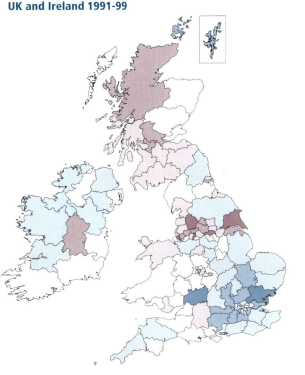

Map **2.16**

**Testis: incidence* by health authority
UK and Ireland 1991-99**

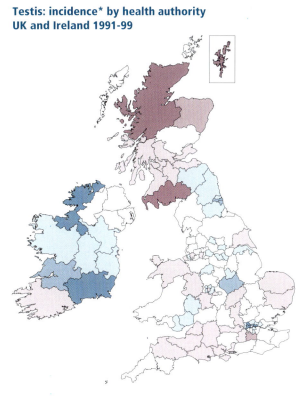

* Ratio of directly age-standardised rate in health authority to UK and Ireland average

The overall geographical patterns in cancer of the kidney showed some similarity with those for (mortality from) bladder cancer in that there were higher than average rates in Scotland and the formerly heavily industrialised north of England. But (relative to their respective overall averages) rates for kidney cancer were lower than for bladder cancer in London and the south east of England and higher in Wales, Northern Ireland and Ireland. There was only very weak similarity between the geographical patterns in kidney and lung cancer – rates for lung cancer were noticeably below average in most of England outside the major conurbations, especially for females, and were quite low in most of Northern Ireland and Ireland. The differences between cancers of the kidney and lung in their time trends, birth cohort trends, male-to-female ratio, relationship with socio-economic deprivation and their geographical patterns suggest that – as with bladder cancer – smoking is not a dominant risk factor.

Brain cancer and malignant melanoma of the skin

Summary

- There was little geographical variation in the incidence of brain cancer in either males or females. Although both incidence and mortality were slightly higher in the more affluent groups, this may reflect access to diagnostic services.

- The incidence of melanoma of the skin was 50 per cent higher in women than in men, but mortality rates were similar. The geographical patterns in incidence were unlike those for any of the other 20 major cancers included in this atlas. In England, most of the above average rates were in the south. Incidence was generally below average in Wales, but consistently above average in Scotland, Northern Ireland and Ireland.

- Variations in the ascertainment of cases of melanoma between the cancer registries may partly account for regional variations in recorded incidence. Higher incidence in Scotland, Northern Ireland and Ireland may be due to better and earlier detection there.

The only established causes of brain cancer are high-dose ionising radiation and heritable syndromes, but both of these factors affect only very small proportions of the population; several occupations have been linked to a raised risk. The incidence of brain cancer appeared to increase during the 1970s and 1980s, but the pattern in the trends in the age-specific rates suggests that part of the rise was due to improvements in diagnostic techniques.[49] The reported increase in incidence, together with advances in molecular biology and immunology, has prompted an intensive search for environmental causal factors. There is no strong evidence linking brain cancer with either power frequency electromagnetic fields or the relatively recent huge expansion in the use of mobile phones.[50,51] Both incidence and mortality are slightly higher in the more affluent groups, as is survival, but this may simply reflect access to diagnostic services and lead-time bias.

There was relatively little geographical variability in the incidence of brain cancer in either males or females, although rates were slightly above average in most health authorities in Wales in both sexes; elsewhere there was little consistency in the patterns in the rates between the sexes (Maps 2.19, 4.1b).

Since the 1970s, the incidence of melanoma of the skin has increased three to four fold in England and Wales – more than for any other cancer. Although it is rare in children, it is one of the few cancers that have an impact on young adults. Survival rates have improved, but could rise further if all patients were treated at an early stage of disease. The unusually strong relationship between (higher) incidence of melanoma and affluence is likely to be related to excessive exposure to the sun on holidays abroad. Melanoma is also unusual in that in the 1990s, there were about 50 per cent more cases in women than in men (Figure 14.1). Mortality was, however, consistently higher in men than in women. This is because survival is much better in females: in men, melanoma occurs more often on the trunk of the body,[52-54] where it has a worse prognosis than elsewhere on the body.[55,56] Also, men are less knowledgeable than women about appropriate prevention measures, respond less well to health education, and present with the disease at a later stage.[57]

The geographical patterns in the incidence of melanoma of the skin are highly unusual, though closely similar in males and females in all the five countries (Maps 2.20 and 14.1a). There were (relative to most other cancers) very few health authorities that had rates that were within 10 per cent above or below the UK and Ireland average (the white areas), and several health authorities had rates that were more than 50 per cent above or below the average (the dark purple and blue areas). Although this is to some extent to be expected because the numbers of cases at the health authority level are fairly small compared with most of the other cancers covered in this atlas, there are large groups of adjoining health authorities with above or below average rates, and these patterns cross administrative and cancer registry boundaries. Rates were higher than average in much of the south of England (but not in London) and in Scotland, Northern Ireland and Ireland. Rates

Map **2.17**

**Bladder: mortality* by health authority
Males, UK and Ireland 1991-2000**

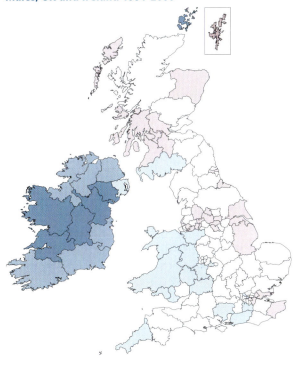

Map **2.18**

**Kidney: mortality* by health authority
Males, UK and Ireland 1991-2000**

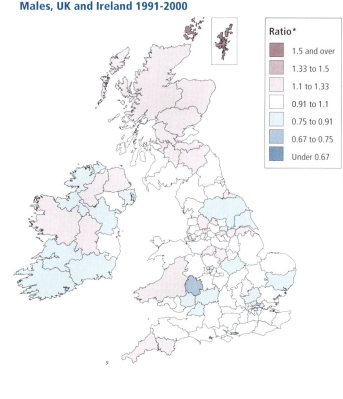

Ratio*

	1.5 and over
	1.33 to 1.5
	1.1 to 1.33
	0.91 to 1.1
	0.75 to 0.91
	0.67 to 0.75
	Under 0.67

Map **2.19**

**Brain: incidence* by health authority
Males, UK and Ireland 1991-99**

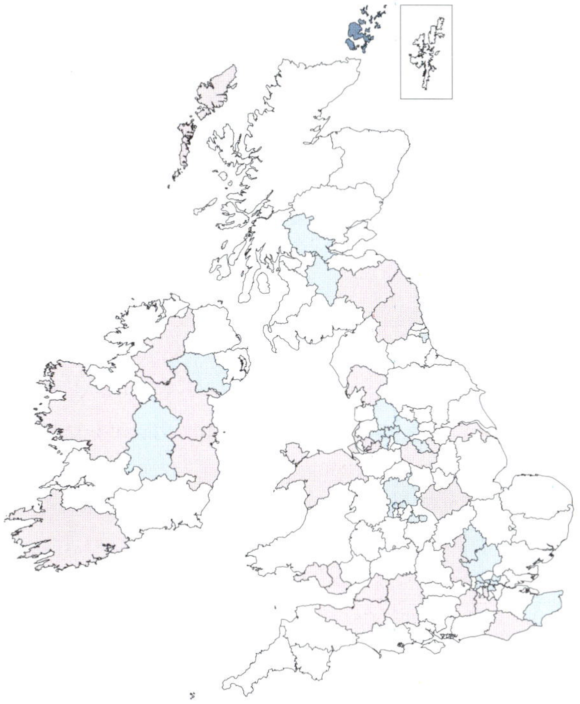

Map **2.20**

**Melanoma of skin: incidence* by health authority
Females, UK and Ireland 1991-99**

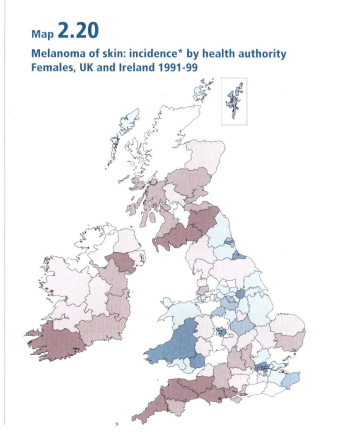

* Ratio of directly age-standardised rate in health authority to UK and Ireland average

were below average in London, in the former industrialised areas of the West Midlands and north of England, and in Wales. In Scotland, Northern Ireland and Ireland, which had generally higher than average incidence, the mortality rates were consistently below average. The maps on an absolute scale (Map 2.31 and Appendix E) clearly illustrate the higher rates in the south of England below a line running north-eastwards across the country from Bristol (apart from London) and in Scotland, Northern Ireland and Ireland.

Variations in ascertainment between the cancer registries may partly account for regional variations in recorded incidence of melanoma, and for the similar patterns in males and females. In Scotland particularly, as well as in Northern Ireland and Ireland, better and earlier detection may have boosted the incidence rates. The apparent geographical variations in incidence and mortality may also reflect different proportions of high risk, fair skinned people and/or low risk ethnic communities in the population.

Lymphomas and leukaemias

Hodgkin's disease, non-Hodgkin's lymphoma, multiple myeloma and leukaemia

Summary

- The incidence of Hodgkin's disease was slightly below average in the south of England, in Wales, and in Scotland; and slightly above average in parts of the midlands and north of England and in Ireland. Hodgkin's disease was, however, the least common of the 21 cancers included in this atlas, and the maps were dominated by random variations in the rates at the health authority level.

- The geographical patterns in the incidence of non-Hodgkin's lymphoma were similar in males and females; there was weak evidence of a north-south divide in England with higher than average rates in London and the south, and lower incidence in the midlands and the north. Incidence was also higher than average in Scotland and Northern Ireland. None of the known risk factors could explain these patterns.

- Multiple myeloma is relatively uncommon, and so – as with Hodgkin's disease – the maps were dominated by random variation in the rates. None of the suspected risk factors explain the geographical variations in incidence. Incidence is also not related to socio-economic deprivation.

- The incidence of leukaemia in both males and females was slightly higher than average in some (but not all) parts of the south east of England, Wales, parts of north east England and southern Scotland; and lower than average in parts of the west and north west of England. None of the known risk factors or socio-economic deprivation explain the observed weak geographical patterns.

At the country, and region of England level, the only noticeable variations in the incidence of Hodgkin's disease (HD) that were consistent in both males and females were the below average rates in the Trent region of England. At the health authority level, incidence was slightly below the average in the southern parts of England, in Wales, and in Scotland, and slightly above average in parts of the midlands and north of England and in Ireland (Maps 2.21, 15.1a). But HD is the least common of the cancers included in this atlas, with just over 800 cases each year in males and 600 in females in the 1990s (Table 2.1); there were on average only about 10 cases each year in total at the health authority level. Survival for HD has improved steadily from about 50 per cent in the early 1970s to 80 per cent in the late 1990s,[12-15] and there were on average fewer than 400 deaths each year in total, an average of about three in each health authority. The maps, especially those for mortality, are therefore dominated by random variation.

From the early 1970s to the late 1990s, the incidence of non-Hodgkin's lymphoma (NHL) increased by a factor of about three in both males and females in England and Wales,[1] and the birth cohort trends have increased continuously since the late nineteenth century. Mortality increased gradually from the 1950s to the 1970s, increased more steeply in the 1980s, then levelled off. The diverging trends in incidence and mortality are consistent with the improvements seen in survival.[12,13] In England and Wales, incidence is slightly higher in those living in affluent areas, but there is no relationship between mortality from NHL and deprivation; this is consistent with survival being better in the more affluent.[12,13]

The geographical patterns in the incidence of NHL in males and females were broadly similar (Maps 2.22, 16.1a). Although incidence rates in only a few health authorities were more than one third above or below the UK and Ireland averages, there was a suggestion of a north-south divide in England, with higher than average rates in London and the south, and lower incidence in the midlands and north. Incidence was also higher than average in Scotland and Northern Ireland. It is unlikely that any of the known risk factors for developing NHL could explain the observed geographical variations in incidence.

There were on average about 1,800 cases of multiple myeloma in males and 1,700 in females each year in the UK and Ireland (Table 2.1). The age-standardised incidence is about 50 per cent higher in males than in females (Table 2.5). Incidence rates in England and Wales rose steadily in both males and females in the 1970s and 1980s but stabilised in the 1990s.[1] Although one-year survival improved from about 40 per cent in the early 1970s to over 60 per cent in the early 1990s, ten-year survival has remained low at below 10 per cent.[12,13,14] Consequently, the trends in mortality were similar to those in incidence.

Map **2.21**

**Hodgkin's disease: incidence* by health authority
Females, UK and Ireland 1991-99**

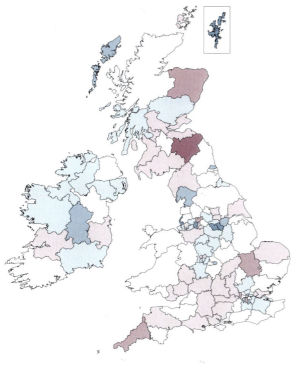

Map **2.22**

**Non-Hodgkin's lymphoma: incidence* by health authority
Females, UK and Ireland 1991-99**

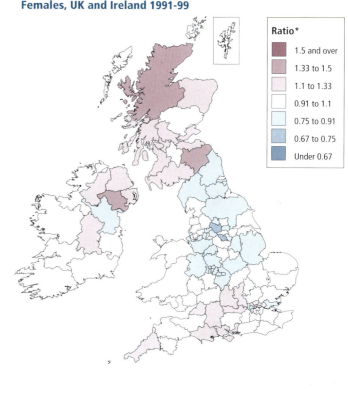

Ratio*	
	1.5 and over
	1.33 to 1.5
	1.1 to 1.33
	0.91 to 1.1
	0.75 to 0.91
	0.67 to 0.75
	Under 0.67

Map **2.23**

**Multiple myeloma: incidence* by health authority
Females, UK and Ireland 1991-99**

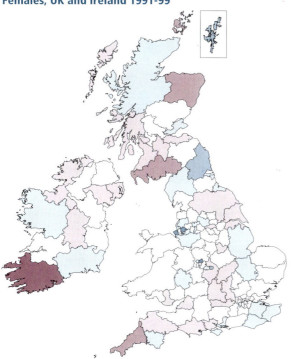

Map **2.24**

**Leukaemia: incidence* by health authority
Females, UK and Ireland 1991-99**

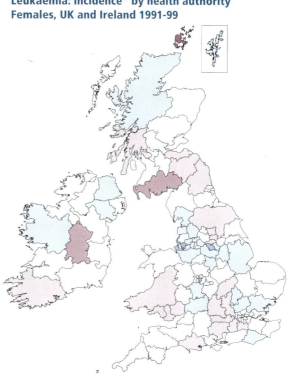

* Ratio of directly age-standardised rate in health authority to UK and Ireland average

Map **2.25**

**Lung: incidence by health authority
Females, UK and Ireland 1991-99**

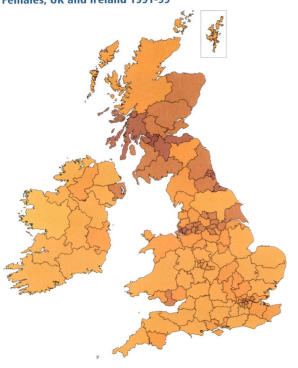

Map **2.26**

**Colorectal: incidence by health authority
Females, UK and Ireland 1991-99**

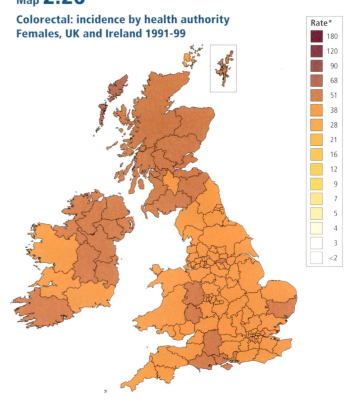

Rate*

	180
	120
	90
	68
	51
	38
	28
	21
	16
	12
	9
	7
	5
	4
	3
	<2

Map **2.27**

**Stomach: incidence by health authority
Females, UK and Ireland 1991-99**

Map **2.28**

**Ovary: incidence by health authority
UK and Ireland 1991-99**

* Rate per 100,000 population directly age standardised using the European standard population.

Map **2.29**

**Cervix: incidence by health authority
UK and Ireland 1991-99**

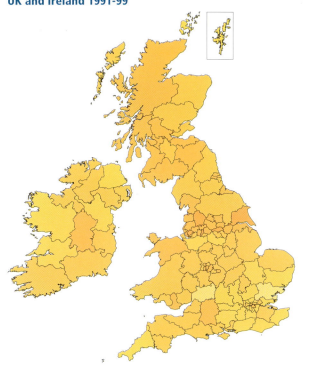

Map **2.30**

**Bladder: mortality by health authority
Females, UK and Ireland 1991-2000**

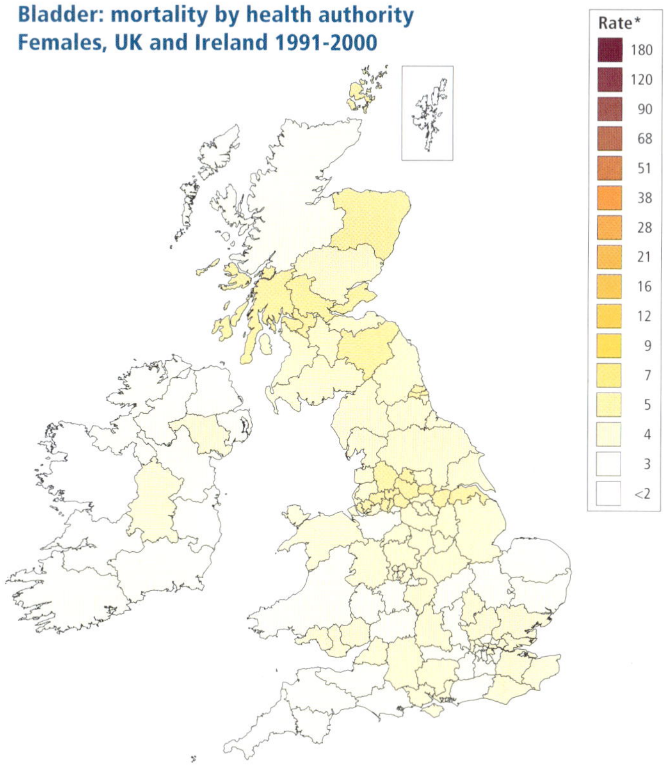

Rate*

	180
	120
	90
	68
	51
	38
	28
	21
	16
	12
	9
	7
	5
	4
	3
	<2

Map **2.31**

**Melanoma of skin: incidence by health authority
Females, UK and Ireland 1991-99**

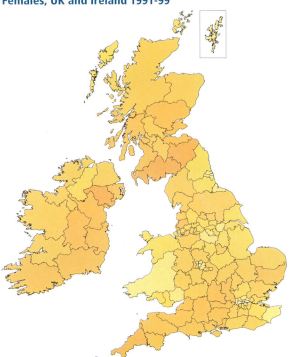

Map **2.32**

**Leukaemia: incidence by health authority
Males, UK and Ireland 1991-99**

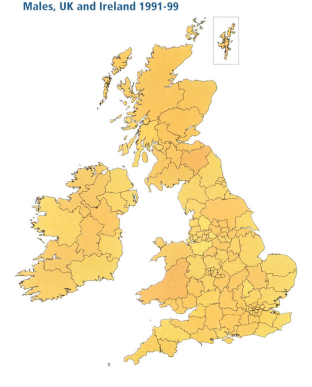

* Rate per 100,000 population directly age standardised using the European standard population.

There was relatively little variation in the incidence of multiple myeloma at the country and regional level, except that rates in both males and females were noticeably higher than average in Northern Ireland and Ireland in males, and lower than average in the North West of England in both sexes. There was some similarity between incidence in males and females at the health authority level (Maps 2.23, 15.1a), but the apparent geographical patterns are mainly due to random variation because of the fairly small numbers of cases. Most of the suspected risk factors for multiple myeloma affect only small proportions of the population and would not account for geographical variations in incidence. Incidence is also not related to socio-economic deprivation, and hence also not to any factors, including lifestyle, for which deprivation may be a marker.

There are five main types of leukaemia, but even when all the sub-types of leukaemia are combined, the numbers of cases are not large – about 6,800 in total on average each year in the UK and Ireland (Table 2.1). The male-to-female ratio (1.7:1) is slightly higher than for the other cancers in this group (Table 2.5). The incidence of leukaemia in England and Wales increased gradually during the 1970s and 1980s, but levelled off in the 1990s.[1] The birth cohort trends have shown gradual increases from the late nineteenth century up to the 1940s, but little change subsequently. Mortality rose gradually in the 1950s and 1960s, but due to improving survival[12] the trends flattened in the 1970s and 1980s – despite the increasing incidence – and fell gradually from the late 1980s onwards.[1] Incidence is slightly higher in those living in affluent areas; but as survival is higher in the affluent, there is no relationship between mortality from leukaemia and socio-economic deprivation.

Since the early 1980s there have been many investigations of the clustering of leukaemia (as a general phenomenon) and of alleged clusters in specific geographic locations. Most of these have been of cases of childhood leukaemia,[58] of which acute lymphoblastic leukaemia (ALL) is the most common. Cases of leukaemia in children (0-14 years) make up only around 5 per cent of all leukaemias.

The geographical patterns in the incidence of leukaemia were broadly similar in males and females (Maps 2.24 and 11.1a). Incidence was slightly higher than average in some (but not other) parts of the south east of England, Wales, parts of north east England and southern Scotland; and lower than average in parts of the west and north west of England. The maps on an absolute scale (Map 2.32 and Appendix E) illustrate that for leukaemia the geographical variations in both incidence and mortality were generally small – there is little variety in the colour of each map. None of the known risk factors or socio-economic deprivation explain the observed weak geographical variations.

Potential for prevention of cancer cases and deaths

Summary

- For several major cancers that have well-defined risk factors, there are wide geographical variations in incidence and mortality. It should theoretically be possible to reduce incidence and mortality rates everywhere at least to the levels found in the areas with the lowest values.

- Smoking tobacco is still a major cause of morbidity and mortality from cancer (as well as from heart disease and stroke). On the above basis, it should be possible to prevent about 6,200 cases of lung cancer in males and 4,000 in females each year in the UK and Ireland. The corresponding numbers of preventable deaths would be about 5,400 in males and 3,200 in females. The largest reductions would occur in the Northern and Yorkshire and North West regions of England, and in Scotland – and within these areas, the bulk of the reductions would occur in just 15 health authorities.

- Smoking tobacco, along with drinking alcohol and poor diet, is also a risk factor for cancers of the bladder; lip, mouth and pharynx; oesophagus; pancreas; and stomach. Reducing incidence for these cancers to the levels found in health authorities with the lowest rates would prevent about 6,100 cases in males and 2,600 in females each year in the UK and Ireland. The corresponding numbers of preventable deaths would be about 3,900 in males and 1,900 in females.

- For the ten cancers considered, on the basis of the observed geographical variations it should theoretically be possible to prevent in total about 25,600 cases and 17,500 deaths each year in the UK and Ireland.

- Far larger numbers of cancer deaths would be prevented by reducing the geographical or socio-economic differentials in incidence and mortality than by reducing the socio-economic differentials in survival.

The incidence of, and mortality from, all types of cancer varies around the world[27,59,60] and – as shown in this atlas – within the UK and Ireland. Even if risk factors for different types of cancer were not known, there would therefore be potential for preventing both cases and deaths, because it should theoretically be possible to reduce the incidence and mortality rates everywhere at least to those in the areas with among the lowest values. Estimates of preventable cases and deaths made on this basis are given in Table 2.6 for ten of the major cancers; details of the methods used are given in Appendix H.

Despite recent restrictions on the advertising and sale of cigarettes, smoking is still a major cause of morbidity and mortality, and (given the birth cohort trends) will continue to be for some considerable time. About 90 per cent of lung cancer cases are attributable to smoking. It should therefore be possible to reduce the annual numbers of lung cancer cases in the UK and Ireland from 27,000 in males and 15,000 in females by 24,300 and 13,500, respectively, to about 3,000 and 1,500, respectively. More realistically, reducing the rates everywhere to the lowest in all but a few health authorities would prevent about 6,200 newly diagnosed cases in males and 4,000 in females; the corresponding numbers of preventable deaths would be about 5,400 and 3,200, respectively. The largest reductions in the numbers of cases would occur in the Northern and Yorkshire and the North West regions of England, and in Scotland. Within these areas, almost 55 per cent of the total in Northern and Yorkshire would occur in just 5 of the 13 health authorities: County Durham, Gateshead and South Tyneside, Newcastle and North Tyneside, Sunderland, and Tees; in the North West region, 45 per cent of the total would occur in just 4 of the 16 health authorities: Liverpool, Manchester, Salford and Trafford, and St Helens and Knowsley; and in Scotland, almost 80 per cent of the total would occur in just 6 of the 15 health authorities: Argyle and Clyde, Ayrshire and Arran, Forth, Greater Glasgow, Lanark, and Lothian. In total, reductions in the 15 health authorities named above would constitute almost one third of the numbers in the whole of the UK and Ireland. In the two regions of England (Northern and Yorkshire, and North West) and in Scotland, the mortality from lung cancer in females exceeds that from breast cancer in all but a few health authorities.

Smoking is also a risk factor for cancers of the bladder; larynx; lip, mouth and pharynx; oesophagus; pancreas; and stomach. Drinking alcohol and poor diet are the other main risk factors for these cancers. Reducing their respective incidence rates to those in health authorities with the lowest rates would prevent about 6,100 cases in males and 2,600 in females each year; the corresponding numbers of deaths would be about 3,900 in males and 1,900 in females.

Table 2.6

Potential for prevention of cancer cases and deaths each year[1], UK and Ireland

Cancer	Cases			Deaths		
	Males	Females	Total	Males	Females	Total
Lung						
Northern and Yorkshire	1,030	690	1,720	920	590	1,510
Trent	560	270	830	460	210	670
West Midlands	550	220	770	480	180	660
North West	1,110	700	1,810	960	580	1,540
Eastern	170	140	310	150	110	260
London	740	520	1,260	630	420	1,050
South East	320	250	570	230	180	410
South West	60	70	130	60	40	100
England	4,540	2,860	7,400	3,890	2,310	6,200
Wales	320	180	500	220	130	350
Scotland	1,180	790	1,970	1,020	620	1,640
Northern Ireland	120	90	210	110	60	170
Ireland	40	90	130	120	120	240
UK and Ireland	6,200	4,010	10,210	5,360	3,240	8,600
Bladder[2]	1,350	510	1,860	610	250	860
Larynx	690	..	690	290	..	290
Lip, mouth and pharynx	1,210	400	1,610	550	170	720
Oesophagus	820	370	1,190	820	370	1,190
Pancreas	310	400	710	320	420	740
Stomach	1,750	900	2,650	1,260	660	1,920
Sub-total	6,130	2,580	8,710	3,850	1,870	5,720
Colorectal	3,370	2,220	5,590	1,570	800	2,370
Melanoma	..	..	..	170	140	310
Cervix	:	1,070	1,070	:	450	450
Total	15,700	9,880	25,580	10,950	6,500	17,450

.. Not estimated   : Not applicable
1 Estimates made on the basis that rates everywhere could be reduced to those in the health authorities with among the lowest rates (see Appendix H for details of calculations). Estimates have been rounded individually and then summed.
2 Because of differences among the cancer registries in the coding of bladder cancer (see Chapter 3), the estimates of preventable cases were based on the estimates of preventable deaths divided by appropriate mortality-to-incidence ratios.

Although neither the incidence of, nor mortality from, colorectal cancer varies markedly with socio-economic deprivation[1] there were some geographical variations in both incidence and mortality, with above average rates in the urban areas of the midlands and north of England, and in Scotland, Northern Ireland and Ireland. As this cancer is the third most common in males and the second most common in females, reducing the rates everywhere to those in the health authorities with the lowest rates – possibly through improvements in diet – would prevent large numbers of cases: about 3,400 in males and 2,200 in females; about one third of these would be in Scotland, Northern Ireland and Ireland. The corresponding numbers of deaths would be about 1,600 in males and 800 in females. Reductions in mortality may eventually be achieved following the introduction of faecal occult blood screening[61] – any reductions due to such screening would take some time to be achieved because five-year relative survival from colorectal cancer is around 50 per cent and so for several years many of the deaths from colorectal cancer would occur in patients diagnosed before screening started. The initial results from a trial of screening by flexi-sigmoidoscopy are also encouraging.[62]

In the early 1970s, the incidence of melanoma of the skin was only about 25 per cent of the rates at the end of the twentieth century. It should therefore be possible to reduce the current rates by 75 per cent, thereby preventing around 4,400 cases each year. This seems unrealistic; and there is evidence that higher rates of incidence, as occur for example in Scotland due in part to higher awareness as a result of public health campaigns and to earlier detection of tumours at an earlier stage of development, are associated with better survival. If mortality rates were reduced everywhere to those in the health authorities with among the lowest rates, this would prevent about 300 deaths each year.

Despite the undoubted success of the re-organised national cervical screening programme in reducing the incidence of cervical cancer (and hence also the mortality),[7-9] there are wide geographical variations in incidence. As noted above, the patterns were not related to differences in the uptake or effectiveness of the cervical screening programme, as similar geographical patterns existed in the 1970s (although the rates were much higher). Reducing the incidence and mortality rates everywhere to those in the health authorities with the lowest rates would prevent over 1,000 cases and about 450 deaths each year.

There was little geographical variation in the incidence of breast cancer in females, and virtually none at all in mortality; and most of the risk factors relate to a woman's reproductive history and are not amenable to primary prevention.[63] About one third of the reduction in breast cancer mortality expected to result directly from the national breast screening programme had been seen in the late 1990s,[6] so by around 2008 there should be a further 900 or so fewer deaths each year in women aged 55-69. The extension of screening to women aged up to 70 will bring additional reductions in deaths from breast cancer, mostly in women aged 70-75.

Much of the geographical variation in the incidence of prostate cancer may be due to variations in the local availability and uptake of PSA testing, so it is difficult to say to what levels the incidence rates might be reduced. Specific causal factors for prostate cancer have not been identified, although incidence is slightly higher in the most affluent groups than in the most deprived.[1]

As with breast cancer in females, there was almost no geographical variation in mortality from prostate cancer. It is not yet known whether earlier diagnosis achieved using PSA testing will result in reduced mortality from prostate cancer. Survival from the disease may appear to improve solely because of increased lead-time, without any lengthening of life, or because the asymptomatic cancers detected by screening tend to grow more slowly than the symptomatic, and such patients appear to survive longer. There are several strong arguments against screening: high prevalence rates at post mortem; high biopsy rates, over-diagnosis and over-treatment;[64] one third of screen-detected cases are incurable; no clear benefit of treatment; side effects of prostatectomy include impotence in a large proportion of cases and incontinence in a smaller proportion; costs, both of screening, and of follow-up and treatment (much of which may be unnecessary) are potentially huge; in many elderly patients there are few years of life to gain; and – crucially – no consequent reduction in mortality has been demonstrated.[65,66] Despite the widespread use of such tests in the USA, and apparent incidence rates almost three times higher than in the UK, mortality in the USA is (and has been for many years) almost the same as in the UK (see Figure 4 in reference 31) and other European countries. Large randomised controlled trials are underway in Europe[67] and the USA[68] to determine the effectiveness of PSA testing as a screening tool. If recent decreases in US prostate cancer mortality are due to early detection as a result of PSA testing, the randomised trials will show early evidence of a mortality benefit.[69] If a randomised controlled trial of screening for prostate cancer were to demonstrate a consequent worthwhile reduction in mortality, a large number of social, psychological, organisational and economic factors would still have to be considered before the introduction of a population-based screening programme.[70]

The cancers not considered here: brain, Hodgkin's disease, kidney, leukaemia, multiple myeloma, non-Hodgkin's lymphoma, ovary, testis and uterus are either less common; or have risk factors which are unknown or not amenable to primary prevention or affect only small proportions of the population; or show little geographical variation. Nevertheless, there must still be some potential for preventing cases of, and deaths from, these cancers. Trials of screening for ovarian cancer are in progress.[71]

In total for the ten cancers considered in Table 2.6, about 25,600 cases, 15,700 in males and 9,900 in females, should, in theory, be preventable; the corresponding numbers of deaths are 17,500 in total, with 11,000 in males and 6,500 in females.

In the *Cancer Trends* book[1] estimates were made of the numbers of preventable cases and deaths in England and Wales from various cancers for which the incidence or morality rates were related to socio-economic deprivation (based on the Carstairs index).[10] For ten cancers that had higher rates in the most deprived groups than in the most affluent, it was assumed that the rates in all groups of society could be reduced to those in the affluent. Based on data for the early 1990s, it was estimated that about 20,000 cases and 16,600 deaths each year from these cancers could be prevented by eliminating the socio-economic differentials in their rates.

The analyses of survival presented in the *Cancer Survival Trends* book[12] showed that for 44 of 47 major cancers in adults, there was either strong or limited evidence that survival rates were lower in those living in the most deprived areas compared with the most affluent. Estimates were made of deaths attributable to cancer that might be avoidable if patients in all groups of society were to have the same survival (at five years from diagnosis) as those in the most affluent groups. In total, just over 2,500 deaths each year were avoidable, 40 per cent of those in patients living in the 20 per cent of areas that were most deprived.

Although the numbers of avoidable deaths resulting from inequalities in cancer survival are substantial, much larger numbers of cases and deaths from cancer could be prevented by reducing the incidence and mortality rates in all patients to those observed in the most affluent group. Possibly even larger numbers – about 25,600 cases and 17,500 deaths – could be prevented by reducing incidence and mortality rates across the UK and Ireland to those in the health authorities with among the lowest values.

References

1. Quinn MJ, Babb PJ, Brock A, Kirby L, Jones J. *Cancer Trends in England and Wales 1950-1999.* Studies on Medical and Population Subjects No. 66. London: The Stationery Office, 2001.

2. General Register Office for Scotland. *Scotland's Population 2001: The Registrar General's Annual Review of Demographic Trends.* Edinburgh: General Register Office for Scotland, 2002.

3. Office of Population Censuses and Surveys. *Mortality statistics 1984: Cause, England and Wales.* Series DH2 No.11 11. London: HMSO, 1985.

4. ONS. *Mortality Statistics: cause 1993 (revised) and 1994: Cause, England and Wales.* Series DH2 No.21. London: HMSO, 1996.

5. Rooney C, Devis T. Mortality trends by cause of death in England and Wales 1980-94: the impact of introducing automated cause coding and related changes in 1993. *Population Trends* 1996: 29-35.

6. Blanks RG, Moss SM, McGahan CE, Quinn MJ et al. Effect of NHS breast screening programme on mortality from breast cancer in England and Wales, 1990-8: comparison of observed with predicted mortality. *British Medical Journal* 2000; 321: 665-669.

7. Quinn M, Babb P, Jones J, Allen E. Effect of screening on incidence of and mortality from cancer of cervix in England: evaluation based on routinely collected statistics. *British Medical Journal* 1999; 318: 904-908.

8. Sasieni P, Adams J. Effect of screening on cervical cancer mortality in England and Wales: analysis of trends with an age period cohort model. *British Medical Journal* 1999; 318: 1244-1245.

9. Sasieni P, Adams J. Analysis of cervical cancer mortality and incidence data from England and Wales: evidence of a beneficial effect of screening. *Journal of the Royal Statistical Society Series A* 2000; 163: 191-209.

10. Carstairs V, Morris R. Deprivation and mortality: an alternative to social class? *Community Medicine* 1989; 11: 213-219.

11. Griffiths C, Fitzpatrick J. *Geographic Variations in Health.* Decennial Supplements No. 16. London: The Stationery Office, 2001.

12. Coleman MP, Babb P, Damiecki P, Grosclaude P et al. *Cancer Survival Trends in England and Wales, 1971-1995: Deprivation and NHS Region.* Studies on Medical and Population Subjects No. 61. London: The Stationery Office, 1999.

13. Coleman MP, Rachet B, Woods LM, Mitry E et al. Trends and socioeconomic inequalities in cancer survival in England and Wales up to 2001. *British Journal of Cancer* 2004; 90: 1367-1373.

14. ONS. Cancer Survival: England and Wales, 1991-2001. March 2004. Available at *http://www.statistics.gov.uk/StatBase/Product.asp?vlnk=1 0821&Pos=2&ColRank=1&Rank=272.*

15. Black R, Brewster D, Brown H, Fraser L et al. *Trends in Cancer Survival in Scotland 1971-1995.* Edinburgh: Information and Statistics Division, 2000.

16. Higginson J, Muir CS, Munoz N. *Human Cancer: Epidemiology and Environmental Causes.* Cambridge Monographs on Cancer Research. Cambridge: Cambridge University Press, 1992.

17. Potter, J. D. *Food, Nutrition and the Prevention of Cancer: a Global Perspective.* Washington: World Cancer Research Fund in association with American Institute for Cancer Research, 1997.

18. Bingham SA, Day NE, Luben R, Ferrari P et al. Dietary fibre in food and protection against colorectal cancer in the European Prospective Investigation into Cancer and Nutrition (EPIC): an observational study. *Lancet* 2003; 361: 1496-1501.

19. Sandhu MS, White IR, McPherson K. Systematic review of the prospective cohort studies on meat consumption and colorectal cancer risk: a meta-analytical approach. *Cancer Epidemiology Biomarkers and Prevention* 2001; 10: 439-446.

20. Norat T, Lukanova A, Ferrari P, Riboli E. Meat consumption and colorectal cancer risk: dose-response meta-analysis of epidemiological studies. *International Journal of Cancer* 2002; 98: 241-256.

21. Lichtenstein P, Holm NV, Verkasalo PK, Iliadou A et al. Environmental and heritable factors in the causation of cancer - analyses of cohorts of twins from Sweden, Denmark, and Finland. *New England Journal of Medicine* 2000; 343: 78-85.

22. Anderson KE, Potter JD, Mack TM. Pancreatic Cancer. In: Schottenfeld D, Fraumeni JF, Jr. (eds) *Cancer Epidemiology and Prevention, second edition.* New York: Oxford University Press, 1996.

23. Kuper H, Boffetta P, Adami HO. Tobacco use and cancer causation: association by tumour type. *Journal of Internal Medicine* 2002; 252: 206-224.

24. Twigg L, Moon G, Walker S. *The Smoking Epidemic in England.* London: Health Development Agency, 2004.

25. Seidell JC, Flegal KM. Assessing obesity: classification and epidemiology. *British Medical Bulletin* 1997; 53: 238-252.

26. Newnham A, Quinn MJ, Babb P, Kang JY et al. Trends in oesophageal and gastric cancer incidence, mortality and survival in England and Wales 1971-1998/1999. *Alimentary Pharmacology and Therapeutics* 2003; 17: 655-664.

27. Parkin DM, Whelan SL, Ferlay J, Teppo L et al. *Cancer Incidence in Five Continents Vol. VIII.* IARC Scientific Publications No. 155. Lyons: International Agency for Research on Cancer, 2000.

28. Majeed A, Babb P, Jones J, Quinn M. Trends in prostate cancer incidence, mortality and survival in England and Wales 1971-1998. *BJU International* 2000; 85: 1058-1062.

29. Quinn M, Babb P. Patterns and trends in prostate cancer incidence, survival, prevalence and mortality. Part II: individual countries. *BJU International* 2002; 90: 174-184.

30. Evans HS, Moller H. Recent trends in prostate cancer incidence and mortality in southeast England. *European Urology* 2003; 43: 337-341.

31. Cartwright RA, Elwood PC, Birch J, Tyrell C et al. Aetology of testicular cancer: association with congenital abnormalities, age at puberty, infertility, and exercise. *British Medical Journal* 1994; 308: 1393-1399.

32. National Cancer Institute. Surveillance, Epidemiology and End Results (SEER). 2005. Available at *http://seer.cancer.gov/.*

33. Wynder E, Stellman S. Environmental factors in the causation of bladder cancer. In: Connolly J (ed) *Carcinoma of the Bladder.* New York: Raven Press, 1981.

34. Ross RK, Paganini-Hill A, Hendersen BE. Epidemiology of bladder cancer. In: Skinner DG, Lieskovsky G (eds) *Diagnosis and Management of Genitourinary Cancer.* Philadelphia: W.B. Saunders Co., 1988.

35. Doll R, Peto R. Mortality in relation to smoking: 20 years' observations on male British doctors. *British Medical Journal* 1976: 1525-1536.

36. Hartge P, Silverman D, Hoover R, Schairer C et al. Changing cigarette habits and bladder cancer risk: a case-control study. *Journal of the National Cancer Institute* 1987; 78: 1119-1125.

37. Hartge P, Silverman DT, Schairer C, Hoover RN. Smoking and bladder cancer risk in blacks and whites in the United States. *Cancer Causes and Control* 1993; 4: 391-394.

38. Bartsch H, Caporaso N, Coda M, Kadlubar F et al. Carcinogen hemoglobin adducts, urinary mutagenicity, and metabolic phenotype in active and passive cigarette smokers. *Journal of the National Cancer Institute* 1990; 82: 1826-1831.

39. McLaughlin JK, Silverman DT, Hsing AW, Ross RK et al. Cigarette smoking and cancers of the renal pelvis and ureter. *Cancer Research* 1992; 52: 254-257.

40. Jensen OM, Knudsen JB, McLaughlin JK, Sorensen BL. The Copenhagen case-control study of renal pelvis and ureter cancer: role of smoking and occupational exposures. *International Journal of Cancer* 1988; 41: 557-561.

41. McCredie M, Stewart JH. Risk factors for kidney cancer in New South Wales. I. Cigarette smoking. *European Journal of Cancer* 1992; 28A: 2050-2054.

42. McLaughlin JK, Mandel JS, Blot WJ, Schuman LM et al. A population-based case-control study of renal cell carcinoma. *Journal of the National Cancer Institute* 1984; 72: 275-284.

43. La Vecchia C, Negri E, D'Avanzo B, Franceschi S. Smoking and renal cell carcinoma. *Cancer Research* 1990; 50: 5231-5233.

44. Kreiger N, Marrett LD, Dodds L, Hilditch S et al. Risk factors for renal cell carcinoma: results of a population-based case-control study. *Cancer Causes and Control* 1993; 4: 101-110.

45. Mellemgaard A, Engholm G, McLaughlin JK, Olsen JH. Risk factors for renal cell carcinoma in Denmark. I. Role of socioeconomic status, tobacco use, beverages, and family history. *Cancer Causes and Control* 1994; 5: 105-113.

46. Bergstrom A, Pisani P, Tenet V, Wolk A et al. Overweight as an avoidable cause of cancer in Europe. *International Journal of Cancer* 2001; 91: 421-430.

47. Mellemgaard A, Engholm G, McLaughlin JK, Olsen JH. Risk factors for renal-cell carcinoma in Denmark. III. Role of weight, physical activity and reproductive factors. *International Journal of Cancer* 1994; 56: 66-71.

48. McCredie M, Stewart JH. Risk factors for kidney cancer in New South Wales, Australia. II. Urologic disease, hypertension, obesity, and hormonal factors. *Cancer Causes and Control* 1992; 3: 323-331.

49. Swerdlow AJ, dos SS, I, Reid A, Qiao Z et al. Trends in cancer incidence and mortality in Scotland: description and possible explanations. *British Journal of Cancer* 1998; 77 Suppl 3: 1-54.

50. Cook A, Woodward A, Pearce N, Marshall C. Cellular telephone use and time trends for brain, head and neck tumours. *New Zealand Medical Journal* 2003; 116: U457.

51. Lonn S, Ahlbom A, Hall P, Feychting M. Mobile phone use and the risk of acoustic neuroma. *Epidemiology* 2004; 15: 653-659.

52. MacKie RM, Bray CA, Hole DJ, Morris A et al. Incidence of and survival from malignant melanoma in Scotland: an epidemiological study. *Lancet* 2002; 360: 587-591.

53. Coebergh JWW, Janssen-Heijenen MLG, Louwman WJ, Voogd AC. *Cancer incidence and survival in the south of the Netherlands, 1955-1999 and incidence in the north of Belgium, 1996-1998.* Eindhoven: Comprehensive Cancer Centre South, 2001.

54. Bulliard JL. Site-specific risk of cutaneous malignant melanoma and pattern of sun exposure in New Zealand. *International Journal of Cancer* 2000; 85: 627-632.

55. Garbe C, Buttner P, Bertz J, Burg G et al. Primary cutaneous melanoma. Prognostic classification of anatomic location. *Cancer* 1995; 75: 2492-2498.

56. Balch CM, Soong SJ, Gershenwald JE, Thompson JF et al. Prognostic factors analysis of 17,600 melanoma patients: validation of the American Joint Committee on Cancer melanoma staging system. *Journal of Clinical Oncology* 2001; 19: 3622-3634.

57. Streetly A, Markowe H. Changing trends in the epidemiology of malignant melanoma: gender differences and their implications for public health. *International Journal of Epidemiology* 1995; 24: 897-907.

58. Alexander FE, Boyle P. *Methods for Investigating Localised Clustering of Disease.* IARC Scientific Publications No.135. Lyon: International Agency for Research on Cancer, 1996.

59. Ferlay J, Parkin DM, Pisani P. *GLOBOCAN 1: Cancer Incidence and Mortality Worldwide.* Lyon: IARC Press, 1998.

60. Parkin DM. Global cancer statistics in the year 2000. *The Lancet Oncology* 2001; 2: 533-543.

61. Steele RJ, Parker R, Patnick J, Warner J et al. A demonstration pilot trial for colorectal cancer screening in the United Kingdom: a new concept in the introduction of healthcare strategies. *Journal of Medical Screening* 2001; 8: 197-202.

62. UK Flexible Sigmoidoscopy Screening Trial Investigators. Single flexible sigmoidoscopy screening to prevent colorectal cancer: baseline findings of a UK multicentre randomised trial. *The Lancet* 2002; 359: 1291-1300.

63. McPherson K, Steel CM, Dixon JM. Breast cancer - epidemiology, risk factors, and genetics. *British Medical Journal* 1994; 309: 1003-1006.

64. Ciatto S, Zappa M, Bonardi R, Gervasi G. Prostate cancer screening: the problem of overdiagnosis and lessons to be learned from breast cancer screening. *European Journal of Cancer* 2000; 36: 1347-1350.

65. Selley S, Donovan J, Faulkner A, Coast J et al. Diagnosis, management and screening of early localised prostate cancer. *Health Technology Assessment* 1997; 1: 1-96.

66. Chamberlain J, Melia J, Moss S, Brown J. The diagnosis, management, treatment and costs of prostate cancer in England and Wales. *Health Technology Assessment* 1997; 1: i-53.

67. Schroder FH, Damhuis RA, Kirkels WJ, De Koning HJ et al. European randomized study of screening for prostate cancer--the Rotterdam pilot studies. *International Journal of Cancer* 1996; 65: 145-151.

68. Gohagan JK, Prorok PC, Kramer BS, Cornett JE. Prostate cancer screening in the prostate, lung, colorectal and ovarian cancer screening trial of the National Cancer Institute. *Journal of Urology* 1994; 152: 1905-1909.

69. Tarone RE, Chu KC, Brawley OW. Implications of stage-specific survival rates in assessing recent declines in prostate cancer mortality rates. *Epidemiology* 2000; 11: 167-170.

70. Wilson JMG, Junger G. *Principles and Practice of Screening for Disease*. WHO Public Health Paper 34. Geneva: World Health Organisation, 1968.

71. Lewis S, Menon U. Screening for ovarian cancer. *Expert Review of Anticancer Therapy* 2003; 3: 55-62.

Chapter 3
Bladder

Nicola Cooper, Ray Cartwright

Summary

- In the UK and Ireland, bladder cancer accounted for around 1 in 20 cases of all cancers and 1 in 30 cancer deaths in the 1990s.

- Geographical patterns in incidence are obscured by known differences among the countries and regions of England in the classification and registration of bladder tumours.

- Mortality in males was over three times that in females. Mortality in both males and females was higher than average in Scotland and in a band across the north of England, and markedly lower in Northern Ireland and Ireland.

- Smoking, which is related to socio-economic deprivation, is an established risk factor for bladder cancer. But neither the incidence nor the mortality trends in bladder cancer correspond with those in lung cancer in either sex, by either time period or birth cohort; and there is only a weak relationship between bladder cancer and deprivation. There is also much less geographical variation in bladder cancer mortality than for lung cancer, both between and within countries and the regions of England.

- Occupational exposure to chemicals, predominantly in male workers in the dye and rubber industries, may explain some of the observed geographical patterns.

Introduction

There are well-recognised differences in the classification and registration of some bladder tumours, which are recorded as malignant by some cancer registries and as non-malignant (benign) by others. These differences in the coding of non-invasive transitional cell papillomas in the bladder will be resolved through the implementation by all registries of the recommendations made by the European Network of Cancer Registries. Registries with a high percentage of all bladder tumours recorded as malignant will have defined transitional cell papillomas as malignant tumours. For the period 1995-99, 77 per cent of bladder tumours were classified as malignant in Scotland and 89 per cent in Wales; the equivalent figures for Northern Ireland and Ireland were 55 per cent and 94 per cent,

respectively. In England, over 90 per cent of bladder cancers were recorded as malignant in five of the nine regional cancer registries. The remaining four registries in England – Northern and Yorkshire, North Western, East Anglia and Thames – classified between 60 and 77 per cent of their bladder tumours as malignant.[1] This difference in the coding and classification of bladder tumours contributes considerably to some of the apparent geographical variation in the incidence of bladder cancer that is illustrated in the charts and maps in this chapter.

Incidence and mortality

In the UK and Ireland, there were about 9,700 cases of bladder cancer in males and 3,900 in females diagnosed each year during the 1990s. Bladder cancer was the fourth most common cancer in males, while in females it ranked eighth. It accounted for 7 per cent of all male cancers and 3 per cent of all female cancers. The age-standardised incidence rates were 29.2 per 100,000 in males and 8.3 per 100,000 in females, a ratio of around 3.5:1. Incidence rates vary considerably by age – bladder cancer is rare in those aged under 40 and rises steeply from age 60 to peak in those aged 85 and over in both sexes. The lifetime risk[2,3] of being diagnosed with bladder cancer, based on England and Wales data for 1997, is 1 in 30 (3.3 per cent) for males and 1 in 79 (1.3 per cent) for females.[4]

In the 1990s, around 3,600 males and 1,800 females died from bladder cancer each year in the UK and Ireland, representing 4 per cent of deaths from all cancers in males and 2 per cent in females, respectively. Bladder cancer was the sixth most common cause of cancer death in males and the tenth most common in females. The age-standardised mortality rates were 10.5 per 100,000 in males and 3.3 per 100,000 in females, a male-to-female ratio of 3.2:1. The mortality rates by age show the same pattern as incidence rates, in that they increase steeply with age.

Incidence and mortality trends

Incidence up to the 1980s increased markedly in both males and females in Scotland with rates for males and females in 1970 of 29.6 and 7.4 per 100,000, respectively, and in 1985 of 40.5 and 13.3 per 100,000.[5] In England and Wales, between 1971 and 1998, directly age-standardised incidence increased by 16 per cent in males and 37 per cent in females.[6] The larger rise in females, particularly in the younger birth cohort, suggests at least partly differing aetiology between the sexes.

Mortality, however, reached a peak in the UK and Ireland in the 1970s and has since declined in all five countries. In England and Wales, directly age-standardised mortality fell by 26 per cent in males between 1971 and 1998, and showed little change in females over the same period.[6] The lifetime risk of bladder cancer in England and Wales for males aged 30-74 years and born in 1900 was 8.7 per 1,000 and 4.8 for those born in 1940; the figures for females were 2.1 and 1.5, respectively.[5] The decline in mortality in males resembles that for lung cancer but is of a lower magnitude.[5]

Survival

Bladder cancer is one of the few cancers in which men have a substantial survival advantage over women. In England and Wales, relative survival rates at one year after diagnosis were 81 per cent and 73 per cent, in men and women, respectively, and after five years were 64 and 56 per cent, respectively, for patients diagnosed in 1996-99.[7] Survival had improved similarly in both males and females, by approximately 4 and 6 percentage points, respectively, every five years during the 1970s and early 1980s,[8] but there was less improvement in survival thereafter. In the early 1990s, survival from bladder cancer in England, Wales and Scotland was within 2 percentage points of the average for Europe for males; survival for females in England and Wales was within 3 percentage points of the average, but in Scotland it was over 7 percentage points lower.[9] There was also variation in the survival rates in the regions of England;[10] some of this variation may be the consequence of the differences in the classification and registration of some bladder tumours mentioned above. Survival rates for transitional cell papillomas are very high, so where cancer registries define these as malignant, the survival rate also appears to be higher than in regions where they are not so defined.

Geographical patterns in incidence

The incidence rates for bladder cancer appear to show wide variability among the countries of the UK and Ireland. Scotland and Wales both had rates that were around 10 and 24 per cent higher than the average for the UK and Ireland for both sexes. Northern Ireland and Ireland had similar rates that were between 20-30 per cent lower than the average for both sexes (Figure 3.1 and Table B3.1). There was less variation between the regions of England than between the countries. Trent and West Midlands had the highest incidence rates of around 32 and 9 per 100,000 population for males and females, respectively, and Eastern the lowest with rates of 24 and 6 per 100,000. Incidence by health authority areas showed more variation, but typically the results within the countries and English regions reflect their overall average but with some outliers (Figure 3.3 and Table B3.1).

The maps were internally consistent for all of Ireland, showing lower rates overall. For Wales, the incidence rates were generally slightly above average. For England and Scotland the

patterns were more complex. There were wider variations in incidence for females than for males. The highest rates in females occurred in the central east coast of Scotland, urban north west of England around Liverpool (St. Helens – males 38.2 and females 12.6 per 100,000), a band across urban Yorkshire and Humberside, and a further band from Buckinghamshire through to Dorset in the south of England. However, there were some similarities between the male and female patterns (which is clearer at the finer geographical level) in that the south east of England, rural north of England and rural Scotland were similar for both sexes and have lower than average rates. In England, the lowest rates were recorded in Cambridgeshire in the Eastern region: 18.4 per 100,000 in males and 4.9 per 100,000 in females.

Incidence rates in females were one third of those in males and the resulting low numbers in some areas could account for the apparent greater fluctuation around the average. In summary, the geographical patterns were complex with some of the differences in incidence attributable to differences in the registration and classification of bladder tumours by the cancer registries. There were similar geographical patterns in the male and female rates, irrespective of the magnitude of the rates.

Geographical patterns in mortality

The mortality rates for bladder cancer among the countries of the UK and Ireland (Figure 3.2) did not show such wide variability as did incidence. Scotland had mortality rates for males and females that were 10 and 24 per cent, respectively, higher than the average for the UK and Ireland – as were its incidence rates. Wales, although having similar incidence rates to Scotland, had mortality rates for males and females of 11 and 6 per cent, respectively, lower than the UK and Ireland average. Northern Ireland and Ireland had mortality rates that were 18-34 per cent lower than average, as were their incidence rates.

The regions of England showed little variation in mortality rates for bladder cancer compared with that among the other four countries. The Northern and Yorkshire, Trent and North West regions all had age-standardised mortality rates slightly above the UK and Ireland average for both males and females: just over 11 per 100,000 in males and around 3.5 per 100,000 in females. The South West and West Midlands regions both had mortality rates that were slightly below average.

Overall, there was little variability in bladder cancer mortality in males at the health authority level (Map 3.2). Within Scotland, slightly higher than average rates of bladder cancer mortality for males occurred in Grampian, and in Glasgow and its surrounding areas.

(continued on page 48)

Figure **3.1**

**Bladder: incidence by sex, country, and region of England
UK and Ireland 1991-99[1]**

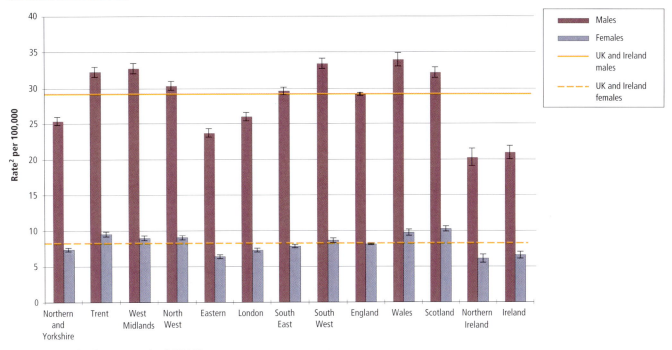

1 Northern Ireland 1993-99, Ireland 1994-99

2 Age standardised using the European standard population, with 95% confidence interval

Figure **3.2**

**Bladder: mortality by sex, country, and region of England
UK and Ireland 1991-2000[1]**

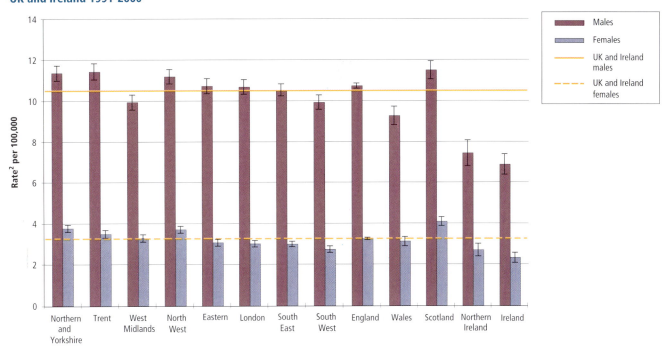

1 Scotland 1991-99, Ireland 1994-2000

2 Age standardised using the European standard population, with 95% confidence interval

Figure **3.3a**

**Bladder: incidence by health authority within country, and region of England
Males, UK and Ireland 1991-99¹**

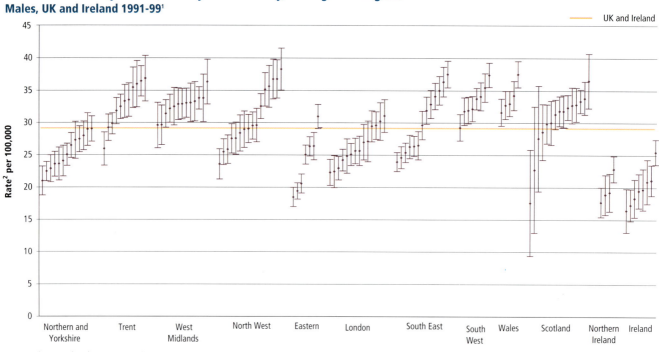

1 Northern Ireland 1993-99, Ireland 1994-99

2 Age standardised using the European standard population, with 95% confidence interval

Figure **3.3b**

**Bladder: incidence by health authority within country, and region of England
Females, UK and Ireland 1991-99¹**

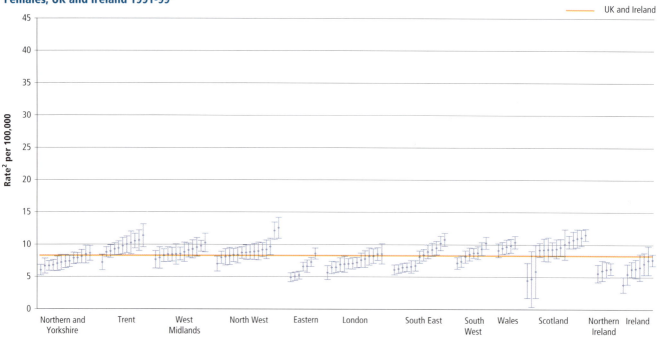

1 Northern Ireland, 19913-99, Ireland 1994-99

2 Age standardised using the European standard population, with 95% confidence interval

Figure **3.4a**

**Bladder: mortality by health authority within country, and region of England
Males, UK and Ireland 1991-2000[1]**

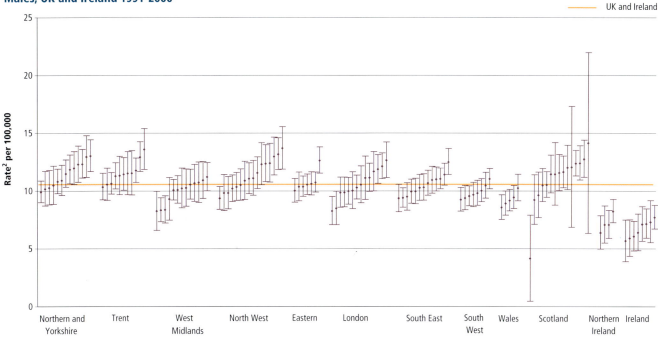

1 Scotland 1991-99, Ireland 1994-2000

2 Age standardised using the European standard population, with 95% confidence interval

Figure **3.4b**

**Bladder: mortality by health authority within country, and region of England
Females, UK and Ireland 1991-2000[1]**

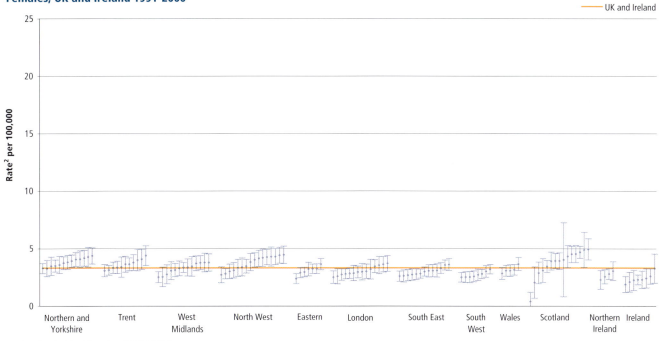

1 Scotland 1991-99, Ireland 1994-2000

2 Age standardised using the European standard population, with 95% confidence interval

Map **3.1a**

**Bladder: incidence* by health authority
Males, UK and Ireland 1991-99**

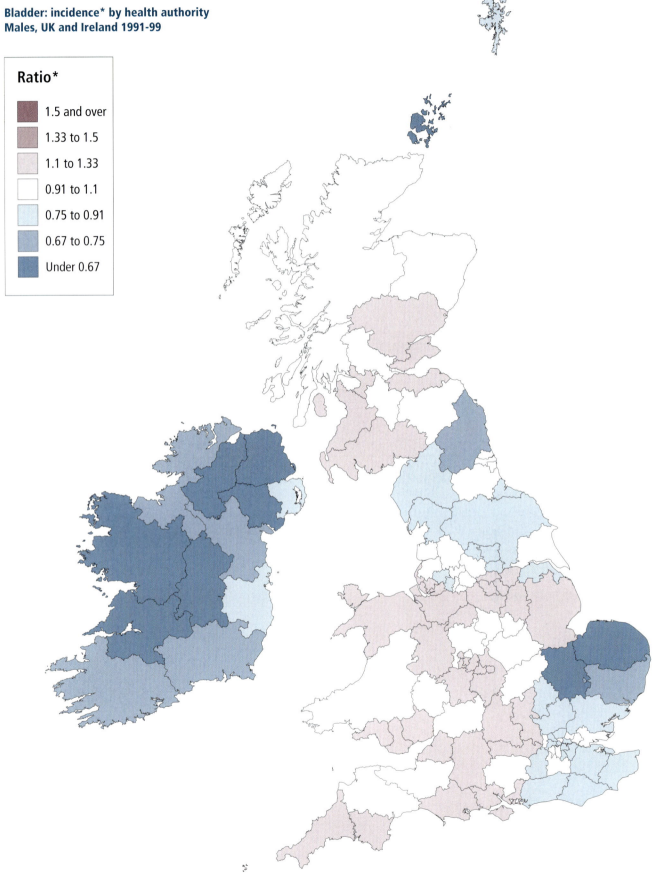

Ratio*

	1.5 and over
	1.33 to 1.5
	1.1 to 1.33
	0.91 to 1.1
	0.75 to 0.91
	0.67 to 0.75
	Under 0.67

*Ratio of directly age-standardised rate in health authority to UK and Ireland average

Map **3.1b**

**Bladder: incidence* by health authority
Females, UK and Ireland 1991-99**

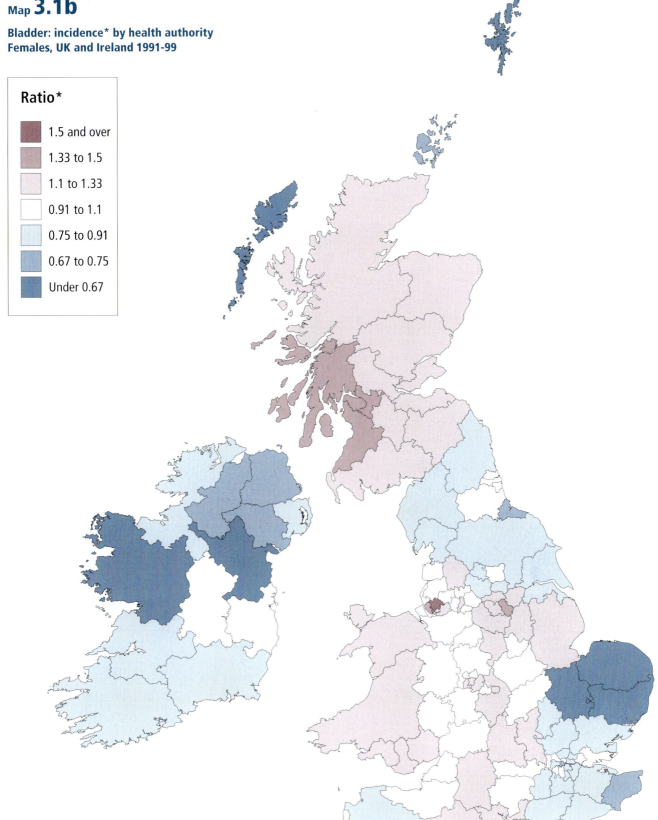

Ratio*

	1.5 and over
	1.33 to 1.5
	1.1 to 1.33
	0.91 to 1.1
	0.75 to 0.91
	0.67 to 0.75
	Under 0.67

*Ratio of directly age-standardised rate in health authority to UK and Ireland average

Map **3.2a**

Bladder: mortality* by health authority
Males, UK and Ireland 1991-2000

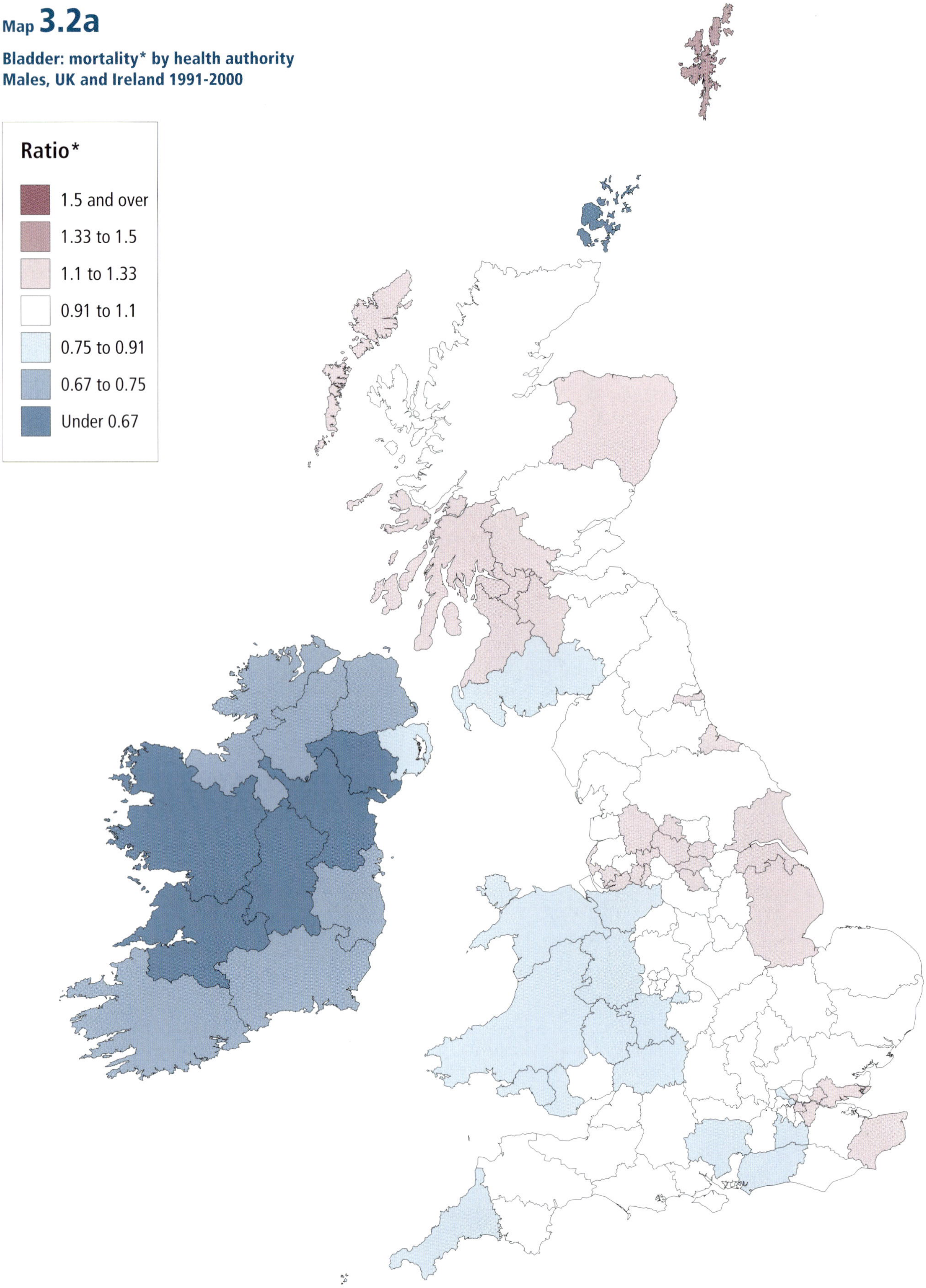

Ratio*

■	1.5 and over
■	1.33 to 1.5
■	1.1 to 1.33
□	0.91 to 1.1
■	0.75 to 0.91
■	0.67 to 0.75
■	Under 0.67

Ratio of directly age-standardised rate in health authority to UK and Ireland average

Map **3.2b**

Bladder: mortality* by health authority
Females, UK and Ireland 1991-2000

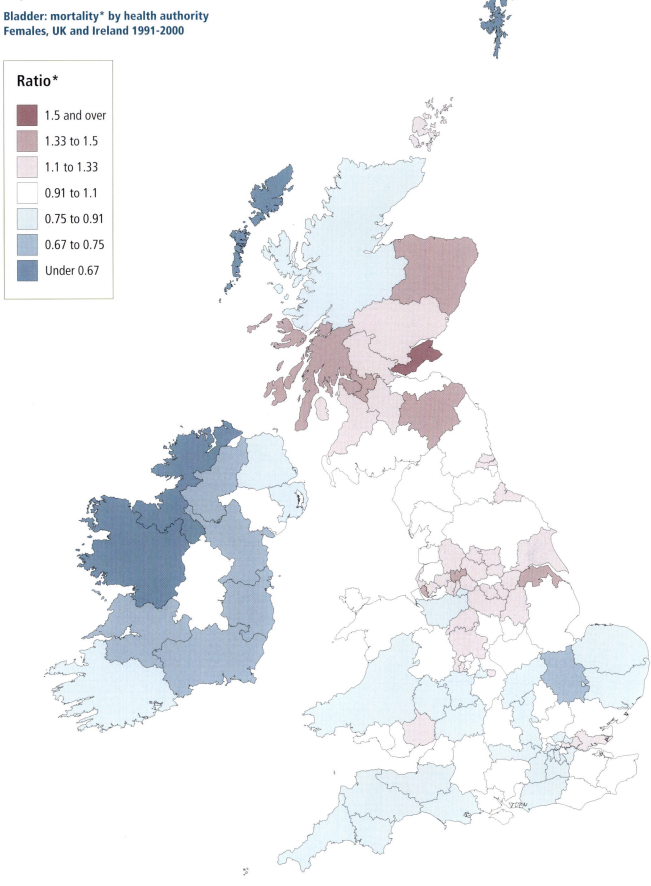

Ratio*

	1.5 and over
	1.33 to 1.5
	1.1 to 1.33
	0.91 to 1.1
	0.75 to 0.91
	0.67 to 0.75
	Under 0.67

Ratio of directly age-standardised rate in health authority to UK and Ireland average

The areas of Shetland, Orkney and Western Isles all had very small average numbers of deaths per year (Table B3.2) and the rates for these areas should be treated with caution (Table B3.1). In England, the areas of Newcastle, Middlesborough, and a band from Liverpool eastward across the Pennines to Hull had slightly higher than average mortality. Slightly higher mortality was also apparent to the east of London along the Thames estuary, including areas of South Essex and East Kent. For males, all the areas within Northern Ireland and Ireland and all but one in Wales had lower than average mortality rates. There was also an apparent band of lower mortality adjacent to Wales, from South Cheshire southward to Gloucestershire.

The map for bladder cancer mortality for females showed more variability than that for males (Map 3.2 and Table B3.1). High rates of bladder cancer mortality in females paralleled the high rates in males around many of the major cities. In the whole of Northern Ireland, Ireland and Wales only one area had a mortality rate that was more than 10 per cent above the UK and Ireland average. In England, except for the areas east of London, the map showed an apparent north-south divide, with the majority of areas south of Birmingham having lower than average mortality rates.

Risk factors and aetiology

Substantial epidemiological evidence supports a relationship between bladder cancer and cigarette smoking. It has been suggested that up to 40 per cent of all male and 10 per cent of female cases might be ascribable to this exposure.[11] The relative risks are around 2-3 fold.[12] The causative links were established as the result of positive associations from at least 8 cohort and over 15 case-control studies, the best known cohort study being that of British doctors.[13] The striking consistency of their findings, the dose-response relationships[14,15] and the identification of at least two known bladder carcinogens in cigarette smoke (2-naphthylamine and 4-aminobiphenyl) as well as the identification of aromatic amine based DNA-adducts,[16] all give confidence that there is a causative link. However, there are only weak overall correlations in incidence between the sexes and very little agreement with the geographical pattern of lung cancer in the UK and Ireland (Chapter 13) or with long-term trends in lung cancer in either sex. The geographical distribution of bladder cancer, despite the links with cigarette smoking, does not show a marked association with areas of higher tobacco consumption, as does that for lung cancer. This seems to indicate that cigarette smoking is not a dominant factor in the aetiology of the condition and raises the possibility that the associations with smoking may be confounded with other factors. The relative risks found in some studies may be related to the kind of tobacco smoked. Black tobacco (composed of air-cured tobacco) used more in southern Europe produces more 4-aminobiphenyl in the mainstream smoke than blond (flue-cured) tobacco.[17,18] The risk of bladder cancer is 2 to 3 times higher among smokers of black tobacco than among smokers of blond tobacco.[19]

Bladder cancer is strongly linked to occupational and environmental exposure to chemicals. Occupational exposure is estimated to be the cause of around 20 per cent of current bladder cancer cases. Studies in the dye intermediates industry and the rubber industry in the 1950s have indicated that arylamines such as 2-naphthylamine, benzidine and (in the USA) 4-aminobiphenyl are all human bladder carcinogens.[20,21] Consequently, the industrial use of 2-naphthylamine and of benzidine was banned in the UK in 1950 and 1962, respectively. However, deaths from occupational cancer may take place several decades after initial causal exposures. Long latent periods of up to 40 years or more from first industrial exposure are observed and therefore, despite removal of known carcinogens, occupationally caused bladder cancers may continue to be diagnosed and account for the excess bladder mortality that occurs to workers in the chemical industry. Also, recent studies still show excess risk of bladder cancer in workers of the rubber industry with no recorded exposure to 2-naphthylamine. This indicates that other agents in this industry may be associated with the occurrence of bladder cancer among rubber workers.[22]

The source of other possible chemical carcinogens causing bladder cancer is more controversial, being based largely on case-control studies only. It is possible that leather workers,[23] painters,[24,25] truck drivers,[26] aluminium workers,[27] and those in jobs with a high exposure to printing inks, cutting oils and solder[28] all might be at some slight excess risk of bladder cancer. The risk areas for females had a higher percentage of workers in textile-related occupations.[29] High-risk areas had more male glass workers and female ceramic workers than the national average.[29] Investigation of excess bladder cancer mortality for the period 1968-78 in London using death certificate data showed significant rate ratios for all road transport drivers and leather workers.[30] It is possible therefore, that the observed regional variation in bladder cancer incidence and mortality could in part be a reflection of a number of different occupational exposures from industries concentrated in different areas of the UK and Ireland.

Some treatments for cancer have been attributed to increasing the risk of subsequent bladder cancer. The alkylating agent cyclophosphamide used largely in cancer chemotherapy confers a risk of bladder cancer.[31] Ionising irradiation undoubtedly can cause bladder cancer in people heavily exposed, such as those treated for cervix cancer by external beam radiation.[32] However, this risk cannot provide a geographical explanation of bladder cancer patterns, as these risks apply only to a well-defined and small group of people.

Environmental studies show increasing evidence that arsenic ingestion causes a risk of bladder cancer. Documented causes are exposure to high levels of arsenic in the national environment as occurred in Taiwan,[33] or (unusual) arsenic ingestion of Fowler's solution as a medication.[34] Chlorination of drinking water has been linked to a modestly increased risk of bladder cancer in the USA[35] but the levels of chlorination in the British Isles are low (well within World Health Organisation guidelines) and generally uniform in those houses supplied with mains water. However, the excessive use of phenacetin (acetaminophen) confers a risk of cancer to the urethelial tract generally.[36] Since the analgesic has been banned in the UK, this is at best, a historical risk. Genetic susceptibility is associated with bladder cancer aetiology. It is not known whether this genetic susceptibility varies geographically and so whether it contributes to the overall pattern of distribution. The geographical distribution of bladder cancer in the UK is unlikely to be influenced by any of these factors.

Socio-economic deprivation

In England and Wales, based on 1993 data, incidence of bladder cancer was only very slightly higher in males and females in more deprived areas than in more affluent areas. There was a slightly steeper deprivation gradient in mortality in males than in females, with rates in the more deprived groups around 20 per cent higher than in the more affluent.[37] In Scotland, incidence rates for the period 1986-95 were very slightly higher in deprived areas.[38] In England and Wales, five-year survival for patients diagnosed with bladder cancer during 1996-99 showed a deprivation gradient: the most affluent group of patients had around 6 percentage points better survival at five years after diagnosis than the most deprived.[39] In the West Midlands, analysis of bladder cancer deaths rather than deaths from all causes for patients diagnosed with bladder cancer showed no difference in survival between patients of different socio-economic groups.[40] In the north west of England, women from more deprived areas had worse survival and were more likely to present with advanced tumours than men or women from less deprived areas.[41]

The overall patterns of bladder cancer distribution by area (Maps 3.1 and 3.2) do not correspond closely with those for social deprivation; this is another aspect in which bladder cancer differs from lung cancer. The association between bladder cancer and smoking may explain the slightly higher

incidence and mortality from bladder cancer in the more deprived groups in the population. In 2002, men in routine occupations were more than twice as likely to smoke as professional workers and to start smoking at an earlier age.[42] Apart from smoking, the main determinants of bladder cancer occur from occupational chemical exposures. Risks arising from the dye and rubber industries and lesser documented risks to leather workers, painters, truck drivers and aluminium workers all relate to exposures occurring to male manual workers. These occupational exposures may explain in some part the correlation between bladder cancer and deprivation and hence the observed geographical patterns.

The fact that the strong epidemiological evidence for an association with cigarette smoking is not reflected in the apparent geographical distribution seen in this atlas is puzzling. It may be that this is a result of the variations in practice among the cancer registries or that other aetiological factors play a strong role in the development of bladder cancer alongside cigarette smoking.

References

1. Anderson O, Stephenson J. UKACR *comparison of cancer registrations. Report number 3. Bladder tumours:* Internal report, 2004.

2. Schouten LJ, Straatman H, Kiemeney LALM, Verbeek ALM. Cancer incidence: Life table risk versus cumulative risk. *Journal of Epidemiology and Community Health* 1994; 48: 596-600.

3. ONS. *Cancer Statistics Registrations: Registrations of cancer diagnosed in 2000, England.* Series MB1 No. 31. London: Office for National Statistics, 2003.

4. Quinn MJ, Babb PJ, Kirby L, Jones J. Registrations of cancer diagnosed in 1994-97, England and Wales. *Health Statistics Quarterly* 2000; 7: 71-82.

5. Coleman MP, Esteve J, Damiecki P, Arslan A et al. *Trends in Cancer Incidence and Mortality.* IARC Scientific Publications No. 121. Lyon: International Agency for Research on Cancer, 1993.

6. Hayne D, Arya M, Quinn MJ, Babb PJ et al. Current trends in bladder cancer in England and Wales. *Journal of Urology* 2004; 172: 1051-1055.

7. ONS. Cancer Survival: England and Wales, 1991-2001. March 2004. Available at *http://www.statistics.gov.uk/statbase/ssdataset.asp?vlnk=7899.*

8. Coleman MP, Babb P, Damiecki P, Grosclaude P et al. Cancer Survival Trends in England and Wales, 1971-1995: *Deprivation and NHS Region.* Studies on Medical and Population Subjects No. 61. London: The Stationery Office, 1999.

9. Sant M, Areleid T, Berrino F, Bielska Lasota M et al. EUROCARE-3: survival of cancer patients diagnosed 1990-1994 - results and commentary. *Annals of Oncology* 2003; 14 Suppl 5: v61-v118.

10. ONS. Cancer Survival in England by Strategic Health Authority. April 2004. Available at *http://www.statistics.gov.uk/StatBase/Product.asp?vlnk=11991&Pos=1&ColRank=1&Rank=272.*

11. Wynder E, Stellman S. Environmental factors in the causation of bladder cancer. In: Connolly J (ed) *Carcinoma of the Bladder.* New York: Raven Press, 1981.

12. Ross RK, Paganini-Hill A, Hendersen BE. Epidemiology of bladder cancer. In: Skinner DG, Lieskovsky G (eds) *Diagnosis and Management of Genitourinary Cancer.* Philadelphia: W.B. Saunders Co., 1988.

13. Doll R, Peto R. Mortality in relation to smoking: 20 years' observations on male British doctors. *British Medical Journal* 1976: 1525-1536.

14. Hartge P, Silverman D, Hoover R, Schairer C et al. Changing cigarette habits and bladder cancer risk: a case-control study. *Journal of the National Cancer Institute* 1987; 78: 1119-1125.

15. Hartge P, Silverman DT, Schairer C, Hoover RN. Smoking and bladder cancer risk in blacks and whites in the United States. *Cancer Causes and Control* 1993; 4: 391-394.

16. Bartsch H, Caporaso N, Coda M, Kadlubar F et al. Carcinogen hemoglobin adducts, urinary mutagenicity, and metabolic phenotype in active and passive cigarette smokers. *Journal of the National Cancer Institute* 1990; 82: 1826-1831.

17. Bryant MS, Vineis P, Skipper PL, Tannenbaum SR. Hemoglobin adducts of aromatic amines: associations with smoking status and type of tobacco. *Proceedings of the National Academy of Sciences of the USA* 1988; 85: 9788-9791.

18. Patriankos C, Hoffmann D. Chemical studies of tobacco smoke. LXIV. On the analysis of aromatic amines in cigarette smoke. *Journal of Analytical Chemistry* 1979; 3: 150-154.

19. Vineis P, Esteve J, Hartge P, Hoover R et al. Effects of timing and type of tobacco in cigarette-induced bladder cancer. *Cancer Research* 1988; 48: 3849-3852.

20. Case RA, Hosker ME, McDonald DB, Pearson JT. Tumours of the urinary bladder in workmen engaged in the manufacture and use of certain dyestuff intermediates in the British chemical industry. I. The role of aniline, benzidine, alpha-naphthylamine, and beta-naphthylamine. *British Journal of Industrial Medicine* 1954; 11: 75-104.

21. Case RA, Hosker ME. Tumour of the urinary bladder as an occupational disease in the rubber industry in England and Wales. *British Journal of Preventative and Social Medicine* 1954; 8: 39-50.

22. Kogevinas M, Sala M, Boffetta P, Kazerouni N et al. Cancer risk in the rubber industry: a review of the recent epidemiological evidence. *Occupational and Environmental Medicine* 1998; 55: 1-12.

23. Vineis P, Magnani C. Occupation and bladder cancer in males: a case-control study. *International Journal of Cancer* 1985; 35: 599-606.

24. Bethwaite PB, Pearce N, Fraser J. Cancer risks in painters: study based on the New Zealand Cancer Registry. *British Journal of Industrial Medicine* 1990; 47: 742-746.

25. Steenland K, Palu S. Cohort mortality study of 57,000 painters and other union members: a 15 year update. *Occupational and Environmental Medicine* 1999; 56: 315-321.

26. Hoar SK, Hoover R. Truck driving and bladder cancer mortality in rural New England. *Journal of the National Cancer Institute* 1985; 74: 771-774.

27. Theriault G, De Guire L, Cordier S. Reducing aluminum: an occupation possibly associated with bladder cancer. *Canadian Medical Association Journal* 1981; 124: 419-22, 425.

28. Coggon D, Pannett B, Acheson ED. Use of job-exposure matrix in an occupational analysis of Lung and Bladder cancers, on the basis of death certificates. *Journal of the National Cancer Institute* 1984; 72: 61-65.

29. Dolin PJ. A descriptive study of occupation and bladder cancer in England and Wales. *British Journal of Cancer* 1992; 65: 476-478.

30. Baxter PJ, McDowell ME. Occupation and cancer in London: an investigation into nasal and bladder cancer using the Cancer Atlas. *British Journal of Industrial Medicine* 1986; 43: 44-49.

31. Travis LB, Curtis RE, Glimelius B, Holowaty EJ et al. Bladder and kidney cancer following cyclophosphamide therapy for non-Hodgkin's lymphoma. *Journal of the National Cancer Institute* 1995; 87: 524-530.

32. Boice JD, Jr., Engholm G, Kleinerman RA, Blettner M et al. Radiation dose and second cancer risk in patients treated for cancer of the cervix. *Radiation Research* 1988; 116: 3-55.

33. Chiang HS, Guo HR, Hong CL, Lin SM et al. The incidence of bladder cancer in the black foot disease endemic area in Taiwan. *British Journal of Urology* 1993; 71: 274-278.

34. Cuzick J, Sasieni P, Evans S. Ingested arsenic, keratoses, and bladder cancer. *American Journal of Epidemiology* 1992; 136: 417-421.

35. Wilkins JR, III, Comstock GW. Source of drinking water at home and site-specific cancer incidence in Washington County, Maryland. *American Journal of Epidemiology* 1981; 114: 178-190.

36. Piper JM, Tonascia J, Matanoski GM. Heavy phenacetin use and bladder cancer in women aged 20 to 49 years. *New England Journal of Medicine* 1985; 313: 292-295.

37. Quinn MJ, Babb PJ, Brock A, Kirby L et al. *Cancer Trends in England and Wales 1950-1999.* Studies on Medical and Population Subjects No. 66. London: The Stationery Office, 2001.

38. Harris V, Sandridge AL, Black RJ, Brewster DH et al. *Cancer Registration Statistics Scotland, 1986-1995.* Edinburgh: ISD Publications, 1998.

39. Coleman MP, Rachet B, Woods LM, Mitry E et al. Trends and socioeconomic inequalities in cancer survival in England and Wales up to 2001. *British Journal of Cancer* 2004; 90: 1367-1373.

40. Begum G, Dunn JA, Bryan RT, Bathers S et al. Socio-economic deprivation and survival in bladder cancer. *BJU International* 2004; 94: 539-543.

41. Moran A, Sowerbutts AM, Collins S, Clarke N et al. Bladder cancer: worse survival in women from deprived areas. *British Journal of Cancer* 2004; 90: 2142-2144.

42. Rickards L, Fox K, Roberts C, Fletcher L et al. *Living in Britain: Results from the 2002 General Household Survey.* London: The Stationery Office, 2004.

Chapter 4

Brain

Paul Silcocks, John Steward, Helen Wood

Summary

- In the UK and Ireland, in the 1990s, brain cancer accounted for about 1 in 60 newly diagnosed cancers and 1 in 50 cancer deaths. Brain cancer accounted for about a fifth of childhood cancers.

- Incidence rates were highest in Wales and Ireland (males) and, within England, in the south (excluding London) and the more urban areas of the north, and lowest in London, and parts of the midlands and north of England.

- The geographical pattern of mortality rates did not match closely with that for incidence. In particular, rates were lower than might have been expected from the incidence in many parts of the north of England and higher in parts of Scotland.

- There is a weak (negative) relationship between incidence and mortality and socio-economic deprivation, with lower than average rates in parts of London and some of the former heavily industrialised areas in the midlands and north of England, and higher rates in predominantly rural areas and/or in the south.

- There is no obvious link between the geographical variation in the incidence of brain cancer and any possible risk factors.

Introduction

This chapter concerns primary malignant neoplasms of the brain. In making comparisons with results in other publications, their inclusion criteria for these tumours should be scrutinised carefully. In particular, benign tumours may be included in some studies, especially those covering childhood. Cancers in other organs often metastasise to the brain – these secondary tumours have been excluded here.

Incidence and mortality

In the 1990s, on average 2,400 males and 1,900 females were diagnosed with brain cancer each year in the UK and Ireland. These cases represented 1.8 per cent of male and 1.4 per cent of female cancers, a male-to-female ratio of 1.3:1. The age-standardised incidence rates in males and females were 7.9 and 5.3 per 100,000, respectively (a ratio of 1.5:1 – slightly higher than the ratio of cases).

Although predominantly a disease affecting adults, brain cancer accounts for about a fifth of childhood cancers. In England and Wales, after a small peak in childhood, the age-specific incidence rates in both men and women rose with age, then declined after 75 years or so, the peak being greater in men.[1]

During the same period 1,900 males and 1,400 females died on average each year from brain cancers in the UK and Ireland, a male-to-female ratio of 1.4:1, slightly higher than the ratio for incident cases. These deaths represented 2 per cent of all cancer deaths, the proportion being slightly higher in males (2.2 per cent) than in females (1.8 per cent). The age-standardised mortality rates were 6.1 and 3.9 per 100,000, respectively (a ratio of 1.6:1).

Incidence and mortality trends

Age-specific incidence and mortality rates in both sexes have shown upward trends since the 1970s and the 1950s respectively, although analysis by birth cohort shows that for England and Wales at least, mortality has declined in cohorts born since the 1920s.[1,2] A similar pattern is evident in the USA, with declining mortality in adults.[3] There is evidence of increasing incidence of brain cancer in childhood in all countries.[4-6] There has also been an increase in incidence in the elderly.

The interpretation of trends in brain cancer incidence and mortality over time is complicated by the introduction of diagnostic techniques such as CT and MRI scans, which have made diagnosis more accurate in the elderly but also reduced the proportion of histologically-verified tumours, risk of misclassified secondary tumours, and the need for autopsy. This might also influence geographical variations within the UK and Ireland. These issues of interpretation are considered in detail by Swerdlow et al.[2]

Survival

Survival from brain cancer is quite low and has improved little in recent decades. One-year relative survival for patients diagnosed in England and Wales during 1996-99 was about 32 per cent in both sexes, while at five years it was about 13 per cent in men and 15 per cent in women.[7] Comparable figures were seen for patients diagnosed in the 1990s in Northern Ireland[8] and Scotland.[9] Across Europe for patients diagnosed in 1990-94, average relative survival rates were 38 per cent at one year and 18 per cent at five years, with the exception of some Nordic countries such as Finland, where rates were markedly higher.[10]

Brain cancer encompasses different histological subtypes with different responses to treatment. The WHO classification is based on morphology,[11] reflected in ICDO2 rather than ICD10. For the malignant gliomas, the commonest subtype in adults, survival depends critically on grade, as well as other factors such as feasibility of complete surgical excision. There is a dichotomy such that five-year survival can be as high as 65 per cent for grades I-II and usually below 10 per cent for grades III-IV.[12] Relative survival decreases with age, as with most other tumour sites.[13] There is little difference in survival between males and females, with females having a marginally higher rate at five years. There has been only a modest increase in five-year survival over recent decades, of about 2 percentage points every five years.

Tumours of the central nervous system are the second most common form of cancer in children aged 0-14 years, constituting group 3 in the International Classification of Childhood Cancer.[14] It should be noted that this group includes some benign tumours such as pituitary adenomas and craniopharyngiomas. However, unlike most other sites, because of their special location, tumours in the brain may prove fatal even if judged to be histologically benign. It is often difficult to differentiate between malignant and benign histology so both are normally registered and included in specialist childhood cancer publications.

Geographical patterns in incidence

Within the countries of the UK and Ireland, the highest rates for incidence were found in Wales where the age-standardised rate for males was 14 per cent higher than the average, and that for females was 17 per cent higher (Figure 4.1). Rates were also higher than average for males in Ireland and South West England, and for both sexes in the South East. Incidence rates were lower than average in London and the West Midlands, and in Eastern England for males only.

At the health authority level there were few areas where incidence rates differed markedly from the average (Figure 4.3. Table B4.1). On the maps (Map 4.1) there were a large number of white areas, where rates were less than 10 per cent different from the average, although the health authority areas that were coloured purple or blue tended to be in small clusters. There were areas of relatively low incidence for both sexes in London, the West Midlands, North West England and the west of Scotland. For males there were clusters of relatively high incidence in Wales, South East and North West England, and

the south of Scotland; for females there were clusters in South East, South West and North West England, Wales and Ireland. There was close similarity between males and females in the pattern of areas with relatively low incidence, whereas there was less similarity in the pattern of areas with relatively high incidence.

Geographical patterns in mortality

The geographical pattern of mortality rates did not match closely with that for incidence (Figure 4.2). Mortality rates were highest in Ireland for both males and females, but as the mortality-to-incidence ratios (Table B4.1) were particularly high for this country; this may have been due to the exclusion of death certificate only (DCO) cases. Mortality rates were slightly higher than average in the South East of England, as were incidence rates, and below average in the North West, and in London (females only). The rates were not significantly different from the UK and Ireland average in any of the other countries, or regions of England.

At the health authority level, as with incidence, the mortality rates differed from the overall average in only a handful of areas (Figure 4.4, Table B4.1). As for incidence, health authorities with low mortality were clustered around London and the North West, but the pattern of areas with high mortality was less clear than that for incidence. The map for mortality (Map 4.2) also confirmed the relatively small amount of variability at the health authority level. A striking difference between the incidence and mortality maps was the cluster of areas with particularly high mortality, for both males and females, in Ireland. In males, mortality was 23 per cent higher than average in Ireland, whereas incidence was only 11 per cent higher. As noted earlier, this may relate to the exclusion of cases registered solely from a death certificate.

Risk factors and aetiology

The reported increase in the incidence of brain cancer in all age groups, linked with advances in molecular biology and immunology, has prompted an intensive search for environmental causal factors in recent decades.[15,16] However, there have been few positive findings to date. One problem is that brain cancer comprises a heterogeneous group of diseases, most likely with different aetiologies. There seems to be some evidence emerging for a link between brain cancers and certain infections, possibly mediated through an effect on the immune system.

The only two established causes for primary brain cancer are high-dose ionising radiation and heritable syndromes such as Li-Fraumeni syndrome.

(continued on page 60)

Figure 4.1

Brain: incidence by sex, country, and region of England
UK and Ireland 1991-99[1]

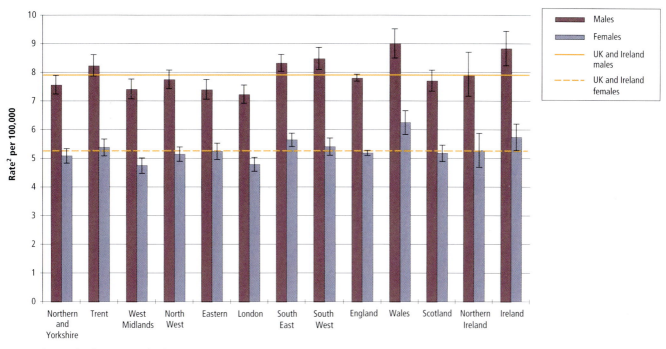

1 Northern Ireland 1993-99, Ireland 1994-99

2 Age standardised using the European standard population, with 95% confidence interval

Figure 4.2

Brain: mortality by sex, country, and region of England
UK and Ireland 1991-2000[1]

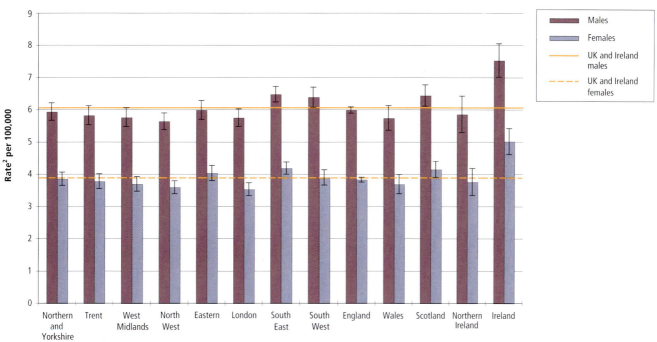

1 Northern Ireland 1991-99, Ireland 1994-2000

2 Age standardised using the European standard population, with 95% confidence interval

Figure **4.3a**

Brain: incidence by health authority within country and region of England
Males, UK and Ireland 1991-99[1]

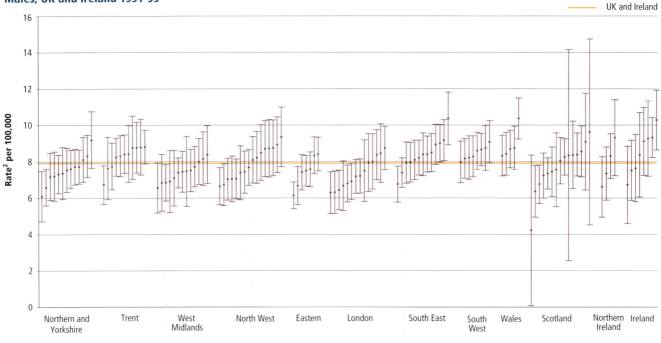

1 Northern Ireland 1993-99, Ireland 1994-99

2 Age standardised using the European standard population, with 95% confidence interval

Figure **4.3b**

Brain: incidence by health authority within country, and region of England
Females, UK and Ireland 1991-99[1]

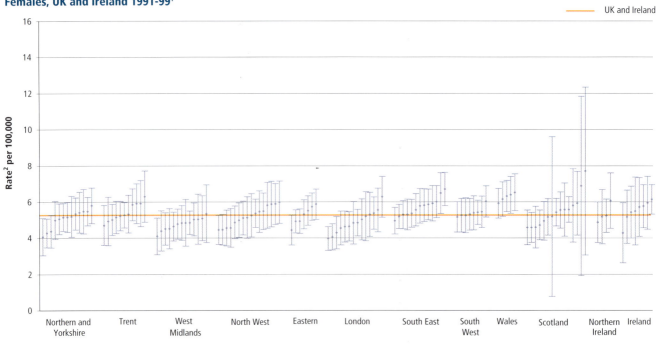

1 Northern Ireland 1993-99, Ireland 1994-99

2 Age standardised using the European standard population, with 95% confidence interval

Figure **4.4a**

**Brain: mortality by health authority within country and region of England
Males, UK and Ireland 1991-2000[1]**

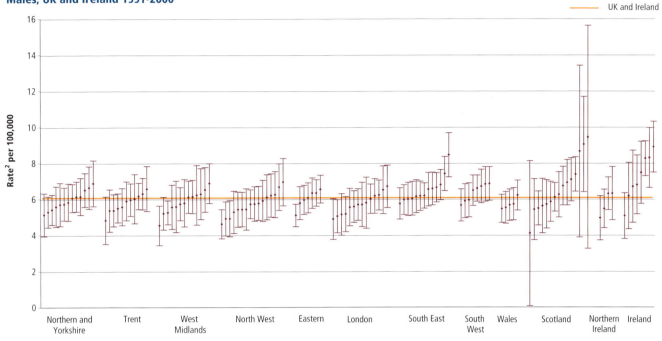

1 Scotland 1991-99, Ireland 1994-2000

2 Age standardised using the European standard population, with 95% confidence interval

Figure **4.4b**

**Brain: mortality by health authority within country, and region of England
Females, UK and Ireland 1991-2000[1]**

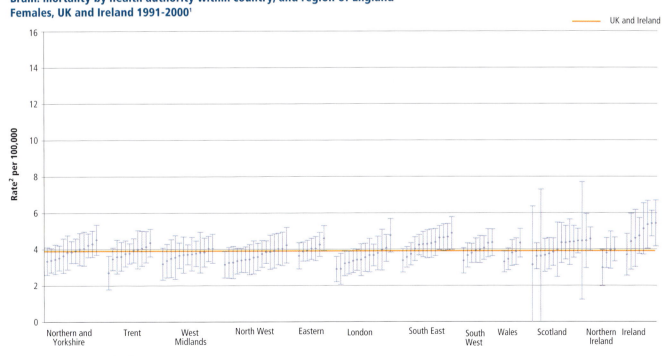

1 Scotland 1991-99, Ireland 1994-2000

2 Age standardised using the European standard population, with 95% confidence interval

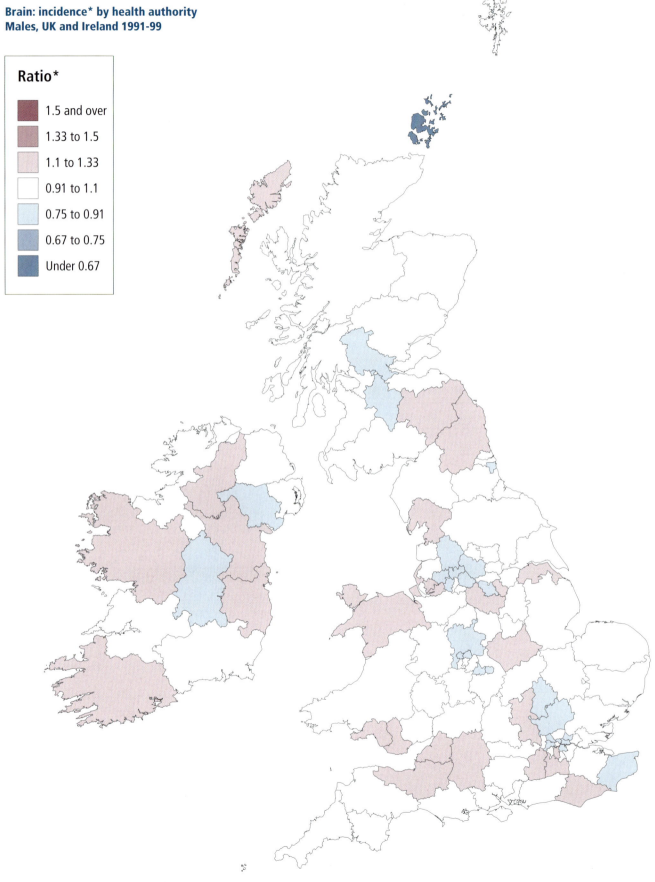

Map **4.1a**

**Brain: incidence* by health authority
Males, UK and Ireland 1991-99**

Ratio*

	1.5 and over
	1.33 to 1.5
	1.1 to 1.33
	0.91 to 1.1
	0.75 to 0.91
	0.67 to 0.75
	Under 0.67

*Ratio of directly age-standardised rate in health authority to UK and Ireland average

Map **4.1b**

Brain: incidence* by health authority
Females, UK and Ireland 1991-99

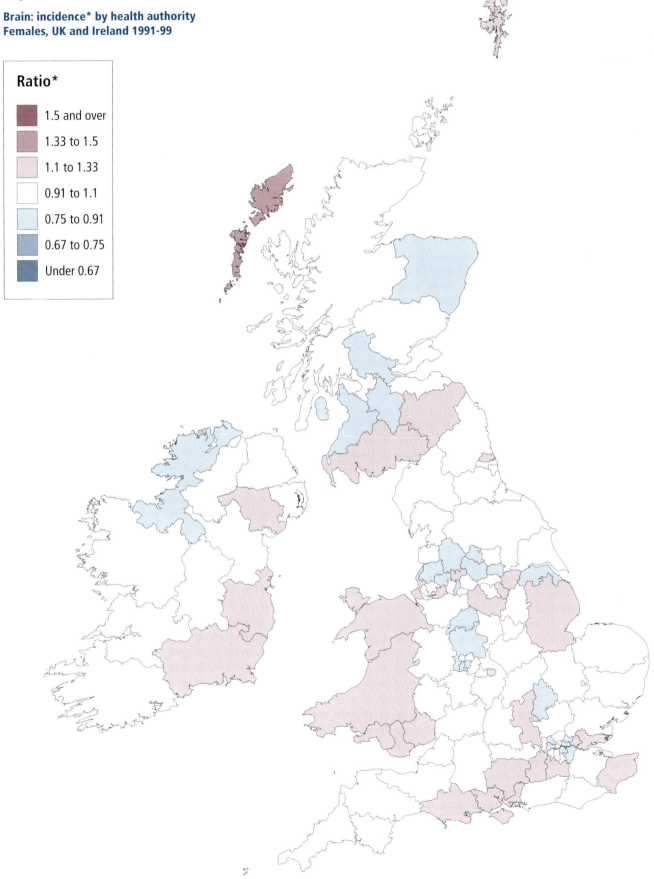

Ratio*

- 1.5 and over
- 1.33 to 1.5
- 1.1 to 1.33
- 0.91 to 1.1
- 0.75 to 0.91
- 0.67 to 0.75
- Under 0.67

*Ratio of directly age-standardised rate in health authority to UK and Ireland average

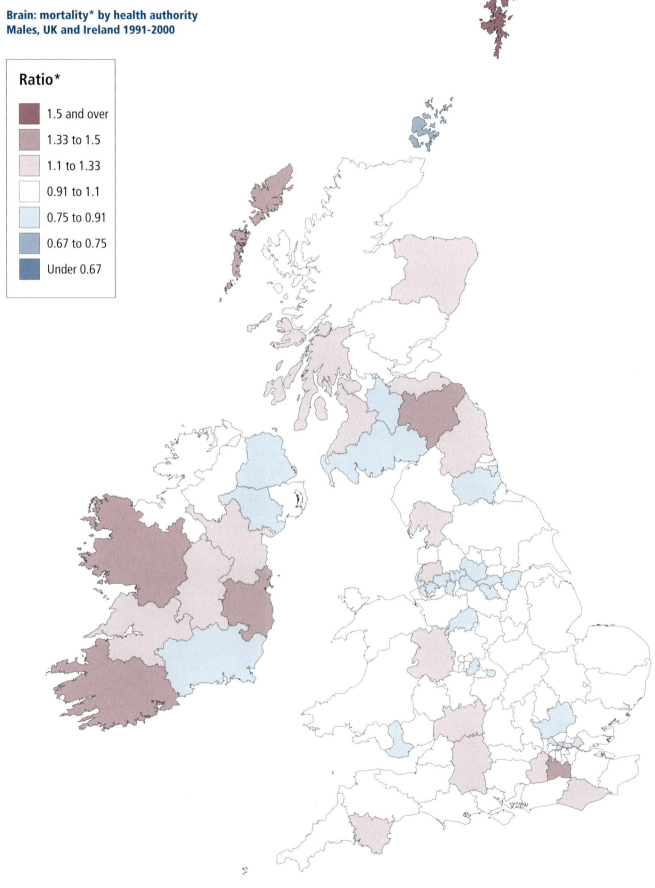

Map **4.2a**

**Brain: mortality* by health authority
Males, UK and Ireland 1991-2000**

Ratio*

- 1.5 and over
- 1.33 to 1.5
- 1.1 to 1.33
- 0.91 to 1.1
- 0.75 to 0.91
- 0.67 to 0.75
- Under 0.67

*Ratio of directly age-standardised rate in health authority to UK and Ireland average

Map **4.2b**

**Brain: mortality* by health authority
Females, UK and Ireland 1991-2000**

Ratio*

	1.5 and over
	1.33 to 1.5
	1.1 to 1.33
	0.91 to 1.1
	0.75 to 0.91
	0.67 to 0.75
	Under 0.67

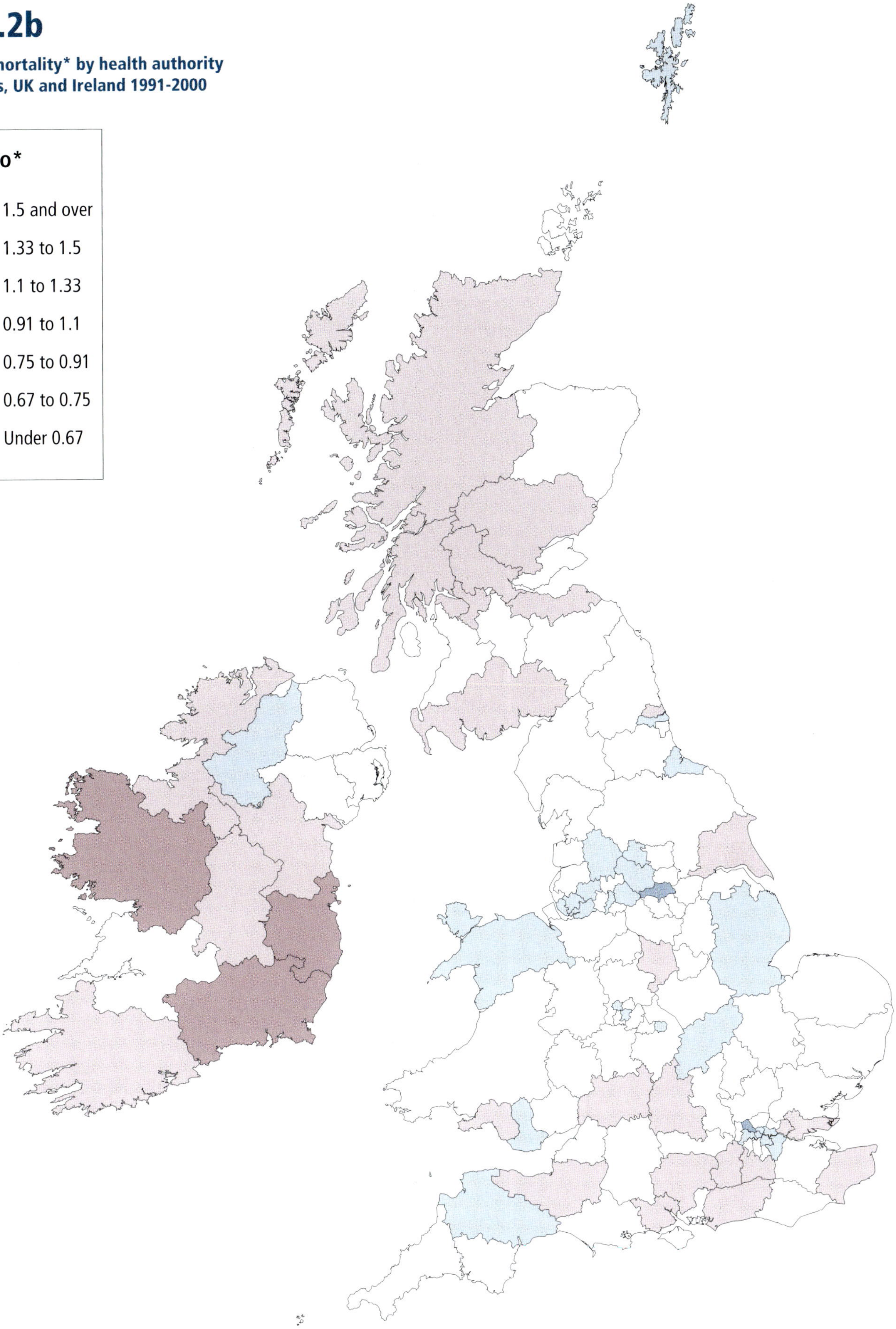

*Ratio of directly age-standardised rate in health authority to UK and Ireland average

The link with radiation has been established through studies of children irradiated for medical conditions such as ringworm, thymic enlargement and tonsil hypertrophy, dental X-rays and of atomic bomb survivors. However, the risk estimates are inconsistent and uncertain compared with the evidence relating to other cancers.[16] The relative risk is greater for benign than malignant tumours, and such exposure is rare.

Studies of international variations in incidence show that incidence rates are higher in western Europe, the USA, Australia and Canada, and lower in eastern Europe, Japan and India. Studies of migrants suggest that lifestyle may be important, since incidence in people who have migrated from countries with low rates to countries with high rates is higher than in the country of origin. However, this relationship is confounded by access to diagnostic facilities.[15]

The role of genetics in the aetiology of brain cancer is becoming better understood.[17] People with certain heritable conditions such as von Recklinghausen's neurofibromatosis have an increased risk of developing central nervous system tumours, including astrocytomas.[18] A case-control study of gliomas revealed that a family history of brain cancer was a risk factor but a history of chickenpox or shingles in the previous three years appeared to be protective.[19] To summarise, less than 5 per cent of brain cancer cases can be attributed to inherited syndromes. However, evidence is emerging that genetic factors in the sense of an inherent predisposition to specific environmental factors might eventually explain a greater proportion than this.[20]

Evidence from occupational studies has suggested a higher risk in various groups including agricultural, electrical, petrochemical and rubber workers, and health professionals. In all these studies there is often a misclassification bias in relation to occupational history and even when this is known the exact exposure may be difficult to assess.[16] There is a slight excess in incidence in rural compared to urban areas, which some have linked to the use of pesticides. A New Zealand study showed a raised risk in livestock farmers.[21] Small case-control studies have made tentative links between householders' use of pesticides and brain cancer incidence. Organophosphate can certainly reduce the effectiveness of the immune system.

The occupational associations with petroleum have suggested that exposure to petrochemicals could be a risk factor, and there have been reports of clusters in workers and residents. However, the evidence is weak and inconsistent and no specific carcinogen has been identified. A recent meta-analysis of cohort studies in petroleum workers failed to find an effect.[22] Evidence from animal experiments has suggested other possible carcinogens, but this does not seem to be supported by the available epidemiological evidence on humans.[23]

The higher risk in health professionals seems to be greatest in those involved in diagnostic services and others with potential exposure to radiation such as dental nurses. Electrical workers have high levels of occupational exposure to electromagnetic fields but electromagnetic radiation is non-ionising and therefore in itself not mutagenic. The evidence linking employment as an electrician or utility worker with brain cancer is also weak and inconsistent. However, some recent work has produced evidence for an interaction between electromagnetic fields and chemicals.[24]

In the general population of England and Wales, exposure to power frequency electromagnetic fields (50 Hz) – for example, domestic appliances and wiring – increased 4.5 fold from the 1950s to a peak around 1970, since when levels of exposure seem to have been relatively constant. Mobile phone use (as measured by number of subscribers) rose exponentially from a negligible number in 1985 to about 19 million in 1999. There is at present no strong evidence to link this with the trend of increasing incidence of brain cancer.[25,26]

There have also been studies examining brain cancer incidence in children living close to electrical power lines and a recent review concluded that there is evidence for an effect, but it is probably not causal (there may be higher incidence in children living close to electrical power lines, but this exposure is not the cause of the cancer).[27] There are difficulties in measuring actual electromagnetic field exposure. One large UK population-based case-control study that attempted to do this, failed to find evidence for an effect of exposure on the incidence of brain cancer.[28]

The evidence for lifestyle factors such as tobacco and alcohol consumption having any association with brain cancer incidence is weak and inconsistent. There is weak evidence for a link with diet, particularly N-nitroso compounds (which have also been linked with stomach cancer).[16]

Socio-economic deprivation

In England and Wales, age-standardised incidence and mortality are both slightly higher in more affluent groups.[1] Survival is also slightly higher in affluent groups[13,29] but the possibility cannot be ruled out that these gradients simply reflect access to diagnostic services and lead-time bias. Although there was relatively little geographical variability in either incidence or mortality for cancer of the brain, in England and Wales the weak inverse relationships with deprivation were reflected in slightly lower than average rates in parts of London and former heavily industrialised areas in the midlands and north; the small numbers of areas with higher rates were predominantly rural and/or in the south. Deprivation, particularly an area-based measure such as the Carstairs index,[30] is of course only a marker for some possible underlying, but unknown, factor(s).

There is no obvious link between the geographical pattern of brain cancer incidence and mortality, as illustrated by the maps (Maps 4.1 and 4.2), and any of the risk factors discussed in the aetiology section.

References

1. Quinn MJ, Babb PJ, Brock A, Kirby L et al. *Cancer Trends in England and Wales 1950-1999*. Studies on Medical and Population Subjects No. 66. London: The Stationery Office, 2001.

2. Swerdlow AJ, dos Santos Silva I, Doll R. *Cancer Incidence and Mortality in England and Wales: Trends and Risk Factors*. Oxford: Oxford University Press, 2001.

3. Legler JM, Ries LA, Smith MA, Warren JL et al. Cancer surveillance series [corrected]: brain and other central nervous system cancers: recent trends in incidence and mortality. *Journal of the National Cancer Institute* 1999; 91: 1382-1390.

4. McKinney PA, Parslow RC, Lane SA, Bailey CC et al. Epidemiology of childhood brain tumours in Yorkshire, UK, 1974-95: geographical distribution and changing patterns of occurrence. *British Journal of Cancer* 1998; 78: 974-979.

5. Smith MA, Freidlin B, Ries LA, Simon R. Trends in reported incidence of primary malignant brain tumors in children in the United States. *Journal of the National Cancer Institute* 1998; 90: 1269-1277.

6. Nishi M, Miyake H, Takeda T, Hatae Y. Epidemiology of childhood brain tumors in Japan. *International Journal of Oncology* 1999; 15: 721-725.

7. ONS. Cancer Survival: England and Wales, 1991-2001. March 2004. Available at *http://www.statistics.gov.uk/statbase/ssdataset.asp?vlnk=7899*.

8. Fitzpatrick D, Gavin A, Middleton R, Catney D. *Cancer in Northern Ireland 1993-2001: A Comprehensive Report*. Belfast: Northern Ireland Cancer Registry, 2004.

9. Black R, Brewster D, Brown H, Fraser L et al. *Trends in Cancer Survival in Scotland 1971-1995*. Edinburgh: Information and Statistics Division, 2000.

10. Sant M, Areleid T, Berrino F, Bielska Lasota M et al. EUROCARE-3: survival of cancer patients diagnosed 1990-1994 - results and commentary. *Annals of Oncology* 2003; 14 Suppl 5: v61-v118.

11. World Health Organisation. *Pathology and Genetics of Tumours of the Central Nervous System*. Lyon: IARC Press, 2000.

12. Souhami R, Tobias J. *Cancer and its Management, fourth edition*. Oxford: Blackwell Science, 2003.

13. Coleman MP, Babb P, Damiecki P, Grosclaude P et al. *Cancer Survival Trends in England and Wales, 1971-1995: Deprivation and NHS Region*. Studies on Medical and Population Subjects No. 61. London: The Stationery Office, 1999.

14. International Agency for Research on Cancer. *International Incidence of Childhood Cancer Volume II*. IARC Scientific Publication No. 144. Lyon: IARC Press, 1999.

15. Preston-Martin S, Mack W. Neoplasms of the Nervous System. In: Schottenfeld D, Fraumeni JF, Jr. (eds) *Cancer Epidemiology and Prevention, second edition*. New York: Oxford University Press, 1996.

16. Savitz D, Trichopoulos D. Brain Cancer. In: Adami HO, Hunter D, Trichopoulos D (eds) *Textbook of Cancer Epidemiology*. Oxford: Oxford University Press, 2002.

17. Preston-Martin S. Epidemiology of primary CNS neoplasms. *Neurologic Clinics* 1996; 14: 273-290.

18. Li FP, Dreyfus MG, Russell TL, Verselis SJ et al. Molecular epidemiology study of a suspected community cluster of childhood cancers. *Medical and Pediatric Oncology* 1997; 28: 243-247.

19. Wrensch M, Lee M, Miike R, Newman B et al. Familial and personal medical history of cancer and nervous system conditions among adults with glioma and controls. *American Journal of Epidemiology* 1997; 145: 581-593.

20. Malmer B, Iselius L, Holmberg E, Collins A et al. Genetic epidemiology of glioma. *British Journal of Cancer* 2001; 84: 429-434.

21. Reif JS, Pearce N, Fraser J. Occupational risks for brain cancer: a New Zealand Cancer Registry-based study. *Journal of Occupational Medicine* 1989; 31: 863-867.

22. Wong O, Raabe GK. A critical review of cancer epidemiology in the petroleum industry, with a meta-analysis of a combined database of more than 350,000 workers. *Regulatory Toxicology Pharmacology* 2000; 32: 78-98.

23. Collins JJ, Strother DE. CNS tumors and exposure to acrylonitrile: inconsistency between experimental and epidemiology studies. *Neuro-Oncology* 1999; 1: 221-230.

24. Navas-Acien A, Pollan M, Gustavsson P, Floderus B et al. Interactive effect of chemical substances and occupational electromagnetic field exposure on the risk of gliomas and meningiomas in Swedish men. *Cancer Epidemiology Biomarkers and Prevention* 2002; 11: 1678-1683.

25. Cook A, Woodward A, Pearce N, Marshall C. Cellular telephone use and time trends for brain, head and neck tumours. *New Zealand Medical Journal* 2003; 116: U457.

26. Lonn S, Ahlbom A, Hall P, Feychting M. Mobile phone use and the risk of acoustic neuroma. *Epidemiology* 2004; 15: 653-659.

27. Ahlbom IC, Cardis E, Green A, Linet M et al. Review of the epidemiologic literature on EMF and Health. *Environmental Health Perspectives* 2001; 109 Suppl 6: 911-933.

28. Skinner J, Mee TJ, Blackwell RP, Maslanyj MP et al. Exposure to power frequency electric fields and the risk of childhood cancer in the UK. *British Journal of Cancer* 2002; 87: 1257-1266.

29. Coleman MP, Rachet B, Woods LM, Mitry E et al. Trends and socioeconomic inequalities in cancer survival in England and Wales up to 2001. *British Journal of Cancer* 2004; 90: 1367-1373.

30. Carstairs V, Morris R. Deprivation and mortality: an alternative to social class? *Community Medicine* 1989; 11: 213-219.

Chapter 5
Breast

Mike Quinn

Summary

- In the UK and Ireland in the 1990s, breast cancer accounted for 1 in 4 cases of cancer, and 1 in 5 deaths from cancer in females.

- There was little geographical variation in incidence and even less for mortality.

- In only a handful of areas in south east England were incidence rates more than 10 per cent above average; in a few areas in the far north of England, and in most of Ireland, rates were more than 10 per cent below average.

- The areas with higher incidence rates tended to be the more affluent ones.

- There is no obvious link between the observed variations in breast cancer incidence and known risk factors for the disease.

Incidence and mortality

Breast cancer is the most common cancer in women worldwide, although cervical cancer is more common in some developing countries.[1] Breast cancer in men is extremely rare[2] and is not considered in this atlas. Breast cancer accounts for about 25 per cent of all malignancies in women; the proportion is higher in women in western, developed, countries. Both incidence and mortality vary considerably around the world.[3-5] Incidence has been rising in many parts of the world, including the USA, Canada, Europe, the Nordic countries, Singapore and Japan.[6,7]

Breast cancer has long been the most common cancer in females in the UK and Ireland, accounting for almost 1 in 3 of all malignant cancers in the 1990s (in females), when there were on average almost 39,000 new cases each year – almost 2.5 times as many as for colorectal cancer (16,300), the second most common cancer in women (see Chapter 2). The age-standardised incidence rate was 108 per 100,000, more than double the rate for colorectal cancer. Age-specific rates were very low in women under 40, increasing rapidly above this age to a peak in women aged 85 and over.

Breast cancer has also been the most common cause of cancer death in women, accounting for almost 1 in 5 of all cancer deaths, and 5 per cent of all deaths in women in the 1990s. There were on average 14,600 deaths each year from breast cancer in women in the UK and Ireland, just under 10 per cent more than for lung cancer (13,400), the second most common cause of cancer death in women. The age-standardised mortality rate for the UK and Ireland was 35 per 100,000, 17 per cent higher than the rate for lung cancer. The overall mortality-to-incidence ratio for breast cancer was 0.33. As for incidence, age-specific mortality increased steeply above age 40, reaching a peak in the oldest age group.

Incidence and mortality trends

During the 1990s, the earlier underlying upward trends in incidence rates in almost all age groups continued, but in Great Britain were overlaid by the effects of the introduction of mammographic screening of women aged 50-64, which started in 1988 and reached full population coverage around 1994.[8] Screening began in Northern Ireland in 1993; there was no organised breast-screening programme in Ireland during the 1990s. Rates were also affected by the increasing numbers of post-menopausal women using hormone replacement therapy.[9] In England and Wales, the first (prevalence) round of screening resulted in increases in incidence of around 25 per cent in women aged 50-64; subsequently, rates in women aged 50-54, many of whom were being screened for the first time, remained at the higher level, but rates in women aged 55-64 returned to the pre-screening trends.[10,11] Incidence in women aged 65-69 fell in the latter part of the period: many cancers in these women would have been detected several years earlier (when they were in the younger age groups) during the prevalence round of screening.

Mortality from breast cancer in the UK was among the highest in the world in the mid-1980s.[11] By the late 1990s, the rate in women aged 55-69 in England and Wales had fallen dramatically, by over 20 per cent; about a third of this was directly due to the beginning of the effect of screening, and two thirds to improved treatment – both chemotherapy and the increasingly widespread use of adjuvant tamoxifen – and to indirect effects of screening such as raised awareness leading to earlier presentation and diagnosis outside the screening programme.[12]

Survival

Even in the late 1980s, five-year (relative) survival from breast cancer was very good in the UK, at around 65 per cent,[13,14] and better than for the other major cancers in women – lung, colorectal and ovary.

(continued on page 68)

Figure **5.1**

**Breast: incidence by country, and region of England
Females, UK and Ireland 1991-99[1]**

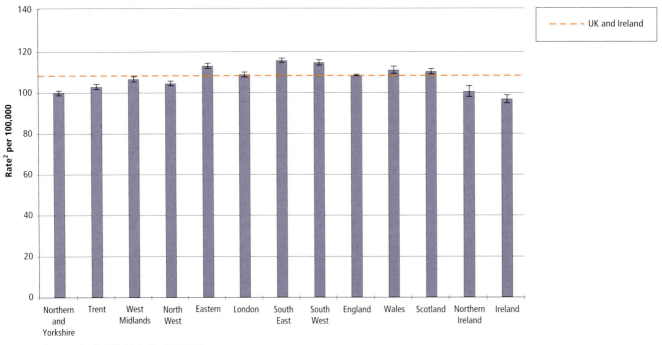

1 Northern Ireland 1993-99, Ireland 1994-99

2 Age standardised using the European standard population, with 95% confidence interval

Figure **5.2**

**Breast: mortality by country, and region of England
Females, UK and Ireland 1991-2000[1]**

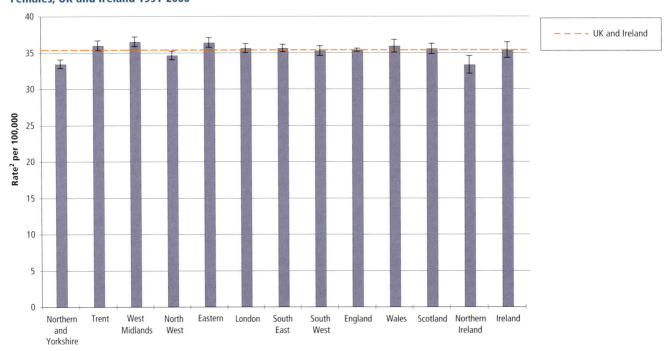

1 Scotland 1991-99, Ireland 1994-2000

2 Age standardised using the European standard population, with 95% confidence interval

Figure **5.3**

Breast: incidence rates by health authority within country, and region of England
Females, UK and Ireland 1991-99[1]

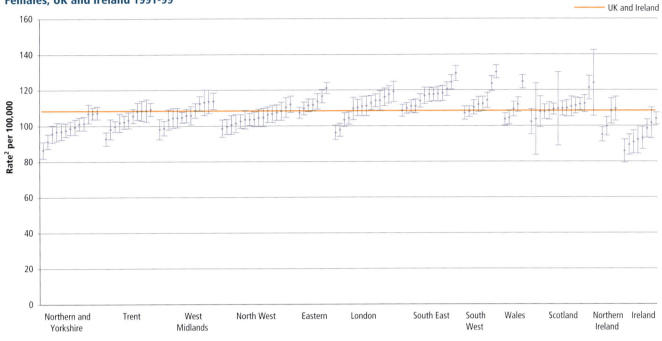

1 Northern Ireland 1993-99, Ireland 1994-99

2 Age standardised using the European standard population, with 95% confidence interval

Figure **5.4**

Breast: mortality by health authority within country, and region of England
Females, UK and Ireland 1991-2000[1]

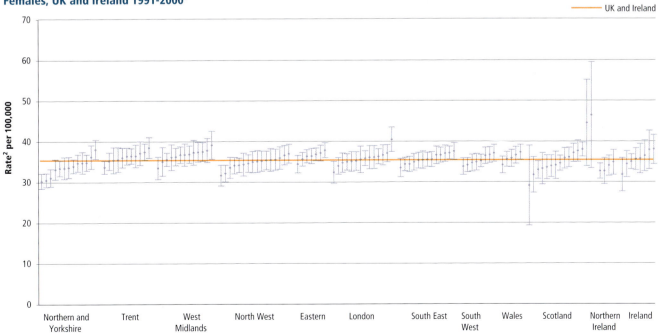

1 Scotland 1991-99, Ireland 1994-2000

2 Age standardised using the European standard population, with 95% confidence interval

Map 5.1

**Breast: incidence* by health authority
Females, UK and Ireland 1991-99**

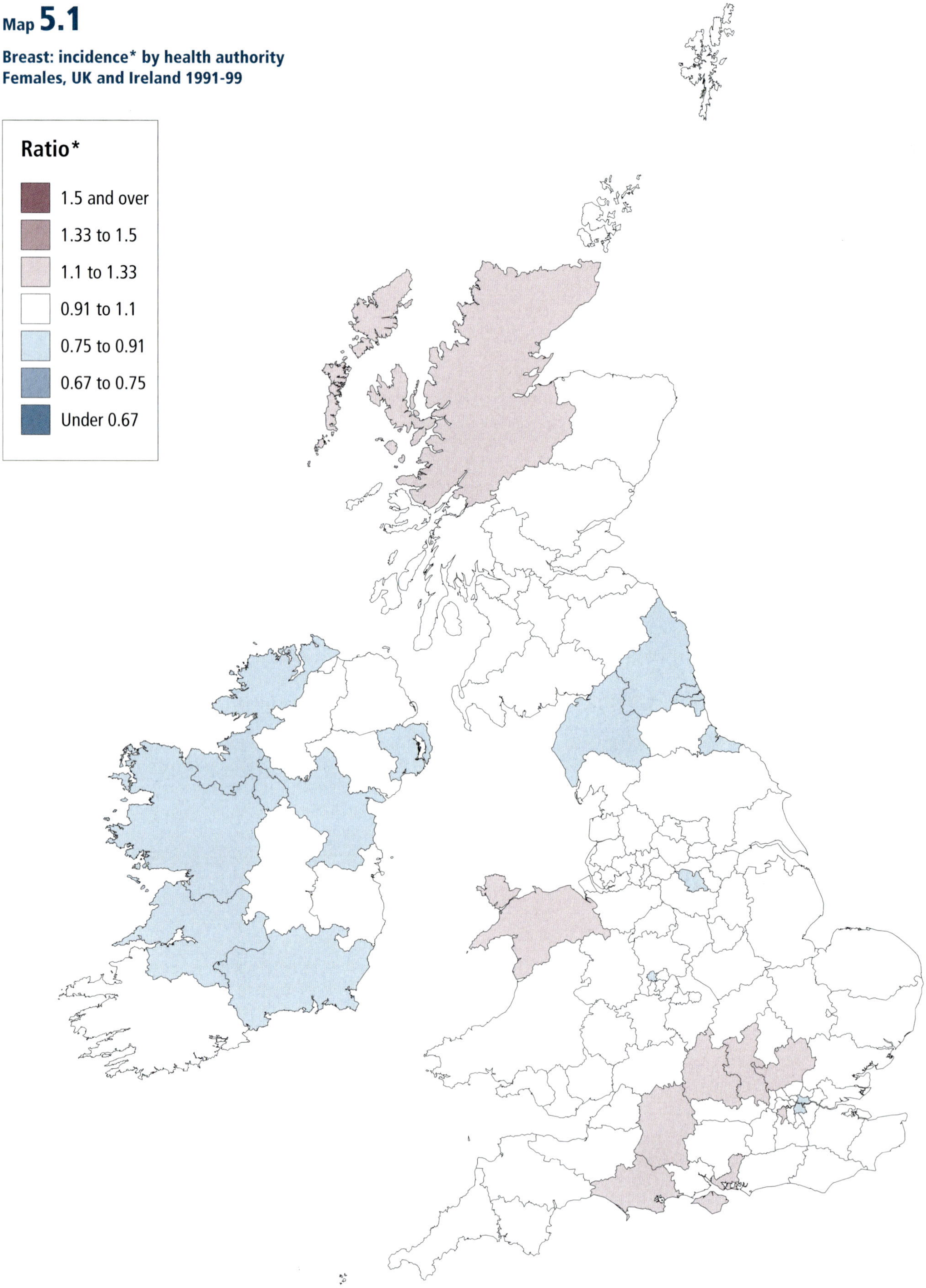

Ratio*

■	1.5 and over
■	1.33 to 1.5
■	1.1 to 1.33
□	0.91 to 1.1
■	0.75 to 0.91
■	0.67 to 0.75
■	Under 0.67

Ratio of directly age-standardised rate in health authority to UK and Ireland average

Map **5.2**

Breast: mortality* by health authority
Females, UK and Ireland 1991-2000

Ratio*

- 1.5 and over
- 1.33 to 1.5
- 1.1 to 1.33
- 0.91 to 1.1
- 0.75 to 0.91
- 0.67 to 0.75
- Under 0.67

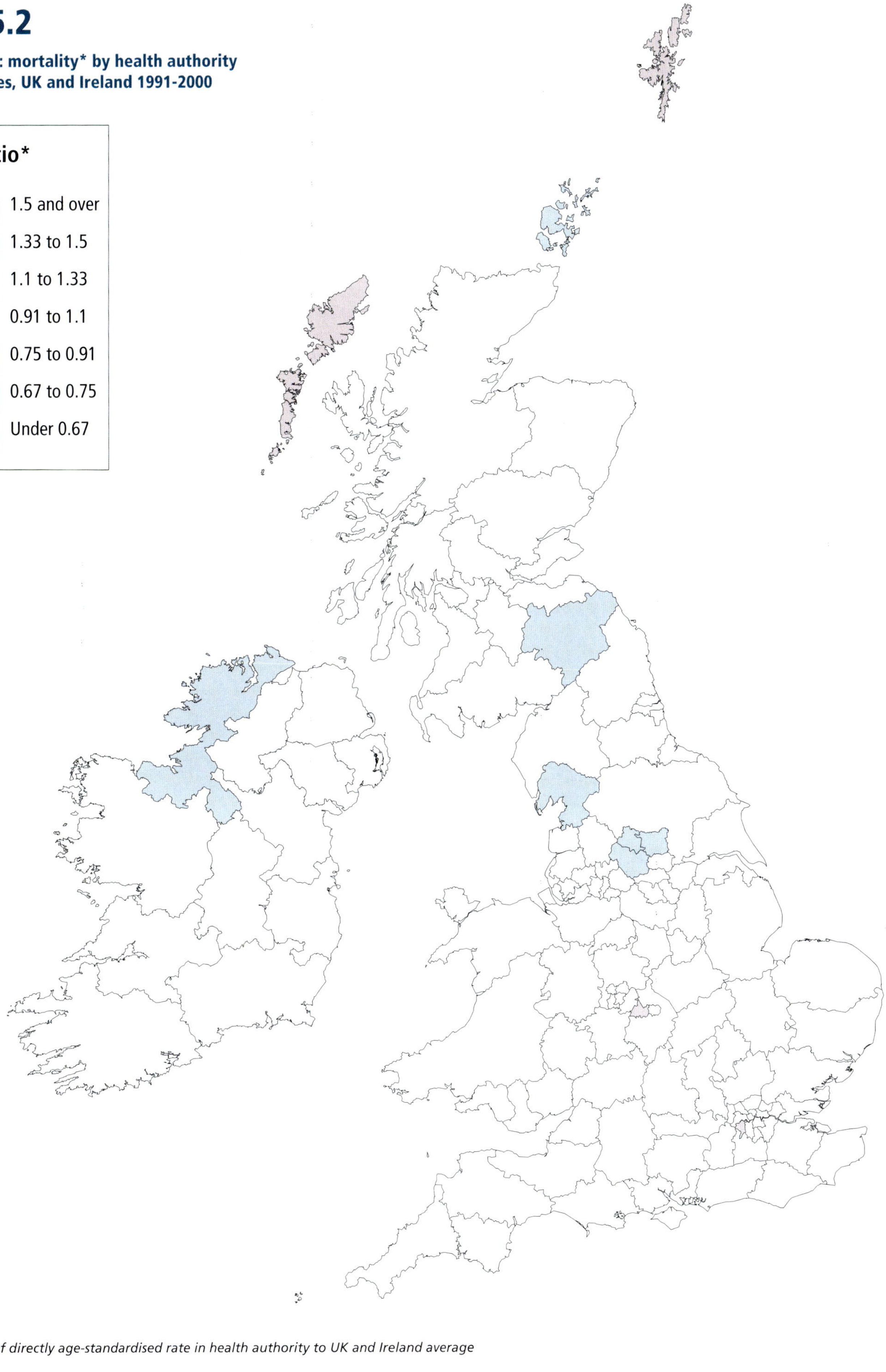

*Ratio of directly age-standardised rate in health authority to UK and Ireland average

During the 1990s, five-year survival increased markedly, partly due to earlier diagnosis (lead-time bias) as a result of mammographic screening, although much of the apparent increases were real – as indicated by the sharp reductions in mortality despite the increases in incidence. By the late 1990s, five-year survival was around 80 per cent (England and Wales data).[15]

Geographical patterns in incidence

There was relatively little variation in the incidence of breast cancer in the UK and Ireland at the country and regional level, with rates only ranging from 97 per 100,000 in Ireland to 116 per 100,000 in the South East of England (less than 20 per cent higher) (Figure 5.1). In England, rates were generally lower than average in the north and slightly higher in the south (except London). There was also relatively little variation in the rates at the level of health authorities (Figure 5.3) – most of the rates in Northern and Yorkshire, in Trent and in Ireland were below average, while most of those in the Eastern, South East and South West regions of England were above. Many of these rates are not very far from the average in either absolute or relative terms, and the large numbers of statistically significantly differences from the overall average (Table B5.1) arise because of the very large numbers of cases. The map of incidence (Map 5.1) clearly shows that the vast majority of rates in the health authorities were within 10 per cent or so of the average (white). In England, only a few areas in the central south east were more than 10 per cent above the average, and a few in the north more than 10 per cent below; rates in Ireland were mostly more than 10 per cent below average.

Geographical patterns in mortality

At the regional and country level there was even less variation in breast cancer mortality than in incidence, with rates ranging only from 33.4 per 100,000 in Northern and Yorkshire to 36.5 per 100,000 in the West Midlands (less than 10 per cent higher) (Figure 5.2). There was also less variation in mortality than in incidence at the health authority level, with very few rates being significantly different from the UK and Ireland average (despite the large numbers of deaths) (Figure 5.4). The map of mortality (Map 5.2) shows that only a handful of areas had mortality rates that were more than 10 per cent above or below the average; the rates in the Western Isles, Orkney, and Shetland are based on relatively small numbers of deaths, have correspondingly wide confidence intervals, and are not significantly different from the overall average (Figure 5.4, Table B5.1).

Risk factors and aetiology

Most of the known risk factors for breast cancer relate to a woman's reproductive history – early menarche (onset of menstrual periods), late first pregnancy, low parity, and late menopause; endogenous hormones, both oestrogens and androgens, probably have an important role. Some types of benign breast disease increase the risk of developing malignant breast cancer. None of these risk factors is currently amenable to primary prevention.[16] Oral contraceptive use and hormone replacement therapy have been linked to increased risk.[9] Alcohol consumption is associated with an increased risk of breast cancer, but cigarette smoking appears not to increase risk.[17] Studies of migrant populations have suggested that differences in incidence between countries are social and environmental, rather than genetic, in origin; only about 5 per cent of breast cancer cases are due to the inheritance of dominant genes, such as BRCA-1 and BRCA-2.[18] Avoidance of obesity may decrease the risk of post-menopausal breast cancer, and switching from a high-fat and low-vegetable diet to a lower-fat, higher-vegetable diet may also contribute to a reduced risk.

Socio-economic deprivation

In England and Wales in the early 1990s, there was an inverse relationship between the incidence of breast cancer and deprivation, as defined by the Carstairs deprivation category:[19] incidence was about 30 per cent higher in the most affluent groups than in the most deprived.[11] In contrast, mortality was not related to deprivation, which implies that survival is better in the more affluent. In fact, survival from breast cancer has consistently been found to be higher in women from affluent areas than in those from deprived areas.[13,20,21] In south east England, women in the most deprived category (again defined by the Carstairs deprivation index) had a 35 per cent greater risk of death than women from the most affluent areas after adjustment for stage at diagnosis, morphological type, and type of treatment. In older women (65-99 years) however, part of the gradient can be explained by patients more often being diagnosed with advanced disease. In England and Wales as a whole, for women diagnosed in the late 1980s there were gaps between the most affluent and most deprived of around 5 percentage points in one-year (relative) survival and 7-8 percentage points in five-year survival.[13] For women diagnosed in the late 1990s, although survival had improved markedly in all groups, the gaps between the most affluent and most deprived remained.[22]

References

1. Parkin DM, Bray FI, Devesa SS. Cancer burden in the year 2000. The global picture. *European Journal of Cancer* 2001; 37 Suppl 8: S4-66.

2. ONS. *Cancer Statistics Registrations: Registrations of cancer diagnosed in 2001, England.* Series MB1 No. 32. London: Office for National Statistics, 2004.

3. Parkin DM, Muir CS, Whelan SL, Gao.Y-T. et al. *Cancer in Five Continents Vol. VI.* IARC Scientific Publications No. 120. Lyon: International Agency for Research on Cancer, 1992.

4. Parkin DM, Whelan SL, Ferlay J, Raymond L et al. *Cancer Incidence in Five Continents Vol. VII.* IARC Scientific Publications No. 143. Lyon: International Agency for Research on Cancer, 1997.

5. Parkin DM, Whelan SL, Ferlay J, Teppo L et al. *Cancer Incidence in Five Continents Vol. VIII.* IARC Scientific Publications No. 155. Lyon: International Agency for Research on Cancer, 2000.

6. Coleman MP, Esteve J, Damiecki P, Arslan A et al. *Trends in Cancer Incidence and Mortality.* IARC Scientific Publications No. 121. Lyon: International Agency for Research on Cancer, 1993.

7. Botha JL, Bray F, Sankila R, Parkin DM. Breast cancer incidence and mortality trends in 16 European countries. *European Journal of Cancer* 2003; 39: 1718-1729.

8. Department of Health and Social Security. *Breast cancer screening: report to the health ministers of England, Wales, Scotland and Northern Ireland (Forrest report).* London: HMSO, 1986.

9. Beral V. Breast cancer and hormone-replacement therapy in the Million Women Study. *Lancet* 2003; 362: 419-427.

10. Quinn M, Allen E. Changes in incidence of and mortality from breast cancer in England and Wales since introduction of screening. United Kingdom Association of Cancer Registries. *British Medical Journal* 1995; 311: 1391-1395.

11. Quinn MJ, Babb PJ, Brock A, Kirby L et al. *Cancer Trends in England and Wales 1950-1999.* Studies on Medical and Population Subjects No. 66. London: The Stationery Office, 2001.

12. Blanks RG, Moss SM, McGahan CE, Quinn MJ et al. Effect of NHS breast screening programme on mortality from breast cancer in England and Wales, 1990-8: comparison of observed with predicted mortality. *British Medical Journal* 2000; 321: 665-669.

13. Coleman MP, Babb P, Damiecki P, Grosclaude P et al. *Cancer Survival Trends in England and Wales, 1971-1995: Deprivation and NHS Region.* Studies on Medical and Population Subjects No. 61. London: The Stationery Office, 1999.

14. Harris V, Sandridge AL, Black RJ, Brewster DH et al. *Cancer Registration Statistics Scotland, 1986-1995.* Edinburgh: ISD Publications, 1998.

15. ONS. Cancer Survival: England and Wales, 1991-2001. March 2004. Available at *http://www.statistics.gov.uk/statbase/ ssdataset.asp?vlnk=7899.*

16. McPherson K, Steel CM, Dixon JM. ABC of breast diseases. Breast cancer - epidemiology, risk factors and genetics. *British Medical Journal* 1994; 309: 1003-1006.

17. Collaborative Group on Hormonal Factors in Breast Cancer. Alcohol, tobacco and breast cancer - collaborative reanalysis of individual data from 53 epidemiological studies, including 58,515 women with breast cancer and 95,067 women without the disease. *British Journal of Cancer* 2002; 87: 1234-1245.

18. Evans DG, Fentiman IS, McPherson K, Asbury D et al. Familial breast cancer. *British Medical Journal* 1994; 308: 183-187.

19. Carstairs V, Morris R. Deprivation and mortality: an alternative to social class? *Community Medicine* 1989; 11: 213-219.

20. Schrijvers CT, Mackenbach JP, Lutz JM, Quinn MJ et al. Deprivation and survival from breast cancer. *British Journal of Cancer* 1995; 72: 738-743.

21. Schrijvers CT, Mackenbach JP, Lutz JM, Quinn MJ et al. Deprivation, stage at diagnosis and cancer survival. *International Journal of Cancer* 1995; 63: 324-329.

22. Coleman MP, Rachet B, Woods LM, Mitry E et al. Trends and socioeconomic inequalities in cancer survival in England and Wales up to 2001. *British Journal of Cancer* 2004; 90: 1367-1373.

Chapter 6
Cervix

Mike Quinn

Summary

- In the UK and Ireland in the 1990s, cancer of the cervix accounted for around 1 in 40 cases of cancer and 1 in 50 deaths from cancer in females.

- Incidence rates were high in a band across the northern part of England, parts of the west midlands and north east of England and in Scotland, and were low in the south of England and in both Northern Ireland and Ireland.

- The geographical pattern of mortality rates was broadly similar to that for incidence.

- There is a clear link between incidence and mortality rates and socio-economic deprivation, with higher rates in the more deprived areas and lower rates in the more affluent.

- The geographical patterns in incidence and mortality in the UK were not related to differences in the uptake or effectiveness of the revised cervical screening programme: the regional variation in incidence was similar in the 1970s and 1980s (although rates were much higher).

Introduction

There were nearly half a million new cases of cervical cancer in the world in 2000, representing about 10 per cent of all cancers in women, among whom it ranked second, after breast cancer.[1] About 80 per cent of cases occur in developing countries, where cervical cancer is often the most common type of cancer.[2] In contrast, cervical cancer accounts for only around 5 per cent of cases in women in North America, Australia and western Europe. Large differences in incidence among different ethnic groups have been reported in some countries. The incidence of the disease had been falling in many developed countries since the second world war[3] – but not, up to the late 1980s, in the UK.[4,5]

Screening

Many cases of invasive cervical cancer appear to arise after the initial development of a precursor premalignant condition – cervical intraepithelial neoplasia (CIN) – and screening for this condition has become a cornerstone of attempts to reduce mortality. The Pap smear was developed over 50 years ago, and screening began in Great Britain, in some of the Nordic countries, and in parts of North America, in the 1960s. Although the effectiveness of cervical screening has never been properly demonstrated in randomised, controlled trials, firm evidence comes from the Nordic countries, where the implementation of widely different policies was followed by sharply contrasting trends in incidence and mortality.[6] Although cervical screening in England started in 1964, for over twenty years it failed to achieve sufficient coverage of women or follow-up of all women with positive smears,[7] and was largely ineffective.[8] During the 1980s, important national policy changes were made which required health districts to provide routine screening at least five-yearly for all women aged 25-64. District call and recall systems had to be in place, and uniform policy and standards were developed covering all aspects of screening and subsequent intervention. The official re-launch of this re-organised NHS cervical screening programme took place in 1988. Financial incentives were first introduced in general practitioner contracts in 1990. The recorded coverage improved dramatically from 42 per cent of the target age group in 1988 to 85 per cent in 1994, a level subsequently maintained. This entailed a workload of some 4.5 million smears annually during the 1990s.[9] Coverage increased in all age groups, but particularly for older women (55 to 64 years).[10] There was no organised cervical screening programme in Ireland during the 1990s.

Incidence and mortality

Cervical cancer always was a relatively uncommon disease. There were on average just over 3,700 new cases of cervical cancer diagnosed in the UK and Ireland in the 1990s, 2.7 per cent of all cancer cases in women, in whom it ranked ninth. The disease is rare below the age of 20, but incidence rates rise rapidly to a first peak in the 35-39 age group; there is then a slight decline in rates, followed by a rise to a second, slightly higher peak in women in their 70s. In England in 1999, cervical cancer was the second most common cancer in women aged 20-34, accounting for 17 per cent of all cancers in this age group (breast cancer was the commonest, accounting for 27 per cent).[11] Overall, the age-standardised incidence rate was 11 per 100,000.

There were on average 1,600 deaths each year from cervical cancer in the UK and Ireland during the 1990s; these represented 2 per cent of cancer deaths in women. Age-specific mortality did not follow the bimodal pattern seen for incidence, but rose progressively to peak in women in their 70s. Overall, the age-standardised mortality rate was 4 per 100,000, and the mortality-to-incidence ratio was 0.39.

(continued on page 76)

Figure **6.1**

**Cervix: incidence by country, and region of England
UK and Ireland 1991-99[1]**

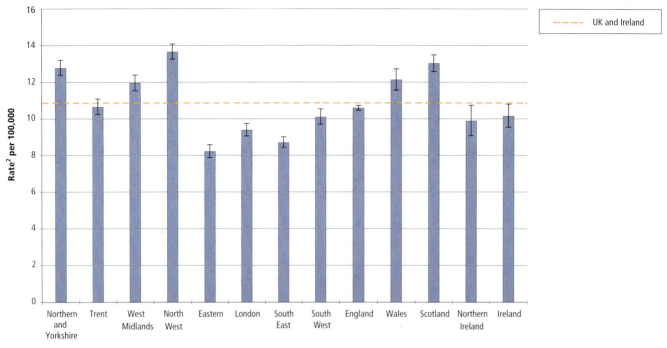

1 Northern Ireland 1993-99, Ireland 1994-99

2 Age standardised using the European standard population, with 95% confidence interval

Figure **6.2**

**Cervix: mortality by country, and region of England
UK and Ireland 1991-2000[1]**

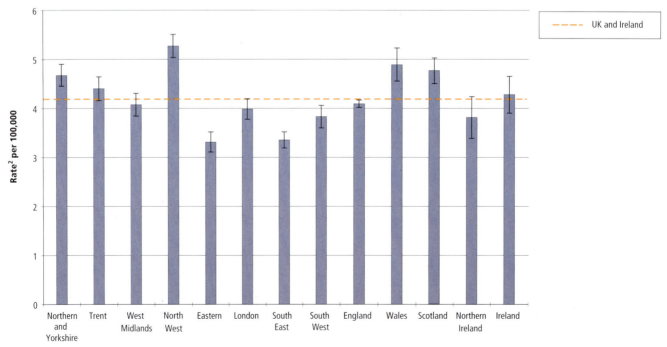

1 Scotland 1991-99, Ireland 1994-2000

2 Age standardised using the European standard population, with 95% confidence interval

Figure **6.3**

**Cervix: incidence by health authority within country, and region of England
UK and Ireland 1991-99[1]**

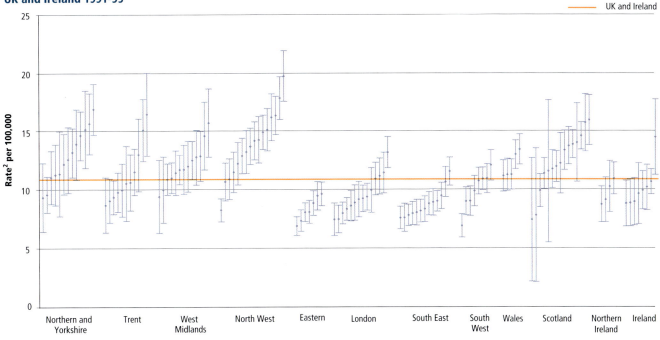

1 Northern Ireland 1993-99, Ireland 1994-99

2 Age standardised using the European standard population, with 95% confidence interval

Figure **6.4**

**Cervix: mortality by health authority within country, and region of England
UK and Ireland 1991-2000[1]**

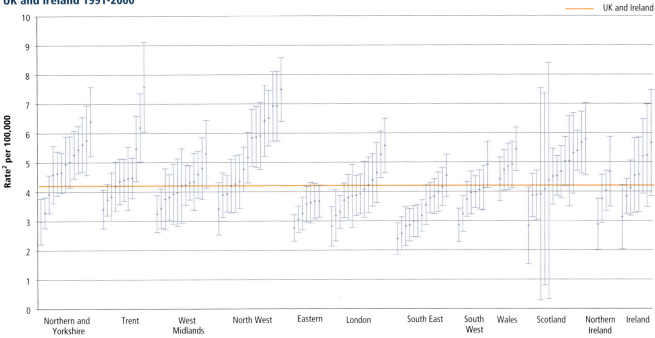

1 Scotland 1991-99, Ireland 1994-2000

2 Age standardised using the European standard population, with 95% confidence interval

Map **6.1**

**Cervix: incidence* by health authority
UK and Ireland 1991-99**

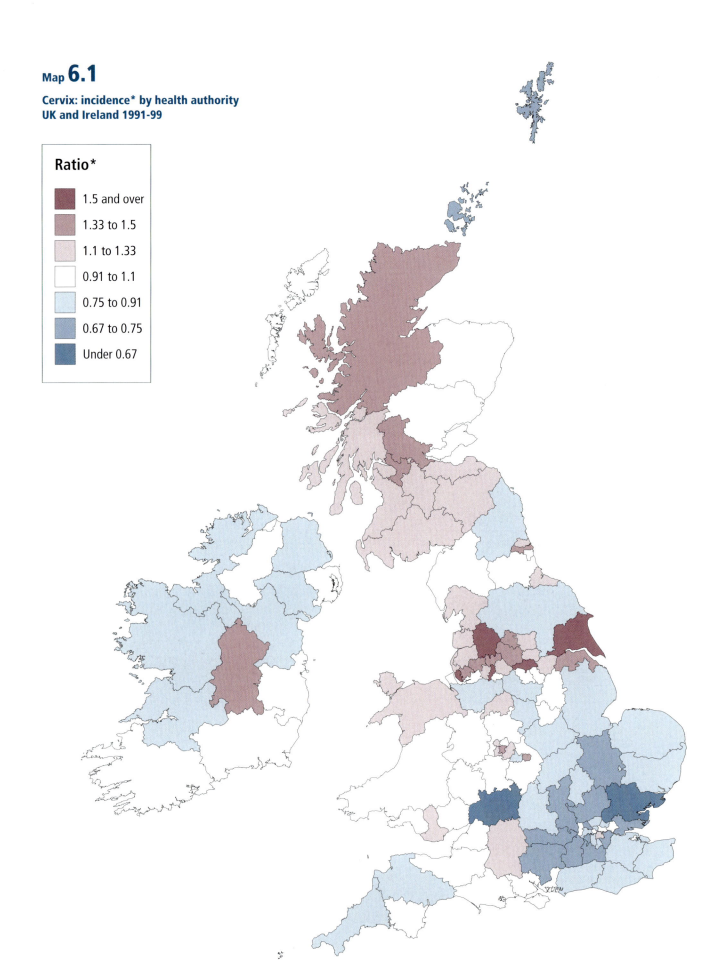

Ratio*

	1.5 and over
	1.33 to 1.5
	1.1 to 1.33
	0.91 to 1.1
	0.75 to 0.91
	0.67 to 0.75
	Under 0.67

Ratio of directly age-standardised rate in health authority to UK and Ireland average

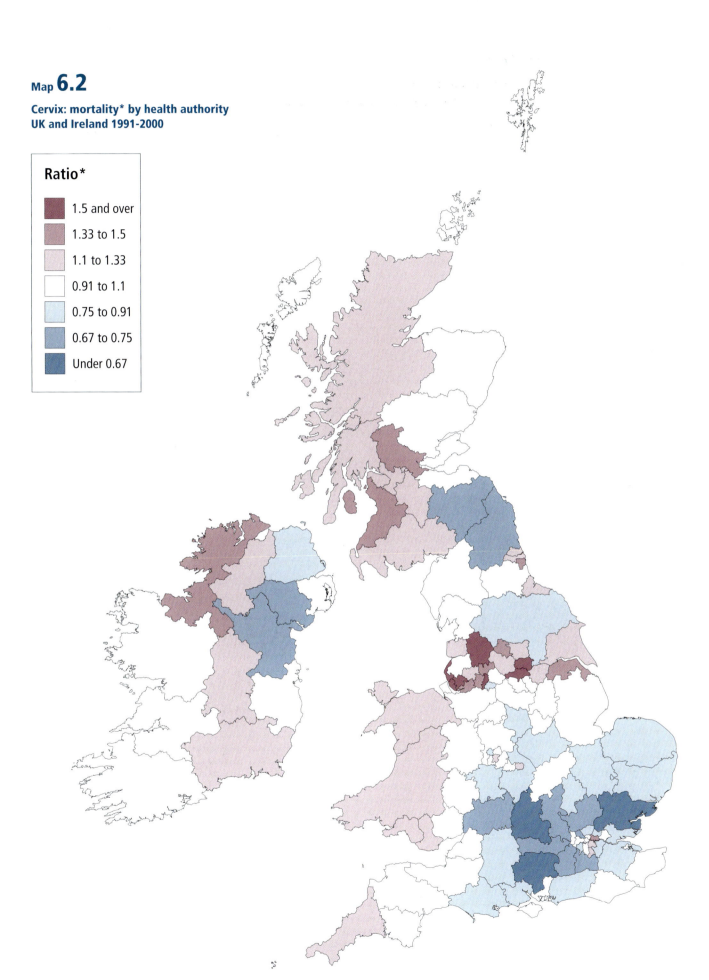

Map **6.2**

**Cervix: mortality* by health authority
UK and Ireland 1991-2000**

Ratio*

	1.5 and over
	1.33 to 1.5
	1.1 to 1.33
	0.91 to 1.1
	0.75 to 0.91
	0.67 to 0.75
	Under 0.67

Ratio of directly age-standardised rate in health authority to UK and Ireland average

Incidence and mortality trends

Since 1971, the numbers of registrations of *in situ* cervical cancer in England and Wales increased broadly in line with the increasing numbers of smears taken, to reach almost 80 per 100,000 (20,000 cases) in the mid-1990s.[10] Since 1987, the trends in registrations in women aged 20-24 and 25-29 have been continually upward, whereas in women aged 30-49 there has been no overall increase. Registrations for older groups were consistently low and declined with age.

From 1971 to the mid-1980s, the incidence of invasive cervical cancer in England and Wales remained between 14 and 16 per 100,000 (on average 3,900 cases a year).[4] This overall stability, however, resulted from a complex dynamic pattern in the underlying age-specific rates resulting from women who were born in different periods having widely different risks of cervical cancer: those women born in the early 1920s and the 1950s onwards have much higher risks than those born in the mid-1930s.[4,12,13] After 1990, incidence in England fell dramatically to just over 9 per 100,000 in 1999,[11] around 40 per cent lower than in the mid-1980s. In 1999, the overall pattern was similar to that in 1990, but the incidence in every age group from 30-34 to 70-74 was substantially (and statistically significantly) lower – by on average around 15 per 100,000 (190 cases).[4]

From 1950 to 1987, mortality from cervical cancer in England and Wales fell steadily by just over 1.5 per cent each year from 11.2 per 100,000 (2,500 deaths) to 6.1 per 100,000 (1,800 deaths).[4] This long-term decline in cervical cancer mortality predates the introduction of screening, and may be due to improvements in hygiene and nutrition; the shifting of childbearing patterns towards smaller family sizes, delayed childbearing and increased mean age at first birth; and a decline in sexually transmitted diseases.[14] Subsequently, the rate of decline trebled and by 1999 mortality had fallen to 3.3 per 100,000 – only 30 per cent of the level in 1950.[4,15] Age-specific mortality, however, changed in different ways in the various age groups. In the youngest women (25-34 years) mortality trebled from around 1 per 100,000 (30 deaths) in the mid-1960s to a plateau of around 3 per 100,000 (100 deaths) in the mid-1980s. Mortality in all other age groups declined, but at different times. Trends in both incidence and mortality in Scotland have been closely similar to those in England and Wales.[16]

Survival

For patients diagnosed in 1986-90, survival from cervical cancer was high: one-year survival was 82 per cent and five-year survival was 61 per cent.[17] Survival declined steeply with age. In 1986-90, five-year survival was around 80 per cent in the youngest age group (under 40) but only 20 per cent in the elderly (80 and over). There was little variation in survival across England and Wales. From the early 1970s to the late 1980s there was an improvement in one-year survival of 7 percentage points, and in five-year survival of 9 percentage points. Although the revised screening programme was highly successful in reducing both incidence and mortality[10,13,18] (see below) there were no significant improvements in the treatment of cervical cancer in the period, and five-year survival for women diagnosed in the late 1990s was unchanged at 61 per cent.[19]

Geographical patterns in incidence

In the 1990s there were considerable variations in the incidence of invasive cervical cancer among the regions of England, the other countries of the UK, and Ireland (Figure 6.1). Rates were highest in the Northern and Yorkshire and North West regions of England, and Scotland; and lowest in the Eastern, South East and London regions of England. Incidence in the North West was about 70 per cent higher than in the Eastern region. There was also considerable variation in the rates in the health authorities within each region and country (Figure 6.3). The highest rates in Scotland were around 60 per cent higher than the lowest (excluding Orkney and Shetland); in the West Midlands the difference between the highest and lowest was around 70 per cent; in Northern and Yorkshire, in Trent and in London it was around 80 per cent; and in the North West it was 140 per cent. Despite the wide ranges in rates at the health authority level, there was little overlap between rates in the health authorities in the north and south of England. Of the 29 health authorities in the Northern and Yorkshire and North West regions, 17 had rates markedly above the average for the UK and Ireland, 11 were not notably different from the average, and only one was notably lower. In Scotland (excluding the Western Isles, Orkney and Shetland) six health authorities had rates that were particularly high, six were not different from the average, and none was markedly low. Of the 42 health authorities in the Eastern, London, South East, and South West regions, 30 had rates that were notably below the UK and Ireland average, 11 were not different, and only one was markedly higher (see Table B6.1).

The map of cervical cancer incidence (Map 6.1) confirms the wide variation at the health authority level, but also reveals patterns that cut across regional boundaries in England. There was a band of high incidence across the northern part of England including very high rates in: Manchester (80 per cent above the average); Liverpool (over 60 per cent higher); East Lancashire; and St Helens and Knowsley (both 50 per cent higher) in the North West; in East Riding (55 per cent higher) and Bradford (45 per cent) in Northern and Yorkshire; and in Barnsley (50 per cent higher) and South Humber (40 per cent) in Trent. Rates were also noticeably above average in Tees and Tyneside. In these same regions, however, there were also some health authorities with relatively low rates, for example: South Cheshire, Stockport, and Wirral in the North West; North Yorkshire, and Northumberland in the Northern and Yorkshire region; and North Derbyshire, Southern Derbyshire, Leicestershire, and Lincolnshire in Trent. Unlike the rates in most of the former highly industrialised northern cities of England, that in Sheffield was not elevated. Rates were markedly lower than average in most of England below the northern band of high rates, with the main exception of a handful of higher rates in the West Midlands, including Coventry and Sandwell. Rates were above average in Scotland apart from Fife, Grampian, and Tayside on the eastern side of the country. Rates in both Northern Ireland and Ireland were generally low. The only noticeably higher than average rate in Ireland was in the Midland area, which does not include Dublin, but the rate is based on around only 13 cases each year.

Geographical patterns in mortality

The pattern in mortality from cervical cancer at the region and country level was closely similar to that in incidence (Figure 6.2) with higher than average rates in the north of England, Wales, and Scotland and lower than average rates in the Eastern, London and South East regions of England. The patterns of mortality within regions and countries were also substantially similar to those in incidence (Figure 6.4).

The map of cervical cancer mortality (Map 6.2) shows closely similar geographical patterns to those in incidence. In England, there was a band of higher than average mortality across the north, with high rates in Tees and Tyneside and parts of the former heavily industrialised parts of the West Midlands. Mortality in much of the rest of central and southern England was well below average. The pattern in mortality in Scotland was also closely similar to that in incidence. Mortality rates in the health authorities in Northern Ireland and Ireland were on average based on fewer than 10 deaths each year, and had correspondingly wide confidence intervals: none was significantly different from the average (see Table B6.1).

Risk factors and aetiology

The risk of developing cervical cancer is closely related to sexual behaviour. Very low rates of the disease occur in nuns. A link between cervical cancer and a sexually transmitted infection was first suggested because it was associated with women who had had many sexual partners (or whose husbands or partners had), and an early age at first intercourse.[12] A number of infections have been considered, including herpes simplex

virus type 2, genital warts, syphilis and gonorrhoea. Evidence of the aetiological role of human papillomavirus (HPV) has accumulated from both molecular and epidemiological studies.[20] The mechanism by which HPV acts is less clear, since many women with the infection do not go on to develop dysplasia, and of those who do, many do not progress to invasive cervical cancer. Other risk factors include smoking, oral contraceptives and high parity. Only the last of these appears to be a risk factor independent of HPV infection. Folate deficiency also requires further investigation for its possible explanation of the effects of parity – pregnancy is associated with the depletion of maternal folate stores. Other possible explanations are cervical trauma during childbirth and hormonal influences of pregnancy.

Changes in these risk factors over time will have affected the incidence of cervical cancer. Over the past 30 years both sexes have had a tendency to have first sexual relationships at earlier ages, and to have more sexual partners than in the past. In addition, younger cohorts of women have had greater exposure to cigarette smoking and oral contraceptives, although any increase in incidence as a result of this may have been attenuated by the reduction in parity.

Socio-economic deprivation

In the early 1990s there was a strongly positive gradient in the incidence of cervical cancer by Carstairs deprivation category,[21] with the rate in the most deprived group about three and a half times that in the most affluent. The relationship, however, was not linear: incidence increased more rapidly with deprivation in categories 12 to 20.[4] The relationship between mortality from cervical cancer and deprivation was closely similar to that for incidence. In England and Wales as a whole, and in most of the regions, there was a significant inverse gradient in survival with deprivation.[17] In 1986-90, the difference in survival rates between patients living in the most affluent areas and the most deprived areas was around 4 percentage points at both one year and five years after diagnosis. For women diagnosed in the late 1990s, there was little change in the deprivation gap, with five-year survival some 5 percentage points higher in the most affluent.[22]

It is important to recognise that the wide geographical variation and the patterns in both the incidence of, and mortality from, cervical cancer are strongly related to deprivation – and not to differences in the effectiveness of the cervical screening programme in the UK. In England and Wales in the 1980s there was the same wide variation in incidence at the regional level as illustrated in Map 6.1 for the 1990s. Following the major changes to the screening programme in the late 1980s, incidence fell dramatically in all regions (as well as in all age groups).[4,10]

References

1. Parkin DM, Bray FI, Devesa SS. Cancer burden in the year 2000. The global picture. *European Journal of Cancer* 2001; 37 Suppl 8: S4-66.

2. Parkin DM, Whelan SL, Ferlay J, Teppo L et al. *Cancer Incidence in Five Continents Vol. VIII*. IARC Scientific Publications No. 155. Lyon: International Agency for Research on Cancer, 2000.

3. Coleman MP, Esteve J, Damiecki P, Arslan A et al. *Trends in Cancer Incidence and Mortality*. IARC Scientific Publications No. 121. Lyon: International Agency for Research on Cancer, 1993.

4. Quinn MJ, Babb PJ, Brock A, Kirby L et al. *Cancer Trends in England and Wales 1950-1999*. Studies on Medical and Population Subjects No. 66. London: The Stationery Office, 2001.

5. Black RJ, Macfarlane GJ, Maisonneuve P, Boyle P. *Cancer Incidence and Mortality in Scotland 1960-89*. Edinburgh: ISD Publications, 1995.

6. Day NE. Screening for cancer of the cervix. *Journal of Epidemiology and Community Health* 1989; 43: 103-106.

7. Farmery E, Gray JAM. *Report of the First Five Years of the NHS Cervical Screening Programme*. Oxford: National Co-ordinating Network, 1994.

8. Murphy MF, Campbell MJ, Goldblatt PO. Twenty years' screening for cancer of the uterine cervix in Great Britain, 1964-84: further evidence for its ineffectiveness. *Journal of Epidemiology and Community Health* 1988; 42: 49-53.

9. Department of Health. *Cervical Screening Programme, England: 2002-03*. Statistical Bulletin 2003 No. 24, 2003.

10. Quinn M, Babb P, Jones J, Allen E. Effect of screening on incidence of and mortality from cancer of cervix in England: evaluation based on routinely collected statistics. *British Medical Journal* 1999; 318: 904-908.

11. ONS. *Cancer Statistics Registrations: Registrations of cancer diagnosed in 1999, England*. Series MB1 No. 30. London: Office for National Statistics, 2002.

12. Beral V. Cancer of the cervix: a sexually transmitted infection? *Lancet* 1974; 1: 1037-1040.

13. Sasieni P, Adams J. Analysis of cervical cancer mortality and incidence data from England and Wales: evidence of a beneficial effect of screening. *Journal of the Royal Statistical Society Series A* 2000; 163: 191-209.

14. Woodman CB, Rollason T, Ellis J, Tierney R et al. Human papillomavirus infection and risk of progression of epithelial abnormalities of the cervix. *British Journal of Cancer* 1996; 73: 553-556.

15. ONS. *Mortality Statistics 1999: Cause*. Series DH2 No. 26. London: The Stationery Office, 2000.

16. Walker JJ, Brewster D, Gould A, Raab GM. Trends in incidence of and mortality from invasive cancer of the uterine cervix in Scotland (1975-1994). *Public Health* 1998; 112: 373-378.

17. Coleman MP, Babb P, Damiecki P, Grosclaude P et al. *Cancer Survival Trends in England and Wales, 1971-1995: Deprivation and NHS Region*. Studies on Medical and Population Subjects No. 61. London: The Stationery Office, 1999.

18. Sasieni P, Adams J. Effect of screening on cervical cancer mortality in England and Wales: analysis of trends with an age period cohort model. *British Medical Journal* 1999; 318: 1244-1245.

19. ONS. Cancer Survival: England and Wales, 1991-2001. March 2004. Available at *http://www.statistics.gov.uk/statbase/ ssdataset.asp?vlnk=7899*.

20. Brinton LA. Epidemiology of cervical cancer - overview. In: Munoz N, Bosch FX, Shah KV, Meheus A (eds) *The Epidemiology of Human Papillomavirus and Cervical Cancer*. IARC Scientific Publications No. 119. Lyon: International Agency for Research on Cancer, 1992.

21. Carstairs V, Morris R. Deprivation and mortality: an alternative to social class? *Community Medicine* 1989; 11: 213-219.

22. Coleman MP, Rachet B, Woods LM, Mitry E et al. Trends and socioeconomic inequalities in cancer survival in England and Wales up to 2001. *British Journal of Cancer* 2004; 90: 1367-1373.

Chapter 7

Colorectal

Steve Rowan, David Brewster

Summary

- In the UK and Ireland, in the 1990s, colorectal cancer accounted for about 1 in 8 newly diagnosed cancers and 1 in 9 deaths from cancer.

- Incidence rates were highest in Scotland, Ireland and Northern Ireland. Rates for males in some of the more urban areas of the midlands and north of England, and for females in parts of the south of England, were slightly above average; rates in most of London and parts of the south east of England were slightly below average.

- Although the geographical pattern of mortality rates was broadly similar to that for incidence, there was more evidence of a north-side divide across England, particularly for males.

- Possible risk factors for colorectal cancer – poor diet, obesity and smoking – are associated with deprivation and may explain some of the geographical variation in incidence.

- The significant difference in five-year survival between affluent and deprived areas observed in England and Wales and in Scotland may in part explain the higher mortality rates, relative to incidence, in some of the areas that have higher levels of socio-economic deprivation.

Introduction

Cancers of the colon and rectum have been considered together here because in many instances it can be difficult to determine in which of these sites within the large bowel a cancer is located. In particular, cancers of the sigmoid colon may be coded to either colon or rectum, since this subsite is also referred to as the 'rectosigmoid junction', which is considered to be part of the rectum.

Incidence and mortality

In the UK and Ireland in the 1990s, there were about 17,600 newly diagnosed cases of colorectal cancer each year in males, of which 62 per cent were in the colon and 38 per cent in the rectum. There were about 16,300 cases in females, of which 71 per cent were in the colon and 29 per cent in the rectum. Colorectal cancer accounted for 13 per cent of all diagnosed cancer cases in males, in whom it was the third most common cancer after lung and prostate cancer, and 12 per cent of cases in females, where it was the second most common after breast cancer.

Overall, the age-standardised incidence rates were 53 and 35 per 100,000 in males and females respectively, giving a male-to-female ratio of about 1.5:1. Colorectal cancer is predominately a disease of the elderly[1] and the age-specific incidence rates rose steeply from about age 50 for both males and females. In England and Wales, the lifetime risk[2] of being diagnosed with colorectal cancer in 1997 was 5.7 per cent (1 in 18) in males and 4.9 per cent (1 in 20) in females.[3]

In the 1990s, about 9,500 males died from colorectal cancer in the UK and Ireland each year; of these deaths, 65 per cent were from colon cancer and 35 per cent from rectal cancer. About 9,100 females died each year from colorectal cancer; of these, 73 per cent were from colon cancer and 27 per cent from rectal cancer. Colorectal cancer accounted for 11 per cent of cancer deaths in males in whom it was the third most common cause of cancer death after lung and prostate cancer, and 12 per cent of cancer deaths in females, where it was also the third most common, after breast and lung cancer.

Overall, the age-standardised mortality rates were 28 and 18 per 100,000 in males and females respectively, giving a male-to-female ratio of about 1.6:1, slightly higher than that for incidence, suggesting slightly better survival for females. The age-specific mortality rates followed a similar pattern to those for incidence, with rates rising steeply in both males and females from about age 50.

Incidence and mortality trends

For males in England and Wales, the age-standardised incidence rates of colon and rectal cancer were broadly similar in the early 1970s. The rates for the two cancers later diverged as the incidence of colon cancer increased by 30 per cent between 1971 and 1997, compared with an increase of only 6 per cent for rectal cancer over the same period. For females, the incidence of colon cancer was about twice that of rectal cancer in the early 1970s. The rise in incidence rates of colon and rectal cancer was more gradual at 5 per cent for both cancers between 1971 and 1997. By 1997, the ratio of colon to rectal cases was similar in males and females.[1,4]

Despite these increases in incidence, there have been declines in the age-standardised mortality rates from both colon and rectal cancer since 1950. The mortality rates from colon cancer fell by 33 per cent in males and 53 per cent in females between 1950 and 1999. Mortality from rectal cancer fell by 56 per cent for both males and females over the same period.

For males in Scotland in 1975, the age-standardised incidence rate of colon cancer was approximately 50 per cent higher than that of rectal cancer, and for females, it was nearly three times higher. Between 1975 and 1997 there was an increase in both cancers in males, particularly from the late 1980s, and more so in rectal cancer. For females, the rise in incidence rates of both cancers was less marked. Mortality trends were broadly similar to those in England and Wales, with the exception of deaths from rectal cancer in males, where the age-standardised mortality rate began rising again from 1990, to reach levels previously seen in the early 1960s.[5]

Survival

Survival from colon and rectal cancers is now moderate, having improved significantly since the early 1970s.[6] For patients diagnosed in England and Wales in 1996–99, five-year relative survival from colon cancer was 47 per cent in males and 48 per cent in females. This compares with 42 per cent for males and 43 per cent for females diagnosed in 1991–95, an improvement of about 5 percentage points. For those diagnosed with rectal cancer in 1996–99, five-year survival was 47 per cent for males and 51 per cent for females. Again, this was an improvement (of more than 6 percentage points) from 1991–95, when survival was 40 per cent in males and 45 per cent in females.[7] Five-year survival from colorectal cancer was around 50 per cent for patients diagnosed in Scotland,[8] Northern Ireland[9] and Ireland,[10] in the 1990s.

Geographical patterns in incidence

For the constituent countries of the UK and Ireland, the incidence rates of colorectal cancer were between 18 and 22 per cent higher than the average for the UK and Ireland for males in Scotland, Ireland and Northern Ireland. Similarly, for females, the incidence rates were 23 and 19 per cent higher in Northern Ireland and Scotland, respectively, and 11 per cent higher in Ireland. The incidence rate for males was also slightly higher than average in Wales.

Within England, the incidence rates for males were slightly above average in the West Midlands, and below average in the London, Eastern, South East and Trent regions. For females, incidence rates were slightly below average in the Trent, Northern and Yorkshire, and Eastern regions, and almost 20 per cent below average in London (Figure 7.1).

Within each country, or region of England, the differences in the incidence rates between the highest and lowest health authorities were generally only 20–30 per cent (Figure 7.3) – this is a very narrow range compared with most of the other major cancers, especially those strongly related to deprivation

(such as cervix [Chapter 6], lip, mouth and pharynx [Chapter 12], larynx [Chapter 10], lung [Chapter 13] and stomach [Chapter 21]).

The maps for incidence (Map 7.1) show relatively high rates in Scotland and Northern Ireland for both males and females, and relatively low rates in parts of the London, South East and Eastern regions of England. For males, there were also high rates in some of the more urban areas of the North West, Northern and Yorkshire, and West Midlands regions and in parts of Ireland including the Eastern health authority, which includes Dublin. For females, rates were relatively high in two of the more rural areas of the West Midlands, and in parts of the South East and South West regions, and lower than average in parts of the Trent, and Northern and Yorkshire regions. The geographical patterns in the incidence of colon and rectal cancer (not shown) were similar to each other, and of course to the patterns for colorectal cancer.

Geographical patterns in mortality

The broad geographical patterns in the age-standardised mortality rates from colorectal cancer were closely similar to those for incidence, with higher than average mortality in Scotland, Ireland and Wales for both males and females. Mortality rates were also higher than average in Northern Ireland for females.

Within England, for males, mortality was slightly above average in the North West, West Midlands, and Northern and Yorkshire regions, and below average in the London, Eastern, South East and South West regions. For females, there was little regional variation in mortality: the largest difference from the average, in London, was only 10 per cent (lower) (Figure 7.2). Within each country, or region of England, the differences in mortality rates between the highest and lowest health authority were similar to those for incidence, being generally only 10–30 per cent (Figure 7.4).

The maps for mortality (Map 7.2) show broadly similar patterns to that for incidence. There were higher than average rates in Scotland for both males and females, in Northern Ireland for females and in much of Ireland for males; the lowest rates occurred in London. For males, the pattern of higher mortality in the more urban areas of the North West, Northern and Yorkshire, and West Midlands regions was more pronounced than for incidence. In addition, there were higher than average rates in south Wales. The lowest rates were in the southern half of England. For females, rates were also relatively high in some of the more urban areas of the north of England, and low in London. The maps for mortality indicate a north-south divide within England that is slightly more marked than for incidence.

(continued on page 88)

Figure **7.1**

Colorectal: incidence by sex, country, and region of England
UK and Ireland 1991-99[1]

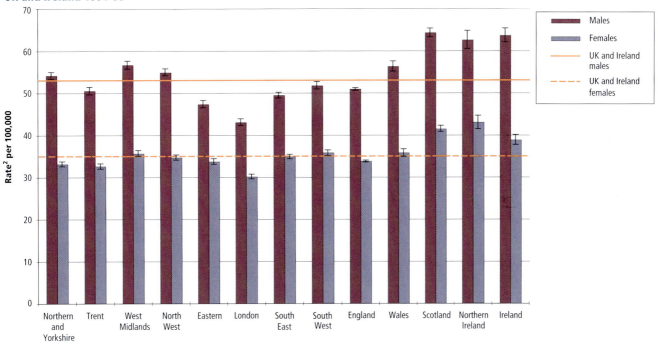

1 Northern Ireland 1993-99, Ireland 1994-99

2 Age standardised using the European standard population, with 95% confidence interval

Figure **7.2**

Colorectal: mortality by sex, country, and region of England
UK and Ireland 1991-2000[1]

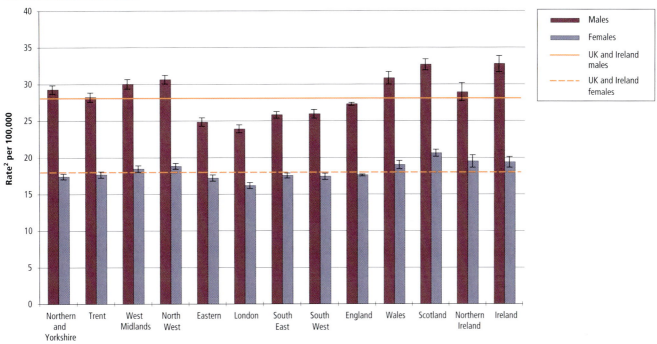

1 Scotland 1991-99, Ireland 1994-2000

2 Age standardised using the European standard population, with 95% confidence interval

Figure **7.3a**

**Colorectal: incidence by health authority within country, and region of England
Males, UK and Ireland 1991-99[1]**

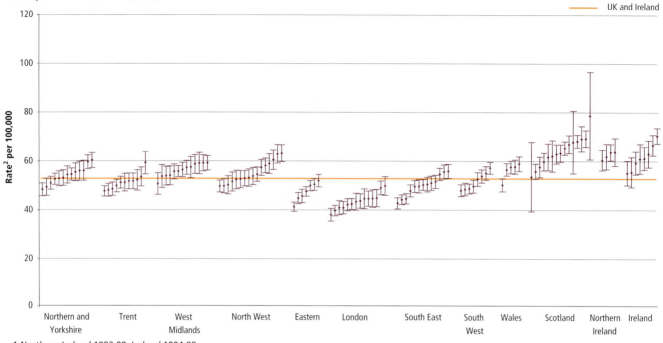

1 Northern Ireland 1993-99, Ireland 1994-99

2 Age standardised using the European standard population, with 95% confidence interval

Figure **7.3b**

**Colorectal: incidence by health authority within country, and region of England
Females, UK and Ireland 1991-99[1]**

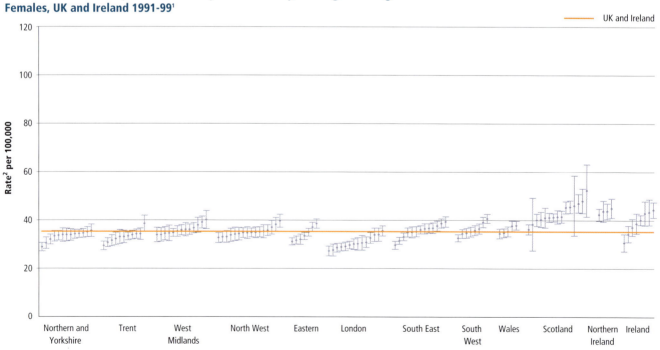

1 Northern Ireland 1993-99, Ireland 1994-99

2 Age standardised using the European standard population, with 95% confidence interval

Figure **7.4a**

**Colorectal: mortality by health authority within country, and region of England
Males, UK and Ireland 1991-2000[1]**

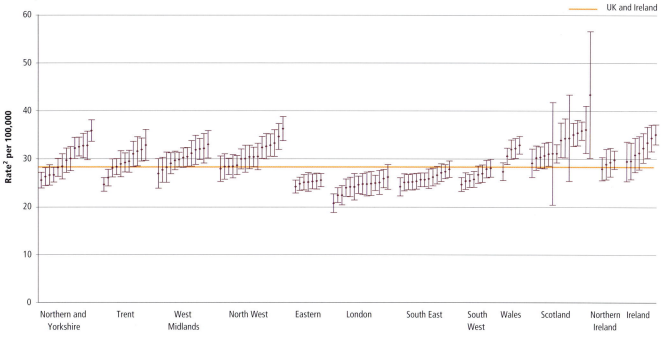

1 Scotland 1991-99, Ireland 1994-2000
2 Age standardised using the European standard population, with 95% confidence interval

Figure **7.4b**

**Colorectal: mortality by health authority within country, and region of England
Females, UK and Ireland 1991-2000[1]**

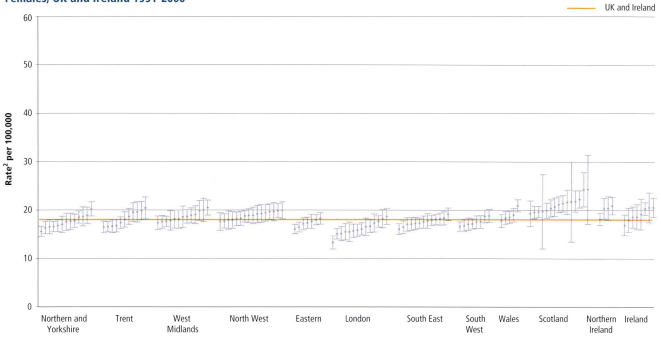

1 Scotland 1991-99, Ireland 1994-2000
2 Age standardised using the European standard population, with 95% confidence interval

Map 7.1a

**Colorectal: incidence* by health authority
Males, UK and Ireland 1991-99**

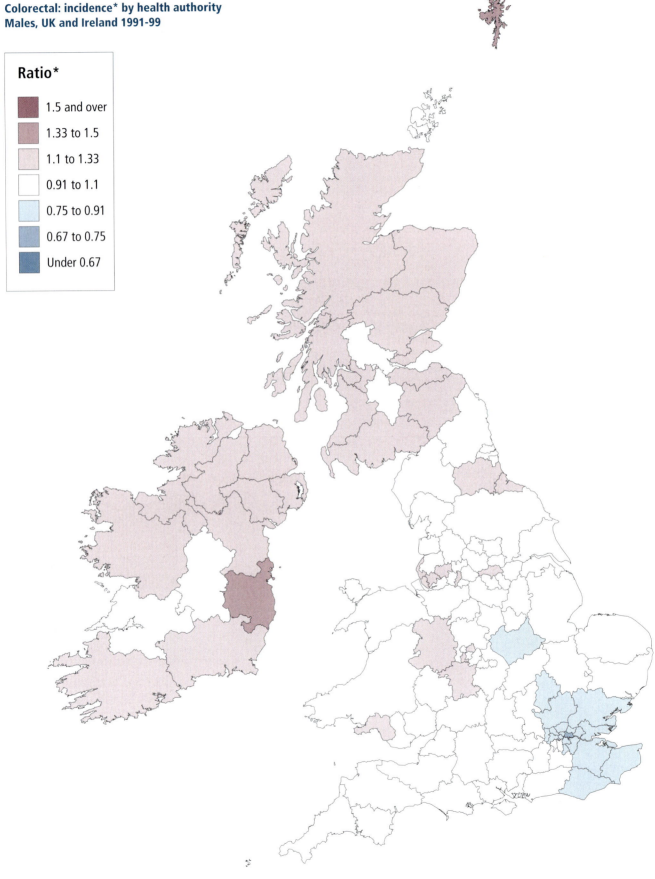

Ratio*

	1.5 and over
	1.33 to 1.5
	1.1 to 1.33
	0.91 to 1.1
	0.75 to 0.91
	0.67 to 0.75
	Under 0.67

*Ratio of directly age-standardised rate in health authority to UK and Ireland average

Map **7.1b**

Colorectal: incidence* by health authority
Females, UK and Ireland 1991-99

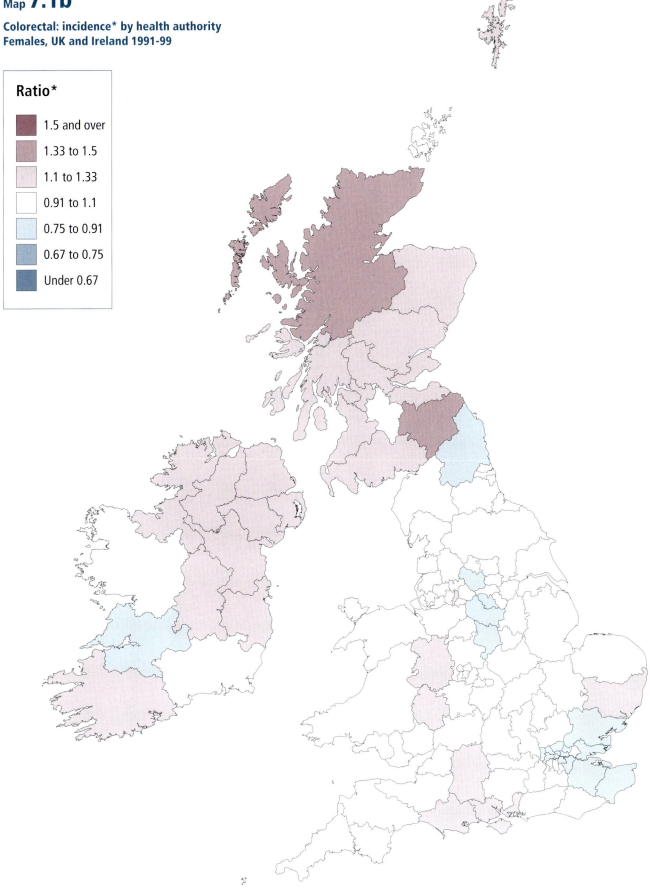

Ratio*

	1.5 and over
	1.33 to 1.5
	1.1 to 1.33
	0.91 to 1.1
	0.75 to 0.91
	0.67 to 0.75
	Under 0.67

*Ratio of directly age-standardised rate in health authority to UK and Ireland average

Map **7.2a**

Colorectal: mortality* by health authority
Males, UK and Ireland 1991-2000

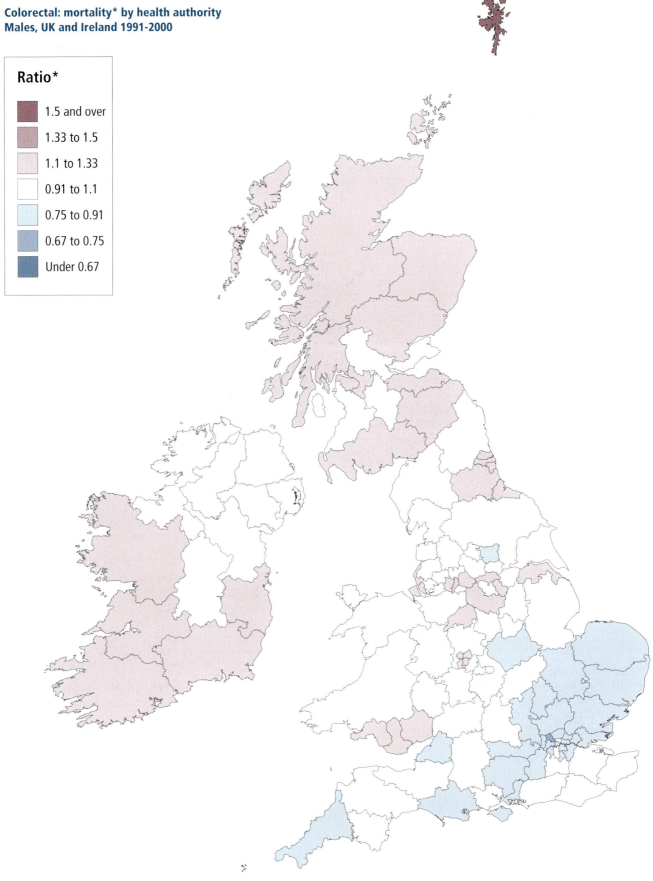

Ratio*

	1.5 and over
	1.33 to 1.5
	1.1 to 1.33
	0.91 to 1.1
	0.75 to 0.91
	0.67 to 0.75
	Under 0.67

*Ratio of directly age-standardised rate in health authority to UK and Ireland average

Map **7.2b**

**Colorectal: mortality* by health authority
Females, UK and Ireland 1991-2000**

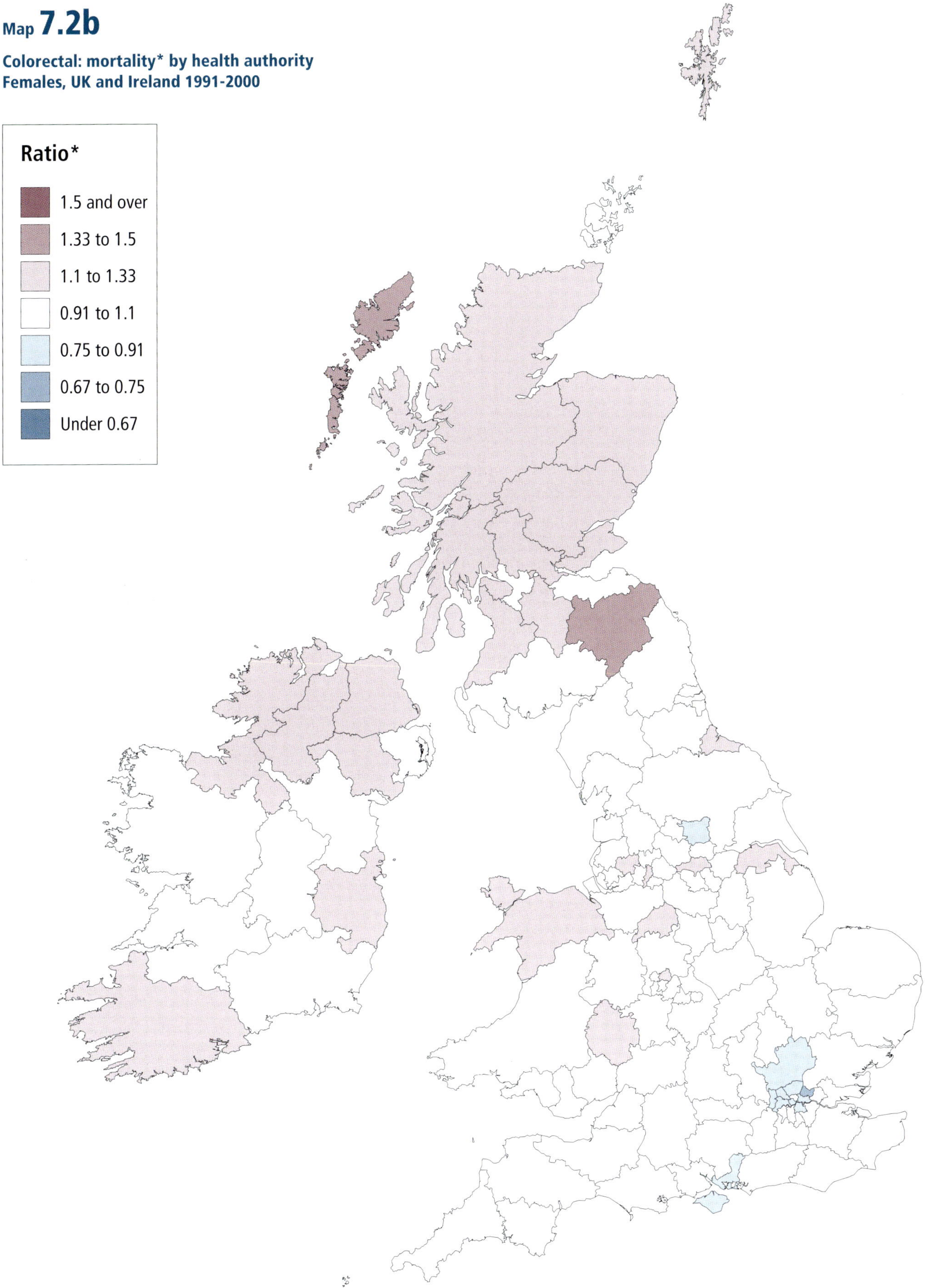

Ratio*

- 1.5 and over
- 1.33 to 1.5
- 1.1 to 1.33
- 0.91 to 1.1
- 0.75 to 0.91
- 0.67 to 0.75
- Under 0.67

*Ratio of directly age-standardised rate in health authority to UK and Ireland average

The maps for colon and rectal cancers (not shown) are closely similar to each other and to those for colorectal cancer, except that in Northern Ireland and Ireland, mortality from rectal cancer is relatively low for both sexes, but mortality from colon cancer is relatively high. These opposing patterns cancel each other out to some extent when the data for the two cancers are combined and so many of the areas on the maps of colorectal cancer show little or no variation from average. Since incidence appears to be relatively high for both colon and rectal cancers in Northern Ireland and Ireland, this suggests that proportionally more deaths were coded to colon than rectal cancer in these countries than in England, Wales and Scotland.

Risk factors and aetiology

The results of a large study of twins in Sweden, Denmark and Finland suggest that approximately 35 per cent of colorectal cancers might be explained by heritable factors.[11] However, geographical, temporal, and migrant studies suggest a more important role for environmental factors, especially diet, in the aetiology of colorectal cancer.[12] In particular, there is evidence that a diet high in vegetables decreases the risk of colorectal cancer, and although the role of dietary fibre intake is unclear,[13] recently published results from the EPIC study (the European Prospective Investigation into Cancer and Nutrition) suggest that dietary fibre is protective.[14] In England, more vegetables are consumed per person than in Scotland, but the overall vegetable consumption in Great Britain has fallen by about 20 per cent between 1974 and 2000.[15] The association between meat consumption and colorectal cancer is inconsistent, and although positive associations have been observed more consistently for red meat and processed meat, these are based on a smaller volume of evidence.[16,17]

There is also evidence that excess body weight is a risk factor for colorectal cancer,[18] and since the prevalence of obesity is increasing in the UK population, this may have contributed to the increasing incidence.[19] Conversely, there is evidence that regular physical activity is protective, especially in males, although this effect seems to be restricted to colon cancer.[13,20]

Hormone replacement therapy has been shown to reduce the risk of colorectal cancer in females in a randomised trial setting.[21] Hormonal factors may explain, at least in part, the lower risk of colorectal cancer in females than males. There is also a large body of evidence suggesting that regular use of aspirin reduces the risk of colorectal cancer, although the evidence relating to other types of non-steroidal anti-inflammatory drug (NSAID) is less substantial.[22]

Although early studies of smoking and colorectal cancer showed no association, more recent studies have shown that long-term smokers are at increased risk.[23] The falling prevalence of smoking in Great Britain[24] appears to be inconsistent with the observed increases in the incidence of colorectal cancer. However, it is important to note that the latent period between exposure and development of disease may be much longer than for some other smoking-related cancers.[23] Certainly, the higher rates of smoking in Scotland compared with England[24] are consistent with the higher incidence of colorectal cancer north of the border.

It is important to note that, as well as reflecting changes in the prevalence of risk and protective factors over time, the apparent increase in incidence of colorectal cancer may be due partly to changing referral practices or increased patient awareness, leading to more colonic investigations being undertaken than previously.[25] Certainly, a proportion of cancers detected early may not otherwise have been diagnosed during the patient's lifetime. Currently, a pilot trial of colorectal cancer screening by faecal occult blood testing (FOBt) is taking place in Coventry, Warwickshire, and the north east of Scotland.[26] However, this cannot have influenced the geographical variation in incidence reported here because screening did not begin until 2000.

Five-year survival from colorectal cancer has improved over time, and despite the increasing incidence, mortality has declined, suggesting earlier detection of symptomatic disease and/or advances in treatment, such as improvements in peri-operative care.

Socio-economic deprivation

There is no substantial variation in the incidence of colorectal cancer across categories of socio-economic deprivation, measured using the Carstairs index,[27] either in England and Wales, or in Scotland.[28] The absence of an obvious association between deprivation and colorectal cancer is also apparent from the incidence maps of the UK and Ireland (Map 7.1). In fact, for females, incidence was higher in some of the more affluent areas of the South East, Eastern and West Midlands regions of England. Within London, incidence rates were lower than average in most areas, and in Scotland, rates were higher than average across the whole country.

Despite increases in survival in England and Wales from colon and rectal cancer in the 1990s, there was a significant difference in survival rates between affluent and deprived patients, from both cancers and for both sexes. For patients diagnosed in 1996–99, the gap in five-year survival between the most deprived and the most affluent patients was 6 and 7 percentage points in males and females respectively, for colon cancer, and 9 and 8 percentage points in males and females respectively, for rectal cancer.[29] This gradient in survival with deprivation, which has also been observed in Scotland,[30] may in part explain the higher mortality rates, relative to incidence, in some of the areas that have higher levels of socio-economic deprivation (Table B7.1).

References

1. Quinn MJ, Babb PJ, Brock A, Kirby L et al. *Cancer Trends in England and Wales 1950–1999.* Studies on Medical and Population Subjects No. 66. London: The Stationery Office, 2001.

2. ONS. *Cancer Statistics Registrations: Registrations of cancer diagnosed in 2001,* England. Series MB1 No. 32. London: Office for National Statistics, 2004.

3. Quinn MJ, Babb PJ, Kirby L, Jones J. Registrations of cancer diagnosed in 1994–97, England and Wales. *Health Statistics Quarterly* 2000; 7: 71–82.

4. Hayne D, Brown RS, McCormack M, Quinn MJ et al. Current trends in colorectal cancer: site, incidence, mortality and survival in England and Wales. *Clinical Oncology* 2001; 13: 448–452.

5. Gray RF, Brewster DH, Kidd J, Burns H. Colorectal Cancer in Scotland: Recent Trends in Incidence and Mortality. *Gastrointestinal Oncology* 2002; 4: 213–222.

6. Coleman MP, Babb P, Damiecki P, Grosclaude P et al. *Cancer Survival Trends in England and Wales, 1971–1995: Deprivation and NHS Region.* Studies on Medical and Population Subjects No. 61. London: The Stationery Office, 1999.

7. ONS. Cancer Survival: England and Wales, 1991–2001. March 2004. Available at http://www.statistics.gov.uk/statbase/ssdataset.asp?vlnk=7899.

8. ISD Scotland. *Trends in Cancer Survival in Scotland, 1977–2001.* Edinburgh: ISD Publications, 2004.

9. Fitzpatrick D, Gavin A, Middleton R, Catney D. *Cancer in Northern Ireland 1993–2001: A Comprehensive Report.* Belfast: Northern Ireland Cancer Registry, 2004.

10. National Cancer Registry of Ireland. *Cancer in Ireland, 1994 to 1998: Incidence, mortality, treatment and survival.* Cork: National Cancer Registry, 2001.

11. Lichtenstein P, Holm NV, Verkasalo PK, Iliadou A et al. Environmental and heritable factors in the causation of cancer - analyses of cohorts of twins from Sweden, Denmark, and Finland. *New England Journal of Medicine* 2000; 343: 78–85.

12. Higginson J, Muir CS, Munoz N. *Human Cancer: Epidemiology and Environmental Causes.* Cambridge Monographs on Cancer Research. Cambridge: Cambridge University Press, 1992.

13. Potter, J. D. *Food, Nutrition and the Prevention of Cancer: a Global Perspective.* Washington: World Cancer Research Fund in association with American Institute for Cancer Research, 1997.

14. Bingham SA, Day NE, Luben R, Ferrari P et al. Dietary fibre in food and protection against colorectal cancer in the European Prospective Investigation into Cancer and Nutrition (EPIC): an observational study. *Lancet* 2003; 361: 1496–1501.

15. Department for Environment Food and Rural Affairs. *National Food Survey.* London: The Stationery Office, 2001.

16. Sandhu MS, White IR, McPherson K. Systematic review of the prospective cohort studies on meat consumption and colorectal cancer risk: a meta-analytical approach. *Cancer Epidemiology Biomarkers and Prevention* 2001; 10: 439–446.

17. Norat T, Lukanova A, Ferrari P, Riboli E. Meat consumption and colorectal cancer risk: dose-response meta-analysis of epidemiological studies. *International Journal of Cancer* 2002; 98: 241–256.

18. Bergstrom A, Pisani P, Tenet V, Wolk A et al. Overweight as an avoidable cause of cancer in Europe. *International Journal of Cancer* 2001; 91: 421–430.

19. Seidell JC, Flegal KM. Assessing obesity: classification and epidemiology. *British Medical Bulletin* 1997; 53: 238–252.

20. International Agency for Research on Cancer. *Weight Control and Physical Activity.* IARC Handbooks of Cancer Prevention, Vol. 6. Lyon: IARC Press, 2002.

21. Rossouw JE, Anderson GL, Prentice RL, LaCroix AZ et al. Risks and benefits of estrogen plus progestin in healthy postmenopausal women: principal results from the Women's Health Initiative randomized controlled trial. *Journal of the American Medical Association* 2002; 288: 321–333.

22. Vainio H, Morgan G, Kleihues P. An international evaluation of the cancer-preventive potential of nonsteroidal anti-inflammatory drugs. *Cancer Epidemiology Biomarkers and Prevention* 1997; 6: 749–753.

23. Giovannucci E. An updated review of the epidemiological evidence that cigarette smoking increases risk of colorectal cancer. *Cancer Epidemiology Biomarkers and Prevention* 2001; 10: 725–731.

24. Walker A, O'Brien M, Traynor J, Fox K et al. *Living in Britain: Results from the 2001 General Household Survey.* London: The Stationery Office, 2002.

25. Rhodes JM. Colorectal cancer screening in the UK: Joint Position Statement by the British Society of Gastroenterology, The Royal College of Physicians, and The Association of Coloproctology of Great Britain and Ireland. *Gut* 2000; 46: 746–748.

26. Steele RJ, Parker R, Patnick J, Warner J et al. A demonstration pilot trial for colorectal cancer screening in the United Kingdom: a new concept in the introduction of healthcare strategies. *Journal of Medical Screening* 2001; 8: 197–202.

27. Carstairs V, Morris R. Deprivation and mortality: an alternative to social class? *Community Medicine* 1989; 11: 213–219.

28. Babb P, Brock A, Jones J, Quinn M. Geographic patterns in cancer incidence. In: Griffiths C, Fitzpatrick J (eds) *Geographic Variations in Health.* Decennial Supplements No. 16. London: The Stationery Office, 2001.

29. Coleman MP, Rachet B, Woods LM, Mitry E et al. Trends and socioeconomic inequalities in cancer survival in England and Wales up to 2001. *British Journal of Cancer* 2004; 90: 1367–1373.

30. Scottish Cancer Intelligence Unit. *Trends in Cancer Survival in Scotland 1971–1995.* Edinburgh: Information and Statistics Division, 2000.

Chapter 8

Hodgkin's disease

Peter Adamson, Richard McNally

Summary

- In the UK and Ireland, Hodgkin's disease accounts for around 1 in 190 diagnosed cases of cancer and 1 in 400 deaths from cancer. In young adults, however, it accounts for one sixth of all cancers.

- Incidence rates were slightly higher than average in the southern parts of England, Wales and Scotland, and slightly lower than average in Ireland and parts of the midlands and north of England.

- There was no clear geographical pattern in mortality rates and little correlation with those for incidence.

- Most of the apparent geographical variations in incidence and mortality at the health authority level are likely to be due to random variation because there are only relatively small numbers of cases and deaths from Hodgkin's disease.

- The two main sub-types of Hodgkin's disease and their associated possible risk factors have a conflicting relationship with socio-economic deprivation. The overall geographical pattern therefore depends on the proportions of the two sub-types within each area or region.

Introduction

Hodgkin's disease (HD) is a rare malignant tumour of the lymphatic system. It is characterised by the presence of Reed-Sternberg giant cells, which replace the normal lymphoid structure. The coding and classification of haematological malignancies, particularly the lymphomas, has changed markedly over the last two decades. Increased biological understanding of the disease process, coupled with the development of modern diagnostic techniques, has led to the recognition of several new disease entities. The composition of many of the major disease groupings has changed. The third edition of the International Classification for Diseases for Oncology (ICDO3) divides HD into the following sub-types: lymphocyte-rich, mixed cellularity, lymphocyte depleted and nodular sclerosing.[1,2] A study from parts of England and Wales

showed that the most common sub-types were nodular sclerosis (55 per cent of cases of HD) and mixed cellularity (22 per cent of cases).[3]

Incidence and mortality

In the UK and Ireland, during the period 1991-99, on average around 820 cases of HD in males and 620 cases in females were diagnosed each year. HD accounted for 0.6 per cent of all cancer cases diagnosed each year for males and 0.5 per cent for females. The age-standardised incidence rate for males was 2.7 per 100,000 and for females it was 1.9, a male-to-female ratio of 1.4:1.

In developed countries HD exhibits a bimodal incidence distribution with low rates in children (aged 0-14 years), a peak in incidence in adolescents and young adults (aged 15-34 years) and another peak for those aged over 60 years. Nodular sclerosis is the predominant sub-type. In contrast, in developing countries there are peaks in incidence in children and older people, and the mixed cellularity and lymphocyte depletion sub-types are predominant.[4] One study, from the UK, has found that the nodular sclerosis sub-type is more common in females than males in the 15-24 age group.[5] However, considering all of the sub-types together, HD is more common in males than females.

In the UK and Ireland, during the period 1991-2000, there were on average around 210 male and 160 female deaths from HD each year. Of all deaths from cancer, 0.2 per cent were due to HD in both males and females. The age-standardised mortality rate for males was 0.7 per 100,000 and for females was 0.4, a male-to-female ratio of 1.8:1, which is higher than that for incidence rates.

Incidence and mortality trends

Hodgkin's disease is a rare malignancy, which ranks as the seventeenth most common cancer for males and females of all ages combined. In young adults, however, it is one of the most common malignancies, accounting for one sixth of all cancers in those aged 15-24 years.

Studies of long-term trends in incidence rates have shown inconsistent results.[6] However, some studies have reported a significant increase in the incidence of HD in adolescents and young adults. This was often due to a large rise in the incidence of the nodular sclerosis sub-type in females.[7-10] Changes to disease classification and improvements in diagnostic technique may explain some of the differences between studies. Indeed, a decrease in incidence described in older adults may be due to changes in diagnostic practice.[7,10]

Trends in mortality have been influenced by improvements in treatments that took place in the 1960s and 1970s. Mortality rates in England and Wales have fallen dramatically.[11] In contrast to many other cancers it is often possible to cure both localised and advanced HD.[12]

Survival

Survival from HD in England and Wales for patients diagnosed during the period 1986-90 was high. One-year relative survival was about 88 per cent and five-year survival was 75 per cent in both men and women.[13] Survival has improved markedly since the early 1970s when the figures were 73 and 53 per cent for one- and five-year survival, respectively. This improvement was due to the development of intensive combined radiotherapy and chemotherapy treatment regimes.[13] For patients diagnosed in England and Wales in 1986-90, the five-year survival rate was lower by 8 percentage points in those living in the most deprived areas compared with those from the most affluent areas.[13]

Geographical patterns in incidence

Within the countries of the UK and Ireland, incidence rates for HD were significantly above the average for males in Wales and Scotland (Figure 8.1), but in females rates were not higher than average in any country. Incidence was lower than average in both males and females in the Trent region of England, and in females only in the West Midlands. At the health authority level, the average number of cases per year ranged from 1-18 in males and 0-17 in females (Table B8.2), hence much of the apparent geographical variation in rates is likely to be due to random variation (Figure 8.3 and Map 8.1), and analysis at this level should be viewed with caution. Looking at the maps for incidence (Map 8.1), rates appear to be above average for both sexes in the southern half of Scotland, south Wales, many areas in southern England, and the Eastern area in Ireland. Incidence appears to be below average for both sexes in most of Ireland, and clusters of areas in the North West, Trent and West Midlands regions of England. In the large majority of areas, however, the rates were not significantly different from the average (Table B8.1 and Figure 8.3).

Geographical patterns in mortality

For most areas in the UK and Ireland, mortality did not differ markedly from the overall average. Mortality appeared to be higher than average in males in Ireland and Wales, and below average in females in the South West and Eastern regions of England (Figure 8.2). However, the average number of deaths each year in each country (excluding England as a whole), or region of England, was very small – ranging from 5-26 in males

and 4-22 in females (Table B8.2). With such small numbers at the regional level it is not possible to interpret the apparent geographical variation in mortality at the health authority level (Figure 8.4 and Map 8.2). The maps for mortality (Map 8.2) show a large number of darkly coloured areas indicating rates more than 50 per cent above or below the UK and Ireland average, but this is most likely due to random variation. There does not appear to be any clear geographical pattern, and there is little correlation with the observed geographical variations in incidence (compare with Map 8.1).

Risk factors and aetiology

Both genetic and environmental factors are likely to be involved in the aetiology of Hodgkin's disease. A number of studies have shown evidence for an association between human leukocyte antigen and increased susceptibility to HD, although a specific genotype remains unidentified.[6,14] Furthermore, there is some evidence for clustering of cases of HD within family groups.[6] An analysis of 28 published studies found that familial HD demonstrated only one peak in incidence, which occurred at ages 15-34 years.[15] This provides support for the involvement of a genetic component in the aetiology of HD in adolescents and young adults, or for an infectious agent affecting young family members living in the same household. However, familial HD is estimated to account for less than 5 per cent of all new cases.[16]

It seems likely that there are at least two distinct aetiologies in the development of HD.[17] A number of studies have reported seasonal variation in the incidence of HD.[6] Also, clustering of cases has been found in both the USA and the UK.[6,18,19] These findings support the role of an environmental factor in the aetiology. They particularly suggest a mechanism that is associated with greater person-to-person contact. Infections have been suggested as a likely agent. Specific candidate viruses include Epstein-Barr virus (EBV), HIV and human herpes virus 6 (HHV-6). EBV has been detected in numerous studies of HD,[6] and has been found to be associated with about one third of cases in developed countries.[20-22] EBV is more often associated with the mixed cellularity sub-type than the nodular sclerosis sub-type. Also, HD in children and older adults is more likely to be EBV-associated than HD in young adults.[23-27] The evidence linking HIV and HHV-6 with HD is more tenuous.[1]

A number of occupational exposures have been suggested as being linked with HD, including occupational exposure to wood and chemicals. However, published studies show inconclusive evidence and a causal link cannot be made.[6] The differences in incidence rates between males and females aged over 30 suggests a role for hormonal factors. Some studies have shown that childbearing has a protective effect in women.[6]

(continued on page 100)

Figure **8.1**

**Hodgkin's disease: incidence by sex, country, and region of England
UK and Ireland 1991-99[1]**

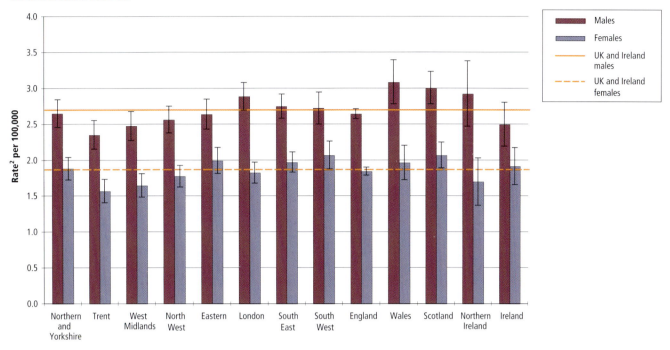

1 Northern Ireland 1993-99, Ireland 1994-99

2 Age standardised using the European standard population, with 95% confidence interval

Figure **8.2**

**Hodgkin's disease: mortality by sex, country, and region of England
UK and Ireland 1991-2000[1]**

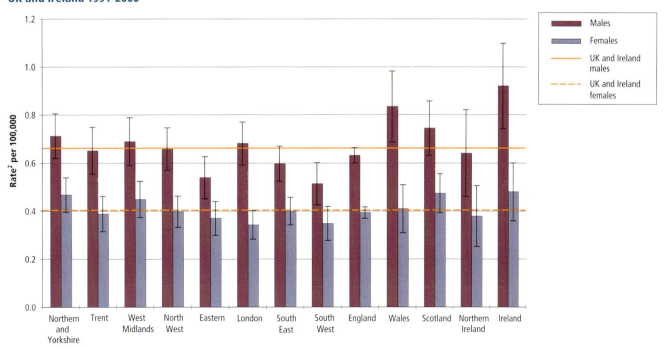

1 Scotland 1991-99, Ireland 1994-2000

2 Age standardised using the European standard population, with 95% confidence interval

Figure **8.3a**

Hodgkin's disease: incidence by health authority within country, and region of England
Males, UK and Ireland 1991-99[1]

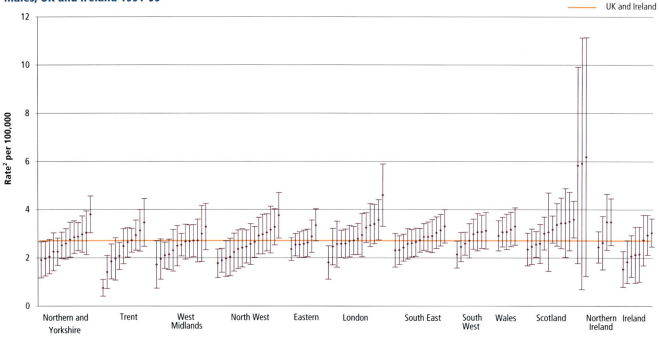

1 Northern Ireland 1993-99, Ireland 1994-99

2 Age standardised using the European standard population, with 95% confidence interval

Figure **8.3b**

Hodgkin's disease: incidence by health authority within country, and region of England
Females, UK and Ireland 1991-99[1]

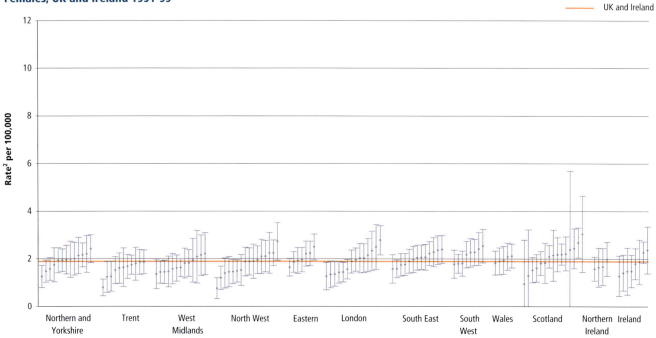

1 Northern Ireland 1993-99, Ireland 1994-99

2 Age standardised using the European standard population, with 95% confidence interval

Figure **8.4a**

Hodgkin's disease: mortality by health authority within country, and region of England
Males, UK and Ireland 1991-2000[1]

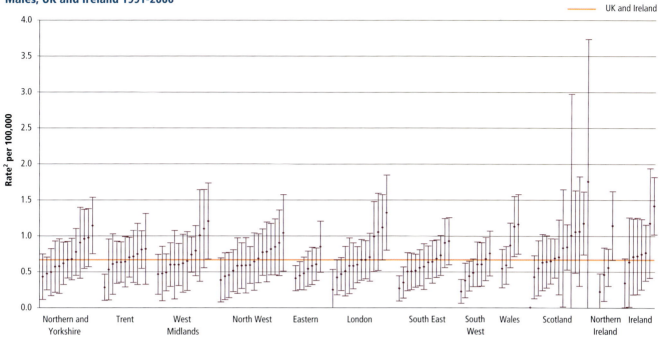

1 Scotland 1991-99, Ireland 1994-2000

2 Age standardised using the European standard population, with 95% confidence interval

Figure **8.4b**

Hodgkin's disease: mortality by health authority within country, and region of England
Females, UK and Ireland 1991-2000[1]

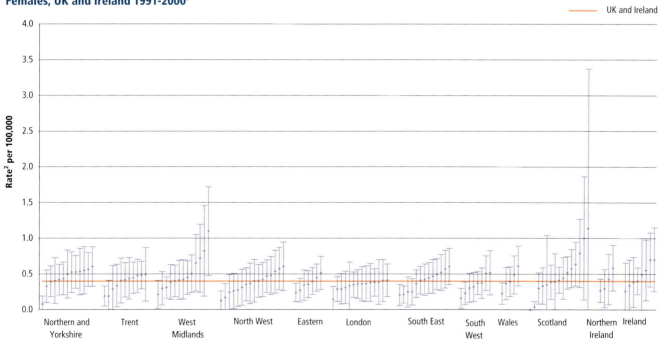

1 Scotland 1991-99, Ireland 1994-2000

2 Age standardised using the European standard population, with 95% confidence interval

Map **8.1a**

**Hodgkin's disease: incidence* by health authority
Males, UK and Ireland 1991-99**

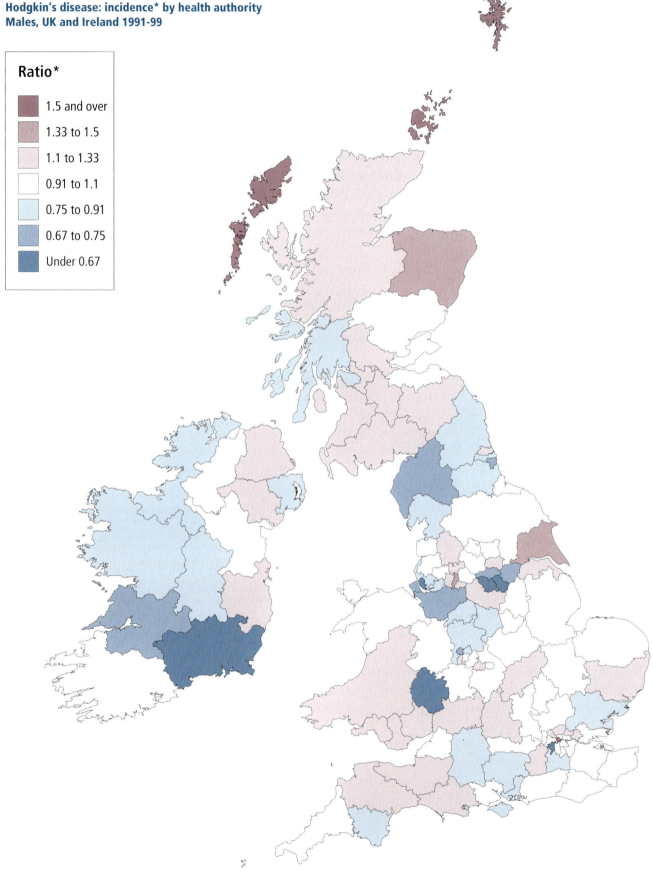

Ratio*

	1.5 and over
	1.33 to 1.5
	1.1 to 1.33
	0.91 to 1.1
	0.75 to 0.91
	0.67 to 0.75
	Under 0.67

Ratio of directly age-standardised rate in health authority to UK and Ireland average

Map **8.1b**

Hodgkin's disease: incidence* by health authority
Females, UK and Ireland 1991-99

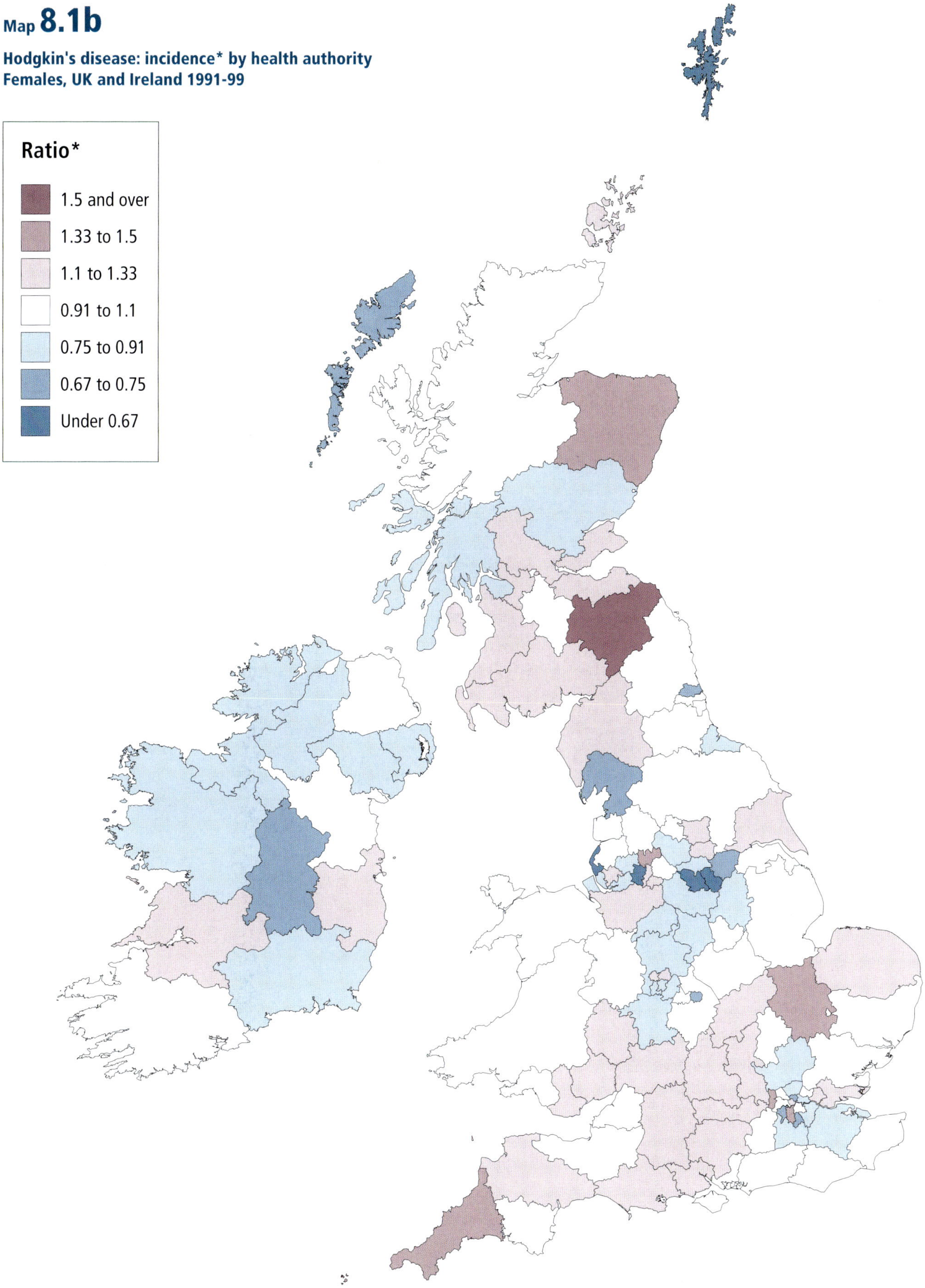

Ratio*

	1.5 and over
	1.33 to 1.5
	1.1 to 1.33
	0.91 to 1.1
	0.75 to 0.91
	0.67 to 0.75
	Under 0.67

*Ratio of directly age-standardised rate in health authority to UK and Ireland average

Map **8.2a**

Hodgkin's disease: mortality* by health authority
Males, UK and Ireland 1991-2000

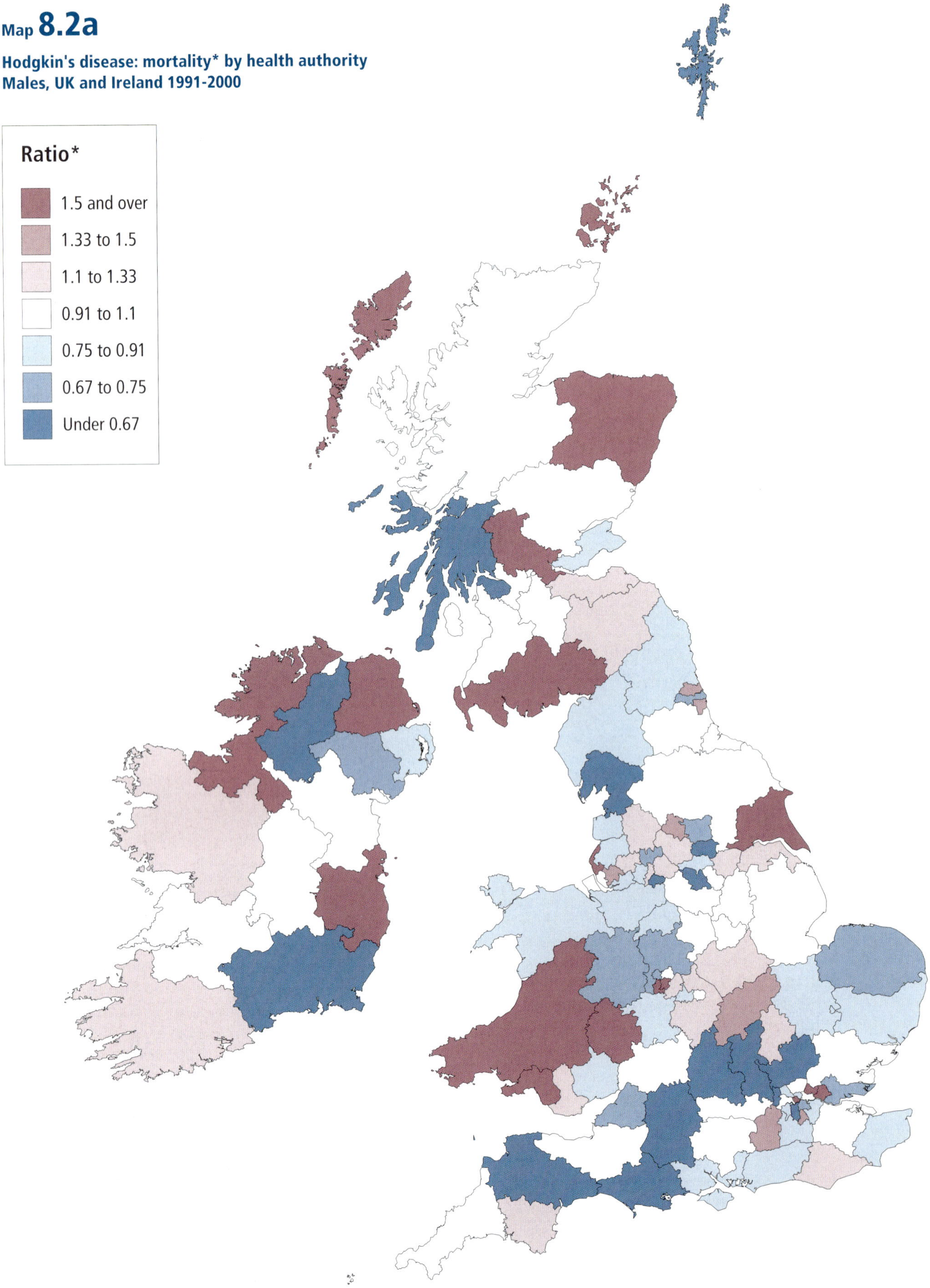

Ratio*

	1.5 and over
	1.33 to 1.5
	1.1 to 1.33
	0.91 to 1.1
	0.75 to 0.91
	0.67 to 0.75
	Under 0.67

*Ratio of directly age-standardised rate in health authority to UK and Ireland average

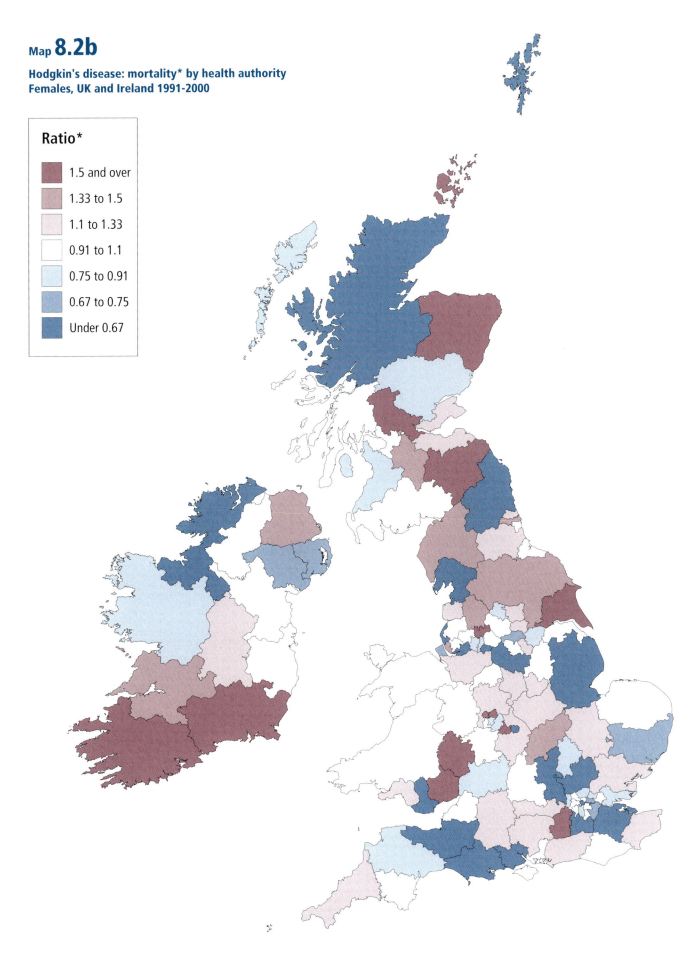

Map 8.2b

**Hodgkin's disease: mortality* by health authority
Females, UK and Ireland 1991-2000**

Ratio*

- 1.5 and over
- 1.33 to 1.5
- 1.1 to 1.33
- 0.91 to 1.1
- 0.75 to 0.91
- 0.67 to 0.75
- Under 0.67

*Ratio of directly age-standardised rate in health authority to UK and Ireland average

Socio-economic deprivation

There is conflicting evidence regarding the incidence of HD and deprivation.[6] Some studies have found an association between higher incidence of HD in children and young adults and affluence, using small-area-based measures.[28,29] Another ecological analysis of children, from north west England, has found a strong association between higher incidence of mixed cellularity HD and deprivation at the small area level, related to higher levels of unemployment and household overcrowding.[30] There have also been conflicting findings from studies of older adults [6] which may reflect the different aetiological sub-types or opportunities for exposure to a relevant aetiological agent.

The geographical distribution of HD incidence throughout the UK and Ireland is likely to depend on both genetic predisposition and environmental exposures. It may also vary by sub-type. Whilst the pattern of incidence presents a complex picture, it is very likely to depend on deprivation. Higher incidence of the mixed cellularity sub-type would be expected in the more deprived areas and lower incidence of this sub-type in the more affluent areas reflecting the increased opportunity for exposure to an aetiological agent, such as EBV, in the more deprived areas. In contrast, the pattern for the nodular sclerosis sub-type is likely to vary by sex, and for females may depend on birth rates and the age of women when first giving birth, as well as exposure to putative aetiological agents for both males and females. For females, higher incidence of the nodular sclerosis sub-type might be predicted in the more affluent areas, with lower incidence in the more deprived areas. The overall incidence pattern for all HD together will depend mainly on the proportions of nodular sclerosis and mixed cellularity cases within each geographical area or region, while that of mortality from HD will depend on both the underlying incidence rates and the survival rates. Since survival is worse in more deprived areas, mortality rates are likely to be higher in these locations.

References

1. World Health Organisation. *International Classification of Diseases for Oncology: Morphology of Neoplasms, third edition*. Geneva: WHO, 2000.

2. Chan JK. The new World Health Organisation classification for lymphomas: the past, the present and the future. *Haematological Oncology* 2001; 19: 129-150.

3. Cartwright RA, Alexander FE, McKinney PA, Ricketts TJ. *Leukaemia and Lymphoma. An atlas of distribution within areas of England and Wales 1984-1988*. London: Leukaemia Research Fund, 1990.

4. Correa FP, O'Connor GT. Epidemiological patterns of Hodgkin's disease. *International Journal of Cancer* 1971; 8: 192.

5. Cartwright R, McNally R, Roman E, Simpson J et al. Incidence and time trends in Hodgkin's disease: from parts of the United Kingdom (1984-1993). *Leukemia and Lymphoma* 1998; 31: 367-377.

6. Cartwright RA, Watkins G. Epidemiology of Hodgkin's disease: a review. *Hematological Oncology* 2004; 22: 11-26.

7. Hjalgrim H, Askling J, Pukkala E, Hansen S et al. Incidence of Hodgkin's disease in Nordic countries. *Lancet* 2001; 358: 297-298.

8. Yeole BB, Jussawalla DJ. Descriptive epidemiology of lymphatic malignancies in Greater Bombay. *Oncology Reports* 1998; 5: 771-777.

9. Chen YT, Zheng T, Chou MC, Boyle P et al. The increase of Hodgkin's disease incidence among young adults. Experience in Connecticut, 1935-1992. *Cancer* 1997; 79: 2209-2218.

10. Liu S, Semenciw R, Waters C, Wen SW et al. Time trends and sex patterns in Hodgkin's disease incidence in Canada, 1970-1995. *Canadian Journal of Public Health* 2000; 91: 188-192.

11. Swerdlow AJ. Epidemiology of Hodgkin's disease and non-Hodgkin's lymphoma. *European Journal of Nuclear Medicine and Molecular Imaging* 2003; 30 Suppl 1: S3-12.

12. Yung L, Linch D. Hodgkin's lymphoma. *Lancet* 2003; 361: 943-951.

13. Coleman MP, Babb P, Damiecki P, Grosclaude P et al. *Cancer Survival Trends in England and Wales, 1971-1995: Deprivation and NHS Region*. Studies on Medical and Population Subjects No. 61. London: The Stationery Office, 1999.

14. Bodmer WF. *Cancer Genetics*. Oxford: Oxford University Press, 1982.

15. Ferraris AM, Racchi O, Rapezzi D, Gaetani GF et al. Familial Hodgkin's disease: a disease of young adulthood? *Annals of Hematology* 1997; 74: 131-134.

16. Kerzin-Storrar L, Faed MJ, MacGillivray JB, Smith PG. Incidence of familial Hodgkin's disease. *British Journal of Cancer* 1983; 47: 707-712.

17. MacMahon B. Epidemiology of Hodgkin's disease. *Cancer Research* 1966; 26: 1189-1201.

18. Smith PG. Current assessment of 'case-clustering' of lymphomas and leukaemias. *Cancer* 1978; 42: 1026-1034.

19. Alexander FE, Williams J, McKinney PA, Ricketts TJ et al. A specialist leukaemia/lymphoma registry in the UK. Part 2: Clustering of Hodgkin's disease. *British Journal of Cancer* 1989; 60: 948-952.

20. International Agency for Research on Cancer. *Epstein-Barr Virus and Kaposi's Sarcoma Herpesvirus/Human Herpesvirus 8*. IARC Monographs on the Evaluation of Carcinogenic Risks to Humans. No. 70. Lyon: IARC Press, 1997.

21. Glaser SL, Jarrett RF. The epidemiology of Hodgkin's disease. In: Diehl V (ed) *Hodgkin's Disease*. London: Bailliere Tindall, 1996.

22. Jarrett RF. Epstein-Barr virus and Hodgkin's disease. *Epstein Barr Virus Report* 1998; 5: 77-85.

23. Jarrett RF, Armstrong AA, Alexander FE. Epidemiology of EBV and Hodgkin's lymphoma. *Annals of Oncology* 1996; 7: S5-S10.

24. Pallesen G, Hamilton-Dutoit SJ, Rowe M, Young LS. Expression of Epstein-Barr virus latent gene products in tumour cells of Hodgkin's disease. *Lancet* 1991; 337: 320-322.

25. Armstrong AA, Alexander FE, Paes RP, Morad NA et al. Association of Epstein-Barr virus with pediatric Hodgkin's disease. *American Journal of Pathology* 1993; 142: 1683-1688.

26. Glaser SL, Lin RJ, Stewart SL, Ambinder RF et al. Epstein-Barr virus-associated Hodgkin's disease: epidemiologic characteristics in international data. *International Journal of Cancer* 1997; 70: 375-382.

27. Preciado MV, Diez B, Grinstein S. Epstein Barr virus in Argentine pediatric Hodgkin's disease. *Leukemia and Lymphoma* 1997; 24: 283-290.

28. Alexander FE, McKinney PA, Williams J, Ricketts TJ et al. Epidemiological evidence for the 'two-disease hypothesis' in Hodgkin's disease. *International Journal of Epidemiology* 1991; 20: 354-361.

29. Alexander FE, Ricketts TJ, McKinney PA, Cartwright RA. Community lifestyle characteristics and incidence of Hodgkin's disease in young people. *International Journal of Cancer* 1991; 48: 10-14.

30. McNally RJ, Alston RD, Cairns DP, Eden OB et al. Geographical and ecological analyses of childhood acute leukaemias and lymphomas in north-west England. *British Journal of Haematology* 2003; 123: 60-65.

Chapter 9

Kidney

Steve Rowan, Robert Haward, David Forman, Caroline Brook

Summary

- Each year in the UK and Ireland, kidney cancer* accounts for around 1 in 55 diagnosed cases of cancer and 1 in 55 deaths from cancer.

- There was a clear north-south divide in incidence – particularly for females – across Great Britain with relatively high rates in Scotland, parts of Wales and in the north of England (females) and low rates in the south east of England (including London) and the midlands (females).

- The geographical pattern of mortality rates was broadly similar to that for incidence.

- There appears to be an association between both incidence and mortality and socio-economic deprivation – more apparent in females – reflected in higher rates in the more urban areas of the north of England and lower rates in the south east.

- Possible risk factors for kidney cancer include smoking and obesity, both of which are associated with deprivation and may explain some of the geographical variation observed.

* In this chapter, data for the UK and Ireland refer to renal cell carcinomas (which make up about 85 per cent of all kidney cancers). However, where reference has been made to trends or survival in England and Wales, the data pertain to all kidney cancers (including cancers of the renal pelvis and ureter, see Table 9.1).

Incidence and mortality

In the UK and Ireland in the 1990s, there were about 3,100 newly diagnosed cases of kidney cancer each year in males and about 1,900 cases in females. It accounted for 2.3 per cent of all newly diagnosed cancer cases in males and 1.4 per cent in females. Overall, the age-standardised incidence rates were 9.8 and 4.8 per 100,000 in males and females respectively, a male-to-female ratio of about 2:1. Kidney cancer is predominately a disease of the elderly with the age-specific incidence rates rising steeply, especially in males, from about age 50. For males in England and Wales, the lifetime risk[3] of being diagnosed with kidney cancer was 1.1 per cent (1 in 90) in males, compared with 0.6 per cent (1 in 160) in females.[4]

In the 1990s, about 1,800 males and 1,200 females died from kidney cancer in the UK and Ireland each year. Kidney cancer accounted for 2.1 per cent of cancer deaths in males and 1.5 per cent in females. Overall, the age-standardised mortality rates were 5.6 and 2.6 per 100,000 in males and females respectively, a male-to-female ratio of about 2.2:1, slightly higher than that for incidence. The age-specific mortality rates followed a similar pattern to those for incidence, with rates rising steeply, particularly in males, from about age 50.

Incidence and mortality trends

In England and Wales, the age-standardised incidence rates of kidney cancer nearly doubled in both males and females, from 5.8 and 2.9 per 100,000, respectively, in 1971 to 11.1 and 5.1 per 100,000 in 1997. In line with incidence, the age-standardised mortality rates in males and females rose steadily from 3.1 and 1.5 per 100,000, respectively, in 1950 to 5.9 and 2.6 per 100,000 in 1999. The increases in both incidence and mortality were greatest in the older age groups.[4]

Table 9.1

Cases of and deaths from kidney cancer by sub-site, 1999[1,2]

Percentages

Cancer site	ICD10 code	ICD9 code	Cases (England)		Deaths (England and Wales)	
			Males	Females	Males	Females
Kidney (renal cell carcinoma)	C64	189.0	88.2	87.8	95.9	95.6
Renal pelvis	C65	189.1	5.6	6.2	0.5	0.6
Ureter	C66	189.2	4.3	3.6	2.9	2.7
Other and unspecified urinary organs	C68	189.3–189.9	2.0	2.4	0.7	1.1
Total number of cases/deaths			2,946	1,914	1,696	1,033

Percentages may not add to 100 due to rounding.

Survival

Survival from kidney cancer is modest, although there has been some improvement over time. For patients diagnosed in England and Wales in 1996–99, five-year age-standardised relative survival from kidney cancer was 45 per cent in males and 44 per cent in females. This compares with 41.5 per cent for males and 40 per cent for females diagnosed in 1991–95, an improvement of around 4 percentage points in both sexes.[5] The European average for five-year survival from kidney cancer was slightly higher, being around 54 per cent for men and 57 per cent for women diagnosed in the early 1990s.[6] In England and Wales, survival from kidney cancer has improved markedly since the early 1970s, when it was around 30 per cent for both sexes.[7]

Geographical patterns in incidence

For the constituent countries of the UK and Ireland, the age-standardised incidence rates of kidney cancer were above the UK and Ireland average in Scotland and Wales in both sexes, and in Northern Ireland in females only (Figure 9.1, Table B9.1). Rates were close to the average for males in Northern Ireland. Within England, incidence rates for males were noticeably lower than average in both sexes in the Eastern and London regions (Figure 9.1). Within each country, or region of England, the differences in the incidence rates between the highest and lowest health authorities were generally 30–50 per cent in males and 30–90 per cent in females (Figure 9.3).

The maps for incidence (Map 9.1) show the higher than average rates in Scotland and parts of Wales for both males and females and the lower rates in parts of the Eastern and London regions of England. There were also high rates in the more urban areas of the North West, Northern and Yorkshire, and Trent regions, particularly in females, suggesting a north-south divide in incidence.

Geographical patterns in mortality

Age-standardised mortality rates for kidney cancer were markedly higher than average only in Scotland, and were close to average for both males and females in the other four countries. Within England, the pattern of mortality rates by region was broadly similar to that for incidence, being below average in the Eastern and London regions in both males and females (Figure 9.2). Within each country, or region of England, the variability in mortality rates between the highest and lowest health authorities was broadly similar to that for incidence, with differences generally of 20–50 per cent in males and 30–60 per cent in females (Figure 9.4).

The maps for mortality (Map 9.2) show generally similar patterns to those for incidence. There were high mortality rates in some, but not all parts of Scotland for both males and females, as well as in the more urban areas of the North West, Northern and Yorkshire, and Trent regions, especially for females. Rates were low in the south of Ireland, parts of London, and the Eastern region. For females, rates were also low in parts of the South East and West Midlands regions, as well as in southern Scotland. The north-south divide seen on the maps appears to be more marked than for incidence.

Risk factors and aetiology

About 85 per cent of kidney cancers are renal cell carcinomas, with most of the remaining cases being transitional cell carcinomas of the renal pelvis and ureter. Renal cell carcinomas have a tendency to spread to other parts of the body via the blood stream rather than the lymph nodes, giving rise to metastases, particularly in the lungs.

For cancer of the renal pelvis and ureter, studies have found that cigarette smoking is a major risk factor,[8–10] although for renal cell carcinoma, the association with smoking is weaker.[10-14] The use of phenacetin-containing analgesics is a major risk factor for cancer of the renal pelvis and ureter,[15] but there are no current UK Marketing Authorisations (product licenses) for any products containing phenacetin as an active substance.

There is evidence of increased risk of renal cell carcinoma with excess body weight [13,16–18] and since the prevalence of obesity is increasing in the UK population, it may have contributed to the increasing incidence.[19] In addition, medications related to the treatment of hypertension or the severity of hypertension itself are also possible risk factors.[20]

For patients who undergo renal dialysis, there is increased incidence of acquired cystic disease of the kidney, which pre-disposes to kidney cancer, particularly in males.[21] While inherited factors such as von Hippel-Lindau Syndrome[22] are very rare, these patients develop multiple types of cancer and are thus at greater risk of developing kidney cancer.

Although the main treatment is surgery, kidney cancer responds to biological treatments, the main ones being interleukin-2 and interferon. Kidney cancer is, however, not very responsive to chemotherapy or radiotherapy.

(continued on page 110)

Figure **9.1**

**Kidney: incidence by sex, country, and region of England
UK and Ireland 1991-99[1]**

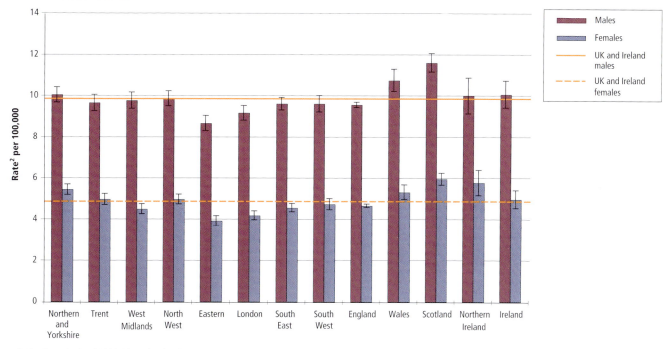

1 Northern Ireland 1993-99, Ireland 1994-99

2 Age standardised using the European standard population, with 95% confidence interval

Figure **9.2**

**Kidney: mortality by sex, country, and region of England
UK and Ireland 1991-2000[1]**

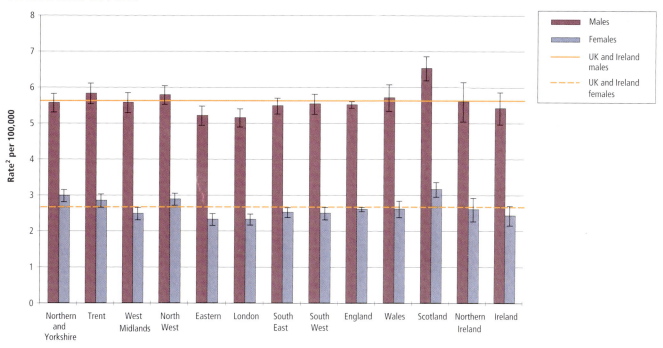

1 Scotland 1991-1999, Ireland 1994-2000

2 Age standardised using the European standard population, with 95% confidence interval

Figure 9.3a

**Kidney: incidence by health authority within country, and region of England
Males, UK and Ireland 1991-99[1]**

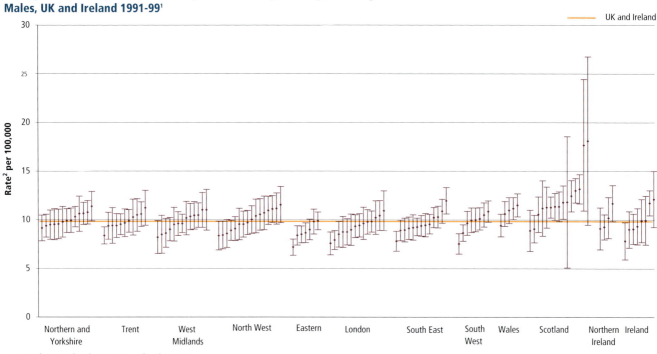

1 Northern Ireland 1993-99, Ireland 1994-99

2 Age standardised using the European standard population, with 95% confidence interval

Figure 9.3b

**Kidney: incidence by health authority within country, and region of England
Females, UK and Ireland 1991-99[1]**

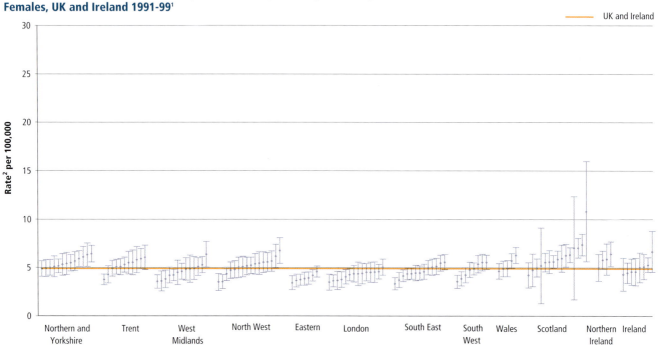

1 Northern Ireland 1993-99, Ireland 1994-99

2 Age standardised using the European standard population, with 95% confidence interval

Figure **9.4a**

**Kidney: mortality by health authority within country, and region of England
Males, UK and Ireland 1991-2000[1]**

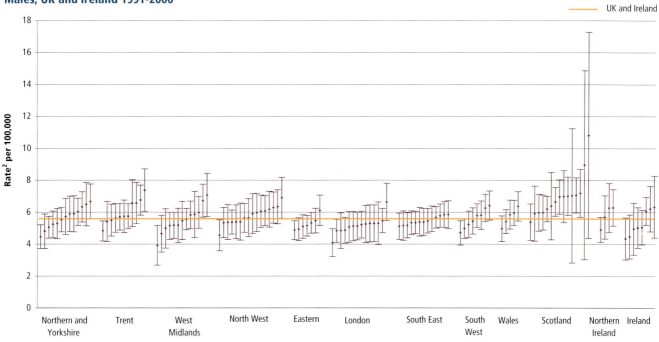

1 Scotland 1991-1999, Ireland 1994-2000

2 Age standardised using the European standard population, with 95% confidence interval

Figure **9.4b**

**Kidney: mortality by health authority within country, and region of England
Females, UK and Ireland 1991-2000[1]**

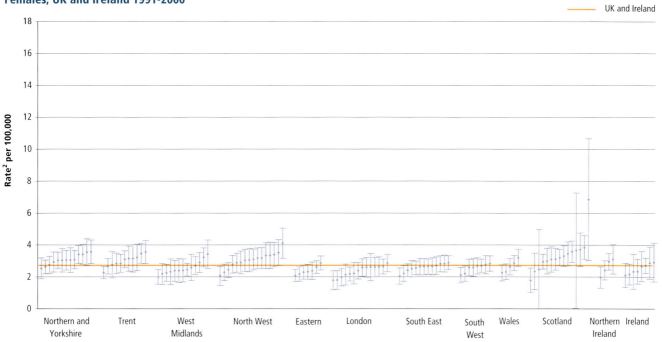

1 Scotland 1991-99, Ireland 1994-2000

2 Age standardised using the European standard population, with 95% confidence interval

Map **9.1a**

**Kidney: incidence* by health authority
Males, UK and Ireland 1991-99**

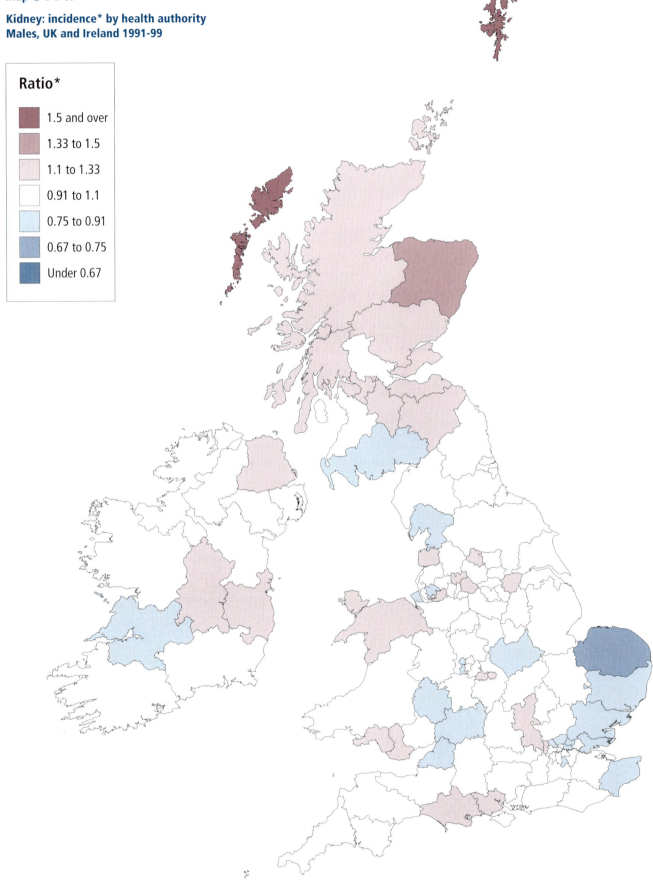

Ratio*

- 1.5 and over
- 1.33 to 1.5
- 1.1 to 1.33
- 0.91 to 1.1
- 0.75 to 0.91
- 0.67 to 0.75
- Under 0.67

*Ratio of directly age-standardised rate in health authority to UK and Ireland average

Map **9.1b**

**Kidney: incidence* by health authority
Females, UK and Ireland 1991-99**

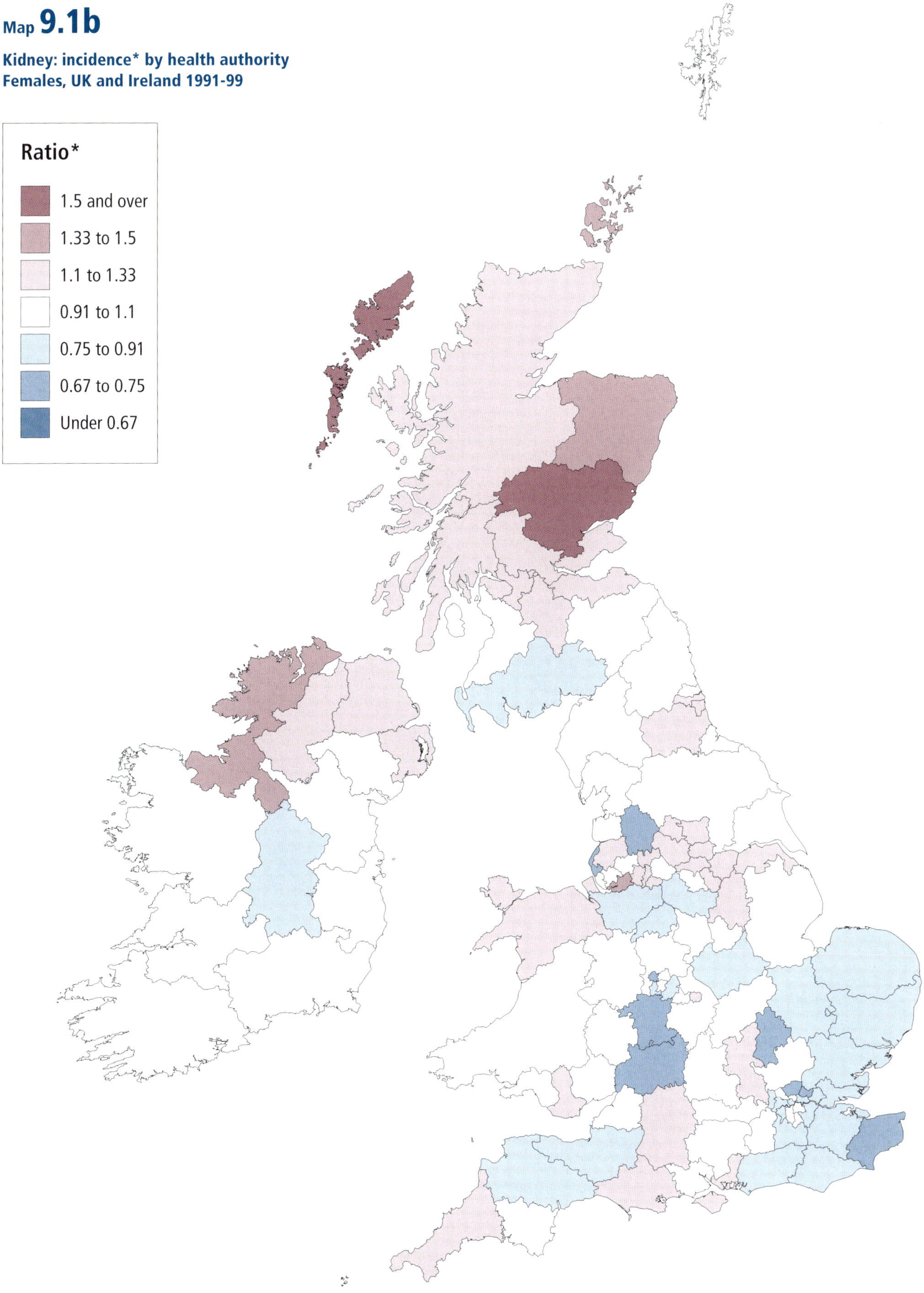

Ratio*

	1.5 and over
	1.33 to 1.5
	1.1 to 1.33
	0.91 to 1.1
	0.75 to 0.91
	0.67 to 0.75
	Under 0.67

*Ratio of directly age-standardised rate in health authority to UK and Ireland average

Map **9.2a**

**Kidney: mortality* by health authority
Males, UK and Ireland 1991-2000**

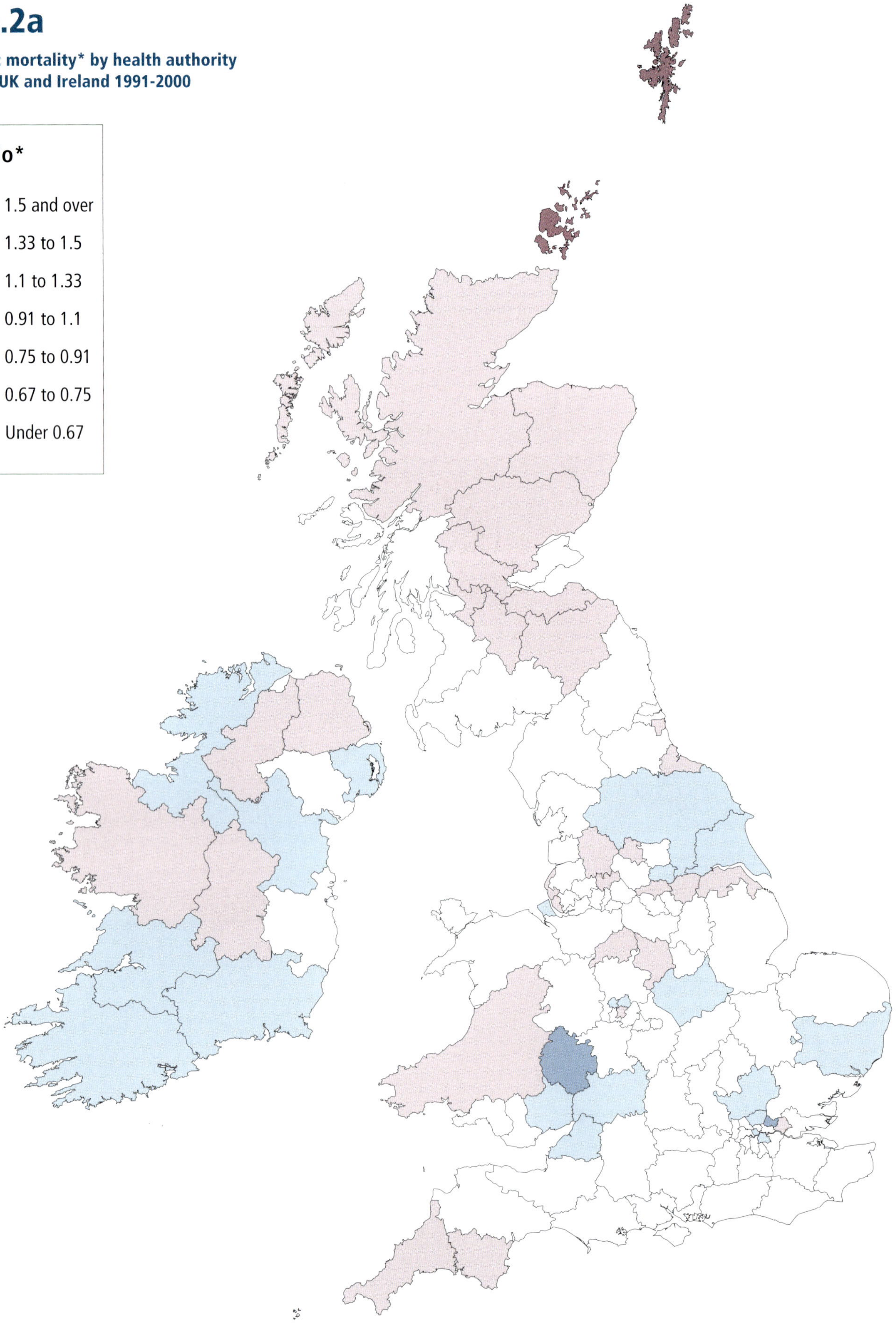

Ratio*

	1.5 and over
	1.33 to 1.5
	1.1 to 1.33
	0.91 to 1.1
	0.75 to 0.91
	0.67 to 0.75
	Under 0.67

Ratio of directly age-standardised rate in health authority to UK and Ireland average

Map **9.2b**

Kidney: mortality* by health authority
Females, UK and Ireland 1991-2000

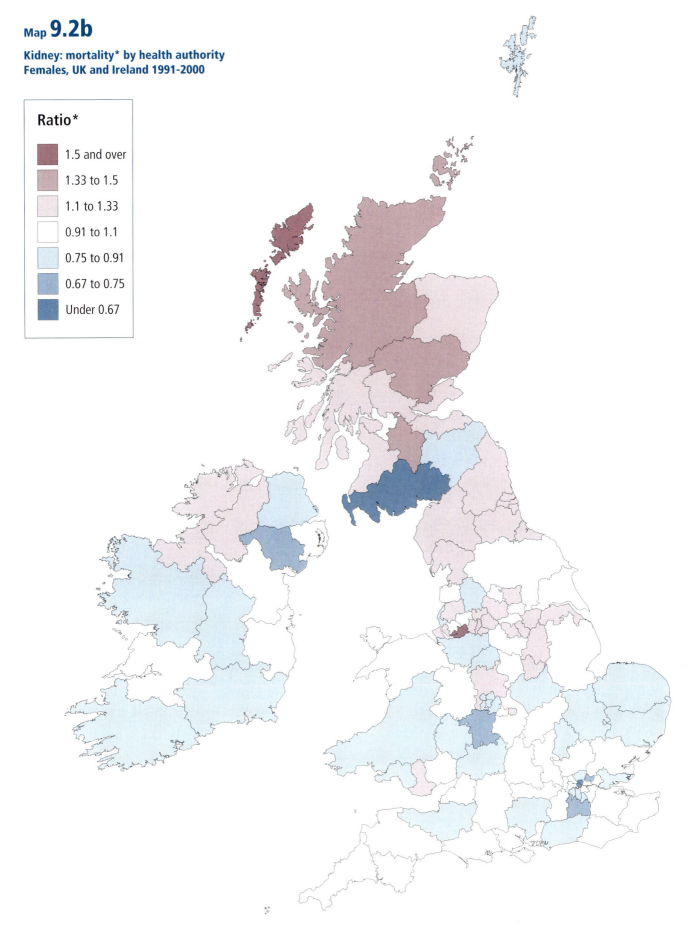

Ratio*

- 1.5 and over
- 1.33 to 1.5
- 1.1 to 1.33
- 0.91 to 1.1
- 0.75 to 0.91
- 0.67 to 0.75
- Under 0.67

*Ratio of directly age-standardised rate in health authority to UK and Ireland average

Socio-economic deprivation

For males diagnosed with kidney cancer in 1990–93 in England and Wales, there was no apparent gradient in incidence across categories of socio-economic deprivation, measured using the Carstairs index,[23] whereas for females, incidence was slightly higher among people living in more deprived areas than among those living in more affluent areas. The pattern for mortality was similar to that for incidence.[4] Comparing the maps for incidence (Map 9.1) with that illustrating deprivation using the Carstairs index (see Appendix F), there appears to be some association between deprivation and kidney cancer in males, but the deprivation gap is again more apparent in females, particularly in the more deprived areas of the North West, Northern and Yorkshire, and Trent regions of England. In Scotland, rates were high across the majority of the country for both males and females.

Despite increases in survival from kidney cancer in England and Wales in the 1990s, there was a significant deprivation gap for males diagnosed in 1996–99, with a difference in five-year survival of 6 percentage points between patients from the most deprived and the most affluent areas. There was, however, no deprivation gap for females diagnosed in the same period.[24]

The maps for mortality (Map 9.2) show higher rates in parts of the constituent countries which are known to have higher levels of socio-economic deprivation (see Appendix F), but as with the incidence maps – and in contrast to the results for survival – the relationship with deprivation is more apparent for females than males.

References

1. ONS. *Cancer Statistics Registrations: Registrations of cancer diagnosed in 1999, England.* Series MB1 No. 30. London: Office for National Statistics, 2002.

2. ONS. *Mortality Statistics 1999: Cause.* Series DH2 No. 26. London: The Stationery Office, 2000.

3. Schouten LJ, Straatman H, Kiemeney LALM, Verbeek ALM. Cancer incidence: Life table risk versus cumulative risk. *Journal of Epidemiology and Community Health* 1994; 48: 596–600.

4. Quinn MJ, Babb PJ, Brock A, Kirby L et al. *Cancer Trends in England and Wales 1950–1999.* Studies on Medical and Population Subjects No. 66. London: The Stationery Office, 2001.

5. ONS. Cancer Survival: England and Wales, 1991–2001. March 2004. Available at http://www.statistics.gov.uk/statbasessdataset.asp?vlnk=7899.

6. Sant M, Areleid T, Berrino F, Bielska Lasota M et al. EUROCARE-3: survival of cancer patients diagnosed 1990–1994 – results and commentary. *Annals of Oncology* 2003; 14 Suppl 5: v61–v118.

7. Coleman MP, Babb P, Damiecki P, Grosclaude P et al. *Cancer Survival Trends in England and Wales, 1971–1995: Deprivation and NHS Region.* Studies on Medical and Population Subjects No. 61. London: The Stationery Office, 1999.

8. McLaughlin JK, Silverman DT, Hsing AW, Ross RK et al. Cigarette smoking and cancers of the renal pelvis and ureter. *Cancer Research* 1992; 52: 254–257.

9. Jensen OM, Knudsen JB, McLaughlin JK, Sorensen BL. The Copenhagen case-control study of renal pelvis and ureter cancer: role of smoking and occupational exposures. *International Journal of Cancer* 1988; 41: 557–561.

10. McCredie M, Stewart JH. Risk factors for kidney cancer in New South Wales. I. Cigarette smoking. *European Journal of Cancer* 1992; 28A: 2050–2054.

11. McLaughlin JK, Mandel JS, Blot WJ, Schuman LM et al. A population-based case-control study of renal cell carcinoma. *Journal of the National Cancer Institute* 1984; 72: 275–284.

12. La Vecchia C, Negri E, D'Avanzo B, Franceschi S. Smoking and renal cell carcinoma. *Cancer Research* 1990; 50: 5231–5233.

13. Kreiger N, Marrett LD, Dodds L, Hilditch S et al. Risk factors for renal cell carcinoma: results of a population-based case-control study. *Cancer Causes and Control* 1993; 4: 101–110.

14. Mellemgaard A, Engholm G, McLaughlin JK, Olsen JH. Risk factors for renal cell carcinoma in Denmark. I. Role of socioeconomic status, tobacco use, beverages, and family history. *Cancer Causes and Control* 1994; 5: 105–113.

15. McCredie M, Stewart JH, Day NE. Different roles for phenacetin and paracetamol in cancer of the kidney and renal pelvis. *International Journal of Cancer* 1993; 53: 245–249.

16. Bergstrom A, Pisani P, Tenet V, Wolk A et al. Overweight as an avoidable cause of cancer in Europe. *International Journal of Cancer* 2001; 91: 421–430.

17. Mellemgaard A, Engholm G, McLaughlin JK, Olsen JH. Risk factors for renal-cell carcinoma in Denmark. III. Role of weight, physical activity and reproductive factors. *International Journal of Cancer* 1994; 56: 66–71.

18. McCredie M, Stewart JH. Risk factors for kidney cancer in New South Wales, Australia. II. Urologic disease, hypertension, obesity, and hormonal factors. *Cancer Causes and Control* 1992; 3: 323–331.

19. Seidell JC, Flegal KM. Assessing obesity: classification and epidemiology. *British Medical Bulletin* 1997; 53: 238–252.

20. Heath CW, Jr., Lally CA, Calle EE, McLaughlin JK et al. Hypertension, diuretics, and antihypertensive medications as possible risk factors for renal cell cancer. *American Journal of Epidemiology* 1997; 145: 607–613.

21. Ishikawa I. Development of adenocarcinoma and acquired cystic disease of the kidney in hemodialysis patients. *Princess Takamatsu Symposium* 1987; 18: 77–86.

22. Latif F, Tory K, Gnarra J, Yao M et al. Identification of the von Hippel-Lindau disease tumor suppressor gene. *Science* 1993; 260: 1317–1320.

23. Carstairs V, Morris R. Deprivation and mortality: an alternative to social class? *Community Medicine* 1989; 11: 213–219.

24. Coleman MP, Rachet B, Woods LM, Mitry E et al. Trends and socio-economic inequalities in cancer survival in England and Wales up to 2001. *British Journal of Cancer* 2004; 90: 1367–1373.

Chapter 10

Larynx

David Brewster, Henrik Møller

Summary

- In the UK and Ireland in the 1990s, cancer of the larynx accounted for around 1 in 70 diagnosed cases of cancer and 1 in 110 deaths from cancer in males.

- Incidence rates were highest in Scotland and the urban areas of the north and midlands of England (although not in inner London) and lowest in the south and east of England.

- The geographical pattern of mortality rates was broadly similar to that for incidence, although mortality in Ireland was higher than might be expected from its incidence.

- Many of the areas with the highest incidence and mortality rates were also areas with high levels of socio-economic deprivation.

- The geographical variations in incidence are likely to reflect the corresponding variation in the historical patterns of prevalence of smoking and alcohol consumption, both of which are established risk factors.

- There are clear similarities between the patterns of distribution of laryngeal cancer and those of cancers of the lip, mouth and pharynx, lung, and oesophagus.

Incidence and mortality

The majority of the data presented in this chapter are restricted to males because the relatively small numbers of cases and deaths in females yielded unstable estimates of incidence and mortality rates in regional comparisons. Data for females are given only at the national level.

In the 1990s there were about 2,000 newly diagnosed cases of laryngeal cancer each year in males and 460 in females. Overall, the age-standardised incidence rate was 6.3 per 100,000 in males (1.2 per 100,000 in females). However, cancer of the larynx is a disease that mainly affects elderly people.[1-4] Incidence was considerably lower than the overall rate in those aged under 50, and then rose rapidly to peak in those aged 70–74.

In the UK and Ireland, around 770 males and 210 females died from cancer of the larynx each year in the 1990s. The age-standardised mortality rate was 2.3 per 100,000 in males (0.5 per 100,000 in females), but as for incidence, rates were higher in elderly people, and reached a peak in those aged 85 and over.

Incidence and mortality trends

In males in England and Wales, age-standardised incidence rates have fluctuated around 6 per 100,000 between the early 1970s and the mid-1990s, rising slightly in the late 1980s and subsequently declining.[1] In contrast, there has been a fairly sustained increase in incidence rates among males in Scotland, from 5 to 9 per 100,000 between 1960 and 1994,[5] although more recent data show some evidence of a plateau.[6] Age-standardised mortality rates in England and Wales decreased substantially between the early 1950s and the late 1990s, from around 4.5 to just over 2 per 100,000 (in males), with particularly marked decreases up to the early 1970s.[1] In Scotland, in contrast, mortality rates have shown an upward trend, at least over the last two decades, rising from around 3 to 3.5 per 100,000 (in males) between 1979 and the early part of the twenty-first century.[6]

Survival

The survival prospects for patients with laryngeal cancer are moderately favourable. The most recently available data for patients in England and Wales show that for men diagnosed in 1996–99, the age-standardised relative survival at one year after diagnosis was 84 per cent, and at five years was 63 per cent.[7] One- and five-year survival for patients diagnosed in Scotland in the 1990s was very similar.[8] In contrast to the findings for many other cancers, survival from cancer of the larynx was close to or above the European average for all the constituent countries of the UK included in the EUROCARE-3 study.[9] This may be due, in part, to a more favourable mix of anatomical sub-sites in the UK compared with other European countries,[10] since tumours of the glottis tend to present earlier and have a better prognosis.[5] One- and five-year relative survival increased by 4 percentage points for men diagnosed in England and Wales in 1986–90 compared with 1971–77.[11] Although one-year relative survival increased by about 5 percentage points in Scottish men over the same period, five-year survival changed little.[12]

Geographical patterns in incidence

Within the countries of the UK, the highest incidence rates for cancer of the larynx occurred in Scotland, where the age-standardised incidence in males was 49 per cent higher than the average (Figure 10.1).

(continued on page 116)

Figure **10.1**

**Larynx: incidence by country, and region of England
UK and Ireland 1991-99[1]**

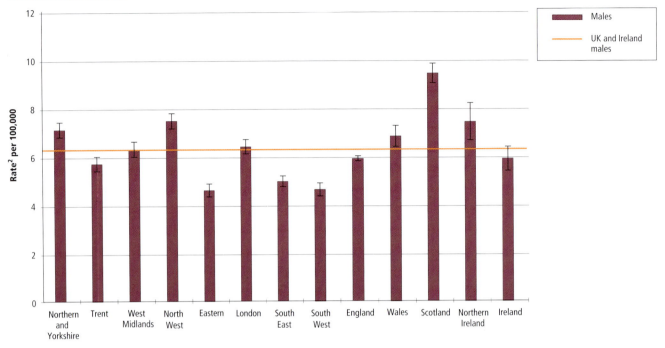

1 Northern Ireland 1993-99, Ireland 1994-99

2 Age standardised using the European standard population, with 95% confidence interval

Figure **10.2**

**Larynx: mortality by country, and region of England
UK and Ireland 1991-2000[1]**

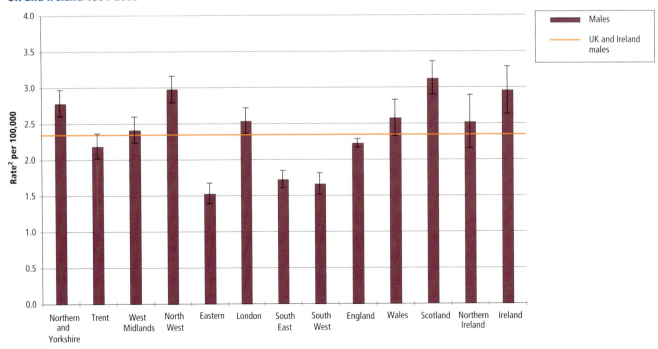

1 Scotland 1991-99, Ireland 1994-2000

2 Age standardised using the European standard population, with 95% confidence interval

Figure **10.3**

**Larynx: incidence by health authority within country, and region of England
Males, UK and Ireland 1991-99[1]**

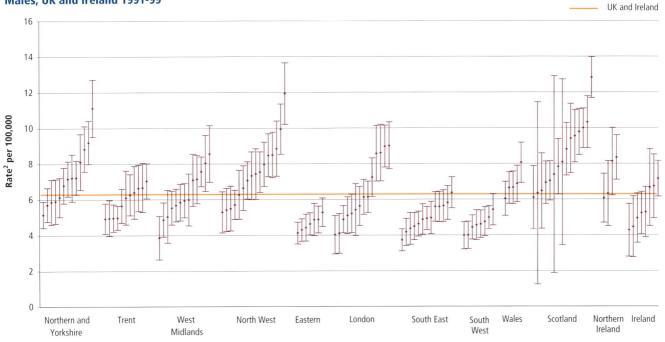

1 Northern Ireland 1993-99, Ireland 1994-99

2 Age standardised using the European standard population, with 95% confidence interval

Figure **10.4**

**Larynx: mortality by health authority within country, and region of England
Males, UK and Ireland 1991-2000[1]**

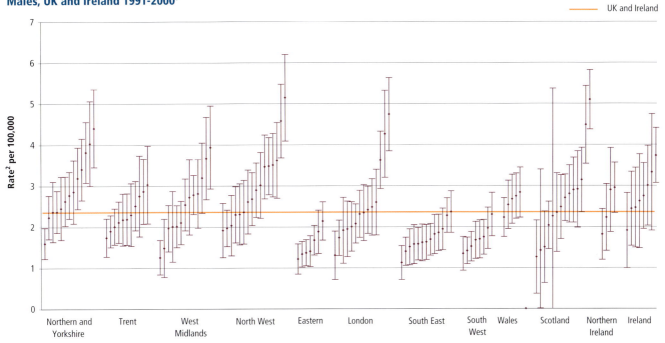

1 Scotland 1991-99, Ireland 1994-2000

2 Age standardised using the European standard population, with 95% confidence interval

Map **10.1**

**Larynx: incidence* by health authority
UK and Ireland 1991-99**

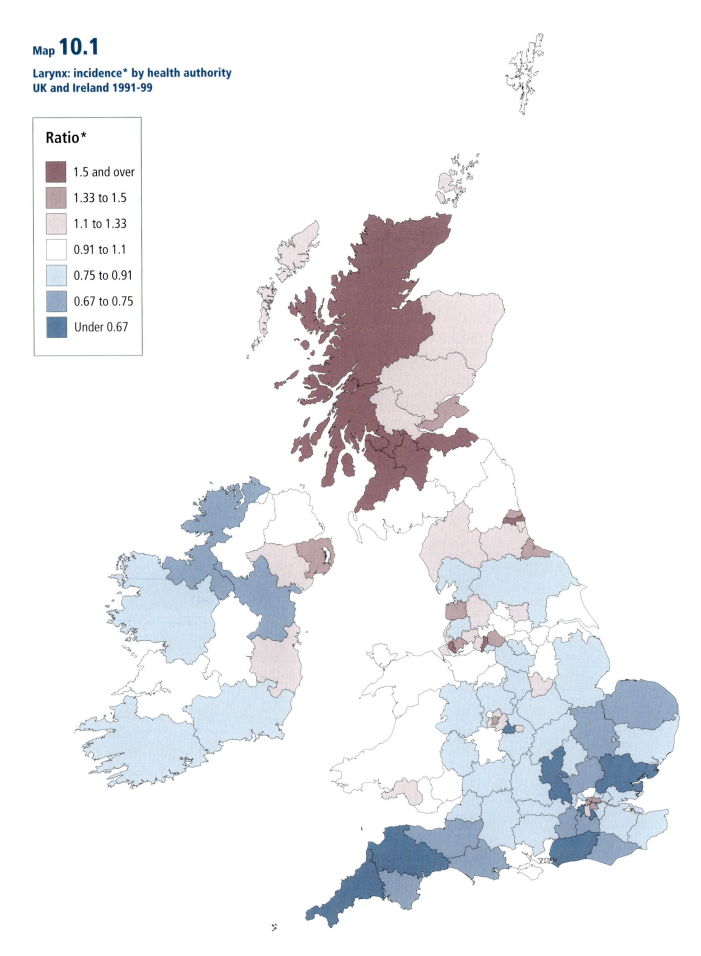

Ratio*

	1.5 and over
	1.33 to 1.5
	1.1 to 1.33
	0.91 to 1.1
	0.75 to 0.91
	0.67 to 0.75
	Under 0.67

*Ratio of directly age-standardised rate in health authority to UK and Ireland average

Map 10.2

**Larynx: mortality* by health authority
UK and Ireland 1991-2000**

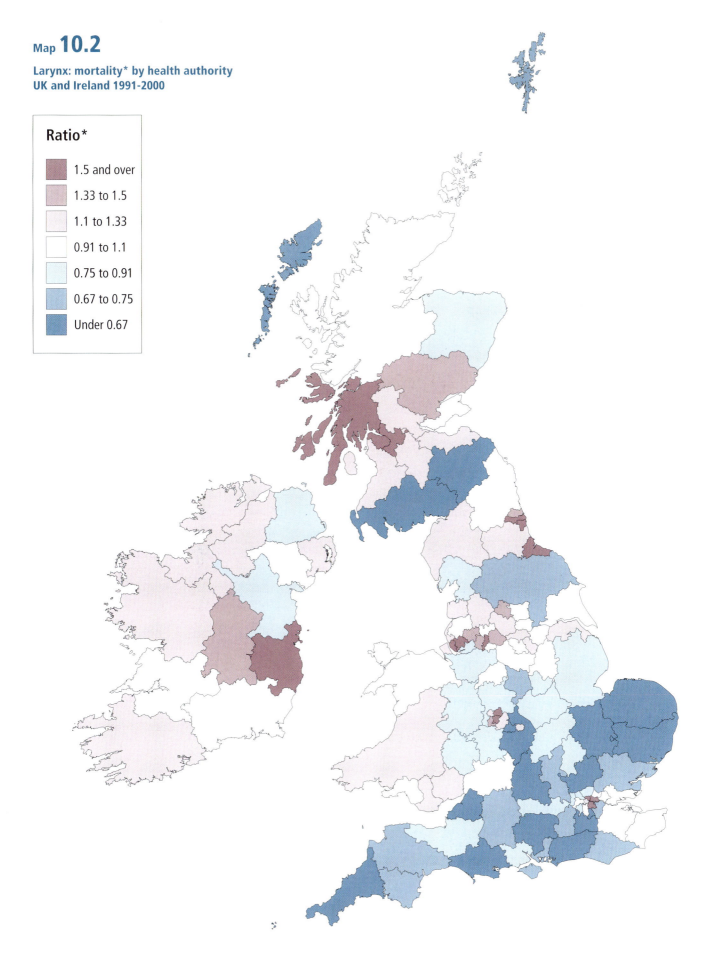

Ratio*

	1.5 and over
	1.33 to 1.5
	1.1 to 1.33
	0.91 to 1.1
	0.75 to 0.91
	0.67 to 0.75
	Under 0.67

Ratio of directly age-standardised rate in health authority to UK and Ireland average

Incidence rates were closer to the average in the other constituent countries. Within England, there was a very clear north-south divide, with incidence rates being higher than average in the North West, and Northern and Yorkshire regions, and well below average in the Eastern, South West, and South East regions.

Variability in the incidence of laryngeal cancer was generally higher between the health authorities within a country or region than between the constituent countries of the UK and regions of England, although this observation may simply reflect greater variability associated with smaller numbers of cases (Figure 10.3). The highest incidence occurred in males in Greater Glasgow, but the next highest rates were seen outside Scotland, in Manchester, and in Gateshead and South Tyneside (Table B10.1).

Incidence rates were markedly below average in all the health authorities in the Eastern region, in most of the health authorities in the South West, and in more than half those in the South East. In the regions with lower incidence, the range between the highest and lowest rates was not as wide as in Scotland and the north of England. In Wales, rates were clustered around the average, but were notably higher in Morgannwg. In Northern Ireland and Ireland, the highest rates were observed in the Eastern areas which include, respectively, Belfast and Dublin.

Unfortunately, the relatively high percentage of tumours registered with unspecified sub-site[11] rules out the possibility of undertaking any meaningful investigation of geographical variation in incidence by sub-site across the UK and Ireland.

Geographical patterns in mortality

Patterns of mortality across the UK and Ireland were similar to those of incidence with the exception of Ireland, which in contrast to incidence, had a mortality rate markedly higher than the average (Figure 10.2). The mortality-to-incidence ratio was also higher in Ireland than elsewhere, owing to the exclusion of death certificate only (DCO) cases, and possibly to below-average survival prospects (Table B10.1). The highest mortality rate was seen in Scotland, and rates were also higher than average in the North West, and Northern and Yorkshire regions of England. Mortality rates were well below average in the Eastern, South West, and South East regions of England.

At the level of health authorities, the geographic patterns of mortality (Figure 10.4) were similar to those of incidence. In relative terms, however, the range of mortality rates within a given region was generally greater than the range of incidence rates, reflecting the considerably smaller numbers of deaths compared with cases. Mortality rates of 4 per 100,000 or

greater were seen in East London and The City; Liverpool; Argyll and Clyde; Gateshead and South Tyneside; Camden and Islington; and Sunderland, with the highest rates observed in Greater Glasgow, and Manchester (Table B10.1).

The maps of incidence and mortality (Maps 10.1 and 10.2) clearly show high rates in much of Scotland (bearing in mind that a sizeable majority of the population is concentrated in the central belt, including Glasgow and Edinburgh), and also in parts of the north of England, the West Midlands, London, and Wales, and in the east of both Northern Ireland and Ireland. Rates were low in large parts of the south of England, mainly rural parts of Ireland (especially incidence), and some largely rural areas of Scotland (mortality only).

Risk factors and aetiology

The majority of laryngeal cancers are squamous cell carcinomas. The main established environmental risk factors for cancer of the larynx are tobacco smoking and alcohol consumption, and the relative importance of these may differ according to the sub-site of the tumour.[13] Both factors act independently, but when combined have a synergistic effect. The role of other potential risk factors, such as diet, gastro-oesophageal reflux, human papillomavirus (HPV), and exposure to asbestos or other occupational hazards is less certain.[13] The geographical distribution of laryngeal cancer in the UK and Ireland is likely to reflect corresponding variation in the historical patterns of prevalence of smoking and alcohol consumption. For example, both the prevalence of smoking and alcohol consumption are known to be higher in Scotland than in England and Wales.[14] Despite reductions in the prevalence of smoking among men of all age groups in Britain since the late 1940s,[15] increases in laryngeal cancer among young men in Scotland up to the 1990s might be due to increased alcohol consumption, and/or an increase in the number of individuals who both smoke and drink heavily.[16] Across the UK as a whole, alcohol consumption has been rising since the middle of the last century.[17]

Socio-economic deprivation

In England and Wales, there is a strong relationship between the incidence of laryngeal cancer and socio-economic deprivation, measured using the Carstairs index.[1,18] In the years around the 1991 Census, incidence in males living in the most deprived areas was almost three times that in males living in the most affluent areas, and a similar gradient was observed in Scotland.[2] The pattern of laryngeal cancer mortality with deprivation was similar to that for incidence, both in England and Wales,[1] and during the early to mid-1990s in Scotland.[12] Survival was also lower among deprived compared with affluent groups,[1,5,11,12] and this may help to explain why the range of mortality rates across different geographical areas appeared to be greater than the range of incidence rates (see above). Social class, based on occupation, has also been related to the risk of dying from laryngeal cancer, although there was no definite pattern in the early part of the twentieth century.[19]

Many of the areas with the highest level of deprivation, as measured by the Carstairs index (see Appendix F), corresponded to areas with high incidence of, and mortality from, laryngeal cancer, including Greater Glasgow, Manchester, and Gateshead and South Tyneside. As noted above, most of the geographical variation in the risk of laryngeal cancer is likely to reflect historical differences in the prevalence and patterns of smoking and alcohol consumption. The prevalence of smoking has, for some time, been substantially higher in unskilled manual workers than in professional people.[14] In contrast, there is currently no clear socio-economic gradient in relation to alcohol consumption among men,[14] although in Scotland, mean weekly consumption is slightly higher in the manual than the non-manual social classes,[20] and men living in the most deprived areas are seven times more likely to die an alcohol-related death or be admitted to hospital with an alcohol-related illness than those in the least deprived areas.[21]

References

1. Quinn MJ, Babb PJ, Brock A, Kirby L et al. *Cancer Trends in England and Wales 1950–1999*. Studies on Medical and Population Subjects No. 66. London: The Stationery Office, 2001.

2. Harris V, Sandridge AL, Black RJ, Brewster DH et al. *Cancer Registration Statistics Scotland, 1986–1995*. Edinburgh: ISD Publications, 1998.

3. National Cancer Registry of Ireland. *Cancer in Ireland, 1996. Incidence and Mortality*. Cork: National Cancer Registry, 1999.

4. Gavin AT, Reid J. *Cancer Incidence in Northern Ireland, 1993–95*. Belfast: The Stationery Office, 1999.

5. Scott N, Gould A, Brewster D. Laryngeal cancer in Scotland, 1960-1994: trends in incidence, geographical distribution and survival. *Health Bulletin* 1998; 56: 749–756.

6. NHS Scotland. Cancer Data. August 2004. Available at http://www.isdscotland.org/isd/info3.jsp?pContentID=402&p_applic=CCC&p_service=Content.show&.

7. ONS. Cancer Survival: England and Wales, 1991-2001. March 2004. Available at http://www.statistics.gov.uk/statbase/ssdataset.asp?vlnk=7899.

8. ISD Scotland. *Trends in Cancer Survival in Scotland, 1977–2001*. Edinburgh: ISD Publications, 2004.

9. Sant M, Areleid T, Berrino F, Bielska Lasota M et al. EUROCARE-3: survival of cancer patients diagnosed 1990–1994 - results and commentary. *Annals of Oncology* 2003; 14 Suppl 5: v61–v118.

10. Berrino F, Gatta G. Variation in survival of patients with head and neck cancer in Europe by the site of origin of the tumours. *European Journal of Cancer* 1998; 34: 2154–2161.

11. Coleman MP, Babb P, Damiecki P, Grosclaude P et al. *Cancer Survival Trends in England and Wales, 1971–1995: Deprivation and NHS Region*. Studies on Medical and Population Subjects No. 61. London: The Stationery Office, 1999.

12. Scottish Cancer Intelligence Unit. *Trends in Cancer Survival in Scotland 1971–1995*. Edinburgh: Information and Statistics Division, 2000.

13. Austin DF, Reynolds P. Laryngeal Cancer. In: Schottenfeld D, Fraumeni JF, Jr. (eds) *Cancer Epidemiology and Prevention, second edition*. New York: Oxford University Press, 1996.

14. Walker A, O'Brien M, Traynor J, Fox K et al. *Living in Britain: Results from the 2001 General Household Survey*. London: The Stationery Office, 2002.

15. Wald N, Nicolaides-Bouman A. UK *Smoking Statistics, second edition*. London: Oxford University Press, 1991.

16. Swerdlow AJ, dos SS, I, Reid A, Qiao Z et al. Trends in cancer incidence and mortality in Scotland: description and possible explanations. *British Journal of Cancer* 1998; 77 Suppl 3: 1–54.

17. Cabinet Office, Prime Minister's Strategy Unit. *Alcohol harm reduction strategy for England*. London: Cabinet Office, 2004.

18. Carstairs V, Morris R. Deprivation and mortality: an alternative to social class? *Community Medicine* 1989; 11: 213–219.

19. Logan WPD. *Cancer mortality by occupation and social class 1851–1971*. IARC Scientific Publications No. 36. Lyon: International Agency for Research on Cancer, 1982.

20. Shaw, A., McMunn, A., Field, J. *Scottish Health Survey 1998, Volume 1*. Edinburgh: The Scottish Executive, 2000.

21. The Scottish Executive. Statistics on Alcohol in Scotland, 2002. October 2004. Available at http://www.alcoholinformation.isdscotland.org/alcohol_misuse/files/Plan_11.pdf.

Chapter 11

Leukaemia

Peter Adamson

Summary

- In the UK and Ireland, leukaemia accounted for about 1 in 40 diagnosed cases of cancer and 1 in 40 deaths from cancer.

- Incidence was relatively high in the south of Scotland, Wales, parts of south east England, and parts of north east England, and relatively low in the north west and east of England.

- The geographical pattern of mortality rates was broadly similar to that for incidence, although mortality was higher in Ireland and lower in Scotland than would have been expected from their incidence rates.

- There is a suggestion of an inverse association between incidence and socio-economic deprivation, reflected in higher rates in some of the more affluent areas of the south of England and lower rates in the more deprived areas of the north.

- None of the known risk factors explain the observed geographical variations.

Introduction

The five main types of leukaemia are biologically and clinically diverse.[1] In the mid-1990s, almost one third of cases were acute myeloid leukaemia (AML) and around a further third were chronic lymphocytic leukaemia (CLL); chronic myeloid leukaemia (CML) and acute lymphoblastic leukaemia (ALL) each accounted for one tenth of cases, the remainder comprising monocytic leukaemia, other specified leukaemia and unspecified leukaemia.[2]

Incidence and mortality

Even when combined, the numbers of cases of the five main types of leukaemia are not large. In the UK and Ireland in the 1990s there were about 3,800 newly diagnosed cases of leukaemia each year in males, in whom it was the eighth most common cancer, and about 3,000 cases in females, in whom it was the twelfth most common. Leukaemia accounted for 2.8 per cent of all newly diagnosed cases per year in males and 2.2

per cent in females. The ratio of the number of cases in males to females was around 1.3:1. Overall, the age-standardised incidence rates were 11.9 and 7.2 per 100,000 in males and females, respectively, giving a male-to-female ratio of 1.7:1. Age-specific incidence for all leukaemias combined is bimodal, with a first (small) peak in incidence in early childhood followed by a trough in young adulthood (age 15-29), and a steady increase in incidence above the age of 30.[2] This bimodal distribution reflects differences in the types of leukaemia which predominate at different ages: common B-cell precursor acute lymphoblastic leukaemia accounts for the childhood peak, with other forms each becoming more common as age increases.

In the 1990s, 2,300 males and 1,900 females died from leukaemia each year in the UK and Ireland. Of all deaths from cancer, 2.7 per cent were due to leukaemia in males, in whom this was the ninth most common cause of cancer death, and 2.4 per cent in females, in whom it was also the ninth most common. The male-to-female ratio of leukaemia deaths, around 1.2:1, was similar to that for newly diagnosed cases. In males, the age-standardised mortality rate was 6.8 per 100,000 and in females was 4.1 per 100,000, a ratio of 1.7:1. The overall mortality-to-incidence ratio was 0.57 for both males and females (Table B11.1).

Incidence and mortality trends

The age-standardised incidence rates in England and Wales since the 1970s have fluctuated around 5 to 7 per 100,000 for females and between 8 and 12 per 100,000 for males.[2] There has been a steady 1-2 per cent increase per year, similar to trends observed throughout Europe.[1] Overall, mortality has declined steadily since the 1970s, although in the late 1990s it appeared to increase slightly in the elderly.[2]

Survival

One-year relative survival from all leukaemias was around 58 per cent and five-year survival was around 35 per cent for patients diagnosed in the period 1996-99 in England and Wales.[3] There have been substantial changes in survival rates since the 1970s with major improvements occurring in the late 1970s and early 1980s.[2] There are, however, differences in survival for the various subtypes of leukaemia. For example, less than 10 per cent of patients diagnosed with AML will survive five years compared with half those diagnosed with CLL. The two other most common subtypes, ALL and CML, have a five-year survival rate of between 20 and 30 per cent.[2] Survival from the leukaemias was similar in Scotland[4] and Northern Ireland,[5] but appeared to be higher in Ireland,[6] for patients diagnosed in the 1990s.

Geographical patterns in incidence

Within the countries of the UK and Ireland, the highest incidence of leukaemia occurred in Wales, where rates were 29 per cent higher than the average for males and 19 per cent higher for females (Table B11.1). Rates were slightly higher than the average for both sexes in Scotland and in South East England, and for males in South West England (Figure 11.1). The lowest incidence was observed in the North West (18 per cent below average for both sexes). Lower than average rates also occurred in the Eastern region for both sexes, and in Trent and Northern Ireland for females. Rates for Northern and Yorkshire; West Midlands; London; and Ireland were close to the overall average for both sexes.

There was generally wider variation in incidence at the health authority level than between the regions of England and the other countries (Figure 11.3). In Wales, for both males and females, 4 out of 5 health authorities had rates that were markedly higher than the UK and Ireland average, and in the North West of England, 10 and 11 health authorities out of 15 were below average, for males and females, respectively. The maps of incidence in males and females are broadly similar, with areas of higher incidence for both sexes in the south of Scotland, Wales, and South East England, and of lower incidence in the North West and Eastern regions of England (Map 11.1).

Geographical patterns in mortality

At the country and regional level, there was much less variability in mortality than in incidence rates. Mortality from leukaemia was below average for both sexes in Scotland and above average for males in Ireland; rates in the other countries and English regions were close to the average (Figure 11.2). At the health authority level, rates were quite variable within regions and countries, but few were markedly different from the average (Figure 11.4). The mortality-to-incidence ratio was higher than average in the North West of England (where incidence was particularly low), which suggests a possible under-ascertainment of cases, and in Scotland and Wales the ratio was lower than average. The most likely explanation for these apparent differences is random variation, as the number of deaths from leukaemia were quite small (Table B11.2).

Risk factors and aetiology

The aetiology of the leukaemias is only partially understood and involves a wide range of factors. Exposure to ionising radiation has been recognised as a cause for developing leukaemia (with the exception of CLL) for over 50 years, although it can only account for a small proportion of cases. The strongest evidence for an increased risk of developing

leukaemia following exposure to ionising radiation comes from studies of survivors of the atomic bombings of Japan, of patients with malignant or benign conditions who received radiotherapy, and of radiologists and x-ray technicians. Diagnostic x-rays are the largest man-made source of radiation exposure to the general population and it is estimated that diagnostic exposures may be the cause of 1.4 per cent of leukaemias in men and 1.9 per cent in women.[7] Weaker evidence comes from studies of nuclear industry workers. In contrast, studies of other groups who may have been exposed to ionising radiation, for example, radium dial workers, underground miners and those living near nuclear power plants, provide little evidence of an increased risk of leukaemia.[8] Childhood leukaemia has been associated with parental employment in hydrocarbon-related occupations and the chemical industry.[9] Increased risk has also been observed in patients treated with radiotherapy for ankylosing spondilitis and various cancers.

An increased risk of secondary AML has been found in some patients treated for various lymphatic and haematological malignancies, and solid tumours, with alkylating agents (a type of chemotherapy drug).[9] In terms of inherited risk, Down's syndrome is associated with a 20- to 30-fold increased risk of developing acute leukaemia of various types,[9] and 90 per cent of CML cases are associated with the Philadelphia chromosome (resulting from a translocation between chromosomes 9 and 22).[10] None of these risk factors could give rise to the observed geographical patterns in incidence.

It has long been hypothesised that the leukaemias may be caused by an infection, as is the case in non-human leukaemias.[2] There is much circumstantial evidence to support the suggestion that exposure to infective agents in the first few years of life could be a risk factor for the development of acute lymphoblastic leukaemia.[11]

Socio-economic deprivation

Observed age-standardised incidence rates show only slight variation by deprivation category, with higher rates in the more affluent groups compared with the deprived. There are no significant differences in mortality by deprivation category.[2] Five-year survival was significantly worse for men living in deprived areas compared with affluent areas, but the difference for women was not significant, for patients diagnosed in England and Wales in 1996-99.[12]

(continued on page 128)

Figure 11.1

**Leukaemia: incidence by sex, country, and region of England
UK and Ireland 1991-99[1]**

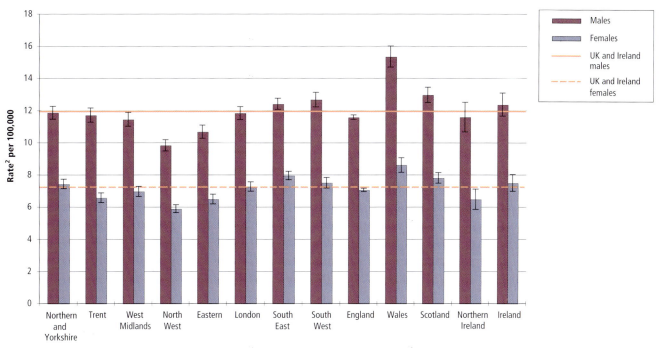

1 Northern Ireland 1993-99, Ireland 1994-99

2 Age standardised using the European standard population, with 95% confidence interval

Figure 11.2

**Leukaemia: mortality by sex, country, and region of England
UK and Ireland 1991-2000[1]**

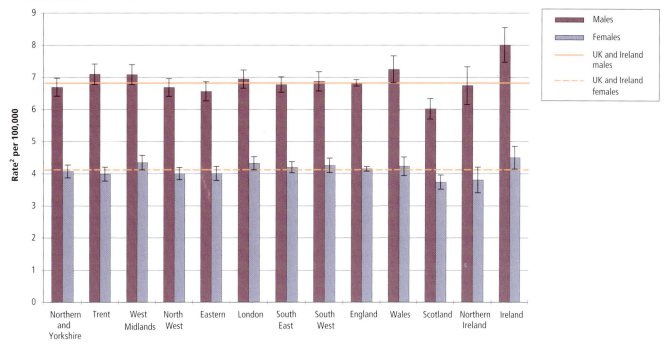

1 Scotland 1991-99, Ireland 1994-2000

2 Age standardised using the European standard population, with 95% confidence interval

Figure **11.3a**

**Leukaemia: incidence by health authority within country, and region of England
Males, UK and Ireland 1991-99[1]**

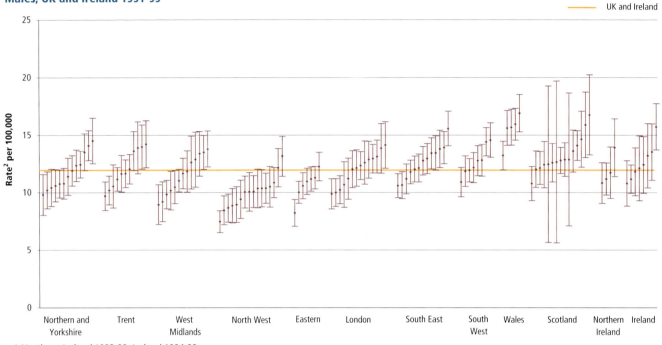

1 Northern Ireland 1993-99, Ireland 1994-99

2 Age standardised using the European standard population, with 95% confidence interval

Figure **11.3b**

**Leukaemia: incidence by health authority within country, and region of England
Females, UK and Ireland 1991-99[1]**

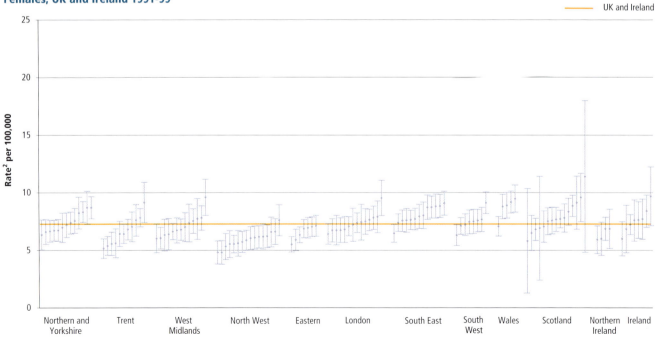

1 Northern Ireland 1993-99, Ireland 1994-99

2 Age standardised using the European standard population, with 95% confidence interval

Figure **11.4a**

Leukaemia: mortality by health authority within country, and region of England
Males, UK and Ireland 1991-2000[1]

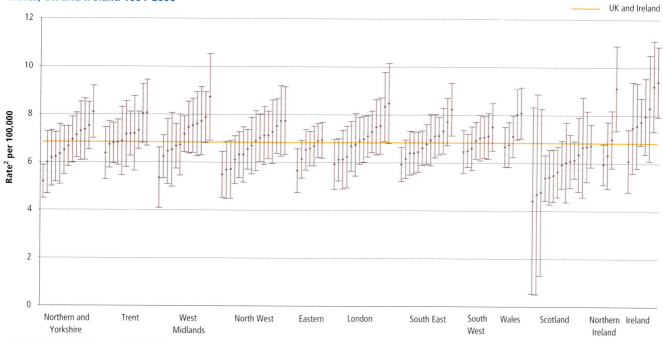

1 Scotland 1991-99, Ireland 1994-2000
2 Age standardised using the European standard population, with 95% confidence interval

Figure **11.4b**

Leukaemia: mortality by health authority within country, and region of England
Females, UK and Ireland 1991-2000[1]

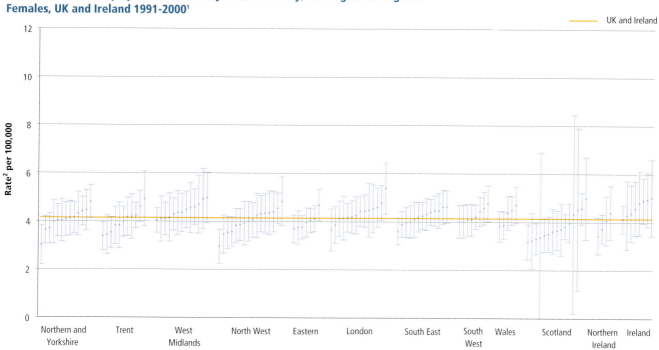

1 Scotland 1991-99, Ireland 1994-2000
2 Age standardised using the European standard population, with 95% confidence interval

Map **11.1a**

**Leukaemia: incidence* by health authority
Males, UK and Ireland 1991-99**

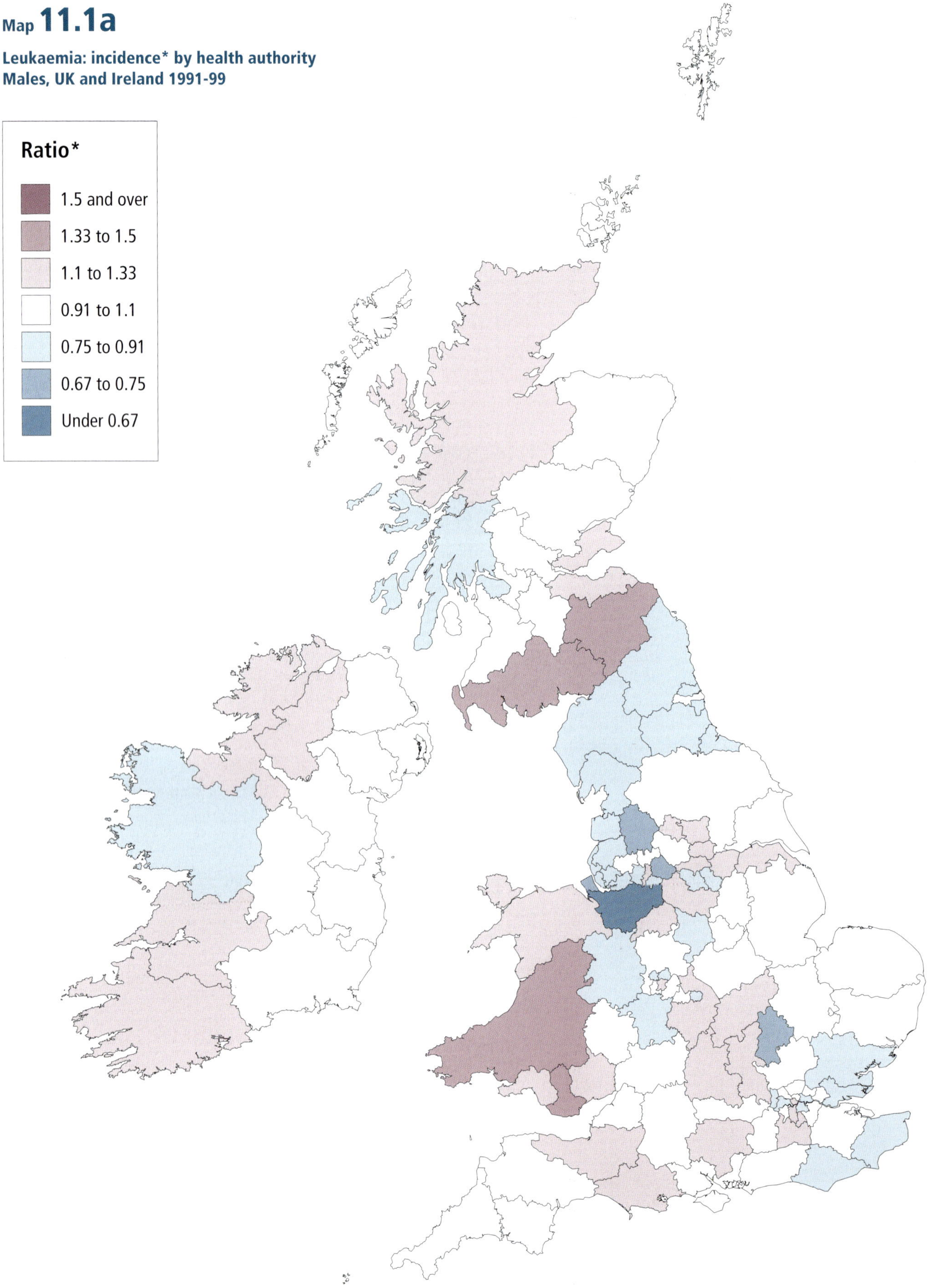

Ratio*

1.5 and over

1.33 to 1.5

1.1 to 1.33

0.91 to 1.1

0.75 to 0.91

0.67 to 0.75

Under 0.67

*Ratio of directly age-standardised rate in health authority to UK and Ireland average

Map **11.1b**

**Leukaemia: incidence* by health authority
Females, UK and Ireland 1991-99**

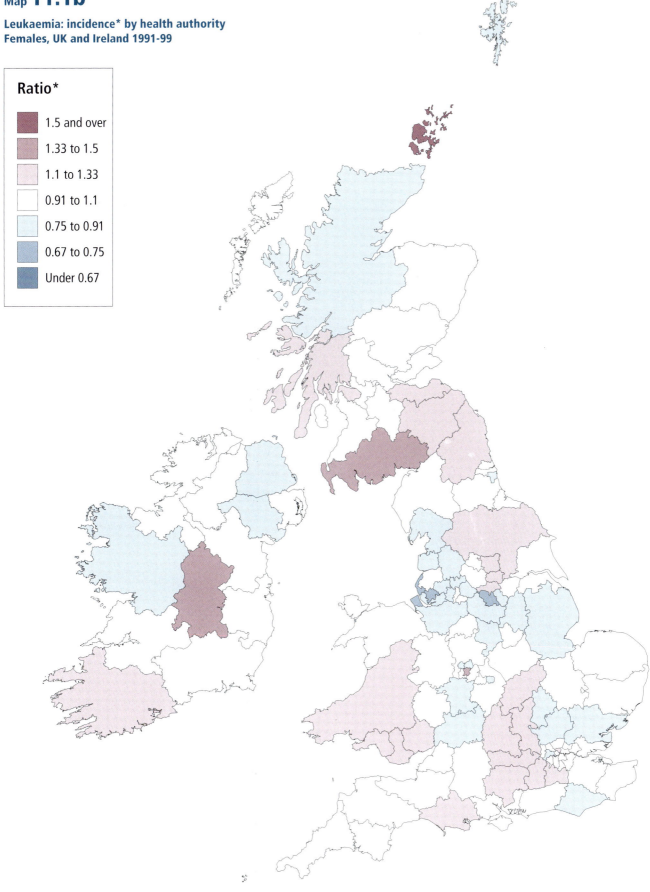

Ratio*

	1.5 and over
	1.33 to 1.5
	1.1 to 1.33
	0.91 to 1.1
	0.75 to 0.91
	0.67 to 0.75
	Under 0.67

*Ratio of directly age-standardised rate in health authority to UK and Ireland average

Map **11.2a**

Leukaemia: mortality* by health authority
Males, UK and Ireland 1991-2000

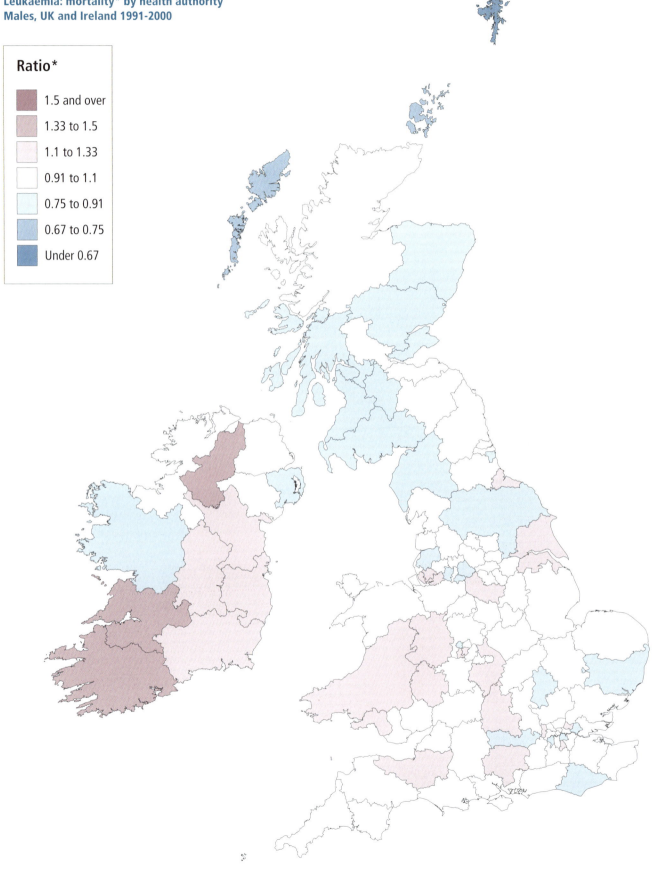

Ratio*

	1.5 and over
	1.33 to 1.5
	1.1 to 1.33
	0.91 to 1.1
	0.75 to 0.91
	0.67 to 0.75
	Under 0.67

Ratio of directly age-standardised rate in health authority to UK and Ireland average

Map **11.2b**

**Leukaemia: mortality* by health authority
Females, UK and Ireland 1991-2000**

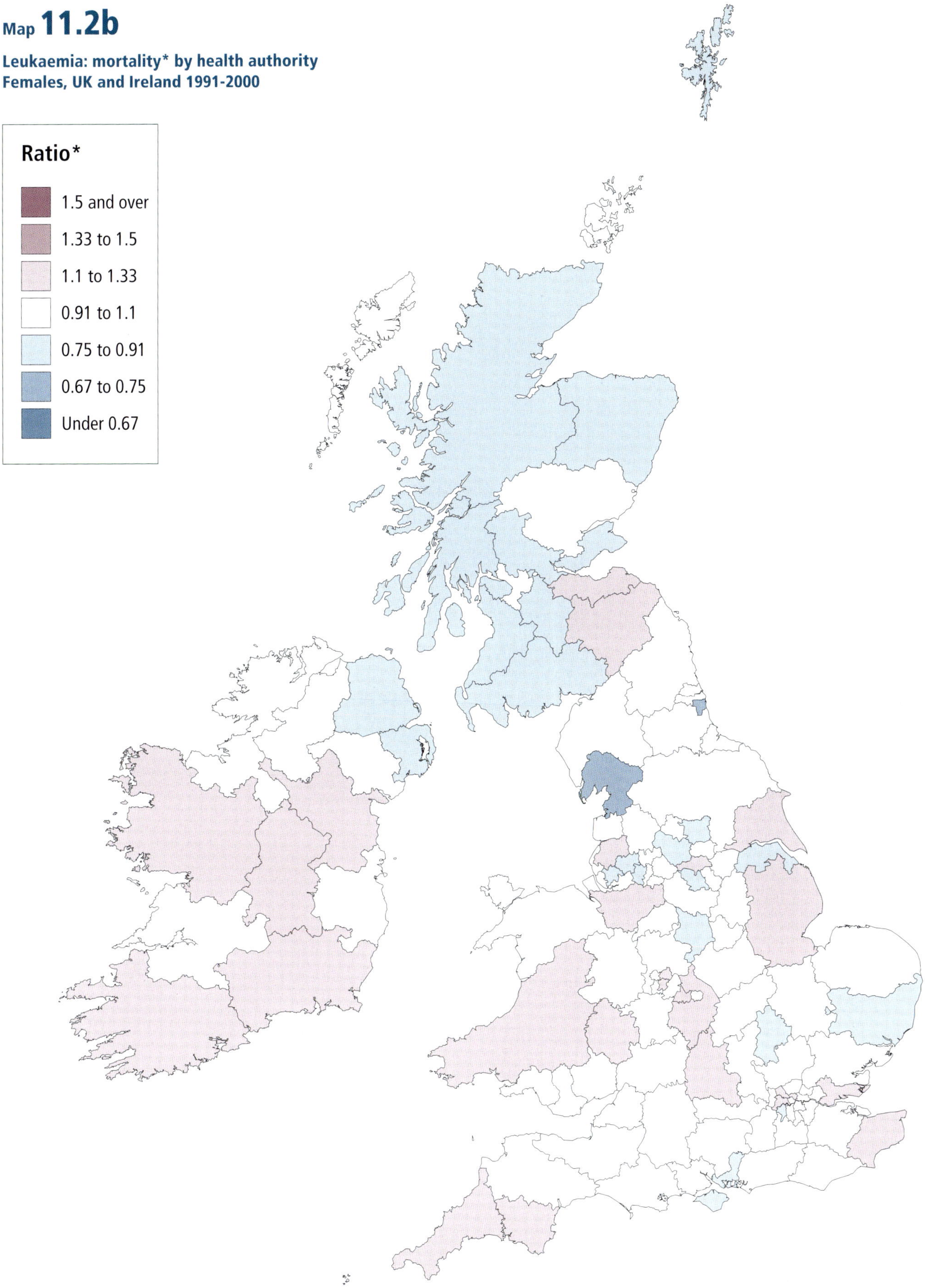

Ratio*

■	1.5 and over
■	1.33 to 1.5
■	1.1 to 1.33
□	0.91 to 1.1
■	0.75 to 0.91
■	0.67 to 0.75
■	Under 0.67

*Ratio of directly age-standardised rate in health authority to UK and Ireland average

References

1. Coleman MP, Babb P, Damiecki P, Grosclaude P et al. *Cancer Survival Trends in England and Wales, 1971-1995: Deprivation and NHS Region.* Studies on Medical and Population Subjects No. 61. London: The Stationery Office, 1999.

2. Quinn MJ, Babb PJ, Brock A, Kirby L et al. *Cancer Trends in England and Wales 1950-1999.* Studies on Medical and Population Subjects No. 66. London: The Stationery Office, 2001.

3. ONS. Cancer Survival: England and Wales, 1991-2001. March 2004. Available at *http://www.statistics.gov.uk/statbase/ ssdataset.asp?vlnk=7899.*

4. Sant M, Areleid T, Berrino F, Bielska Lasota M et al. EUROCARE-3: survival of cancer patients diagnosed 1990-1994 - results and commentary. *Annals of Oncology* 2003; 14 Suppl 5: v61-v118.

5. Fitzpatrick D, Gavin A, Middleton R, Catney D. *Cancer in Northern Ireland 1993-2001: A Comprehensive Report.* Belfast: Northern Ireland Cancer Registry, 2004.

6. National Cancer Registry of Ireland. *Cancer in Ireland, 1994 to 1998: Incidence, mortality, treatment and survival.* Cork: National Cancer Registry, 2001.

7. Berrington de Gonzalez A, Darby S. Risk of cancer from diagnostic X-rays: estimates for the UK and 14 other countries. *Lancet* 2004; 363: 345-351.

8. United Nations Scientific Committee on the Effects of Atomic Radiation. *Sources and effects of ionizing radiation: United Nations Scientific Committee on the Effects of Atomic Radiation (UNSCEAR) 2000 Report to the General Assembly, with Scientific Annexes. Volume II: effects.* New York: United Nations Publications, 2000.

9. Linet MS, Cartwright RA. The Leukaemias. In: Schottenfeld D, Fraumeni JF, Jr. (eds) *Cancer Epidemiology and Prevention, second edition.* New York: Oxford University Press, 1996.

10. Ruddon Jr RW. Molecular and Genetic Events in Neoplastic Transformation. In: Schottenfeld D, Fraumeni JF, Jr. (eds) *Cancer Epidemiology and Prevention, second edition.* New York: Oxford University Press, 1996.

11. Anderson LM, Diwan BA, Fear NT, Roman E. Critical windows of exposure for children's health: cancer in human epidemiological studies and neoplasms in experimental animal models. *Environmental Health Perspectives* 2000; 108 Suppl 3: 573-594.

12. Coleman MP, Rachet B, Woods LM, Mitry E et al. Trends and socioeconomic inequalities in cancer survival in England and Wales up to 2001. *British Journal of Cancer* 2004; 90: 1367-1373.

Chapter 12

Lip, mouth and pharynx

Henrik Møller, David Brewster

Summary

- In the UK and Ireland in the 1990s, cancers of the lip, mouth and pharynx accounted for around 1 in 60 cancer cases and 1 in 80 deaths from cancer.

- Incidence was much higher in males than in females (ratio of rates 2.3:1).

- There was wide variability in incidence in males, with much higher than average rates in most of Scotland, Wales, Northern Ireland and Ireland, and in parts of London and the formerly heavily industrialised north west and north east of England; the pattern for females was similar but less pronounced.

- The geographical pattern of mortality rates was broadly similar to that for incidence.

- There is a clear link between incidence and mortality rates and socio-economic deprivation, with higher rates in the more deprived areas and lower rates in the more affluent.

- The geographical variations in incidence largely reflect variations in the prevalence of the established risk factors – alcohol consumption and smoking.

- There are clear similarities between the geographical patterns in the incidence of cancers of the lip, mouth and pharynx and those of cancers of the larynx, lung and oesophagus.

Introduction

This group includes cancers with different aetiologies. The largest sub-group consists of cancers of the oral cavity and pharynx (68 per cent), followed by cancer of the salivary glands (10 per cent), pyriform sinus and hypopharynx (10 per cent), lip (7 per cent), and nasopharynx (5 per cent) (data for cases diagnosed in the UK and Ireland in 1999).

Incidence and mortality

In the UK and Ireland in the 1990s, cancers of the lip, mouth and pharynx were diagnosed in around 3,100 males and 1,700 females each year. The age-standardised incidence rates were 9.9 and 4.3 per 100,000 in males and females, respectively, a

male-to-female ratio of 2.3:1. As for cancers of the upper gastro-intestinal and respiratory tracts, the age-specific incidence rates of cancers of the lip, mouth and pharynx increased with age, being highest in the oldest age group.

In the 1990s, an average of 1,400 males and 700 females died from lip, mouth and pharynx cancer each year in the UK and Ireland. The corresponding age-standardised mortality rates were 4.2 and 1.6 per 100,000, a male-to-female ratio of 2.6:1. As for incidence rates, age-specific mortality rates increased with age.

Incidence and mortality trends

The age-standardised incidence rates for cancers of the lip, mouth and pharynx in England and Wales decreased from the 1970s to the mid-1980s, but increased thereafter.[1] When analysed by year of birth, the lifetime risk was highest in individuals born before 1900.[1] Studies from Scotland however, have shown marked increases in both incidence and mortality rates in men in the age group 35-64 years since the 1980s, reflecting an increasing risk in successive generations born after 1920.[2] The pattern in women for Scotland is similar, but the increasing risk is less marked. In England and Wales, lip, mouth and pharynx cancer mortality in males fell markedly in the 1950s and 1960s and stabilised from the mid-1970s onwards; in females it decreased gradually over the period 1950-99.[3]

Survival

Five-year relative survival from oral and pharyngeal cancer, for patients diagnosed in 1990-94 was around 40 per cent (England 42, Scotland 37, Wales 40 per cent) in males and around 50 per cent (England 52, Scotland 47, Wales 55 per cent) in females.[4] Between 1971 and 1990 in England, Wales and Scotland there were important increases in five-year relative survival for most sub-sites of oral and pharyngeal cancer.[5,6]

The improvement in survival of patients with head and neck cancer in the 1990s has been analysed in detail in a large patient population from Liverpool.[7] The main explanatory factors in the improvement of survival were thought to be the increasing use of radical post-operative radiotherapy, and a great reduction in nodal recurrences in the neck.

Geographical patterns in incidence

Within the UK and Ireland, the age-standardised incidence rates of lip, mouth and pharynx cancer were particularly high in both males and females in Scotland, Wales, Northern Ireland, and in males in Ireland (Figure 12.1). The highest rates were in Scotland where incidence was 58 per cent higher in males and

42 per cent higher in females than the UK and Ireland averages. Regional rates in the midlands and south of England (except London) were below the average.

At the level of health authority, the incidence of lip, mouth and pharynx cancer was particularly higher than average for both males and females in western Scotland, Manchester, central London, North Wales, and Morgannwg and Bro Taf in the south of Wales (Figure 12.3). In males, exceptionally higher than average rates were also found in Liverpool, the area around Tyneside, the Western and North Western areas of Ireland, and the Western area of Northern Ireland. For males, incidence was lower than the UK and Ireland average in most health authorities in the South West, South East and Eastern regions, and in the West Midlands. In females, variation between health authority areas was less pronounced than in males (Figure 12.3). From the maps for incidence (Map 12.1) it can be seen that, with the exception of Wales and central London, rates in males were higher in the northern parts of the UK and Ireland and lower in the southern parts. The pattern was similar, although less pronounced, for females.

Geographical patterns in mortality

Patterns of regional variation illustrated in figures 12.2 and 12.4 closely resemble those for incidence, although the rates were lower and the variation was generally less pronounced.

Mortality rates were particularly high in Scotland and the North West of England for both males and females, and in Ireland and the Northern and Yorkshire region for males (Figure 12.4). The health authorities with especially high mortality rates corresponded very closely to those with high rates for incidence (western Scotland, the area around Tyneside, Manchester, Liverpool, central London and the west of Ireland). Throughout most of the south and east of England, mortality rates were again lower than average for both males and females. The maps for mortality (Map 12.2) show closely similar patterns to those for incidence.

Risk factors and aetiology

The majority of oral and pharyngeal cancers considered in this chapter comprise a fairly homogeneous group commonly referred to simply as 'oral cancer'. They include predominantly squamous cell carcinoma of the oral cavity and pharynx, and aetiologically are quite well characterised. The individual patterns of risk factors for the less common tumour types do not influence the observed geographical patterns to an appreciable extent. The observed variations largely reflect variations in the risk factors for the dominant group of oral cancers.

Although evidence is accumulating for a causal role for human papillomavirus (HPV) in some head and neck squamous cell carcinomas,[8] tobacco smoking and excessive alcohol drinking are the main causes of oral cancer and, as for oesophageal cancer, the effects are independent and multiplicative.[9] Due to the synergistic action of tobacco and alcohol and to the frequent combination of the two habits in individuals, most cases of oral cancer are attributable to the combined effects of smoking and drinking.

In the context of observed increases in the incidence of oral cancer in young adults in Scotland, the joint distribution of tobacco smoking, alcohol drinking and dietary habits were explored in a relatively small sample of 38 persons.[10] In this study, 68 per cent of the patients were smokers (more than ten cigarettes per day), regular drinkers (more than two glasses per day), or consumed less than one serving of fruit or vegetables per day.

In a study of oral and pharyngeal cancer in the region of the Thames Cancer Registry, oral cancer was highest in the Asian populations and nasopharyngeal cancer was highest in Chinese populations.[11] A study of Bangladeshi people in Tower Hamlets, London found a high prevalence of oral mucosal lesions (40 per cent), the most common of which was leukoplakia (25 per cent).[12]

Consistent with the high incidence in Ireland compared with England, an elevated incidence was observed in men of Irish origin living in England and Wales.[13]

Hindle et al.[14] compared the trends in oral cancer mortality with the corresponding trends in lung cancer (as an indicator of tobacco smoking), and liver cirrhosis (as a marker of alcohol abuse) in England and Wales. The similarity of the trends with oral cancer was much closer for liver cirrhosis than for lung cancer, and it was suggested that rising alcohol consumption since the 1950s, superimposed on continued tobacco usage, could account for the increase in oral cancer among younger males since the 1970s.

The remaining cases, which fall outside this dominant group of oral cancers, are cancers of the lip, for which sun exposure and tobacco smoking are important causes,[15] cancers of the salivary glands, mainly adenocarcinomas of unknown aetiology,[15] and nasopharyngeal cancer, which is associated with Epstein-Barr virus infection and, in high risk populations in Asia, with the consumption of salted fish and other traditional preserved foods, especially in children.[16]

(continued on page 138)

Figure **12.1**

Lip, mouth and pharynx: incidence by sex, country, and region of England
UK and Ireland 1991-99[1]

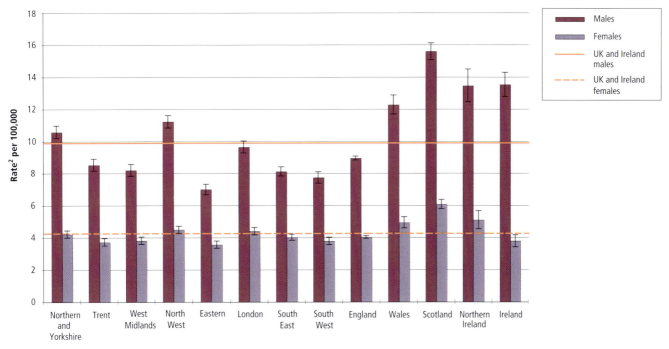

1 Northern Ireland 1993-99, Ireland 1994-99

2 Age standardised using the European standard population, with 95% confidence interval

Figure **12.2**

Lip, mouth and pharynx: mortality by sex, country, and region of England
UK and Ireland 1991-2000[1]

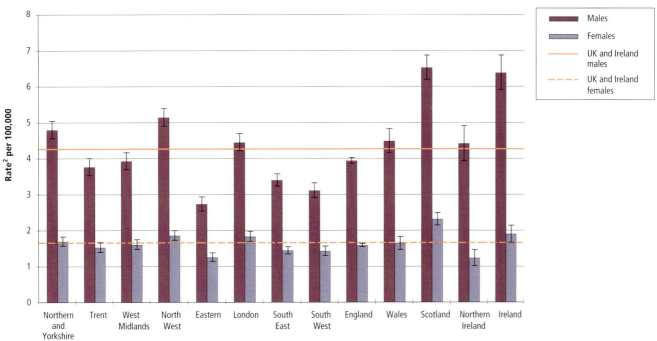

1 Scotland 1991-99, Ireland 1994-2000

2 Age standardised using the European standard population, with 95% confidence interval

Figure **12.3a**

**Lip, mouth and pharynx: incidence by health authority within country, and region of England
Males, UK and Ireland 1991-99[1]**

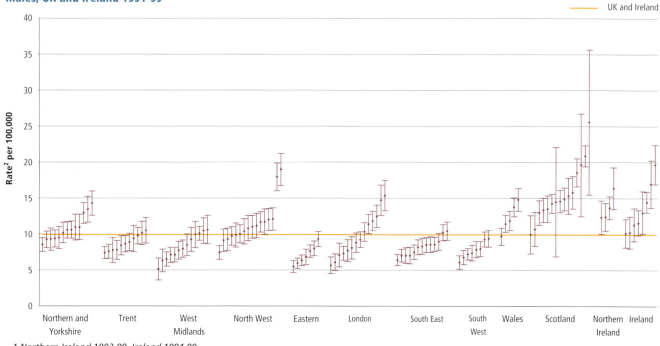

1 Northern Ireland 1993-99, Ireland 1994-99

2 Age standardised using the European standard population, with 95% confidence interval

Figure **12.3b**

**Lip, mouth and pharynx: incidence by health authority within country, and region of England
Females, UK and Ireland 1991-99[1]**

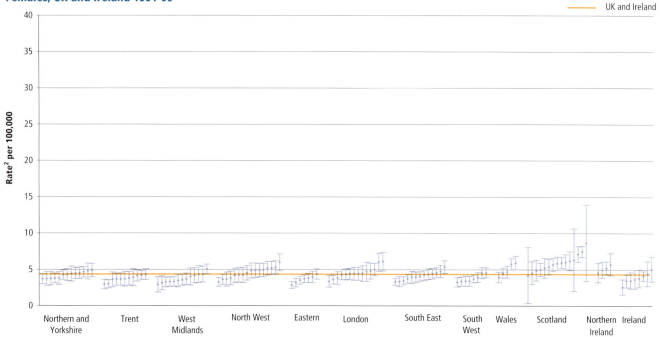

1 Northern Ireland 1993-99, Ireland 1994-99

2 Age standardised using the European standard population, with 95% confidence interval

Figure **12.4a**

Lip, mouth and pharynx: mortality by health authority within country, and region of England
Males, UK and Ireland 1991-2000[1]

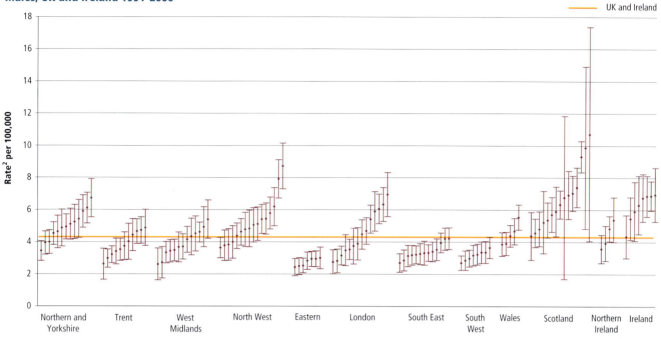

1 Scotland 1991-99, Ireland 1994-2000

2 Age standardised using the European standard population, with 95% confidence interval

Figure **12.4b**

Lip, mouth and pharynx: mortality by health authority within country, and region of England
Females, UK and Ireland 1991-2000[1]

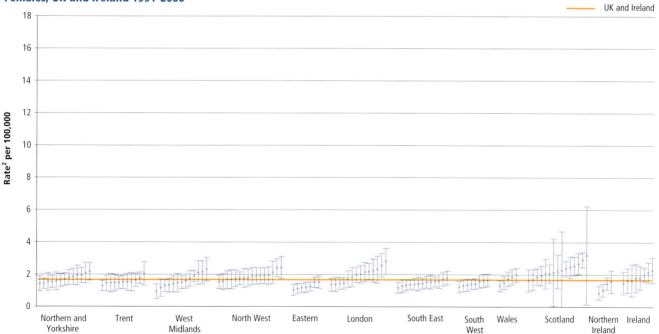

1 Scotland 1991-99, Ireland 1994-2000

2 Age standardised using the European standard population, with 95% confidence interval

Map **12.1a**

Lip, mouth and pharynx: incidence* by health authority
Males, UK and Ireland 1991-99

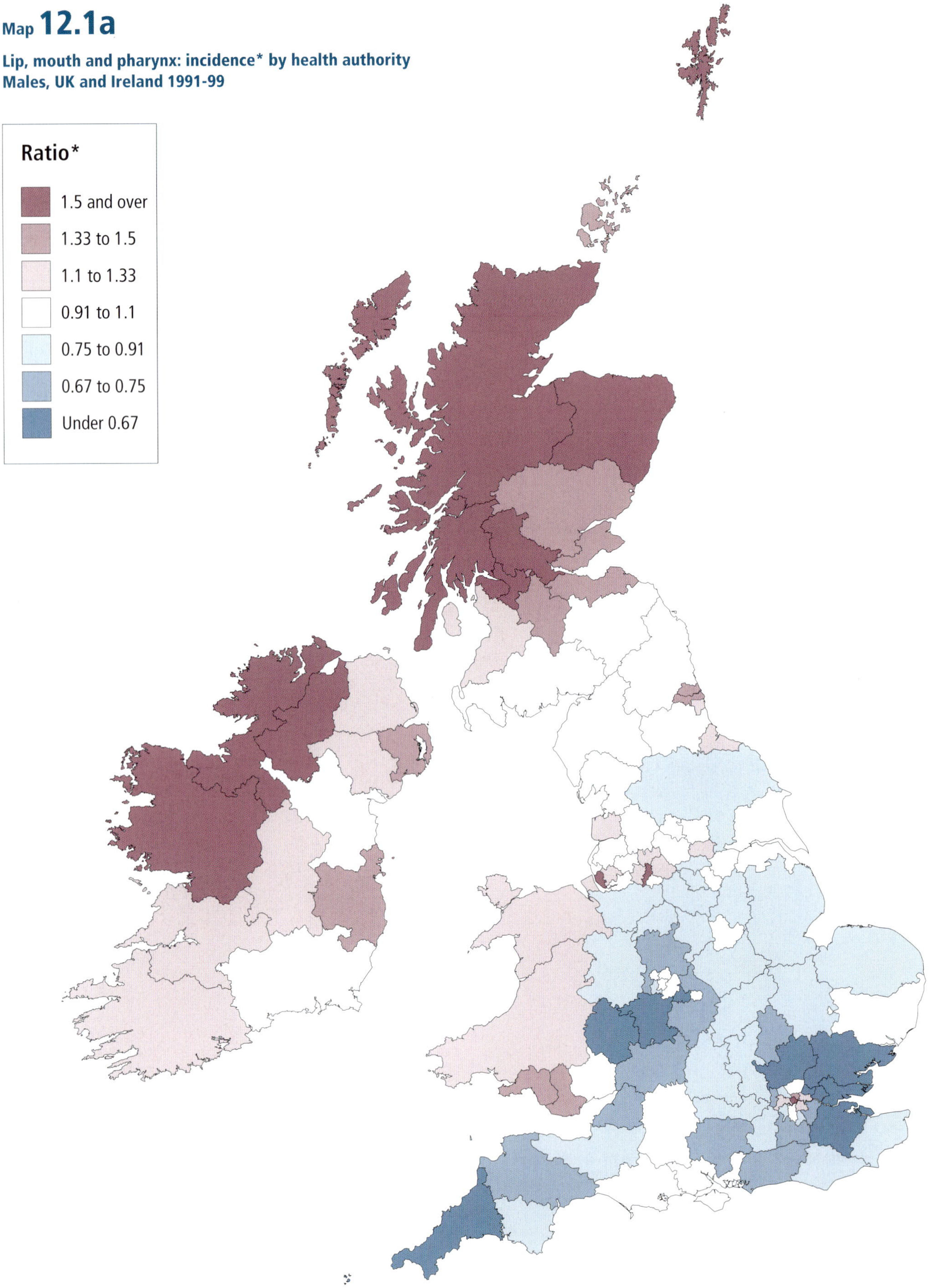

Ratio*

	1.5 and over
	1.33 to 1.5
	1.1 to 1.33
	0.91 to 1.1
	0.75 to 0.91
	0.67 to 0.75
	Under 0.67

*Ratio of directly age-standardised rate in health authority to UK and Ireland average

Map **12.1b**

Lip, mouth and pharynx: incidence* by health authority
Females, UK and Ireland 1991-99

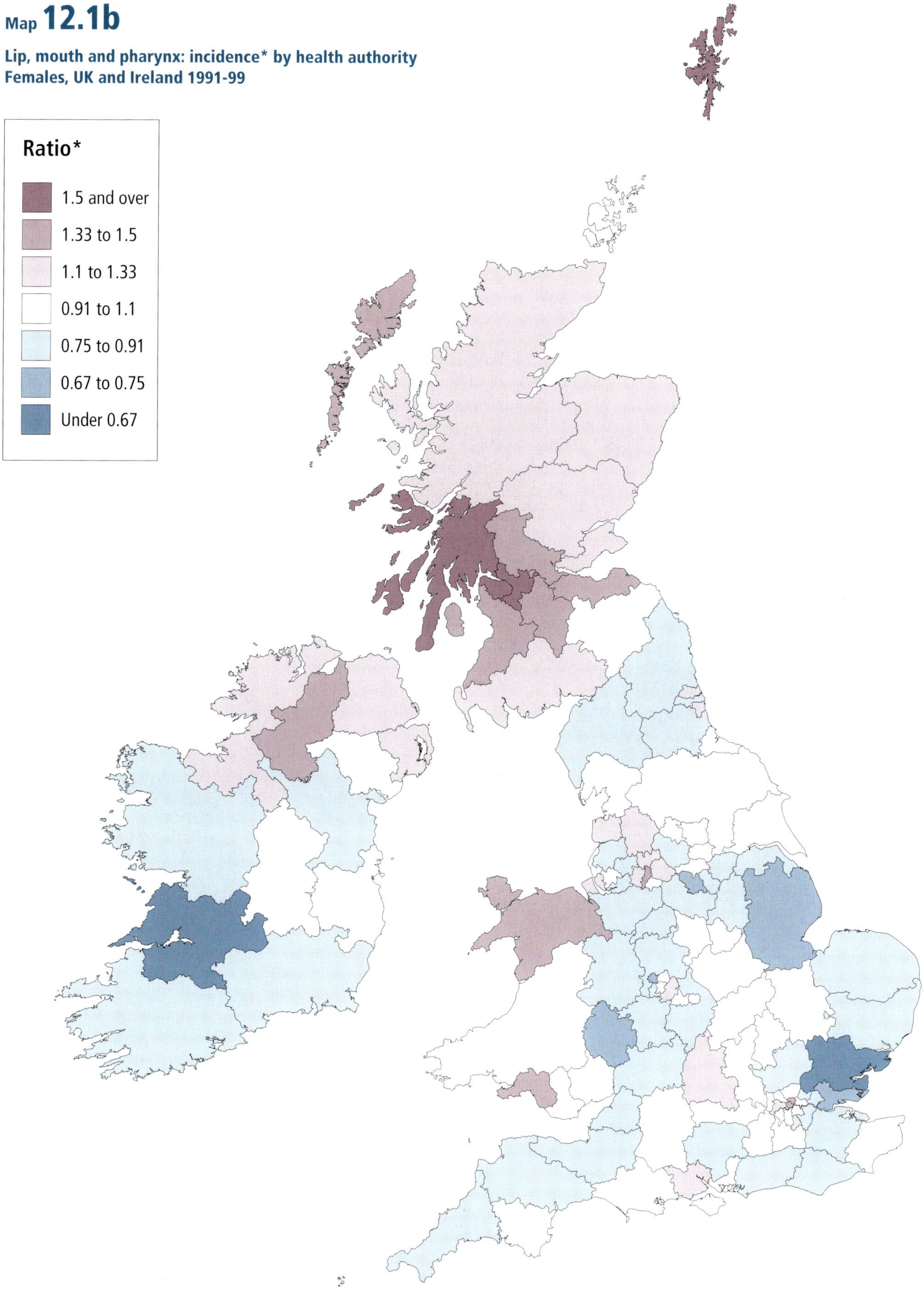

Ratio*

	1.5 and over
	1.33 to 1.5
	1.1 to 1.33
	0.91 to 1.1
	0.75 to 0.91
	0.67 to 0.75
	Under 0.67

*Ratio of directly age-standardised rate in health authority to UK and Ireland average

Map **12.2a**

**Lip, mouth and pharynx: mortality* by health authority
Males, UK and Ireland 1991-2000**

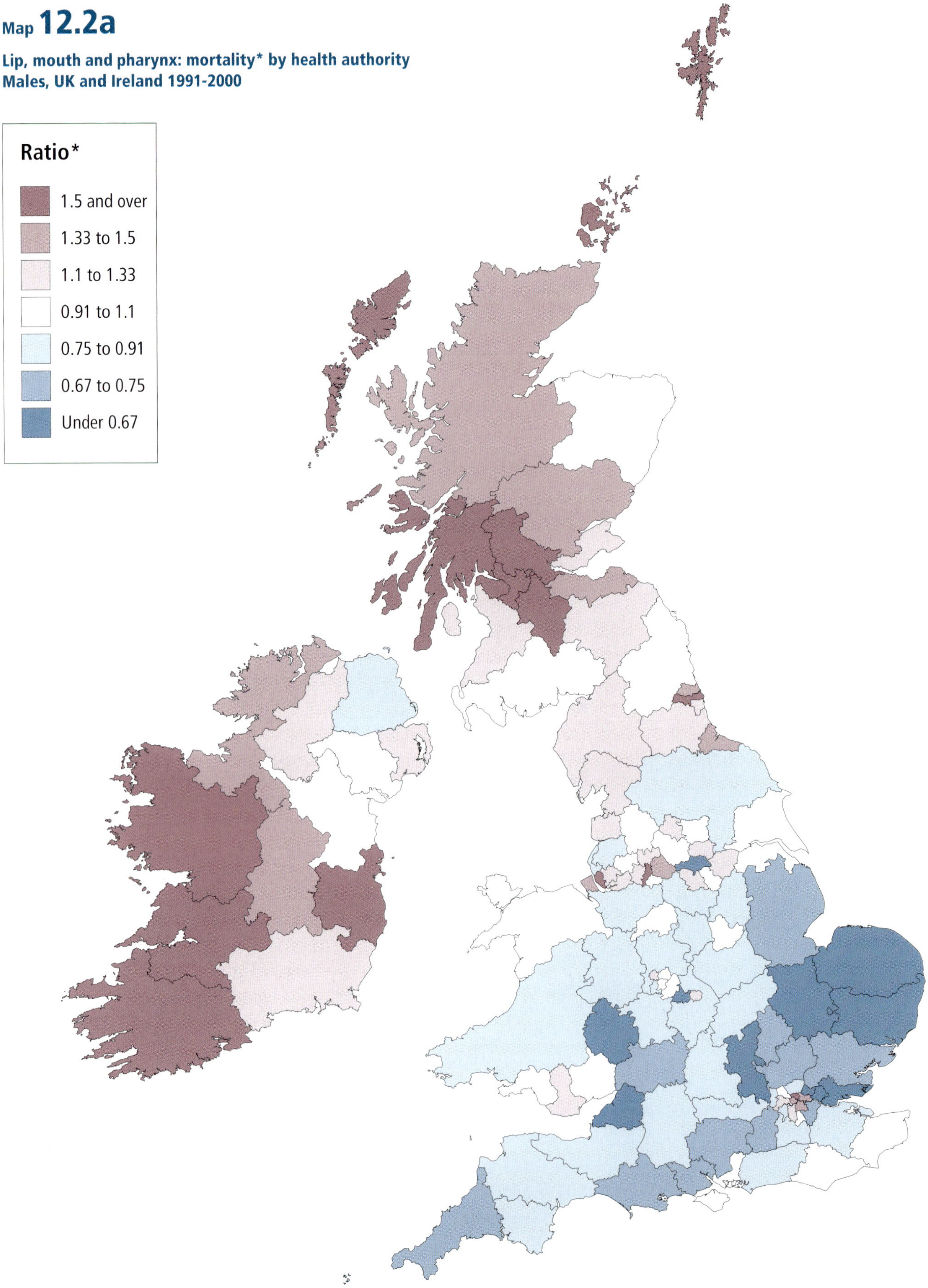

Ratio*

	1.5 and over
	1.33 to 1.5
	1.1 to 1.33
	0.91 to 1.1
	0.75 to 0.91
	0.67 to 0.75
	Under 0.67

*Ratio of directly age-standardised rate in health authority to UK and Ireland average

Map **12.2b**

**Lip, mouth and pharynx: mortality* by health authority
Females, UK and Ireland 1991-2000**

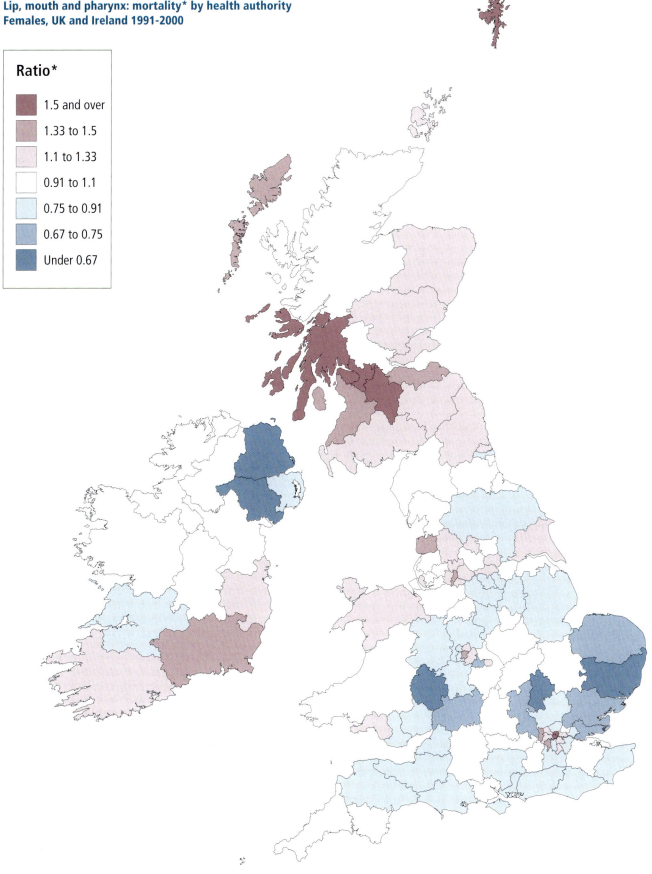

Ratio*

	1.5 and over
	1.33 to 1.5
	1.1 to 1.33
	0.91 to 1.1
	0.75 to 0.91
	0.67 to 0.75
	Under 0.67

Ratio of directly age-standardised rate in health authority to UK and Ireland average

Cancers of the oral cavity and pharynx are predominantly squamous cell carcinomas and share important aetiological characteristics with cancers of the larynx and oesophagus, particularly the very strong associations with tobacco smoking and excessive alcohol drinking. It is therefore not surprising that the maps for lip, mouth and pharynx cancer show very close similarities to those for cancers of the larynx (Chapter 10) and lung (Chapter 13), and to a lesser extent, to the map for cancer of the oesophagus (Chapter 17).

Socio-economic deprivation

Both the incidence of, and mortality from, cancers of the lip, mouth and pharynx are higher in more deprived areas than in affluent areas.[1,17] A study in Newcastle investigated the joint effects of high alcohol consumption, smoking and long-term unemployment on the incidence of oral cancer.[18] All three factors were significant risk factors when considered individually, but when considered in combination the association with long-term unemployment was greatly reduced and no longer significant. This suggests that the gradient in incidence with deprivation is brought about by variation between socio-economic groups with respect to the prevalence of established risk factors (alcohol consumption and smoking) for oral cancer (see Chapter 13 on Lung and Chapter 17 on Oesophagus for further details).

There is also a very strong socio-economic gradient in survival from oral and pharyngeal cancer with the more deprived groups having the lowest survival.[5] This gradient is steeper than for any other type of cancer, with a gap in five-year relative survival of 10-15 percentage points between the highest and the lowest socio-economic group. A reported decrease in survival in persons aged under 65 years in Scotland may reflect an increase in the proportion of cases arising in socio-economically deprived communities.[19]

References

1. Quinn MJ, Babb PJ, Brock A, Kirby L et al. *Cancer Trends in England and Wales 1950-1999*. Studies on Medical and Population Subjects No. 66. London: The Stationery Office, 2001.

2. Robinson KL, Macfarlane GJ. Oropharyngeal cancer incidence and mortality in Scotland: are rates still increasing? *Oral Oncology* 2003; 39: 31-36.

3. Hindle I, Downer MC, Speight PM. The temporal and spatial epidemiology of lip cancer in England and Wales. *Community Dental Health* 2000; 17: 152-160.

4. Sant M, Areleid T, Berrino F, Bielska Lasota M et al. EUROCARE-3: survival of cancer patients diagnosed 1990-1994 - results and commentary. *Annals of Oncology* 2003; 14 Suppl 5: v61-v118.

5. Coleman MP, Babb P, Damiecki P, Grosclaude P et al. *Cancer Survival Trends in England and Wales, 1971-1995: Deprivation and NHS Region*. Studies on Medical and Population Subjects No. 61. London: The Stationery Office, 1999.

6. Black R, Brewster D, Brown H, Fraser L et al. *Trends in Cancer Survival in Scotland 1971-1995*. Edinburgh: Information and Statistics Division, 2000.

7. Jones AS, Houghton DJ, Beasley NJ, Husband DJ. Improved survival in patients with head and neck cancer in the 1990s. *Clinical Otolaryngology and Allied Sciences* 1998; 23: 319-325.

8. Gillison ML, Lowy DR. A causal role for human papillomavirus in head and neck cancer. *Lancet* 2004; 363: 1488-1489.

9. Mackenzie J, Ah-See K, Thakker N, Sloan P et al. Increasing incidence of oral cancer amongst young persons: what is the aetiology? *Oral Oncology* 2000; 36: 387-389.

10. Mucci L, Adami HO. Oral and pharyngeal cancer. In: Adami HO, Hunter D, Trichopoulos D (eds) *Textbook of Cancer Epidemiology*. New York: Oxford University Press, 2002.

11. Warnakulasuriya KA, Johnson NW, Linklater KM, Bell J. Cancer of mouth, pharynx and nasopharynx in Asian and Chinese immigrants resident in Thames regions. *Oral Oncology* 1999; 35: 471-475.

12. Pearson N, Croucher R, Marcenes W, O'Farrell M. Prevalence of oral lesions among a sample of Bangladeshi medical users aged 40 years and over living in Tower Hamlets, UK. *International Dental Journal* 2001; 51: 30-34.

13. Harding S, Rosato M. Cancer incidence among first generation Scottish, Irish, West Indian and South Asian migrants living in England and Wales. *Ethnicity and Health* 1999; 4: 83-92.

14. Hindle I, Downer MC, Moles DR, Speight PM. Is alcohol responsible for more intra-oral cancer? *Oral Oncology* 2000; 36: 328-333.

15. Blot WJ, McLaughlin JK, Devesa SS, Fraumeni JF, Jr. Cancers of the Oral Cavity and Pharynx. In: Schottenfeld D, Fraumeni JF, Jr. (eds) *Cancer Epidemiology and Prevention, second edition*. New York: Oxford University Press, 1996.

16. Yu MC, Hendersen BE. Nasopharyngeal Cancer. In: Schottenfeld D, Fraumeni JF, Jr. (eds) *Cancer Epidemiology and Prevention, second edition*. New York: Oxford University Press, 1996.

17. Edwards DM, Jones J. Incidence of and survival from upper aerodigestive tract cancers in the UK: the influence of deprivation. *European Journal of Cancer* 1999; 35: 968-972.

18. Greenwood M, Thomson PJ, Lowry RJ, Steen IN. Oral cancer: material deprivation, unemployment and risk factor behaviour - an initial study. *International Journal of Oral and Maxillofacial Surgery* 2003; 32: 74-77.

19. Macfarlane GJ, Sharp L, Porter S, Franceschi S. Trends in survival from cancers of the oral cavity and pharynx in Scotland: a clue as to why the disease is becoming more common? *British Journal of Cancer* 1996; 73: 805-808.

Chapter 13

Lung

Helen Wood, Nicola Cooper, Steve Rowan, Mike Quinn

Summary

- In the UK and Ireland in the 1990s, lung cancer accounted for 1 in 6 diagnosed cases of cancer, 1 in 4 deaths from cancer, and 5 per cent of all deaths.

- Geographical variations in incidence and mortality were very similar, as survival from lung cancer is low.

- There was a clear north-south divide across Great Britain, with high incidence in Scotland and northern England, and generally low incidence in Wales, the midlands and southern England.

- Incidence was low in Northern Ireland and Ireland, except in the regions around Belfast and Dublin.

- Incidence was highest in areas traditionally associated with heavy industry and shipbuilding, areas which also have particularly high levels of socio-economic deprivation.

- The greatest risk factor for lung cancer – cigarette smoking – is most prevalent in Scotland and the north of England and is also associated with deprivation.

Incidence and mortality

In the 1990s there were about 27,000 newly diagnosed cases of lung cancer each year in males, in whom it was the most common cancer, and about 15,000 cases in females, in whom it was the third most common, after breast and colorectal cancer. Lung cancer accounted for 16 per cent of all newly diagnosed cases per year; in males the figure was 20 per cent and in females 11 per cent. The ratio of the number of cases in males to females was around 1.8:1. Overall, the age-standardised rates were 81 and 35 per 100,000 in males and females, respectively, a male-to-female ratio of 2.3:1, higher than that for the number of cases. Lung cancer is predominantly a disease of the elderly and incidence was considerably lower in those aged under 50. Above the age of 50, rates increased steeply to peak at age 80-84 in males and age 75-79 in females. The lifetime risk[1,2] of being diagnosed with lung cancer in England and Wales was 8.0 per cent (1 in 13) for males and 4.3 per cent (1 in 23) for females.[3] For non-smokers the risks are much lower than these figures, which were averaged across the whole population, while for smokers the risks are much higher.

In the 1990s, 24,300 males and 13,400 females died from lung cancer each year in the UK and Ireland. Of all deaths from cancer, 23 per cent were due to lung cancer. The figure was 29 per cent in males, in whom this was the commonest cause of cancer death, and 17 per cent in females, in whom it was the second commonest, after breast cancer. In males, the age-standardised mortality rate was 72 per 100,000 and in females it was 30 per 100,000, a male-to-female ratio of 2.4:1. Following the pattern for incidence, mortality rose steeply with age to peak in males aged 80-84 and in females aged 75-79.

Incidence and mortality trends

Lung cancer was a relatively rare disease at the beginning of the twentieth century, but is currently the most common cancer in the world.[4] In males, the age-standardised mortality rate in England and Wales rose steeply through the first half of the century, increasing six fold by the 1950s, and continued to rise to a peak in the 1970s. The rate began to decline in the early 1980s, reaching levels similar to those in the early 1950s by the late 1990s. In females, changes in the mortality rate lag around 20 years behind those in males, and only reached a plateau in the mid-1990s.[3] Trends in the incidence of lung cancer over time are closely similar to those for mortality due to the very low survival. These patterns reflect the changes in the prevalence of cigarette smoking in different birth cohorts of men and women.

Survival

Lung cancer has one of the lowest survival rates of any cancer. In England and Wales, for patients diagnosed in 1996-99, only 23 per cent of men and 24 per cent of women were alive after one year and around 6 per cent of both men and women were alive after five years.[5] The five-year survival figures were similar for cases diagnosed in Scotland,[6] and slightly higher for cases diagnosed in Northern Ireland[7] and Ireland,[8] in the 1990s. The European average survival rates for patients diagnosed in 1990-94 were about 30 per cent after one year and 10 per cent after five years.[9] The low rates of survival from lung cancer are explained by the frequently advanced stage of the disease at diagnosis, the aggressiveness of the disease, and the small number of patients who are eligible for surgery.

Geographical patterns in incidence

The highest incidence rates for lung cancer occurred in Scotland, where the incidence rate in males was 34 per cent higher, and in females was 48 per cent higher, than the average for the UK and Ireland (Figure 13.1). The apparently low rates in Ireland are affected by the exclusion of death certificate only (DCO) cases: the rates were almost the same as the mortality (Table B13.1) whereas elsewhere they were around 10 per cent higher. Within England there was a very clear north-south divide, with incidence rates being higher than average in the North West, and Northern and Yorkshire regions, and below average in the South West, South East and Eastern regions.

Variability in the incidence of lung cancer was generally higher between the health authorities within a country or region than between the constituent countries of the UK and regions of England (Figures 13.1 and 13.3). In those countries and regions with the highest average rates, there was around a two-fold range in incidence among the health authorities.

The highest incidence in both males and females occurred in Greater Glasgow, where rates were markedly higher than in Argyll and Clyde, which had the second highest rates in Scotland. In the Northern and Yorkshire region, in only one area (North Yorkshire) were the rates in both males and females markedly below the average for the UK and Ireland. Incidence was notably higher than average in the north east, especially in Gateshead and South Tyneside; Newcastle and North Tyneside; and Sunderland, in both males and females. There was a band of high incidence across the north of England with particularly high rates in both males and females in Manchester; Liverpool; and St Helen's and Knowsley. In the West Midlands there were 'hotspots' of high incidence in the urban areas around Birmingham, and in North Staffordshire, but these were higher than average only in males.

Incidence rates were below the average for the UK and Ireland in the majority of health authorities in the south and east of England, and in Wales. The exception was in London where incidence was above average in both males and females in several areas, in particular in East London and the City; Lambeth, Southwark and Lewisham; and Camden and Islington. In England, the incidence of lung cancer was lowest in the South West region, with rates 29 and 40 per cent below average in males and females, respectively.

In Northern Ireland, incidence in the Eastern health authority, which includes Belfast, was much higher than elsewhere, with rates in the other areas being below average. The pattern in Ireland was similar, with rates being markedly higher than elsewhere in the Eastern health authority, which includes Dublin.

The overriding picture presented by the maps for incidence (Map 13.1) is of a very clear north-south divide within Great Britain, with higher incidence in Scotland and the north of England and lower incidence across most of the south of England, Wales, Northern Ireland and Ireland.

Geographical patterns in mortality

Survival from lung cancer is very low and varies little (in absolute terms) among the countries of the UK and regions of England. Consequently, almost exactly the same patterns are seen in mortality (Figure 13.2) as in incidence. Mortality rates were very high in Scotland, above average in the north of England and below average in the south and east, and below average in Northern Ireland and Ireland.

As survival from lung cancer also varies little (in absolute terms) among health authorities, the geographic patterns in mortality (Figure 13.4) were closely similar to those in incidence. Areas with notably high incidence, such as Greater Glasgow; Manchester; Liverpool; Gateshead and South Tyneside; Newcastle and North Tyneside; Sunderland; East London and The City; the areas including Belfast and Dublin; urban areas around Birmingham; and North Staffordshire (males only) had correspondingly high mortality. For mortality rates the rank order of health authorities within each country or region was very similar to that for incidence rates (Table B13.1).

In 1999, the percentage of cancer deaths that were due to lung cancer in the Greater Glasgow health authority was 36 per cent in males and 27 in females, compared to 31 and 23 per cent for male and females, respectively, in Scotland overall. In Glasgow City, mortality from lung cancer was 57 per cent higher than the average for Scotland.[10]

The north-south divide in Great Britain was again very clear from the maps for mortality (Map 13.2), with many areas of higher mortality in northern England, and Scotland, and predominantly areas of lower mortality in Wales, southern and eastern England, Northern Ireland and Ireland.

Risk factors and aetiology

By far the greatest risk factor for lung cancer is tobacco smoking, which causes 90 per cent of cases in men and 80 per cent in women.[11,12] Current and historical differences in smoking habits between men and women can explain the observed differences in incidence and mortality between the sexes. Although similar proportions of men and women currently smoke, this has not been the case historically. Cigarette smoking became common in men in the early twentieth century, about 20 years earlier than in women, and was considerably more prevalent in men by the late 1940s (65 per cent in men and 41 in women in 1948). The proportion of smokers fell among males but remained fairly constant among females for the next 20 years.[13]

(continued on page 148)

Figure **13.1**

Lung: incidence by sex, country, and region of England
UK and Ireland 1991-99[1]

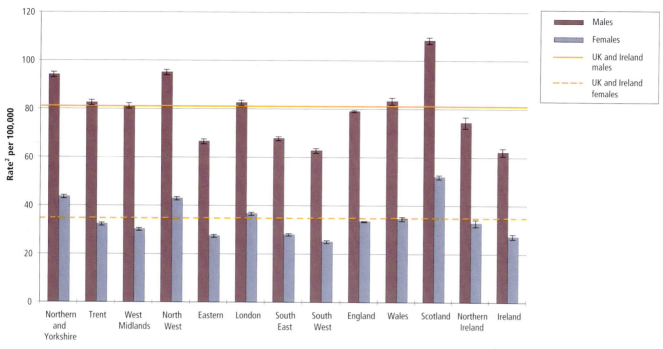

1 Northern Ireland 1993-99, Ireland 1994-99

2 Age standardised using the European standard population, with 95% confidence interval

Figure **13.2**

Lung: mortality by sex, country, and region of England
UK and Ireland 1991-2000[1]

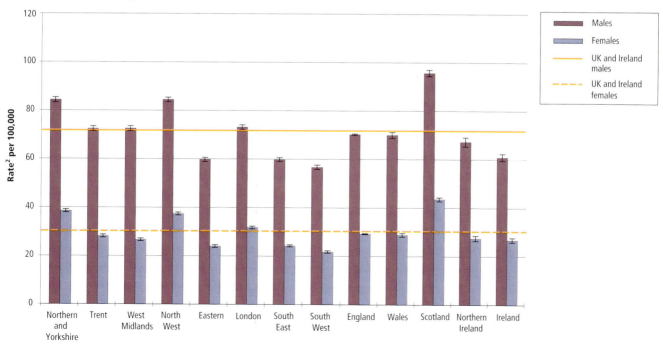

1 Scotland 1991-99, Ireland 1994-2000

2 Age standardised using the European standard population, with 95% confidence interval

Figure 13.3a

**Lung: incidence by health authority within country, and region of England
Males, UK and Ireland 1991-99[1]**

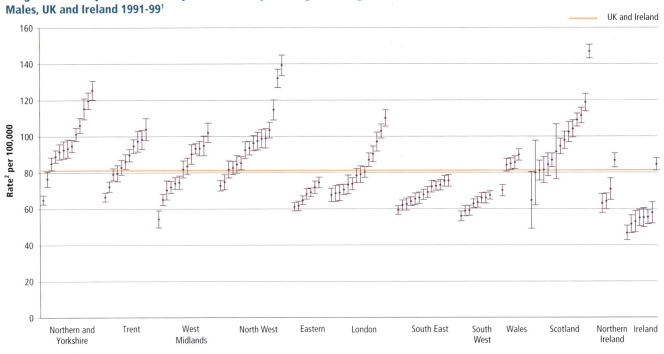

1 Northern Ireland 1993-99, Ireland 1994-99

2 Age standardised using the European standard population, with 95% confidence interval

Figure 13.3b

**Lung: incidence by health authority within country, and region of England
Females, UK and Ireland 1991-99[1]**

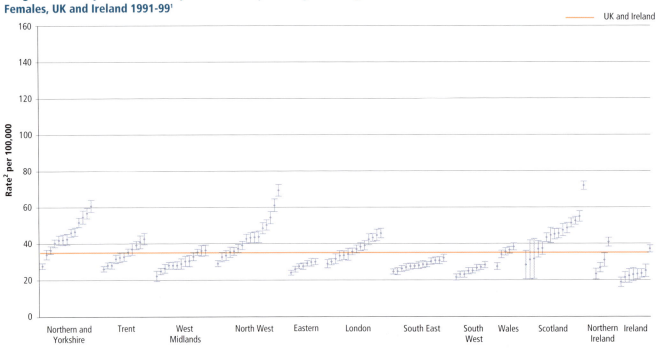

1 Northern Ireland 1993-99, Ireland 1994-99

2 Age standardised using the European standard population, with 95% confidence interval

Figure 13.4a

**Lung: mortality by health authority within country, and region of England
Males, UK and Ireland 1991-2000[1]**

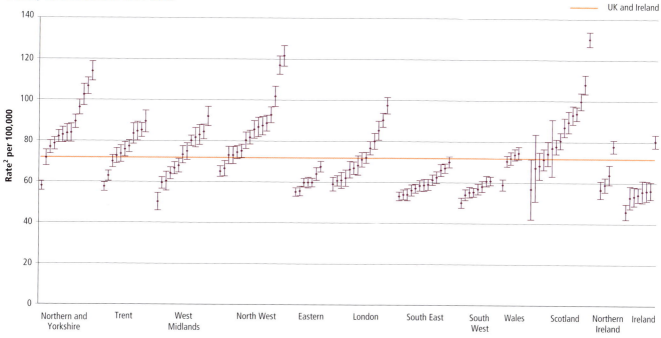

1 Scotland 1991-99, Ireland 1994-2000
2 Age standardised using the European standard population, with 95% confidence interval

Figure 13.4b

**Lung: mortality by health authority within country, and region of England
Females, UK and Ireland 1991-2000[1]**

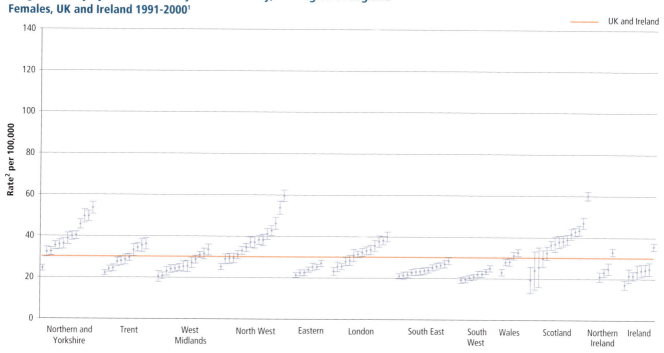

1 Scotland 1991-99, Ireland 1994-2000
2 Age standardised using the European standard population, with 95% confidence interval

Map **13.1a**

**Lung: incidence* by health authority
Males, UK and Ireland 1991-99**

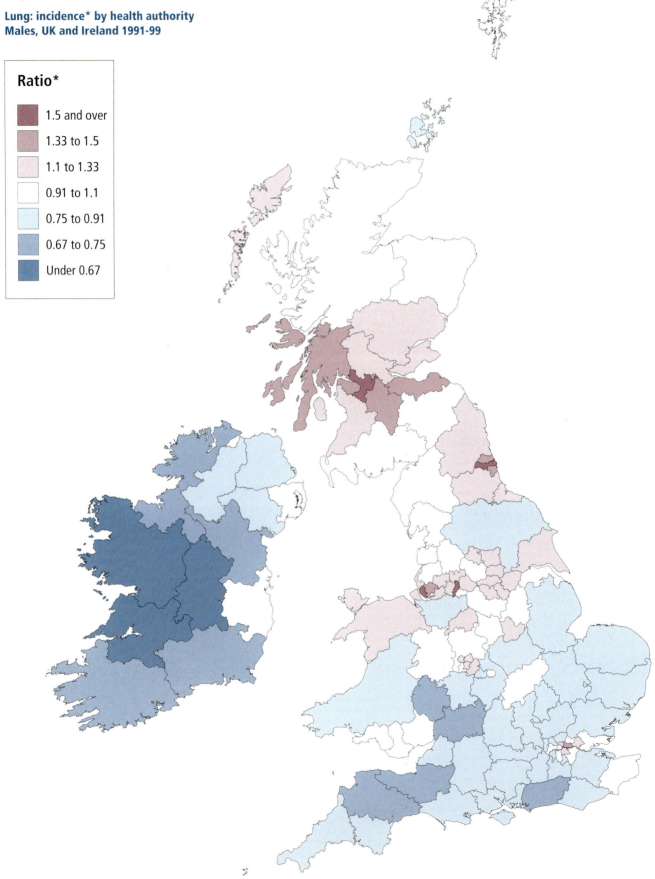

Ratio*

	1.5 and over
	1.33 to 1.5
	1.1 to 1.33
	0.91 to 1.1
	0.75 to 0.91
	0.67 to 0.75
	Under 0.67

*Ratio of directly age-standardised rate in health authority to UK and Ireland average

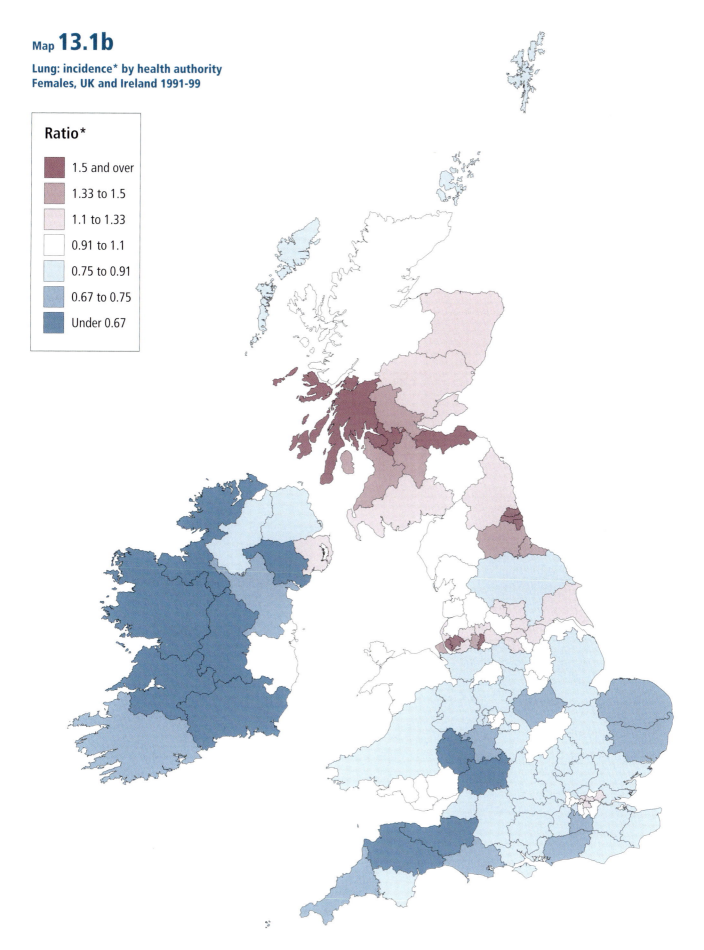

Map 13.1b

Lung: incidence* by health authority
Females, UK and Ireland 1991-99

Ratio*

- 1.5 and over
- 1.33 to 1.5
- 1.1 to 1.33
- 0.91 to 1.1
- 0.75 to 0.91
- 0.67 to 0.75
- Under 0.67

Ratio of directly age-standardised rate in health authority to UK and Ireland average

Map **13.2a**

**Lung: mortality* by health authority
Males, UK and Ireland 1991-2000**

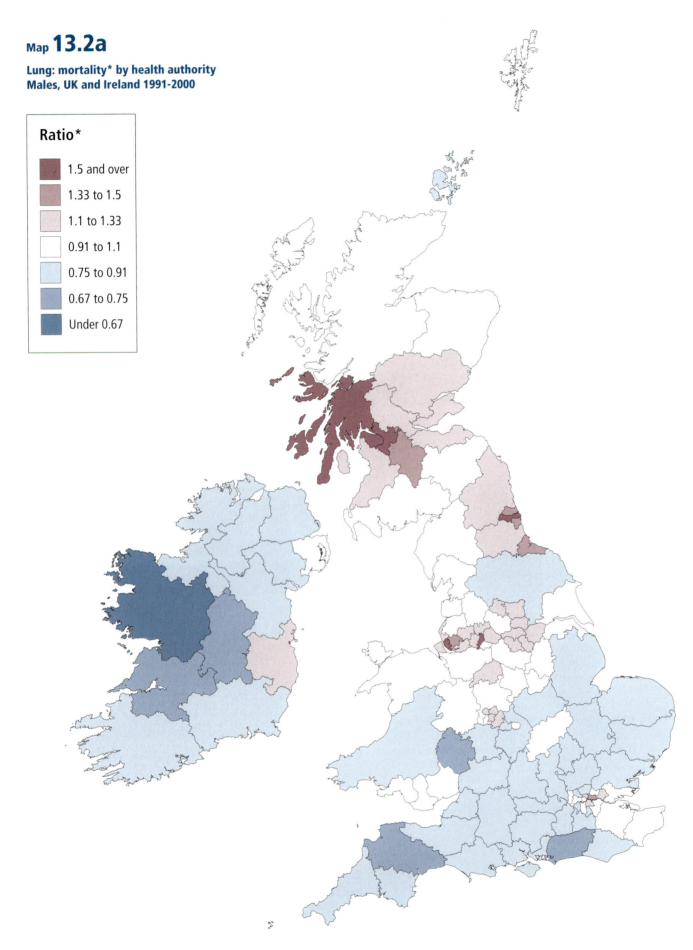

Ratio*

- 1.5 and over
- 1.33 to 1.5
- 1.1 to 1.33
- 0.91 to 1.1
- 0.75 to 0.91
- 0.67 to 0.75
- Under 0.67

*Ratio of directly age-standardised rate in health authority to UK and Ireland average

Map **13.1b**

**Lung: incidence* by health authority
Females, UK and Ireland 1991-99**

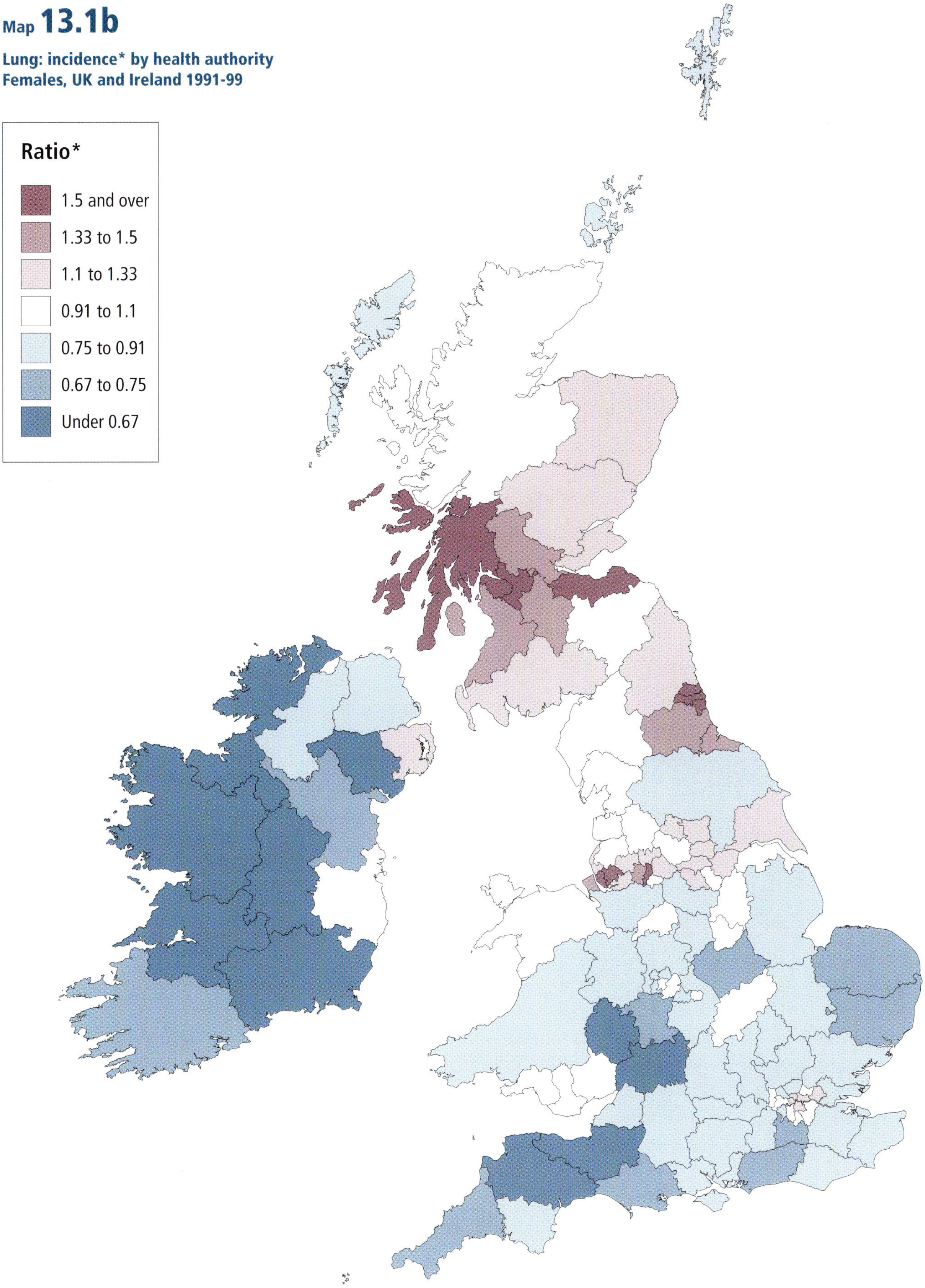

Ratio*

	1.5 and over
	1.33 to 1.5
	1.1 to 1.33
	0.91 to 1.1
	0.75 to 0.91
	0.67 to 0.75
	Under 0.67

*Ratio of directly age-standardised rate in health authority to UK and Ireland average

Map **13.2a**

**Lung: mortality* by health authority
Males, UK and Ireland 1991-2000**

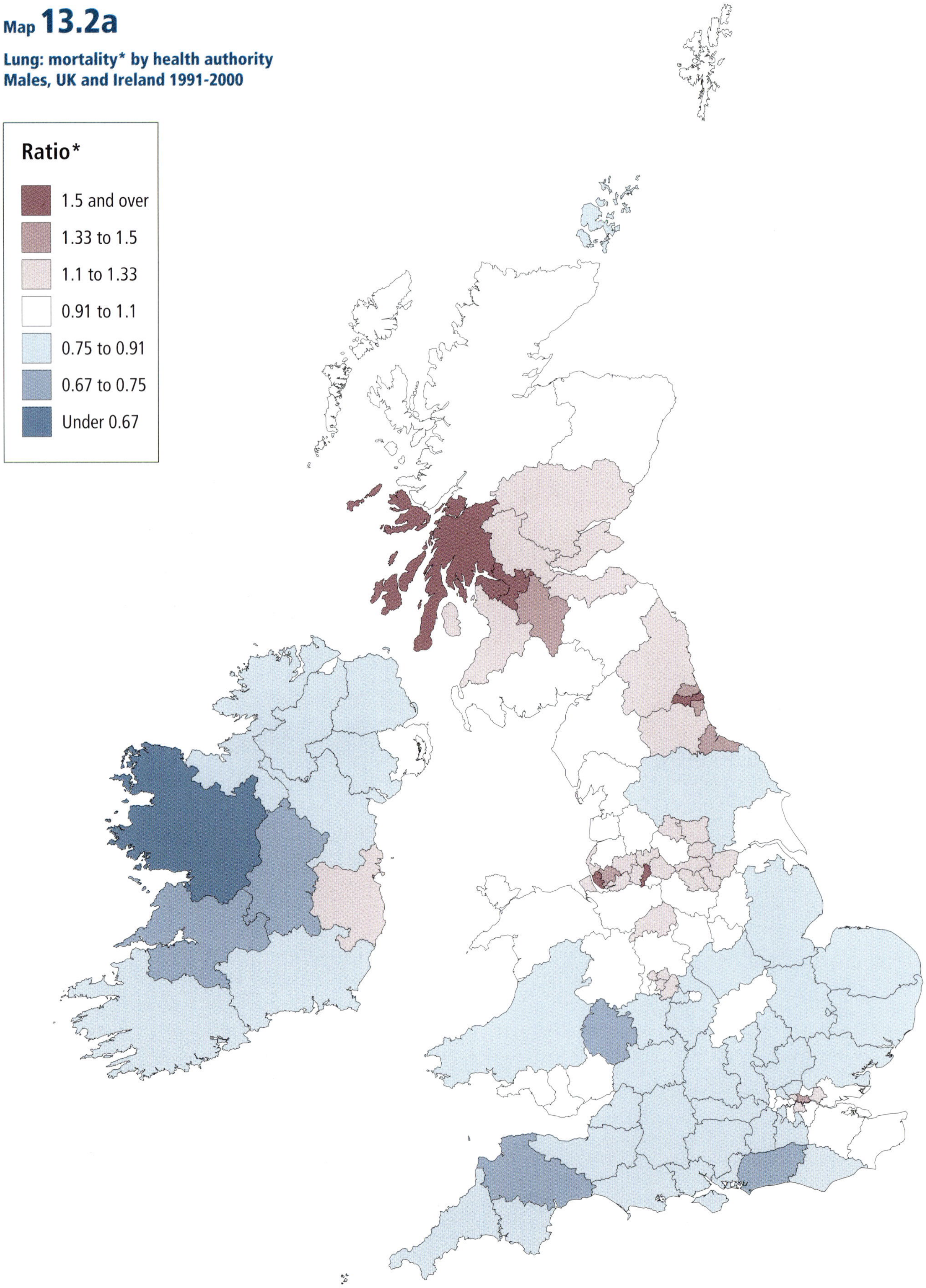

Ratio*

■	1.5 and over
■	1.33 to 1.5
■	1.1 to 1.33
□	0.91 to 1.1
■	0.75 to 0.91
■	0.67 to 0.75
■	Under 0.67

*Ratio of directly age-standardised rate in health authority to UK and Ireland average

146

Map **13.2b**

**Lung: mortality* by health authority
Females, UK and Ireland 1991-2000**

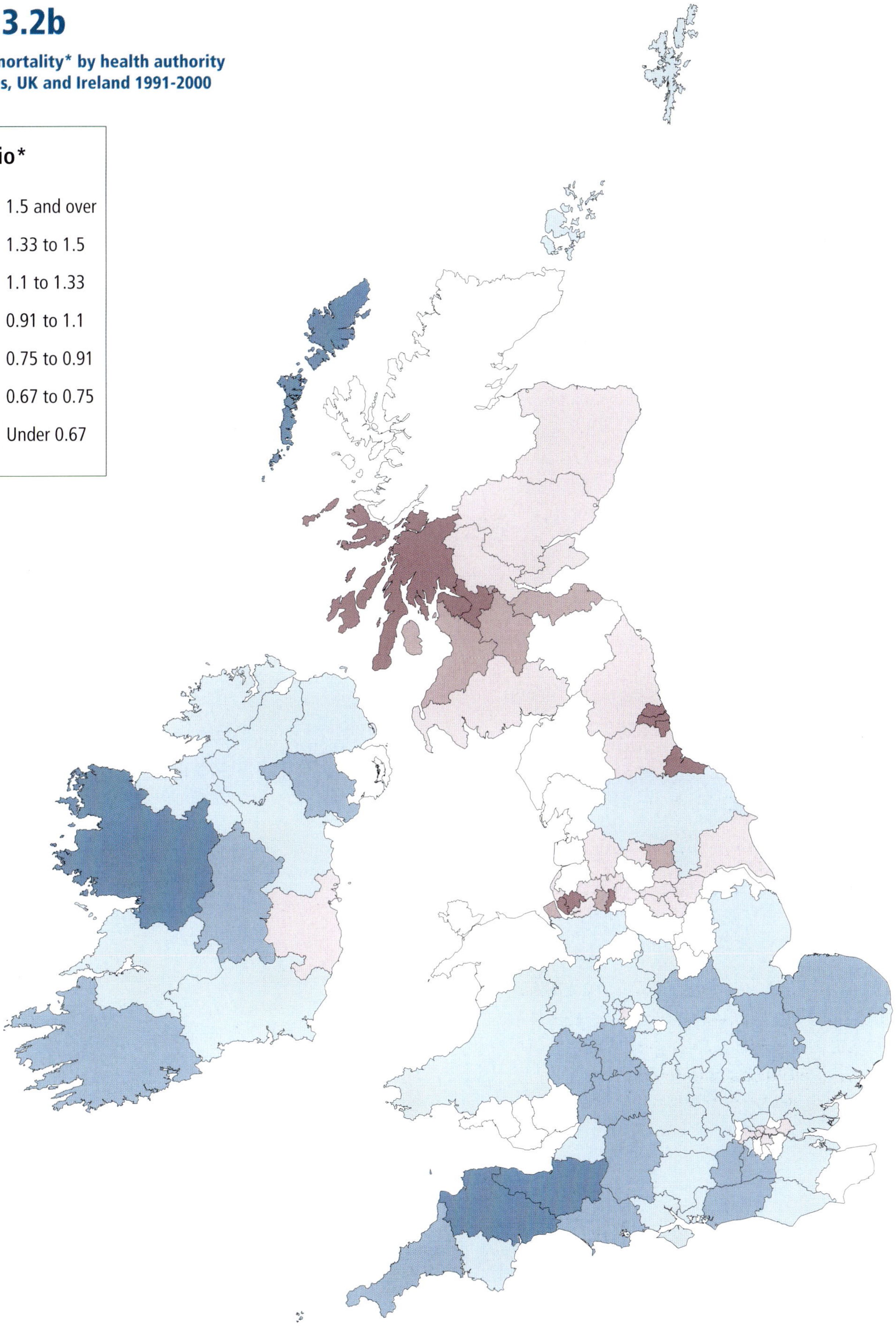

Ratio*

	1.5 and over
	1.33 to 1.5
	1.1 to 1.33
	0.91 to 1.1
	0.75 to 0.91
	0.67 to 0.75
	Under 0.67

*Ratio of directly age-standardised rate in health authority to UK and Ireland average

Only after 1970 did the prevalence among women begin to fall, and as levels continued to fall among men, by the end of the 1990s the prevalence of smoking in men and women was similar (27 and 26 per cent, respectively, in 1998).[14] The latency period for lung cancers attributable to smoking is long – often more than 20 years – as can be seen from the time differences in the peaks of prevalence in smoking and the peaks in incidence and mortality rates.

Scotland has the highest prevalence of smoking in Great Britain, at 28 per cent. In England the highest rates of smoking are in the north, with a prevalence of 28 per cent in the North West region, and 27 per cent in both the North East, and Yorkshire and the Humber regions. Prevalence is lowest in the West Midlands (23 per cent), the East Midlands and London (both 24 per cent).[15] Although these are current figures, the geographical pattern of smoking prevalence has been consistent since the 1970s, and this explains the observed geographical variations in incidence and mortality for lung cancer. Current figures show that in England, at the level of Strategic Health Authority, the prevalence of smoking is highest in Northumberland, Tyne and Wear, at 33 per cent. Levels of smoking are also particularly high in South East, North East and North Central London; South Yorkshire; Greater Manchester; and County Durham and Tees Valley.[12]

Many industrial carcinogens have been conclusively or putatively associated with an increased risk of lung cancer, including asbestos, arsenic, nickel, chromium, zinc, polycyclic hydrocarbons and radon.[11] It has been suggested that up to 15 per cent of cases in men (5 per cent in women) may be attributable to occupational factors in conjunction with smoking.[16] Geographical analysis of deaths from mesothelioma highlights the greatest excesses in areas containing industrial sites where asbestos has been used in the past, particularly areas containing ports or dockyards, and those with a large railway engineering industry.[17] The very high incidence of lung cancer in the west of Scotland is likely to be due in part to occupational exposures, particularly in dockyard workers. Around 6 per cent of male lung cancers in this area of Scotland have been attributed to asbestos exposure.[18] Other possible risk factors include air pollution and indoor radon and its decay products, but without smoking, lung cancer would be an uncommon disease.

Socio-economic deprivation

Lung cancer incidence and mortality rates are strongly associated with deprivation. An analysis of 1993 incidence data for England and Wales by Carstairs deprivation index[19] found that for males, incidence in the most deprived groups was almost two and a half times that in the most affluent, while for females, the ratio was around three.[3] A similar association with deprivation is evident in Scotland, where incidence was around three times higher in both males and females in the most socially deprived groups.[20] Survival was significantly lower among deprived compared to affluent men (a gap in five-year survival of 1.4 percentage points), although the difference in survival between deprived and affluent women was small.[21] Stage of disease at diagnosis was not an important explanatory factor of the association between deprivation and survival.[22] Social class, based on occupation, is also related to the risk of mortality from lung cancer. Results based on individual records from the Longitudinal Study (a one per cent linked sample of census records) showed that in 1986-92 men in the manual classes were twice as likely to die of lung cancer than those in the non-manual classes, while the ratio for women was 2.6.[23]

Many of the areas with the highest level of deprivation, as measured by the Carstairs index, corresponded to areas with high incidence of, and mortality from, lung cancer, including Greater Glasgow; Gateshead and South Tyneside; Newcastle and North Tyneside; Sunderland; Liverpool; Manchester; and East London and The City (see Appendix F). These variations can mainly be explained by differences in the prevalence of smoking. Before the dangers of smoking were widely recognised, the prevalence of smoking varied little by socio-economic group. Today there are clear differences due to the differential decline in smoking by social class. In 2002, male routine workers were more than twice as likely to smoke as professional workers, and also started smoking at a younger age.[15]

References

1. Schouten LJ, Straatman H, Kiemeney LALM, Verbeek ALM. Cancer incidence: Life table risk versus cumulative risk. *Journal of Epidemiology and Community Health* 1994; 48: 596-600.

2. ONS. *Cancer Statistics Registrations: Registrations of cancer diagnosed in 2001, England.* Series MB1 No. 32. London: Office for National Statistics, 2004.

3. Quinn MJ, Babb PJ, Brock A, Kirby L et al. *Cancer Trends in England and Wales 1950-1999.* Studies on Medical and Population Subjects No. 66. London: The Stationery Office, 2001.

4. Parkin DM. Global cancer statistics in the year 2000. *The Lancet Oncology* 2001; 2: 533-543.

5. ONS. Cancer Survival: England and Wales, 1991-2001. March 2004. Available at *http://www.statistics.gov.uk/statbase/ssdataset.asp?vlnk=7899.*

6. ISD Scotland. *Trends in Cancer Survival in Scotland, 1977-2001.* Edinburgh: ISD Publications, 2004.

7. Fitzpatrick D, Gavin A, Middleton R, Catney D. *Cancer in Northern Ireland 1993-2001: A Comprehensive Report.* Belfast: Northern Ireland Cancer Registry, 2004.

8. National Cancer Registry of Ireland. *Cancer in Ireland, 1994 to 1998: Incidence, mortality, treatment and survival.* Cork: National Cancer Registry, 2001.

9. Sant M, Areleid T, Berrino F, Bielska Lasota M et al. EUROCARE-3: survival of cancer patients diagnosed 1990-1994 - results and commentary. *Annals of Oncology* 2003; 14 Suppl 5: v61-v118.

10. Registrar General for Scotland. *Annual Report of the Registrar General of Births, Deaths and Marriages for Scotland, 1999.* Edinburgh: General Register Office for Scotland, 2000.

11. Blot WJ, Fraumeni JF, Jr. Cancers of the Lung and Pleura. In: Schottenfeld D, Fraumeni JF, Jr. (eds) *Cancer Epidemiology and Prevention, second edition.* New York: Oxford University Press, 1996.

12. Twigg L, Moon G, Walker S. *The Smoking Epidemic in England.* London: Health Development Agency, 2004.

13. Wald N, Nicolaides-Bouman A. *UK Smoking Statistics, second edition.* London: Oxford University Press, 1991.

14. ONS. *Smoking-related Behaviour and Attitudes, 1999.* London: The Stationery Office, 2000.

15. Rickards L, Fox K, Roberts C, Fletcher L et al. *Living in Britain: Results from the 2002 General Household Survey.* London: The Stationery Office, 2004.

16. Doll R, Peto R. *The Causes of Cancer. Quantitative estimates of avoidable risks of cancer in the United States today.* Oxford: Oxford University Press, 1981.

17. Health and Safety Executive. *Occupational Health Statistics Bulletin 2002/03,* 2003.

18. De Vos Irvine H, Lamont DW, Hole DJ, Gillis CR. Asbestos and lung cancer in Glasgow and the west of Scotland. *British Medical Journal* 1993; 306: 1503-1506.

19. Carstairs V, Morris R. Deprivation and mortality: an alternative to social class? *Community Medicine* 1989; 11: 213-219.

20. Harris V, Sandridge AL, Black RJ, Brewster DH et al. *Cancer Registration Statistics Scotland, 1986-1995.* Edinburgh: ISD Publications, 1998.

21. Coleman MP, Rachet B, Woods LM, Mitry E et al. Trends and socioeconomic inequalities in cancer survival in England and Wales up to 2001. *British Journal of Cancer* 2004; 90: 1367-1373.

22. Schrijvers CT, Mackenbach JP, Lutz JM, Quinn MJ et al. Deprivation, stage at diagnosis and cancer survival. *International Journal of Cancer* 1995; 63: 324-329.

23. Harding S, Bethune A, Maxwell R, Brown J. Mortality trends using the Longitudinal Study. In: Drever F, Whitehead M (eds) *Health Inequalities.* Decennial Supplement No. 15. London: The Stationery Office, 1997.

Chapter 14

Melanoma of skin

Anna Gavin, Paul Walsh

Summary

- In the UK and Ireland in the 1990s, melanoma of the skin accounted for 1 in 50 diagnosed cases of cancer and 1 in 100 deaths from cancer.

- Incidence was higher than average in the south of England, in Scotland and Ireland, and in Northern Ireland in females, and below average in London, the midlands, Wales, and most of northern England.

- There was less geographic variation in mortality; rates were relatively high in southern England and low across most of the midlands and north of England.

- In contrast to incidence, mortality was relatively low in Scotland, Northern Ireland and Ireland – possibly related to health education campaigns leading to earlier detection.

- Incidence rates were consistently higher in females than in males, but mortality rates were consistently higher in males than in females.

- The areas with higher incidence tended to be the more affluent ones.

- The greatest risk factor for melanoma of the skin is exposure to ultraviolet radiation, mainly from the sun. The relationship between high incidence and affluence is likely to be related to excessive sun exposure on holidays abroad.

Incidence and mortality

In the 1990s there were 2,400 newly registered cases of melanoma each year in males, in whom it was the thirteenth most common cancer, and 3,500 cases in females, in whom it was the eleventh most common. Melanoma accounted for about 2 per cent of all newly diagnosed cases per year (1.8 per cent in males, 2.6 in females). Unlike most other malignancies, melanoma was more common in females than males. The age-standardised incidence rates were 7.7 and 9.7 per 100,000 in males and females, respectively. The ratio of the rates in males to females was around 0.8:1. Melanoma affects younger people more than most cancers, with about 40 per cent of cases in people under 50, although incidence rates were highest above age 75.

In the 1990s, there were around 800 deaths in each sex (one per cent of all cancer deaths) from melanoma of the skin each year in the UK and Ireland. It ranked fourteenth as a cause of cancer death in males, and sixteenth in females. Despite the higher number of cases in females there were almost the same numbers of deaths in males and females, reflecting the poorer average survival for males with this cancer (see below). The age-standardised mortality rate was 2.5 per 100,000 in males and 2.0 per 100,000 in females. Mortality rates were highest in the oldest age group (85 and over).

Incidence and mortality trends

Melanoma of the skin was a very rare disease in the 1960s, but a long-term increase in incidence and mortality has occurred in most white populations across the world over several decades.[1] In England and Wales, both males and females showed a three- to four-fold increase in age-standardised incidence rates between the early 1970s and early 1990s, with large increases seen across most age groups.[2] Mortality rates increased fairly steadily in England and Wales from the 1950s onwards, before stabilising in females (but not males) in the late 1980s.[2] Marked increases in incidence since the 1960s, with less marked increases in mortality, have been seen in Scotland.[3,4] Melanoma of the skin was formerly the most rapidly increasing cancer in the USA, where older men carry the highest risk (to a greater extent than in the UK). The most recent trends in incidence show the biggest rise in older age groups,[5,6] in the UK.[2] Early detection has had an impact on the changes described in the USA as evidenced by a higher rate of increase in localised compared with regional or metastasised melanomas. There were however, increased incidence rates for all stages in males, but only for localised disease in females.[6] Pathological evidence supports the suggestion that increasing incidence rates reflect real change rather than improved diagnosis and ascertainment. The increase was most rapid in the 1970s, when there was little awareness of melanoma, and education to enhance early detection only started in the USA in 1985 at a national level.[7] Similar educational measures in England and Wales in the late 1980s appeared to bring forward diagnosis somewhat, producing a short-term increase in melanoma incidence,[2,8] but a longer-term increase was already underway.

Survival

Relative survival for melanoma patients diagnosed in England and Wales during 1996-99 was about 77 per cent after five years for males and 87 per cent for females.[9] Survival for patients diagnosed in Scotland,[10] Ireland,[11] and Northern Ireland[12] during the 1990s was very similar. Worse survival in men than women has been noted in most datasets from European countries in the EUROCARE-3 project.[13] Melanoma in men presents more often on the trunk,[14-16] where it has a poorer prognosis than melanoma generally.[17,18] Men have been shown to be less knowledgeable than women about appropriate prevention measures, to respond less well to health education, and to present with the disease at later stages.[19]

Geographical patterns in incidence

On a country or regional scale within the UK, the highest incidence rates for melanoma in both males and females occurred in South West England (44 and 42 per cent higher, respectively, than the UK and Ireland average) (Figure 14.1). Ireland also had a markedly high rate for females (37 per cent above average), more so than for males (21 per cent above average). Incidence rates were also particularly high for females in Northern Ireland, Scotland, and South East England (24, 23 and 13 per cent above average, respectively) and males in Scotland and South East England (25 and 17 per cent above average). Incidence rates were markedly below average in Wales; Northern and Yorkshire; Trent; West Midlands; and London for both males and females. At a country and regional scale, there was a strong correlation between male and female incidence rates.

Variability in the incidence of melanoma was generally higher between the health authorities within a country or region than between the countries and regions themselves (Figure 14.3 and Map 14.1). For both males and females, notably high rates (more than 50 per cent above the UK and Ireland average) were recorded in Borders; Dumfries and Galloway; Cornwall and Isles of Scilly; Dorset; and Southampton and South West Hampshire. Similarly high rates were recorded for females in the Southern health authorities in Ireland and Northern Ireland; in Somerset; and North and East Devon; and for males in Isle of Wight, Portsmouth and South East Hampshire; and South and West Devon (Table B14.1). The rank order of health authorities within regions or countries was similar for males and females, and the maps showed many similarities between the sexes.

Variations in ascertainment between registries may partly account for regional variations in recorded melanoma incidence and the similarity between geographic patterns for males and females. For example, it was previously reported that up to 23 per cent of melanoma cases were not registered in the Northern and Yorkshire region, but it is not known if this under-registration has been consistent over time.[20] Relatively high incidence, compared with mortality, in Ireland, Northern Ireland, and Scotland, and to a lesser extent in South West and North West England, may suggest more complete registration in these populations, while data for Wales, and London may

indicate some under-registration. Alternatively, earlier detection in some regions may have boosted incidence rates (in addition to reducing mortality).

Geographical patterns in mortality

At the country and regional level (Figure 14.2), there was far less geographic variation for mortality than for incidence, especially for females. Mortality rates in both males and females were markedly above average in South East and South West England, and notably lower than average in the midlands and north of England, and in both Northern Ireland and Ireland. At the health authority level, variation in mortality appeared fairly substantial (Figure 14.4 and Map 14.2) although due to the small numbers of deaths involved (fewer than 20 annually in most health authorities) many of the rates were not significantly different from the average. Compared to those for incidence (Map 14.1), the maps for mortality showed a much clearer north-south divide across England, with the highest mortality rates near the south and south west coasts, and lower rates in the midlands and north of the country (Map 14.2).

In general, there was a low level of correlation between mortality and incidence rates at the national or regional level, although it was somewhat higher among English regions (Table 14.1). Within most countries or regions, there was again only moderate similarity between the health authority rankings for mortality and incidence rates. Incidence data are probably more susceptible to geographic variations in ascertainment, and are also likely to be influenced by variations in early detection. The latter, as well as potentially inflating the incidence data in some populations, will also tend to reduce mortality rates, further exaggerating disparities between mortality and incidence.

Nevertheless, mortality-to-incidence ratios also reflect survival, and were higher in males (overall average 0.32) than females (0.20), which is consistent with the worse survival of males with melanoma. The ratios were highest, possibly reflecting worse survival, in Wales, and London. The lower mortality and higher incidence in females in Scotland, and Northern Ireland, and perhaps some other regions, possibly reflects health promotion campaigns from the late 1980s onwards.[21] However, an evaluation of the Cancer Research Campaign publicity drive aimed at early detection of melanoma in one Scottish and six English regions in the late 1980s, did not find any significant reduction in mortality associated with the intervention.[22] The similarly low mortality-to-incidence ratio seen in Ireland does not seem to be explained by early detection.

(continued on page 160)

Figure **14.1**

**Melanoma of skin: incidence by sex, country, and region of England
UK and Ireland 1991-99[1]**

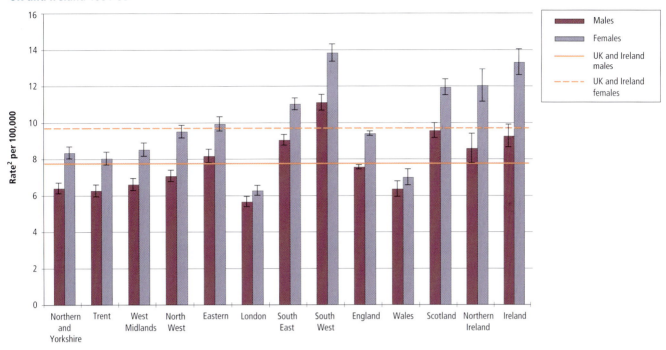

1 Northern Ireland 1993-99, Ireland 1994-99

2 Age standardised using the European standard population, with 95% confidence interval

Figure **14.2**

**Melanoma of skin: mortality by sex, country, and region of England
UK and Ireland 1991-2000[1]**

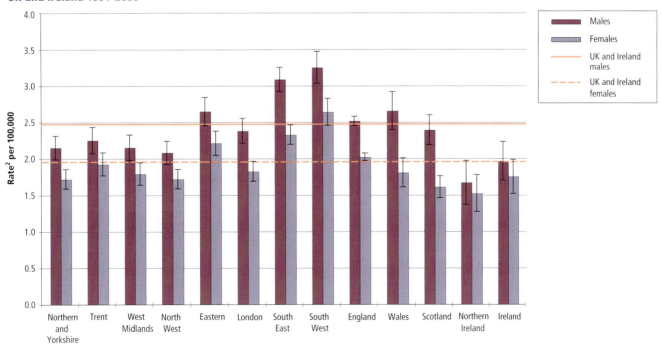

1 Scotland 1991-99, Ireland 1994-2000

2 Age standardised using the European standard population, with 95% confidence interval

Figure **14.3a**

**Melanoma of skin: incidence by health authority within country, and region of England
Males, UK and Ireland 1991-99[1]**

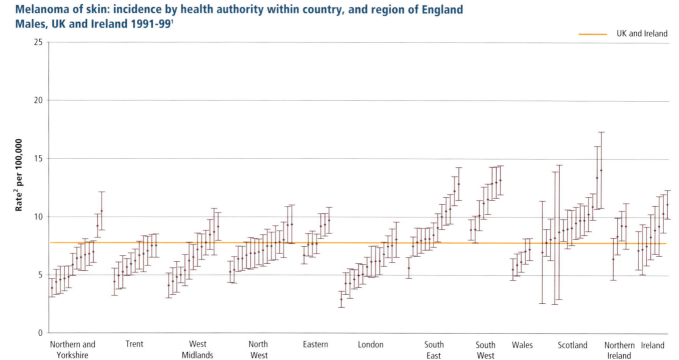

1 Northern Ireland 1993-99, Ireland 1994-99

2 Age standardised using the European standard population, with 95% confidence interval

Figure **14.3b**

**Melanoma of skin: incidence by health authority within country, and region of England
Females, UK and Ireland 1991-99[1]**

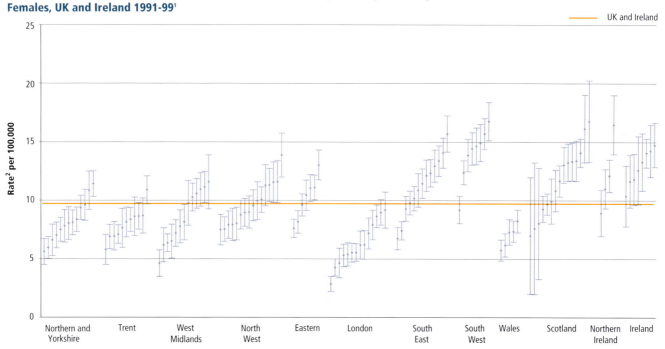

1 Northern Ireland 1993-99, Ireland 1994-99

2 Age standardised using the European standard population, with 95% confidence interval

Figure **14.4a**

Melanoma of skin: mortality by health authority within country, and region of England
Males, UK and Ireland 1991-2000[1]

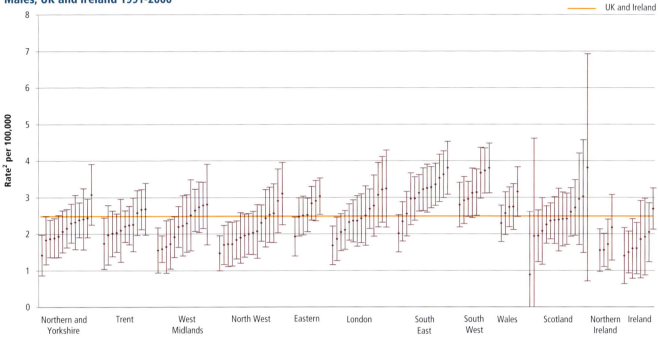

1 Scotland 1991-99, Ireland 1994-2000

2 Age standardised using the European standard population, with 95% confidence interval

Figure **14.4b**

Melanoma of skin: mortality by health authority within country, and region of England
Females, UK and Ireland 1991-2000[1]

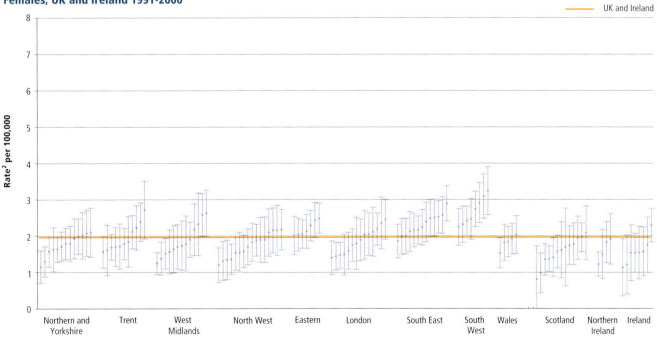

1 Scotland 1991-99, Ireland 1994-2000

2 Age standardised using the European standard population, with 95% confidence interval

Map **14.1a**

**Melanoma of skin: incidence* by health authority
Males, UK and Ireland 1991-99**

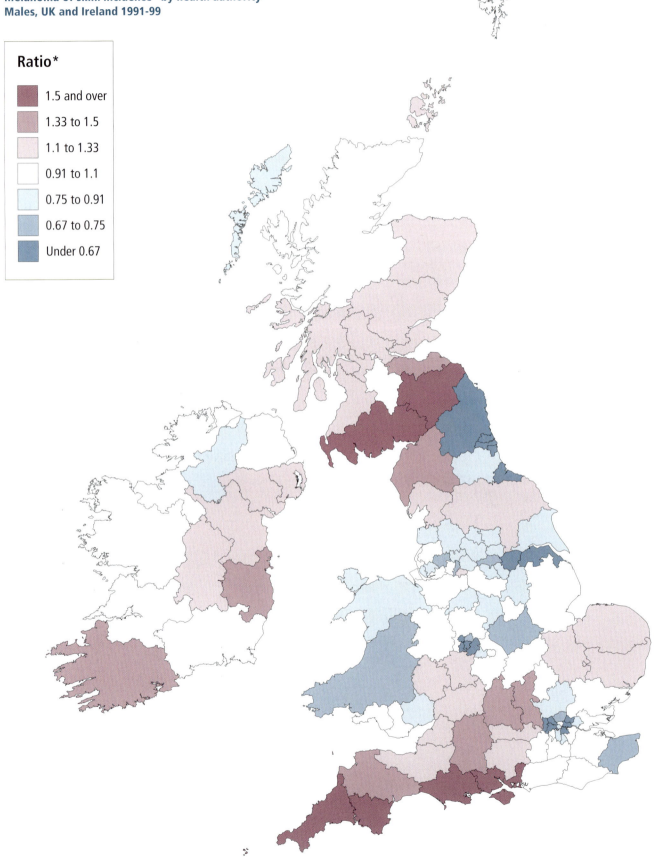

Ratio*

	1.5 and over
	1.33 to 1.5
	1.1 to 1.33
	0.91 to 1.1
	0.75 to 0.91
	0.67 to 0.75
	Under 0.67

*Ratio of directly age-standardised rate in health authority to UK and Ireland average

Map **14.1b**

**Melanoma of skin: incidence* by health authority
Females, UK and Ireland 1991-99**

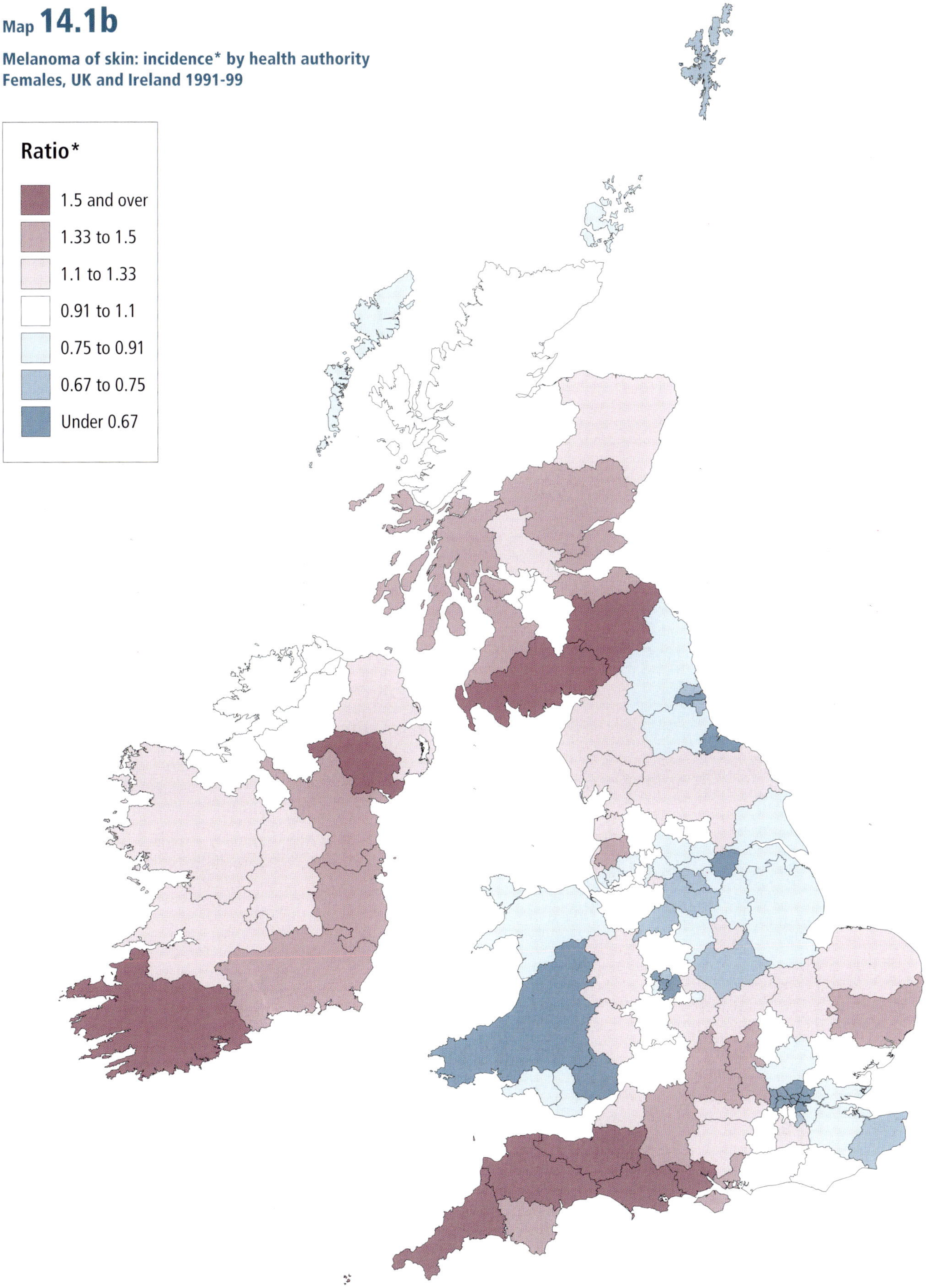

Ratio*

	1.5 and over
	1.33 to 1.5
	1.1 to 1.33
	0.91 to 1.1
	0.75 to 0.91
	0.67 to 0.75
	Under 0.67

Ratio of directly age-standardised rate in health authority to UK and Ireland average

Map **14.2a**

Melanoma of skin: mortality* by health authority
Males, UK and Ireland 1991-2000

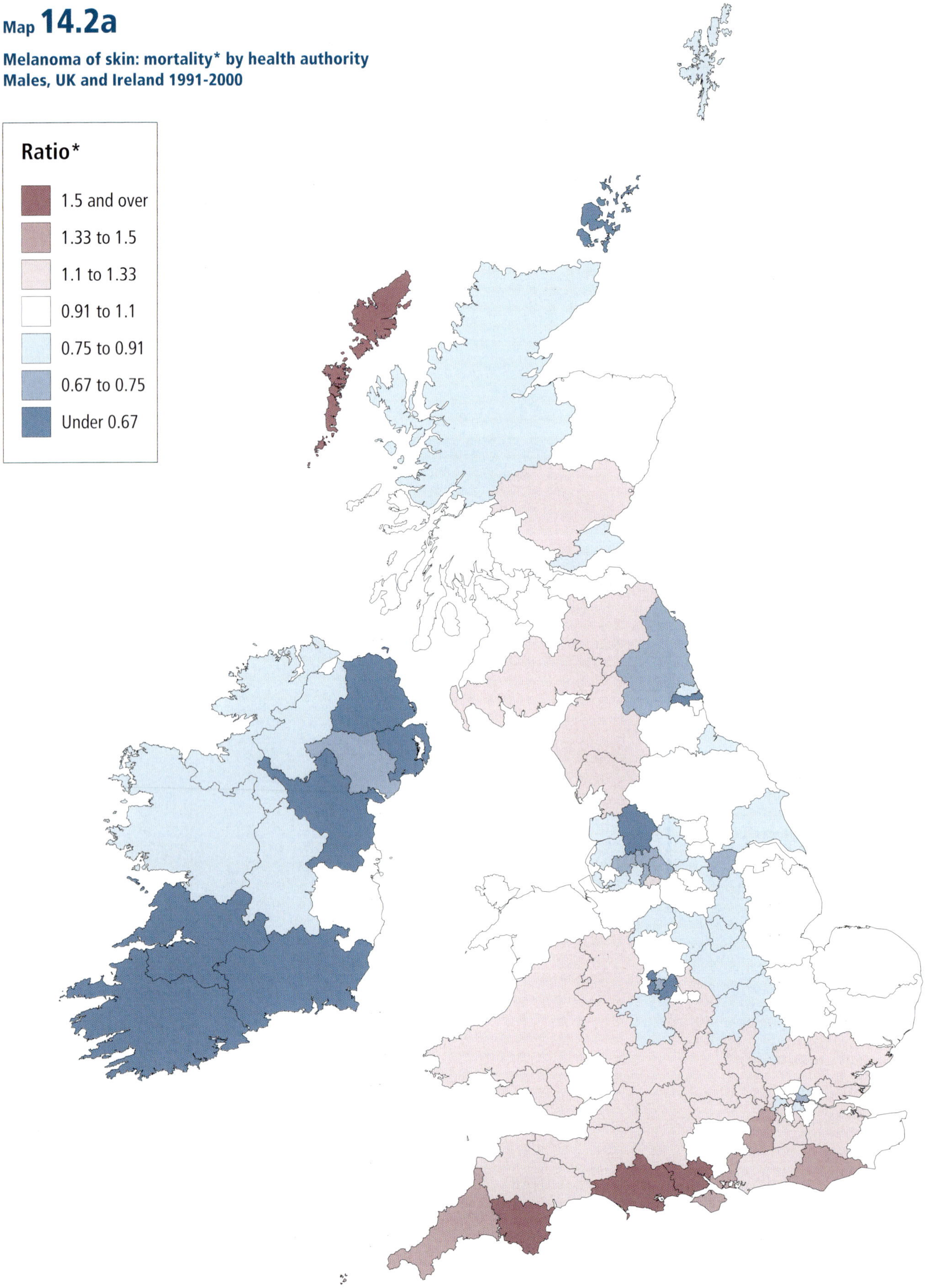

Ratio*

■	1.5 and over
■	1.33 to 1.5
■	1.1 to 1.33
□	0.91 to 1.1
■	0.75 to 0.91
■	0.67 to 0.75
■	Under 0.67

*Ratio of directly age-standardised rate in health authority to UK and Ireland average

Map **14.2b**

**Melanoma of skin: mortality* by health authority
Females, UK and Ireland 1991-2000**

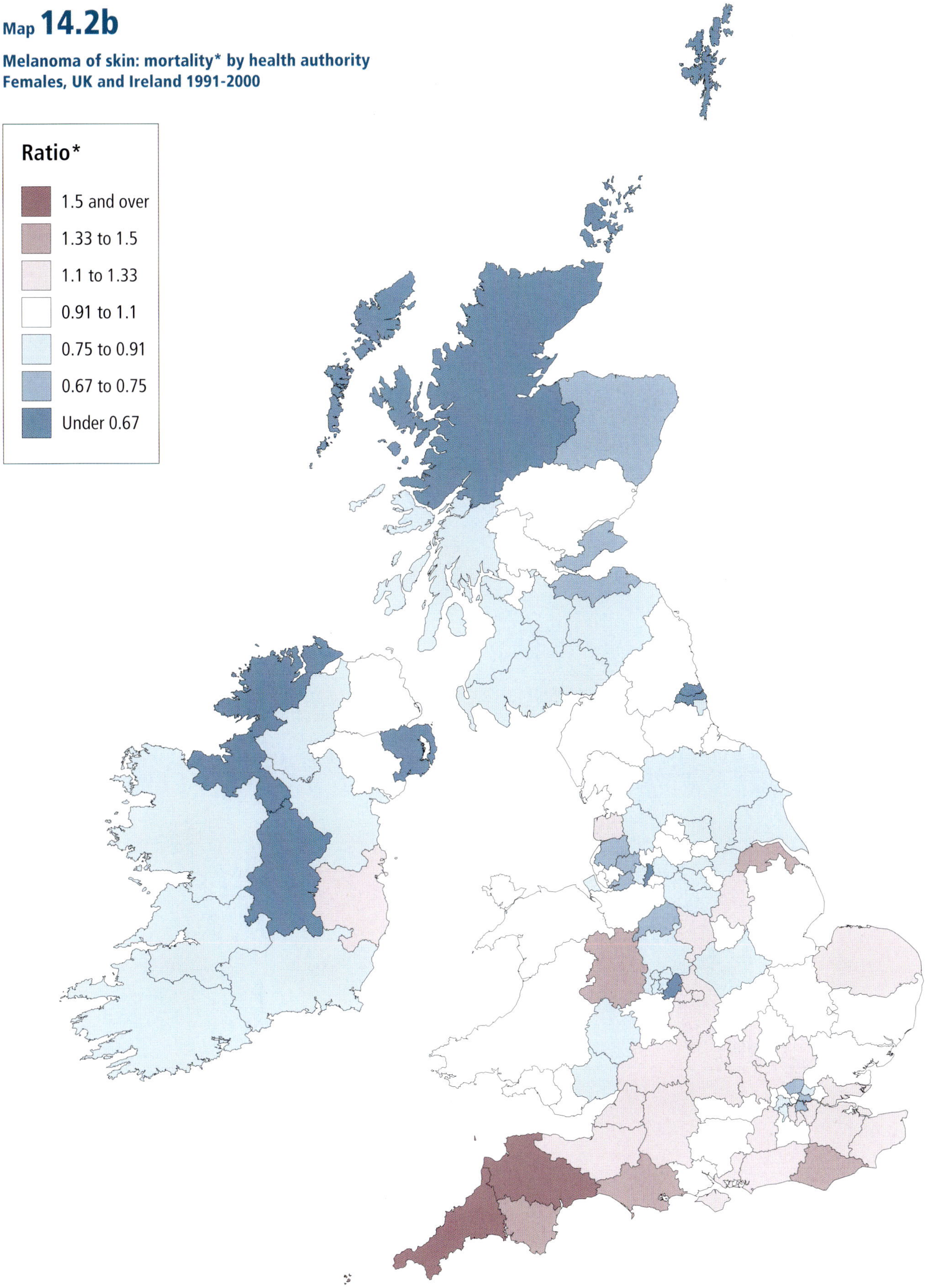

Ratio*

▮	1.5 and over
▮	1.33 to 1.5
▮	1.1 to 1.33
▯	0.91 to 1.1
▮	0.75 to 0.91
▮	0.67 to 0.75
▮	Under 0.67

*Ratio of directly age-standardised rate in health authority to UK and Ireland average

Nevertheless, EUROCARE-3 showed substantial differences in survival among European countries for melanoma and noted that countries with the highest incidence rates also had the highest survival rates, whereas in those countries where incidence was low, survival rates were relatively low. It was speculated that this might be due to higher awareness of melanoma risk in high-incidence areas, resulting in earlier detection and better survival.[13]

Risk factors and aetiology

Light skin type, large number of naevi, atypical naevi, family history of skin cancer and excessive sun exposure (mainly in childhood) are the major risk factors for melanoma of the skin.[23] Melanoma arises from the malignant transformation of melanocytes (skin cells that produce the pigment melanin, which determines skin colour). Ultraviolet radiation can promote the proliferation capacity of melanocytes.[24] Most melanomas are thought to be caused by intermittent rather than chronic exposure to ultraviolet radiation, especially during childhood, although exposure in adulthood certainly also plays a part,[25,26] with melanomas in older people more strongly related to chronic sun exposure.[27] Variation in recreational or holiday exposure to ultraviolet radiation almost certainly contributes strongly to the increased melanoma risk seen in higher socio-economic groups (see below).

When melanoma is detected at an early stage it is curable, but once advanced it is difficult to treat. Limiting exposure to ultraviolet radiation reduces the risk of melanoma and other skin cancers. As the main environmental source of ultraviolet radiation is sun exposure, the European Code Against Cancer advises Europeans to 'reduce their total lifetime exposure, and in particular to avoid extremes of sun exposure and sunburn'. Simple measures such as avoiding the sun between 11am and 3pm, seeking shade, wearing close-weave heavy cotton clothing, and using sunscreen, as well as avoiding the use of sun beds, are recommended.[28]

Socio-economic deprivation

In England and Wales, there is a strong inverse relationship between the incidence of melanoma and social deprivation, measured using the Carstairs Index.[2] In 1988-93, incidence in the most affluent groups was about three times that in the most deprived populations. A similar gradient was seen in Scotland.[3] Variation in mortality with affluence was much less marked than for incidence, but mortality rates in England and Wales were, nevertheless, about 50 per cent higher in the most affluent compared with the most deprived populations.[2] The less marked influence of affluence on mortality rates is consistent with higher survival rates in melanoma patients from

more affluent backgrounds,[29,30] partly, but not wholly, reflecting earlier detection.[31] The maps show some correlation between areas with low incidence of melanoma and areas with high levels of deprivation (see Appendix F). Notably, for both sexes, there were pockets of low melanoma incidence and high levels of deprivation in London; around Wolverhampton and Birmingham; Doncaster; Tees; and Gateshead and South Tyneside (Map 14.1).

It is not clear to what extent (if any) the variations in incidence or mortality in some parts of the UK reflect different proportions of high-risk, fair-skinned people and/or low-risk ethnic minorities. More detailed analysis is needed to quantify the role of such factors, along with variation in risk behaviour (sun exposure), ascertainment, and distribution of stage at diagnosis or other measures of early detection, in explaining the geographic patterns seen in the incidence and mortality of melanoma.

References

1. Coleman MP, Esteve J, Damiecki P, Arslan A et al. *Trends in Cancer Incidence and Mortality.* IARC Scientific Publications No. 121. Lyon: International Agency for Research on Cancer, 1993.

2. Quinn MJ, Babb PJ, Brock A, Kirby L et al. *Cancer Trends in England and Wales 1950-1999.* Studies on Medical and Population Subjects No. 66. London: The Stationery Office, 2001.

3. Harris V, Sandridge AL, Black RJ, Brewster DH et al. *Cancer Registration Statistics Scotland, 1986-1995.* Edinburgh: ISD Publications, 1998.

4. Black RJ, Macfarlane GJ, Maisonneuve P, Boyle P. *Cancer Incidence and Mortality in Scotland 1960-89.* Edinburgh: ISD Publications, 1995.

5. Armstrong BK, Kricker A. Cutaneous melanoma. *Cancer Surveys* 1994; 19-20: 219-240.

6. Jemal A, Devesa SS, Hartge P, Tucker MA. Recent trends in cutaneous melanoma incidence among whites in the United States. *Journal of the National Cancer Institute* 2001; 93: 678-683.

7. Koh HK, Norton LA, Geller AC, Sun T et al. Evaluation of the American Academy of Dermatology's National Skin Cancer Early Detection and Screening Program. *Journal of the American Academy of Dermatology* 1996; 34: 971-978.

8. Melia J, Cooper EJ, Frost T. Cancer Research Campaign health education programme to promote the early detection of cutaneous malignant melanoma. II. Characteristics and incidence of melanoma. *British Journal of Dermatology* 1995; 132: 414-421.

9. ONS. Cancer Survival: England and Wales, 1991-2001. March 2004. Available at *http://www.statistics.gov.uk/statbase/ssdataset.asp?vlnk=7899.*

10. ISD Scotland. *Trends in Cancer Survival in Scotland, 1977-2001.* Edinburgh: ISD Publications, 2004.

11. National Cancer Registry of Ireland. *Cancer in Ireland, 1994 to 1998: Incidence, mortality, treatment and survival.* Cork: National Cancer Registry, 2001.

12. Fitzpatrick D, Gavin A, Middleton R, Catney D. *Cancer in Northern Ireland 1993-2001: A Comprehensive Report.* Belfast: Northern Ireland Cancer Registry, 2004.

13. Sant M, Areleid T, Berrino F, Bielska Lasota M et al. EUROCARE-3: survival of cancer patients diagnosed 1990-1994 - results and commentary. *Annals of Oncology* 2003; 14 Suppl 5: v61-v118.

14. MacKie RM, Bray CA, Hole DJ, Morris A et al. Incidence of and survival from malignant melanoma in Scotland: an epidemiological study. *Lancet* 2002; 360: 587-591.

15. Coebergh JWW, Janssen-Heijenen MLG, Louwman WJ, Voogd AC. *Cancer incidence and survival in the south of the Netherlands, 1955-1999 and incidence in the north of Belgium, 1996-1998.* Eindhoven: Comprehensive Cancer Centre South, 2001.

16. Bulliard JL. Site-specific risk of cutaneous malignant melanoma and pattern of sun exposure in New Zealand. *International Journal of Cancer* 2000; 85: 627-632.

17. Garbe C, Buttner P, Bertz J, Burg G et al. Primary cutaneous melanoma. Prognostic classification of anatomic location. *Cancer* 1995; 75: 2492-2498.

18. Balch CM, Soong SJ, Gershenwald JE, Thompson JF et al. Prognostic factors analysis of 17,600 melanoma patients: validation of the American Joint Committee on Cancer melanoma staging system. *Journal of Clinical Oncology* 2001; 19: 3622-3634.

19. Streetly A, Markowe H. Changing trends in the epidemiology of malignant melanoma: gender differences and their implications for public health. *International Journal of Epidemiology* 1995; 24: 897-907.

20. Northern and Yorkshire Cancer Registry and Information Service. *Registration of skin cancer in Yorkshire. A study on the completeness and validity of cancer registration data.* Leeds: NYCRIS, 2001.

21. Doherty VR, MacKie RM. Experience of a public education programme on early detection of cutaneous malignant melanoma. *British Medical Journal* 1988; 297: 388-391.

22. Melia J, Moss S, Coleman D. The relation between mortality from malignant melanoma and early detection in the Cancer Research Campaign Mole Watcher Study. *British Journal of Cancer* 2001; 14: 803-807.

23. Tucker MA, Goldstein AM. Melanoma etiology: where are we? *Oncogene* 2003; 22: 3042-3052.

24. Holman CD, Armstrong BK, Heenan PJ. A theory of the etiology and pathogenesis of human cutaneous malignant melanoma. *Journal of the National Cancer Institute* 1983; 71: 651-656.

25. Armstrong BK, Kricker A. How much melanoma is caused by sun exposure? *Melanoma Research* 1993; 3: 395-401.

26. Gilchrest BA, Eller MS, Geller AC, Yaar M. The pathogenesis of melanoma induced by ultraviolet radiation. *New England Journal of Medicine* 1999; 340: 1341-1348.

27. Spek-Keijser LM, van der Rhee HJ, Toth G, Van Westering R et al. Site, histological type, and thickness of primary cutaneous malignant melanoma in western Netherlands since 1980. *British Journal of Dermatology* 1997; 136: 565-571.

28. Boyle P, Autier P, Bartelink H, Baselga J et al. European Code Against Cancer and scientific justification: third version (2003). *Annals of Oncology* 2003; 14: 973-1005.

29. MacKie RM, Hole DJ. Incidence and thickness of primary tumours and survival of patients with cutaneous malignant melanoma in relation to socioeconomic status. *British Medical Journal* 1996; 312: 1125-1128.

30. Coleman MP, Rachet B, Woods LM, Mitry E et al. Trends and socioeconomic inequalities in cancer survival in England and Wales up to 2001. *British Journal of Cancer* 2004; 90: 1367-1373.

31. Coleman MP, Babb P, Damiecki P, Grosclaude P et al. *Cancer Survival Trends in England and Wales, 1971-1995: Deprivation and NHS Region.* Studies on Medical and Population Subjects No. 61. London: The Stationery Office, 1999.

Chapter 15

Multiple myeloma

Peter Adamson

Summary

- In the UK and Ireland in the 1990s, multiple myeloma accounted for around 1 in 80 diagnosed cases of cancer and 1 in 70 deaths from cancer.

- There was relatively little geographical variation, although what patterns there were appeared to be similar for incidence and mortality.

- Incidence and mortality were noticeably higher than average in Northern Ireland and Ireland in males, and slightly lower than average in the North West in both sexes.

- There is no apparent link between the observed geographical variations in incidence and any known or suspected risk factor for the disease.

- The geographical variations in incidence also appear not to be related to deprivation, and hence to any factors, including lifestyle, for which deprivation may be a marker.

Introduction

Multiple myeloma is a cancer which affects plasma cells in the bone marrow. In myeloma a single cell becomes malignant and produces a very large number of identical copies. Plasma cells produce antibodies which the body needs to fight infection. In myeloma normal antibody levels are reduced and this can lead to a susceptibility to life-threatening infections. It has some similarities with chronic lymphocytic leukaemia (CLL), but in myeloma, malignant cells rarely move from the bone marrow to enter the bloodstream.

Incidence and mortality

In the 1990s there were about 1,800 newly diagnosed cases of multiple myeloma each year in males, and about 1,700 in females in the UK and Ireland. Myeloma accounted for 1.3 per cent of all cases of cancer per year in males and 1.2 per cent in females. Overall, the age-standardised incidence rates were 5.5 and 3.7 per 100,000 in males and females, respectively, a male-to-female ratio of 1.5:1. Myeloma is a disease of the elderly and rarely affects people under the age of 40. Incidence increased with age, with the highest rates in people in their 70s and 80s. Above the age of 40, all age-specific rates were markedly higher in men than in women.

Multiple myeloma accounted for about 1,300 deaths in both males and females each year in the UK and Ireland in the 1990s. Of all deaths from cancer, in males 1.5 per cent were due to myeloma, and in females 1.7 per cent. The age-standardised mortality rates were 3.7 per 100,000 in males and 2.6 in females, giving a male-to-female ratio of 1.4:1 – similar to that for incidence rates. Age-specific mortality followed the same pattern as for incidence, with the highest rates in those aged over 85, and higher rates in men than women. The average mortality-to-incidence ratio for the UK and Ireland was about 0.7 for both sexes.

Incidence and mortality trends

From 1971 to the mid-1980s, incidence increased steeply in both sexes; the rate of increase slowed down in males and rates levelled off in females in the 1990s. In the mid-1990s, incidence rates were about 80 per cent higher in males and about 70 per cent higher in females than in the early 1970s. This rise was mainly due to large increases in incidence in those aged 75 and over.[1] The pattern for mortality was similar, with steep increases in rates, particularly in the elderly, up to the 1980s, followed by a slower increase or levelling off.[1] The increases in incidence and mortality rates were markedly higher in males than females over the period. The increases in mortality in the elderly can be partially explained by better diagnostic practice (which may also account for some of the increase in incidence) and improvements in the accuracy of death certification.[2]

Survival

Survival from myeloma is remarkable in that there have been improvements in one-year, five-year and median survival but little or no improvement in long-term survival. Relative survival for patients diagnosed in England and Wales in 1996-99 was around 60 and 23 per cent, at one and five years respectively,[3] compared with 42 and 13 per cent for patients diagnosed in 1971-75.[2] Ten-year survival has changed little from the 1970s, increasing from about 5 per cent to 7 per cent at the end of 1995.[2] Median survival improved from around two years in the 1980s and early 1990s to around four years in the late 1990s. The most modern treatments can induce a complete response in patients, but remissions are not durable and the cure rate is low.[4]

Geographical patterns in incidence

As myeloma is an uncommon malignancy and the average numbers of cases and deaths per year are small (Table B15.2), the variations in incidence and mortality, particularly at health authority level, should be interpreted with caution.

Within the countries of the UK and Ireland, the highest incidence rates in males were in Northern Ireland and Ireland (Figure 15.1), where the rates were 15-20 per cent higher than the average for males. Incidence was slightly above average in the South West region of England. In females, incidence rates appeared to be slightly higher than average in Scotland, Northern Ireland, Ireland and the South West region of England. Rates in the North West of England were below average for both sexes. Within countries and regions there was relatively little variation in incidence at the health authority level, although the range of values was wider in males than in females (Table B15.1).

The maps for incidence (Map 15.1) do not show any particularly obvious geographical patterns apart from those already described at the country and regional level. Incidence was generally higher than average for both sexes in Northern Ireland, Ireland (except for the South Eastern area), southern Scotland and some of the health authorities in the northern part of the South East region of England. In much of the North West and North East, and the eastern part of the South East region incidence was generally below average.

Geographical patterns in mortality

At the country and regional level, the pattern of mortality rates was similar to that for incidence (Figure 15.2). Rates were noticeably higher than average in males in Northern Ireland and both sexes in Ireland, and slightly lower than average in both sexes in the North West of England. In all other countries and regions, rates were close to the average. At the health authority level, rates showed a marked difference from the average in very few areas (Figure 15.4), with no clusters of relatively high or low rates except in males in Northern Ireland and Ireland – a similar pattern to that in incidence. As with incidence, there was otherwise no obvious geographical pattern in the variation of mortality rates (Map 15.2), and there was less correlation in mortality between the sexes than for incidence, with the exception of Ireland.

Risk factors and aetiology

People with monoclonal gammopathy of unknown significance (MGUS) are predisposed to developing multiple myeloma. MGUS is a disorder related to myeloma. The majority of people diagnosed with MGUS never develop symptoms but require regular follow-up checks. Each year about 2 per cent of people with MGUS will go on to develop myeloma or a related condition.[5] Additionally, autoimmune disorders, chronic immune stimulation, exposure to ionising radiation, occupational exposures, exposure to hair-colouring products, consumption of alcohol and tobacco, and a family history of myeloma and other diseases have been examined as possible risk factors for multiple myeloma.[5] Evidence from epidemiological studies supports a causal role for autoimmune disorders, exposure to ionising radiation, and occupational exposure to benzene and pesticides in the development of myeloma,[5] although these risk factors would affect very few people and are therefore unlikely to explain the observed geographical variations in incidence and mortality.

Socio-economic deprivation

In England and Wales in the early 1990s there was no relationship between either the incidence of, or mortality from, multiple myeloma and deprivation, measured using the Carstairs index.[1] For patients diagnosed in the 1970s and 1980s, there was virtually no difference in survival between affluent and deprived groups,[2] but for patients diagnosed in the late 1990s, five-year relative survival was higher in the most affluent group by about 5 percentage points in men and 8 in women (for women this deprivation gap was statistically significant).[6] The observed geographical variations in both incidence and mortality are not related to any factor for which deprivation may be a relevant marker.

References

1. Quinn MJ, Babb PJ, Brock A, Kirby L et al. *Cancer Trends in England and Wales 1950-1999*. Studies on Medical and Population Subjects No. 66. London: The Stationery Office, 2001.

2. Coleman MP, Babb P, Damiecki P, Grosclaude P et al. *Cancer Survival Trends in England and Wales, 1971-1995: Deprivation and NHS Region*. Studies on Medical and Population Subjects No. 61. London: The Stationery Office, 1999.

3. ONS. Cancer Survival: England and Wales, 1991-2001. March 2004. Available at *http://www.statistics.gov.uk/statbase/ssdataset.asp?vlnk=7899*.

4. Stewart BW, Kleihues P. *World Cancer Report*. Lyon: IARC Press, 2003.

5. Herrinton LJ, Weiss NS, Olshan AF. Multiple myeloma. In: Schottenfeld D, Fraumeni Jnr JF (eds) *Cancer Epidemiology and Prevention, second edition*. New York: Oxford University Press, 1996.

6. Coleman MP, Rachet B, Woods LM, Mitry E et al. Trends and socioeconomic inequalities in cancer survival in England and Wales up to 2001. *British Journal of Cancer* 2004; 90: 1367-1373.

Figure **15.1**

**Multiple myeloma: incidence by sex, country, and region of England
UK and Ireland 1991-99[1]**

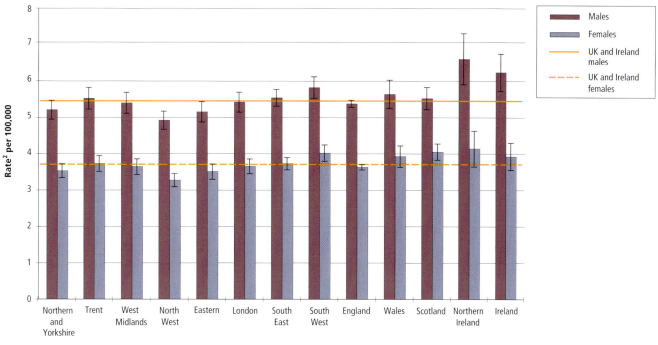

1 Northern Ireland 1993-99, Ireland 1994-99

2 Age standardised using the European standard population, with 95% confidence interval

Figure **15.2**

**Multiple myeloma: mortality by sex, country, and region of England
UK and Ireland 1991-2000[1]**

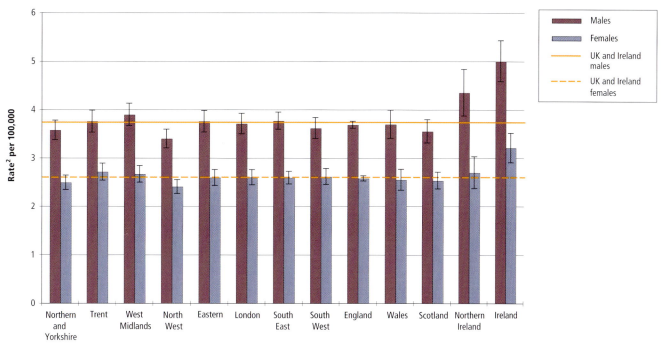

1 Scotland 1991-99, Ireland 1994-2000

2 Age standardised using the European standard population, with 95% confidence interval

Figure **15.3a**

**Multiple myeloma: incidence by health authority within country, and region of England
Males, UK and Ireland 1991-99[1]**

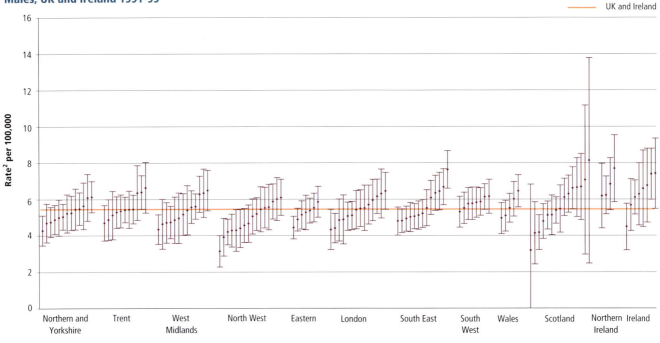

1 Northern Ireland 1993-99, Ireland 1994-99

2 Age standardised using the European standard population, with 95% confidence interval

Figure **15.3b**

**Multiple myeloma: incidence by health authority within country, and region of England
Females, UK and Ireland 1991-99[1]**

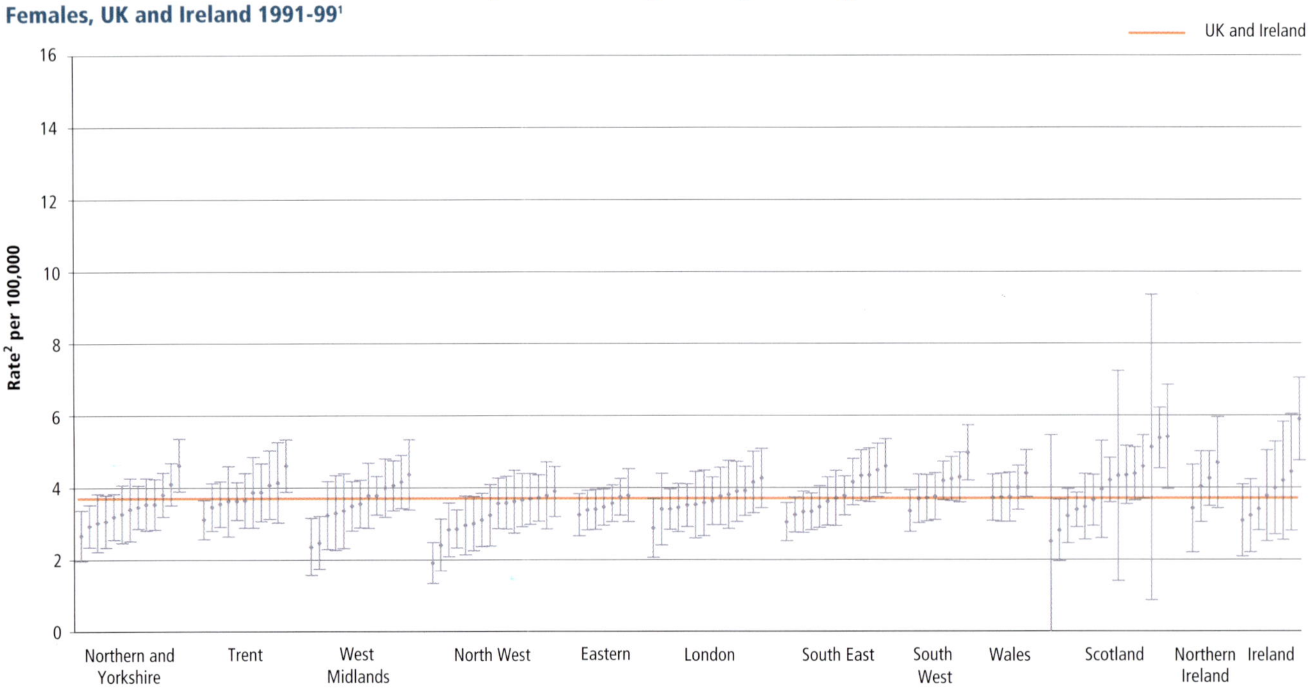

1 Northern Ireland 1993-99, Ireland 1994-99

2 Age standardised using the European standard population, with 95% confidence interval

Figure **15.4a**

Multiple myeloma: mortality by health authority within country, and region of England
Males, UK and Ireland 1991-2000[1]

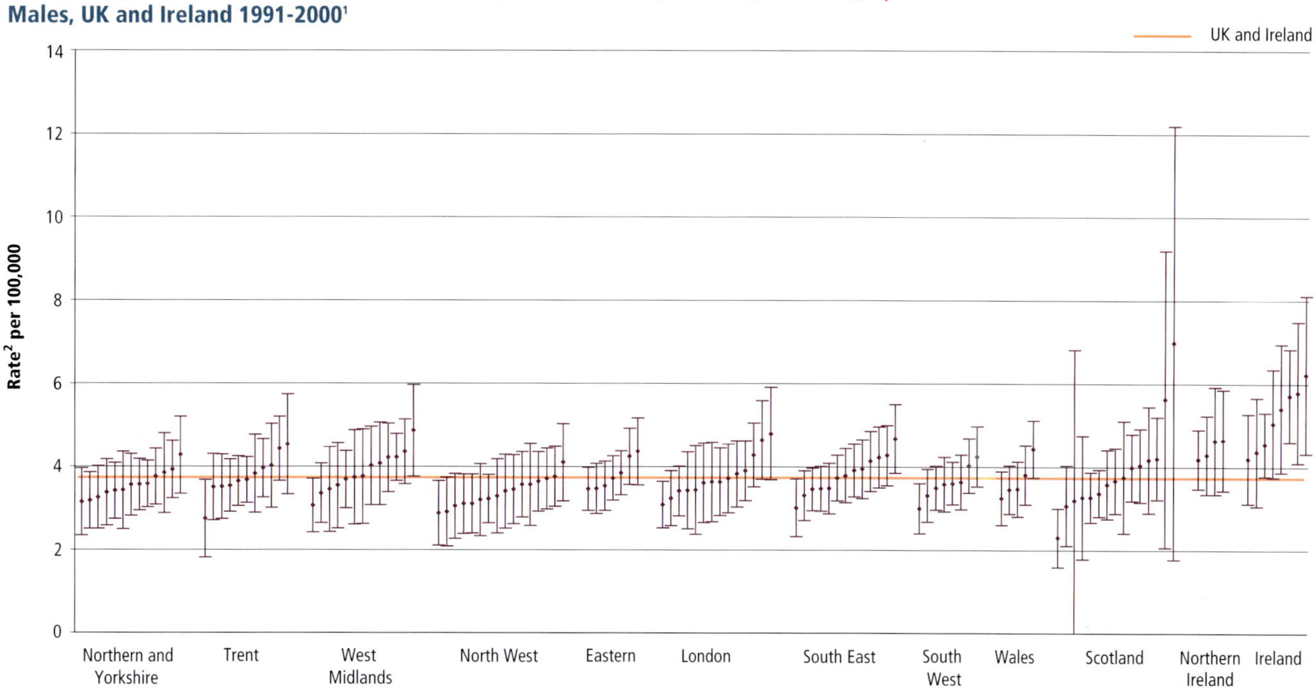

1 Scotland 1991-99, Ireland 1994-2000

2 Age standardised using the European standard population, with 95% confidence interval

Figure **15.4b**

Multiple myeloma: mortality by health authority within country, and region of England
Females, UK and Ireland 1991-2000[1]

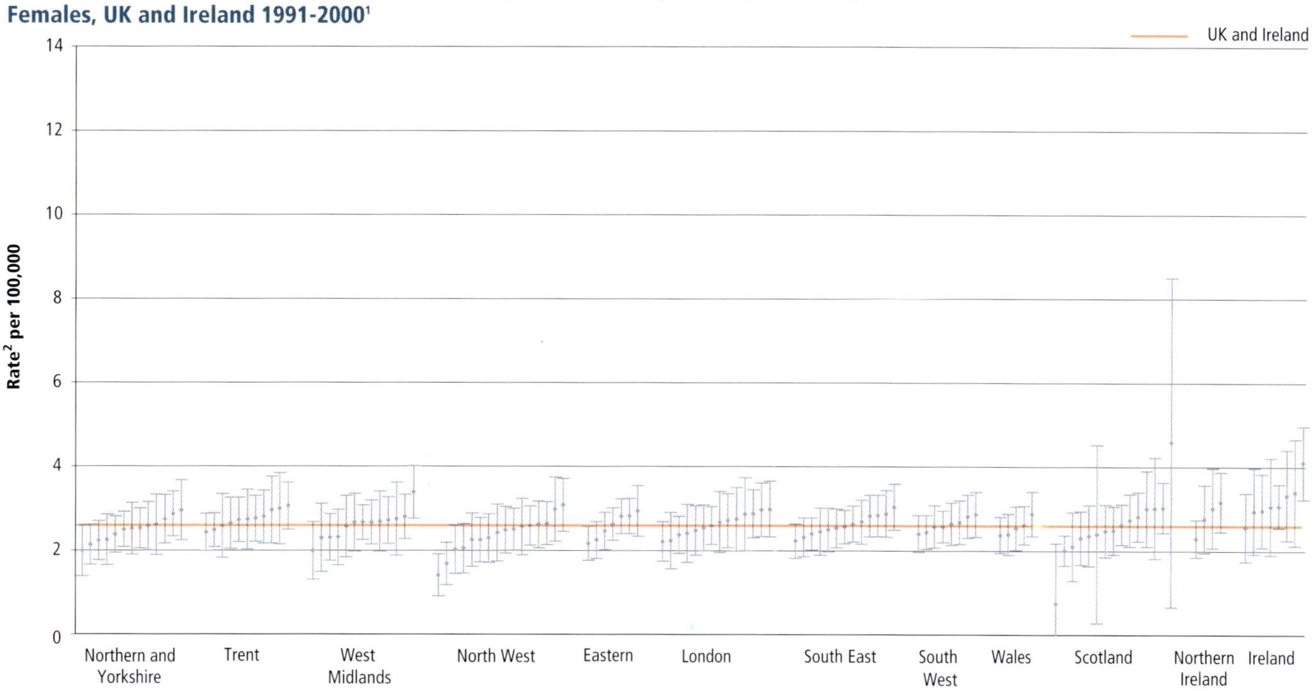

1 Scotland 1991-1999, Ireland 1994-2000

2 Age standardised using the European standard population, with 95% confidence interval

Map **15.1a**

Multiple myeloma: incidence* by health authority
Males, UK and Ireland 1991-99

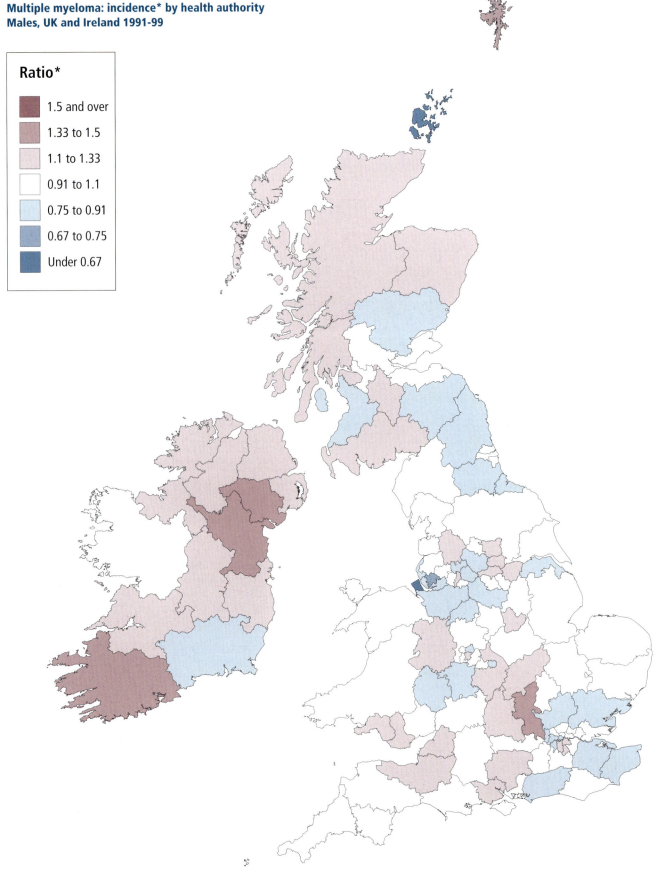

Ratio*

	1.5 and over
	1.33 to 1.5
	1.1 to 1.33
	0.91 to 1.1
	0.75 to 0.91
	0.67 to 0.75
	Under 0.67

*Ratio of directly age-standardised rate in health authority to UK and Ireland average

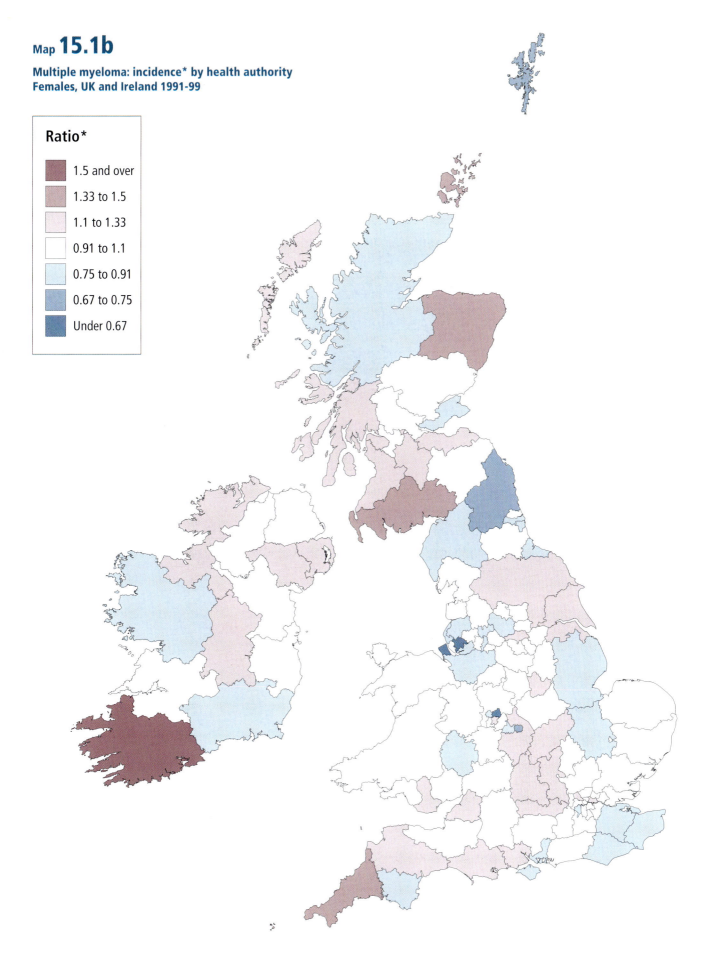

Map **15.1b**

**Multiple myeloma: incidence* by health authority
Females, UK and Ireland 1991-99**

Ratio*

- 1.5 and over
- 1.33 to 1.5
- 1.1 to 1.33
- 0.91 to 1.1
- 0.75 to 0.91
- 0.67 to 0.75
- Under 0.67

*Ratio of directly age-standardised rate in health authority to UK and Ireland average

Map 15.2a

**Multiple myeloma: mortality* by health authority
Males, UK and Ireland 1991-2000**

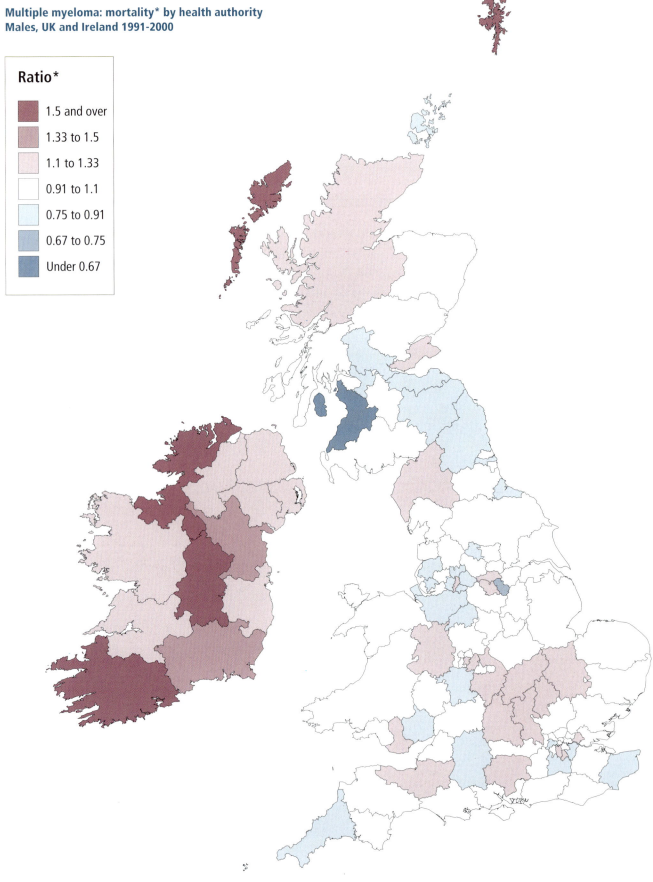

Ratio*

	1.5 and over
	1.33 to 1.5
	1.1 to 1.33
	0.91 to 1.1
	0.75 to 0.91
	0.67 to 0.75
	Under 0.67

*Ratio of directly age-standardised rate in health authority to UK and Ireland average

Map **15.2b**

**Multiple myeloma: mortality* by health authority
Females, UK and Ireland 1991-2000**

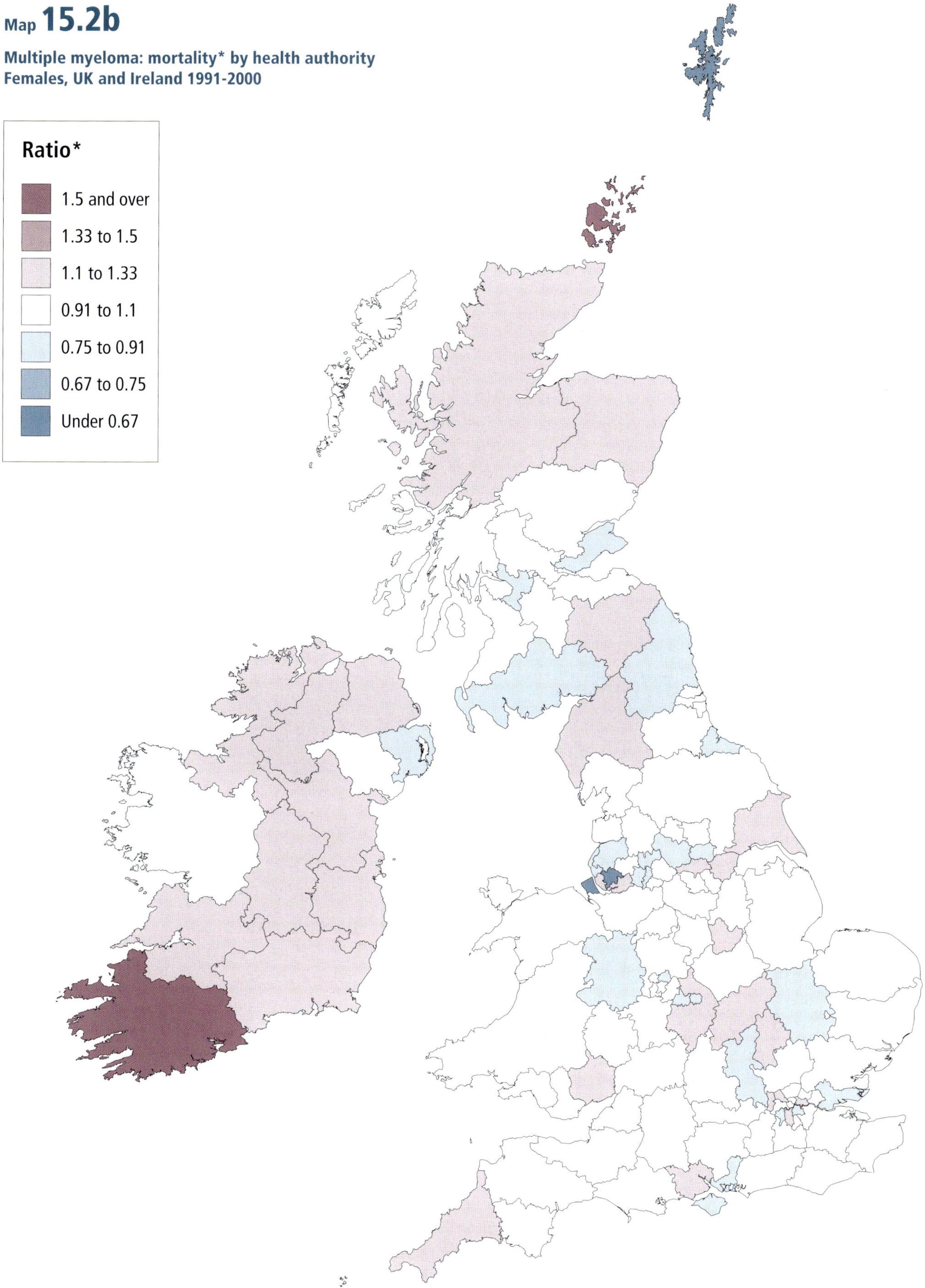

Ratio*

	1.5 and over
	1.33 to 1.5
	1.1 to 1.33
	0.91 to 1.1
	0.75 to 0.91
	0.67 to 0.75
	Under 0.67

*Ratio of directly age-standardised rate in health authority to UK and Ireland average

Chapter 16

Non-Hodgkin's lymphoma

Ray Cartwright, Helen Wood, Mike Quinn

Summary

- In the UK and Ireland in the 1990s, non-Hodgkin's lymphoma accounted for 1 in 30 cases of cancer and 1 in 40 deaths from cancer.

- There was the suggestion of a north-south divide in incidence across England, with higher than average rates in London and the south, and lower incidence in the midlands and north. Incidence was also higher than average in Northern Ireland and Scotland.

- The observed pattern in mortality was similar to, but less clear-cut than that for incidence.

- There appears to be a weak (negative) link between incidence and deprivation, with slightly higher rates in more affluent areas, although there is no known causative factor for which affluence could be a marker.

- It is unlikely that any of the known risk factors for developing NHL could explain the observed geographical variations in incidence.

Introduction

Non-Hodgkin's lymphoma (NHL) is a group of conditions, all of which are malignancies of the cells on the lymphocyte developmental pathway, part of the immune system. The different histological sub-types vary widely in their clinical behaviour, progress and management, and there is some evidence that the different sub-groups of NHL have differing epidemiological features.[1] It is not known to what extent the amalgamation of these different sub-types into a single diagnostic group obscures the geographical distribution of these diseases or confuses the aetiology. NHL is the most common malignancy among the leukaemias and lymphomas. It typically arises in lymph node tissue, but in 15-20 per cent of patients the tumour develops in a site other than a node, for example, in bone, stomach or intestines, brain or breast.[2]

Incidence and mortality

In the UK and Ireland in the 1990s, there were roughly 4,500 new cases of NHL diagnosed in males each year and 4,000 in females (a male-to-female ratio of around 1.1:1). Incidence rates increased markedly with age in both sexes, but at all ages the rates in females were roughly one third lower than in males, particularly in older age groups. The overall age-standardised incidence rates were 14.2 and 9.9 per 100,000 for males and females, respectively (a male-to-female ratio of 1.4:1).

In the 1990s, around 2,400 males and 2,200 females died from NHL each year. In common with the incidence data, the male-to-female ratio of the number of deaths was 1.1:1. The overall age-standardised mortality rates were 7.3 per 100,000 for males and 4.8 for females (a ratio of 1.5:1). The overall mortality-to-incidence ratio for the UK and Ireland was around 0.50 for both sexes (Table B16.1).

Incidence and mortality trends

In England and Wales, age-standardised incidence rates increased around three fold in both sexes between 1971 and the late 1990s. There was a steady increase throughout the 1970s and 1980s, with the largest increases occurring among the elderly.[2] The cumulative risk for 30-74 year-old males from the 1915 and 1940 cohorts in three UK cancer registry areas[3] showed marked increases by cohort of 14, 24 and 50 per cent (Birmingham, Scotland and South Thames, respectively) between 1973 and 1987. There were similar increases in cumulative risk for females. The apparently marked variation in rates of change between regions indicates that geographical patterns of incidence are also likely to have changed over time.

Trends in mortality approximately followed those in incidence, with particularly large increases in the elderly. After a steep rise in the 1980s, mortality rates levelled off in the mid-1990s.[2] Increasing mortality from NHL occurred almost entirely in cohorts born before 1920. There was then little change, until a decline beginning with the cohort born in 1945-49.[4] Mortality increased by over 70 per cent in both sexes in each region between the early 1960s and early 1990s, with no obvious geographical pattern.[4]

The increased incidence of NHL around the world has been documented by numerous publications since the 1980s. It has occurred in mainly white populations in parts of western Europe, North America[5] and Australia, with annual increases in the period 1985-92 of over 4 per cent in parts of Europe.[1]

Closer examination suggests that the changes in incidence over time are not consistent for either the different histological sub-types[1] or the primary site of diagnosis of NHL, with skin lymphoma, for example, showing some of the largest increases.[6] There are some recent indications that these trends may be slowing down. The underlying basis of these trends could be either the changing ability of laboratories to diagnose the condition more accurately, or changes in some unknown underlying biological or environmental processes (see below).

Survival

Relative survival from NHL was around 70 per cent at one year and 50 per cent at five years after diagnosis in patients diagnosed in 1996-99 in England and Wales.[7] Five-year survival was similar in Northern Ireland,[8] Ireland[9] and Scotland[10] for patients diagnosed in the 1990s. Survival rates in England and Wales were close to the European average for patients diagnosed in 1990-94.[11] In England and Wales since the 1970s, five-year survival has improved by around 14 percentage points for both men and women. Prior to the 1970s, most lymphomas were fatal, but the development of effective combination chemotherapy has resulted in the cure of many advanced tumours.[12]

Geographical patterns in incidence

Noticeably higher than average rates of NHL occurred in London, the South West of England and Northern Ireland in males (around 15 per cent above the UK and Ireland average); and in Scotland and Northern Ireland in females (17 and 27 per cent, respectively) (Figure 16.1). Rates were lower than average for both males and females in the Northern and Yorkshire; Trent; North West and West Midlands regions of England.

Within countries and regions there was some variation in incidence rates, among both males and females (Figure 16.3). Variation between the health authorities with the highest and lowest incidence within a country or region ranged from around 20 per cent in Ireland and the Eastern region of England to over 40 per cent in London and Trent (Table B16.1). Most of the rates for health authorities were, however, based on relatively small numbers of cases and it is therefore difficult to interpret this variability (Table B16.2). In both males and females the vast majority of rates at the health authority level did not differ significantly from the average. But in the Northern and Yorkshire; Trent; West Midlands; and North West regions incidence rates in most of the health authorities were below average; and in London, the South East and South West of England, Scotland and Northern Ireland the rates in many health authorities were above average. The maps (Map 16.1) dramatically reflect this pattern with virtually all of the health authorities with slightly raised incidence being in the south and south west of England, and virtually all of those with slightly lower rates being in the midlands and north of the country. The maps also emphasise the generally higher rates in Scotland and Northern Ireland, particularly for females.

Geographical patterns in mortality

Mortality rates by country and region of England broadly reflected those for incidence (Figure 16.2), although there was less variation. As for incidence, mortality rates were below average for both sexes in the Northern and Yorkshire; North West; Trent; and West Midlands regions of England. However, mortality rates were markedly higher than average only in London for males and in Scotland for females.

At the health authority level there was less variation in mortality than incidence, and the rates in most areas were not more than 10 per cent different from the average (Figure 16.4). Reflecting this, the maps for mortality (Map 16.2) show a less clear pattern than those for incidence, although again those areas in England that had slightly higher than average mortality were predominantly in the south and south west, and those areas with slightly below average mortality were in the north. The patterns of health authorities with higher incidence in Scotland and Northern Ireland were reflected, to a lesser degree, in the mortality rates, and the differences between these and the rates in the midlands and north of England were again visible, although less marked than for incidence.

Risk factors and aetiology

There are three broad lines of research linked to studies of the causation of NHL: studies of altered immunity; occupational investigations; and lifestyle studies.

The studies of altered immunity have resulted in some convincing associations. These include an excess of NHL cases occurring in certain inherited syndromes that are characterised by failures of function within the immune system, such as ataxia telangectasia.[13] In addition, people with certain chronic illnesses (for example, those with renal disease or renal transplantation) who receive therapeutic immunosuppressant drugs, have a considerable excess risk of NHL (20 fold and more).[14] Those who suffer from viral immunosuppression also have a higher risk of NHL. These include younger people in parts of Africa with Burkitt's lymphoma, which is partly linked to Epstein-Barr virus infection in endemic malarial areas. In addition, individuals with chronic HIV infection have up to a 60-fold excess risk of developing NHL.[15,16]

Occupational associations are much weaker (2- to 3-fold risk) and are linked to chemicals that may have an adverse effect on the immune system. The most investigated area is that of certain types of agricultural exposures, in particular herbicides and other agrichemicals.[17] Despite numerous studies, no specific association has been found that is thought to be causal.

(continued on page 182)

Figure **16.1**

**Non-Hodgkin's lymphoma: incidence by sex, country, and region of England
UK and Ireland 1991-99[1]**

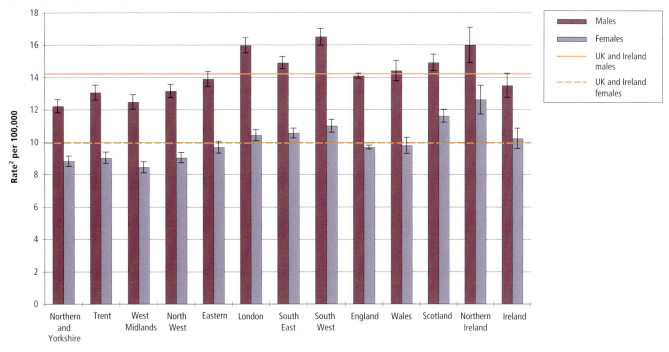

1 Northern Ireland 1993-99, Ireland 1994-99

2 Age standardised using the European standard population, with 95% confidence interval

Figure **16.2**

**Non-Hodgkin's lymphoma: mortality by sex, country, and region of England
UK and Ireland 1991-2000[1]**

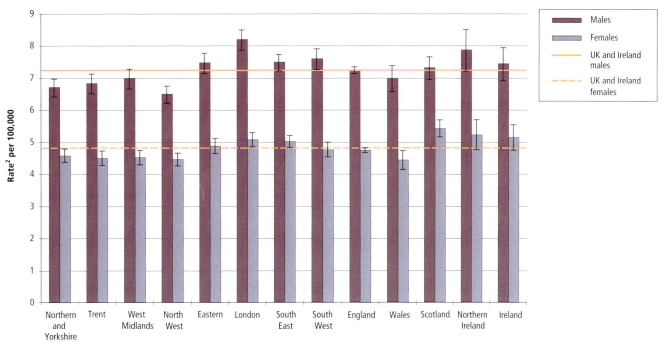

1 Scotland 1991-99, Ireland 1994-2000

2 Age standardised using the European standard population, with 95% confidence interval

Figure **16.3a**

Non-Hodgkin's lymphoma: incidence by health authority within country, and region of England
Males, UK and Ireland 1991-99[1]

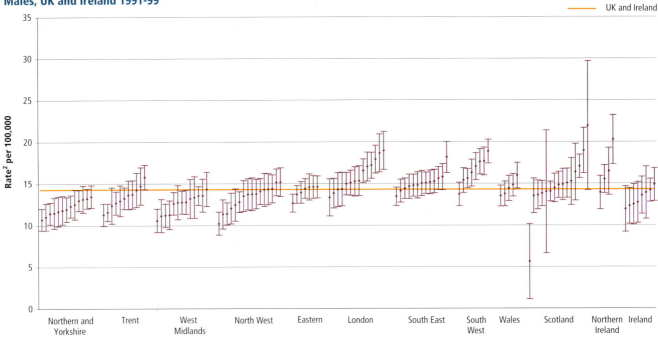

1 Northern Ireland 1993-99, Ireland 1994-99

2 Age standardised using the European standard population, with 95% confidence interval

Figure **16.3b**

Non-Hodgkin's lymphoma: incidence by health authority within country, and region of England
Females, UK and Ireland 1991-99[1]

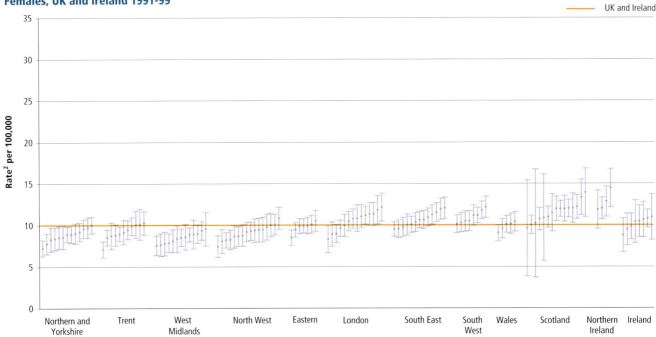

1 Northern Ireland 1993-99, Ireland 1994-99

2 Age standardised using the European standard population, with 95% confidence interval

Figure **16.4a**

Non-Hodgkin's lymphoma: mortality by health authority within country, and region of England
Males, UK and Ireland 1991-2000[1]

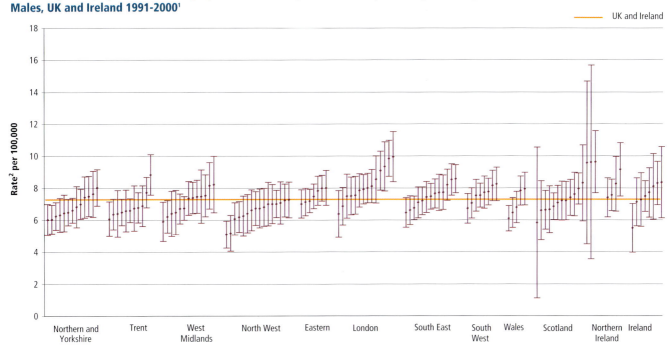

1 Scotland 1991-99, Ireland 1994-2000

2 Age standardised using the European standard population, with 95% confidence interval

Figure **16.4b**

Non-Hodgkin's lymphoma: mortality by health authority within country, and region of England
Females, UK and Ireland 1991-2000[1]

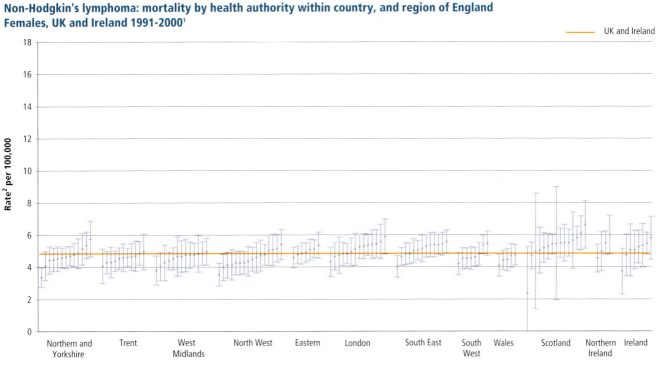

1 Scotland 1991-99, Ireland 1994-2000

2 Age standardised using the European standard population, with 95% confidence interval

Map **16.1a**

**Non-Hodgkin's lymphoma: incidence* by health authority
Males, UK and Ireland 1991-99**

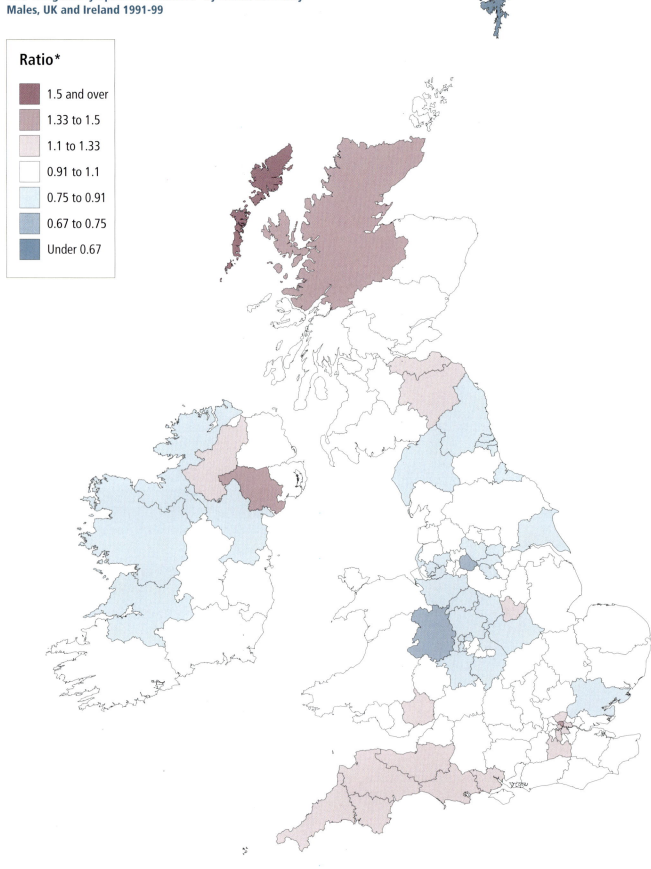

Ratio*

	1.5 and over
	1.33 to 1.5
	1.1 to 1.33
	0.91 to 1.1
	0.75 to 0.91
	0.67 to 0.75
	Under 0.67

*Ratio of directly age-standardised rate in health authority to UK and Ireland average

Map **16.1b**

Non-Hodgkin's lymphoma: incidence* by health authority
Females, UK and Ireland 1991-99

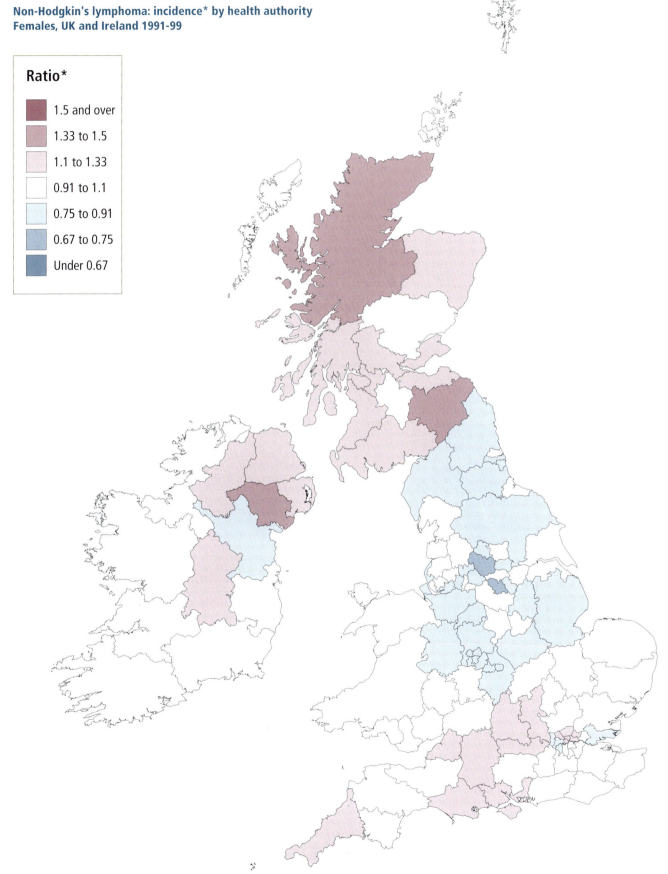

Ratio*

- 1.5 and over
- 1.33 to 1.5
- 1.1 to 1.33
- 0.91 to 1.1
- 0.75 to 0.91
- 0.67 to 0.75
- Under 0.67

*Ratio of directly age-standardised rate in health authority to UK and Ireland average

Map **16.2a**

**Non-Hodgkin's lymphoma: mortality* by health authority
Males, UK and Ireland 1991-2000**

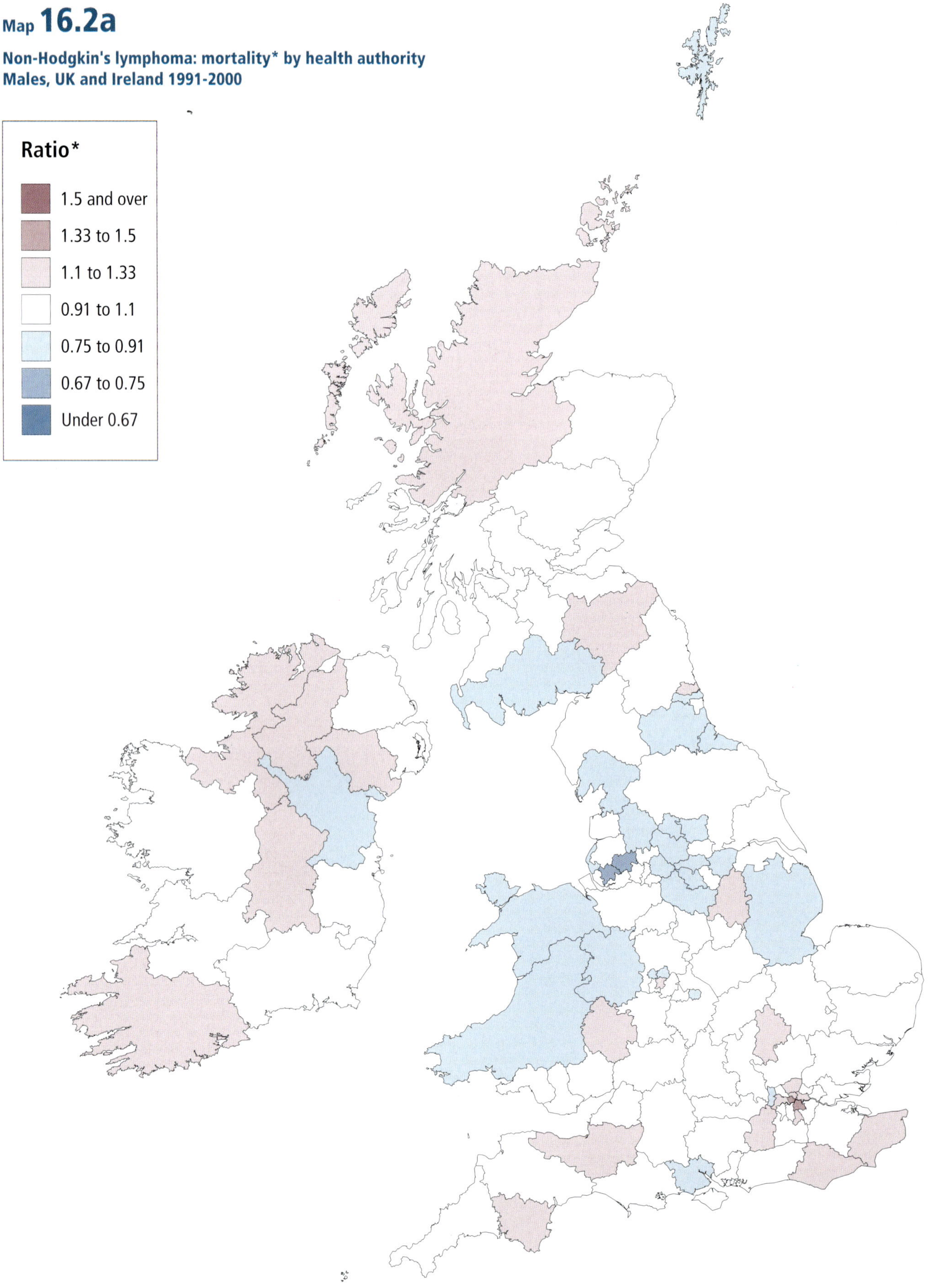

Ratio*

	1.5 and over
	1.33 to 1.5
	1.1 to 1.33
	0.91 to 1.1
	0.75 to 0.91
	0.67 to 0.75
	Under 0.67

*Ratio of directly age-standardised rate in health authority to UK and Ireland average

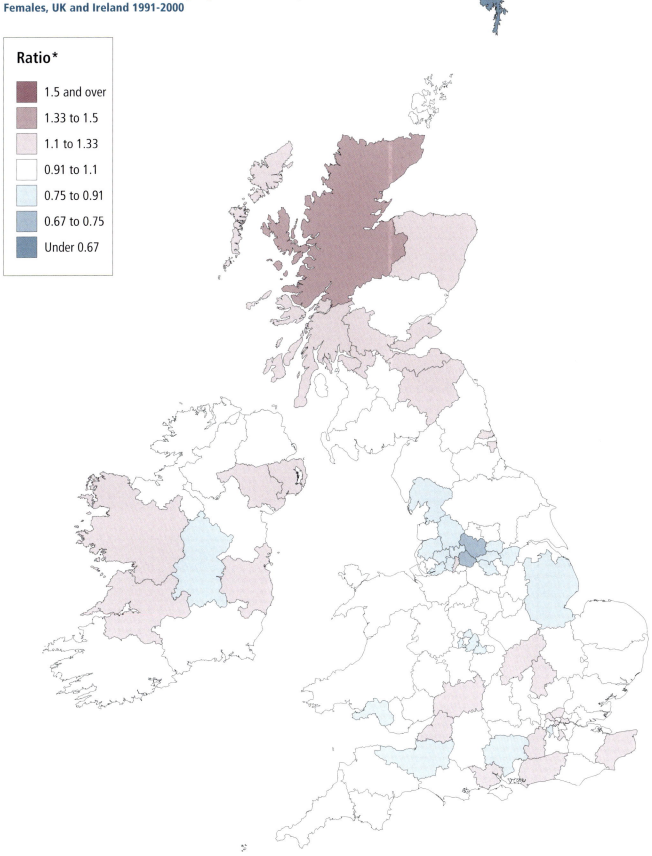

Map **16.2b**

**Non-Hodgkin's lymphoma: mortality* by health authority
Females, UK and Ireland 1991-2000**

Ratio*

	1.5 and over
	1.33 to 1.5
	1.1 to 1.33
	0.91 to 1.1
	0.75 to 0.91
	0.67 to 0.75
	Under 0.67

*Ratio of directly age-standardised rate in health authority to UK and Ireland average

A few studies have shown a link with the petrochemical industry but the results are weaker than those for agrichemicals.[18]

Various lifestyle studies have shown little relationship between exposure to cigarette smoke or to ionising radiation and the risk of NHL. Little else has been published to suggest other possible aetiological factors in the pathogenesis of NHL, although some hypotheses have yet to be investigated. One such hypothesis, for example, is based on the possible adverse immune effects which sunlight exposure might have on lymphocytes circulating in the skin capillaries.[19] However, no strong support has emerged for this in direct studies,[20] although some indirect support is given in ecological studies and one cohort study.[21]

The strong association between immune suppression and NHL relate to a small number of people and so would not have any noticeable impact on the geographical variation in incidence. Similarly, in the UK and Ireland in the 1990s, agricultural exposures would have affected a very small proportion of the total workforce, and they are unlikely to account for the observed geographical distribution of NHL. The impact of HIV on rates of NHL in males has not been evaluated in the UK or Ireland.

Socio-economic deprivation

The incidence of NHL is only very weakly linked to deprivation, with slightly higher rates in the affluent, and there is no association between mortality from NHL and deprivation.[2] Deprivation, or any aetiological factor for which it may be a marker, would therefore not be expected to affect geographical variation to any large degree. This indicates that the dichotomy in incidence rates in England – with virtually all of the higher than average rates occurring in the south and south west, and virtually all of the lower rates occurring in both urban and rural areas in the midlands and north (Map 16.1) – suggests the involvement of factors other than those linked to deprivation. Survival from NHL has consistently been lower among deprived groups:[12] for patients diagnosed in 1996-99, there was a gap of 5-7 percentage points in five-year survival between the least and most affluent groups.[22] This partly explains why less variation was observed in mortality than incidence rates, as the lower incidence of NHL in the midlands and north of England would have been offset by worse survival, resulting in mortality rates that were closer to the average.

References

1. Cartwright R, Brincker H, Carli PM, Clayden D et al. The rise in incidence of lymphomas in Europe 1985-1992. *European Journal of Cancer* 1999; 35: 627-633.

2. Quinn MJ, Babb PJ, Brock A, Kirby L et al. *Cancer Trends in England and Wales 1950-1999.* Studies on Medical and Population Subjects No. 66. London: The Stationery Office, 2001.

3. Coleman MP, Esteve J, Damiecki P, Arslan A et al. *Trends in Cancer Incidence and Mortality.* IARC Scientific Publications No. 121. Lyon: International Agency for Research on Cancer, 1993.

4. Swerdlow AJ, dos Santos Silva I, Doll R. *Cancer Incidence and Mortality in England and Wales: Trends and Risk Factors.* Oxford: Oxford University Press, 2001.

5. Clarke CA, Glaser SL. Changing incidence of non-Hodgkin lymphomas in the United States. *Cancer* 2002; 94: 2015-2023.

6. Cartwright RA, Gilman EA, Gurney KA. Time trends in incidence of haematological malignancies and related conditions. *British Journal of Haematology* 1999; 106: 281-295.

7. ONS. Cancer Survival: England and Wales, 1991-2001. March 2004. Available at *http://www.statistics.gov.uk/statbase/ ssdataset.asp?vlnk=7899.*

8. Fitzpatrick D, Gavin A, Middleton R, Catney D. *Cancer in Northern Ireland 1993-2001: A Comprehensive Report.* Belfast: Northern Ireland Cancer Registry, 2004.

9. National Cancer Registry of Ireland. *Cancer in Ireland, 1994 to 1998: Incidence, mortality, treatment and survival.* Cork: National Cancer Registry, 2001.

10. ISD Scotland. *Trends in Cancer Survival in Scotland, 1977-2001.* Edinburgh: ISD Publications, 2004.

11. Sant M, Areleid T, Berrino F, Bielska Lasota M et al. EUROCARE-3: survival of cancer patients diagnosed 1990-1994 - results and commentary. *Annals of Oncology* 2003; 14 Suppl 5: v61-v118.

12. Coleman MP, Babb P, Damiecki P, Grosclaude P et al. *Cancer Survival Trends in England and Wales, 1971-1995: Deprivation and NHS Region.* Studies on Medical and Population Subjects No. 61. London: The Stationery Office, 1999.

13. Morrell D, Cromartie E, Swift M. Mortality and cancer incidence in 263 patients with ataxia-telangiectasia. *Journal of the National Cancer Institute* 1986; 77: 89-92.

14. Kinlen L. Immunosuppressive therapy and cancer. *Cancer Surveys* 1982; 1: 565-583.

15. Beral V, Peterman T, Berkelman R, Jaffe H. AIDS-associated non-Hodgkin lymphoma. *Lancet* 1991; 337: 805-809.

16. Eltom MA, Jemal A, Mbulaiteye SM, Devesa SS et al. Trends in Kaposi's sarcoma and non-Hodgkin's lymphoma incidence in the United States from 1973 through 1998. *Journal of the National Cancer Institute* 2002; 94: 1204-1210.

17. Blair A, Cantor KP, Zahm SH. Non-Hodgkin's lymphoma and agricultural use of the insecticide lindane. *American Journal of Industrial Medicine* 1998; 33: 82-87.

18. Cartwright RA. Non-Hodgkin's Lymphoma. In: Hancock B, Selby PJ, Maclennon K, Armitage J (eds) *Malignant Lymphoma.* London: Arnold, 2000.

19. Cartwright R, McNally R, Staines A. The increasing incidence of non-Hodgkin's lymphoma (NHL): the possible role of sunlight. *Leukemia and Lymphoma* 1994; 14: 387-394.

20. van Wijngaarden E, Savitz DA. Occupational sunlight exposure and mortality from non-Hodgkin lymphoma among electric utility workers. *Journal of Occupational and Environmental Medicine* 2001; 43: 548-553.

21. Adami J, Gridley G, Nyren O, Dosemeci M et al. Sunlight and non-Hodgkin's lymphoma: a population-based cohort study in Sweden. *International Journal of Cancer* 1999; 80: 641-645.

22. Coleman MP, Rachet B, Woods LM, Mitry E et al. Trends and socioeconomic inequalities in cancer survival in England and Wales up to 2001. *British Journal of Cancer* 2004; 90: 1367-1373.

Chapter 17

Oesophagus

Helen Wood, David Brewster, Henrik Møller

Summary

- In the UK and Ireland in the 1990s, oesophageal cancer accounted for 1 in 40 diagnosed cases and 1 in 25 deaths from cancer.

- Geographical variations in incidence and mortality were very similar because survival from this cancer is very low.

- There was a north-south divide across Great Britain, with high incidence in Scotland and North West England, and low incidence in the south east of England.

- Incidence was highest in western Scotland, urban areas of North West England, and north Wales – areas associated with high levels of deprivation.

- This pattern is likely to be related to known risk factors for oesophageal cancer – smoking, alcohol consumption, poor nutrition and obesity – which are associated with deprivation.

- There are clear similarities between geographical variations in oesophageal cancer and those in cancers of the larynx, lip, mouth and pharynx, and lung.

Incidence and mortality

In the 1990s, oesophageal cancer was diagnosed in around 4,200 males and 2,900 females each year in the UK and Ireland. Males were more than twice as likely to be diagnosed with oesophageal cancer as females. Overall, the age-standardised incidence rate in males was 13 per 100,000, and in females was 5.9, a male-to-female ratio of around 2.2:1. Oesophageal cancer is largely a disease of the elderly, and cases were rare below the age of 50. Age-specific incidence rates increased sharply above the age of 50 to peak in the oldest age groups in both males and females. The lifetime risk[1,2] of being diagnosed with oesophageal cancer was around one per cent in males and 0.5 per cent in females, in England and Wales.[3]

In the 1990s, around 4,200 males and 2,700 females in the UK and Ireland died from oesophageal cancer each year. The overall age-standardised mortality rate was 12.8 per 100,000 for males and 5.3 for females. The male-to-female ratio of the

rates was higher than that for incidence, 2.4:1, consistent with a slightly better survival in females than in males. Following the age-specific pattern for incidence, the mortality rate increased sharply above the age of 50 to peak in those aged 85 and over.

Incidence and mortality trends

The incidence of oesophageal cancer increased during the 1980s and 1990s, although less steeply in females than males. In England and Wales, the age-standardised incidence rate rose by around 70 per cent in males and 35 per cent in females between 1971 and 1998.[4] Trends in mortality over time closely followed those in incidence due to the low survival from oesophageal cancer; the age-standardised mortality rate increased by around 65 per cent in males and 30 per cent in females between 1971 and 1999 in England and Wales.[4]

Survival

Survival from oesophageal cancer in England and Wales for patients diagnosed in 1996-99 was low – around 30 per cent after one year and 8 per cent after five years.[5] Figures for patients diagnosed in Scotland[6] and in Northern Ireland during the 1990s[7] were similar, with the exception that five-year survival was notably higher for women in Northern Ireland (19 per cent). Survival decreased with age and was considerably better than average in the youngest patients (aged 15-39).[5]

Geographical patterns in incidence

By far the highest incidence of oesophageal cancer in both males and females occurred in Scotland (Figure 17.1). The age-standardised rate for males was 28 per cent higher, and that for females was 39 per cent higher, than the UK and Ireland average (Table B17.1). Rates were slightly above average for females in Wales, close to the average for both sexes in Northern Ireland, below average for males in Ireland and both males and females in England. Within England, the North West was the only region in which incidence was higher than average for both sexes. Rates were close to the average in Trent and the West Midlands, and below average in the Northern and Yorkshire; Eastern; London; and South East regions.

In Scotland, incidence rates were above average in 8 out of 15 health authorities, with particularly high rates in the west of the country (Figure 17.3). Incidence was also above average in both males and females in the urban areas of the North West (Liverpool; St Helen's and Knowsley; and Manchester), and in Nottingham and North Wales (Table B17.1).

The maps for incidence (Map 17.1) show notably higher than average rates in Scotland, and a north-south divide in England, with higher rates in the urban areas of the North West and

West Midlands and lower rates in the South East. The notable exception to this pattern was around North Yorkshire where rates were lower than average. Rates were also lower in large parts of Northern Ireland and Ireland. Compared to other countries within Europe, Scotland had the highest incidence of oesophageal cancer in females and the second highest in males, after some areas of northern France. Age-standardised incidence rates were also very high compared to the rest of Europe in the north west of England for both males and females (the area covered by the North Western and Merseyside and Cheshire cancer registries).[8]

Geographical patterns in mortality

The geographical pattern for mortality from oesophageal cancer was very similar to that for incidence, because survival from this cancer is low. Rates were highest in Scotland; in males mortality was 26 per cent higher and in females it was 40 per cent higher than the average for the UK and Ireland (Figure 17.2). There was a north-south divide in England, as for incidence, with lower than average rates in the south. Mortality rates were above average in both sexes in the North West and in males in the West Midlands, and close to average in Ireland, Wales and the Trent region. Mortality was below average in females in Northern and Yorkshire, in males in the South West, and in both sexes in Northern Ireland, and the Eastern, London, and South East regions of England.

The pattern of health authorities in which the mortality rates were either above or below average for both sexes was very similar to that for incidence (Figure 17.4). Mortality was lowest in London, where the rate was below average in males and females in 5 out of 14 health authorities, being 14-17 per cent below the average for the UK and Ireland overall (Table B17.1).

The overall mortality-to-incidence ratio (M:I ratio) for the UK and Ireland was close to one in males and slightly lower in females, consistent with survival being very low in males and slightly higher in females (Table B17.1). It is of note that the M:I ratio was greater than one in Ireland and several regions of England (in males only), which suggests an under-ascertainment of cases or more likely that there has been some misclassification between oesophagus and stomach in the mortality data.[4,9]

The geographical patterns in incidence and mortality for cancer of the oesophagus are closely similar to those for cancers of the larynx (Ch 10), lip, mouth and pharynx (Ch 12), and lung (Ch 13). In particular the north-south divide across Great Britain is striking, with consistently high rates in Scotland and the North West of England and low rates in the south.

Risk factors and aetiology

There are two main histological types of oesophageal cancer; squamous cell carcinoma (SCC), which is most common in the upper two thirds of the oesophagus; and adenocarcinoma, which is most common in the lower third. The distribution of these subtypes and trends in their incidence over time differ between the sexes. In England and Wales, adenocarcinomas accounted for over 40 per cent of oesophageal cancers in males and 20 per cent in females, while the pattern was reversed for SCCs (20 per cent in males and 40 in females);[3] the remaining 40 per cent of tumours in each sex were most likely classified as unspecified carcinomas. The incidence of SCC increased by around 25 per cent in males and 40 per cent in females between 1971 and 1998, so that in 1998 there were similar numbers of SCCs in both sexes (male-to-female ratio of 1.1:1). In contrast, the incidence of adenocarcinoma increased three fold in females and almost six fold in males over the same period, resulting in a male-to-female ratio of 4.8:1 in 1998 (data for England and Wales).[9]

The risk factors also differ between the two main subtypes. For SCC, the main risk factors are excessive alcohol consumption and tobacco smoking, which in combination have an effect that is greater than additive. In a study from France, the risk of developing oesophageal cancer was increased five fold in heavy smokers, more than ten fold in heavy drinkers and about fifty fold in individuals with both habits.[10] Low consumption of fresh fruit, vegetables, meat and dairy products are also associated with SCC. However in developed countries, 90 per cent or more of the risk can be attributed to alcohol and tobacco, and rising alcohol consumption has been suggested as one possible explanation for the increase in oesophageal cancer throughout the twentieth century.[11]

Adenocarcinoma is preceded by a condition called Barrett's oesophagus, which is caused by chronic gastro-oesophageal reflux. A recent study of patients in Northern Ireland with this condition found that it increased the risk of developing oesophageal cancer eight fold, but only in patients who had high-grade dysplasia (specialised intestinal metaplasia). However, only a small proportion can have gone on to develop oesophageal cancer, as it accounted for only 5 per cent of deaths in patients with Barrett's oesophagus.[12] Smoking is associated with an increased risk of adenocarcinoma (as well as with SCC) and overall, over 70 per cent of deaths from cancer of the oesophagus in England in 1998-2002 were estimated to be attributable to smoking.[13] Low consumption of fresh fruit and vegetables is also associated with oesophageal adenocarcinoma, but the main risk factor is obesity. The increasing prevalence of obesity in the UK population may have contributed to the increasing incidence of adenocarcinoma of the oesophagus.[14] (continued on page 192)

Figure **17.1**

**Oesophagus: incidence by sex, country, and region of England
UK and Ireland 1991-99[1]**

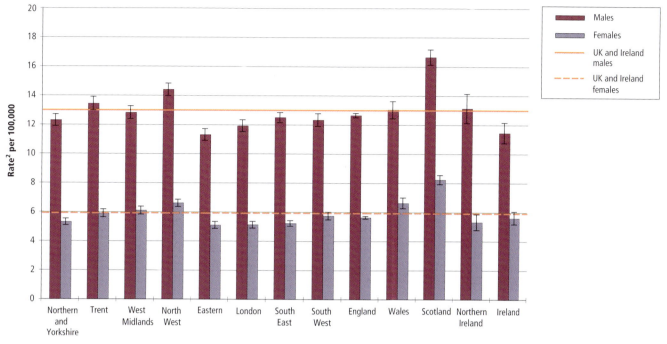

1 Northern Ireland 1993-99, Ireland 1994-99

2 Age standardised using the European standard population, with 95% confidence interval

Figure **17.2**

**Oesophagus: mortality by sex, country, and region of England
UK and Ireland 1991-2000[1]**

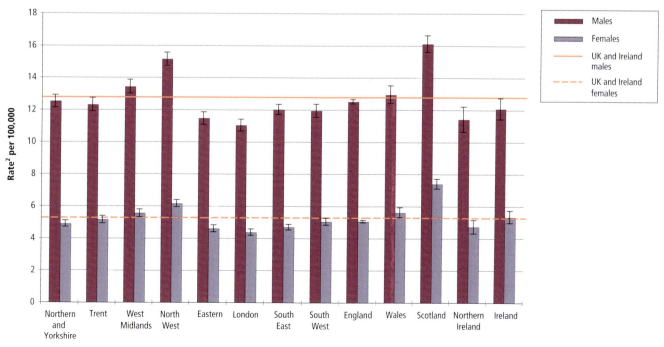

1 Scotland 1991-99, Ireland 1994-2000

2 Age standardised using the European standard population, with 95% confidence interval

Figure **17.3a**

**Oesophagus: incidence by health authority within country, and region of England
Males, UK and Ireland 1991-99[1]**

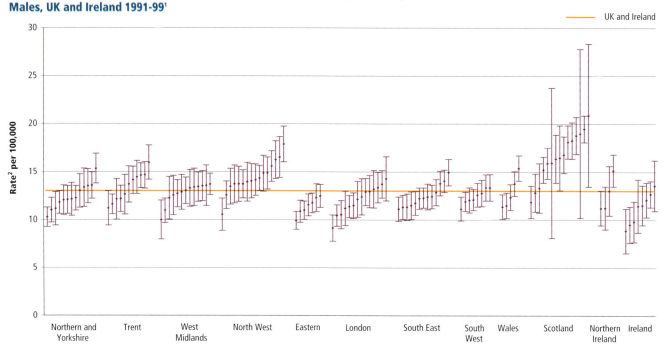

1 Northern Ireland 1993-99, Ireland 1994-99

2 Age standardised using the European standard population, with 95% confidence interval

Figure **17.3b**

**Oesophagus: incidence by health authority within country, and region of England
Females, UK and Ireland 1991-99[1]**

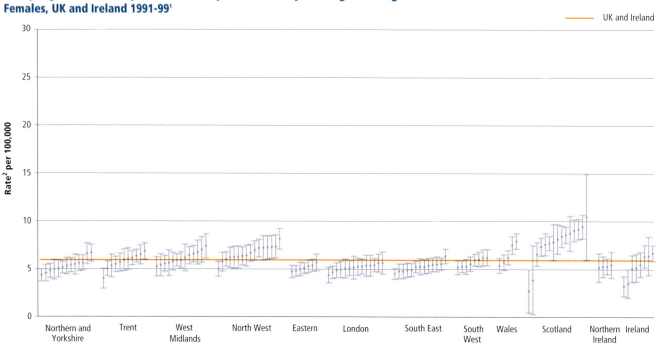

1 Northern Ireland 1993-99, Ireland 1994-99

2 Age standardised using the European standard population, with 95% confidence interval

Figure **17.4a**

Oesophagus: mortality by health authority within country, and region of England
Males, UK and Ireland 1991-2000[1]

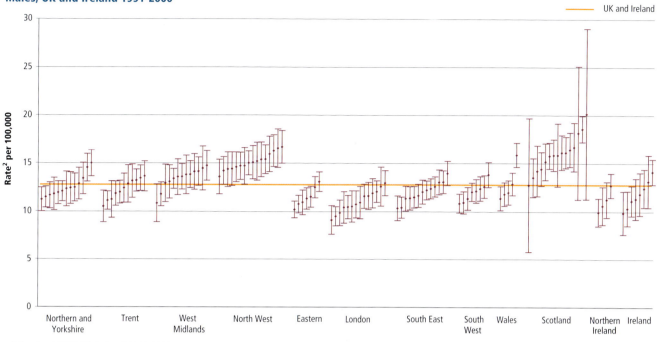

1 Scotland 1991-99, Ireland 1994-2000

2 Age standardised using the European standard population, with 95% confidence interval

Figure **17.4b**

Oesophagus: mortality by health authority within country, and region of England
Females, UK and Ireland 1991-2000[1]

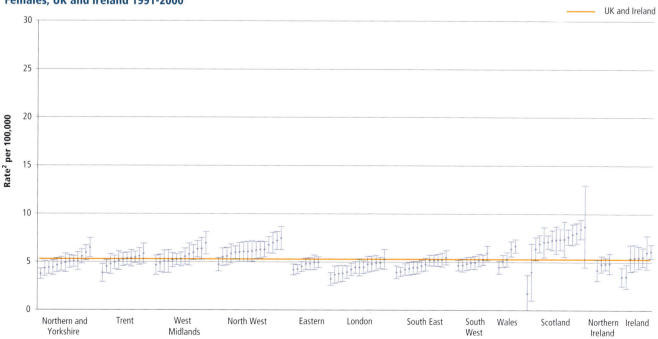

1 Scotland 1991-99, Ireland 1994-2000

2 Age standardised using the European standard population, with 95% confidence interval

Map **17.1a**

Oesophagus: incidence* by health authority,
Males, UK and Ireland, 1991-99

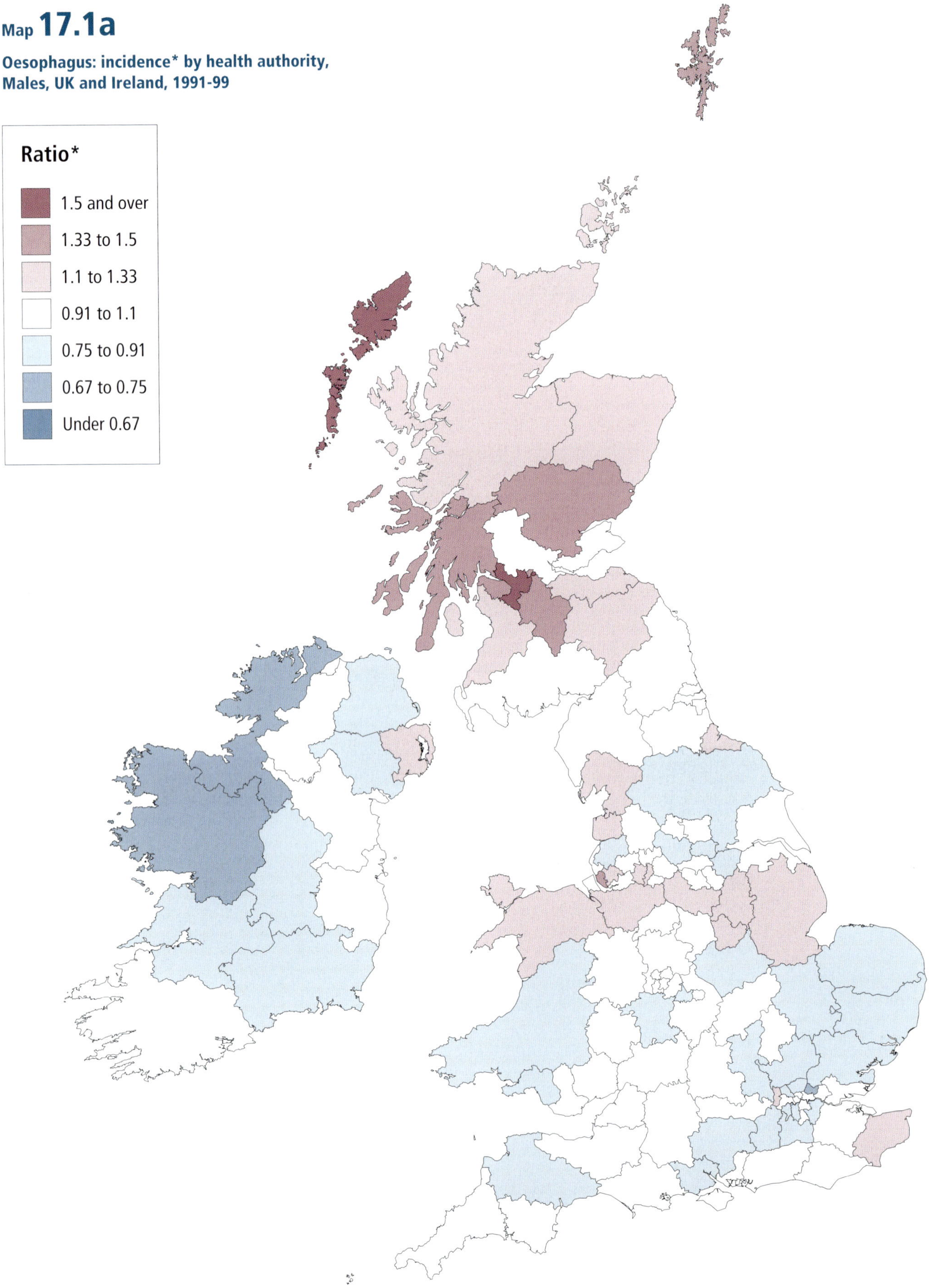

Ratio*

	1.5 and over
	1.33 to 1.5
	1.1 to 1.33
	0.91 to 1.1
	0.75 to 0.91
	0.67 to 0.75
	Under 0.67

*Ratio of directly age-standardised rate in health authority to UK and Ireland average

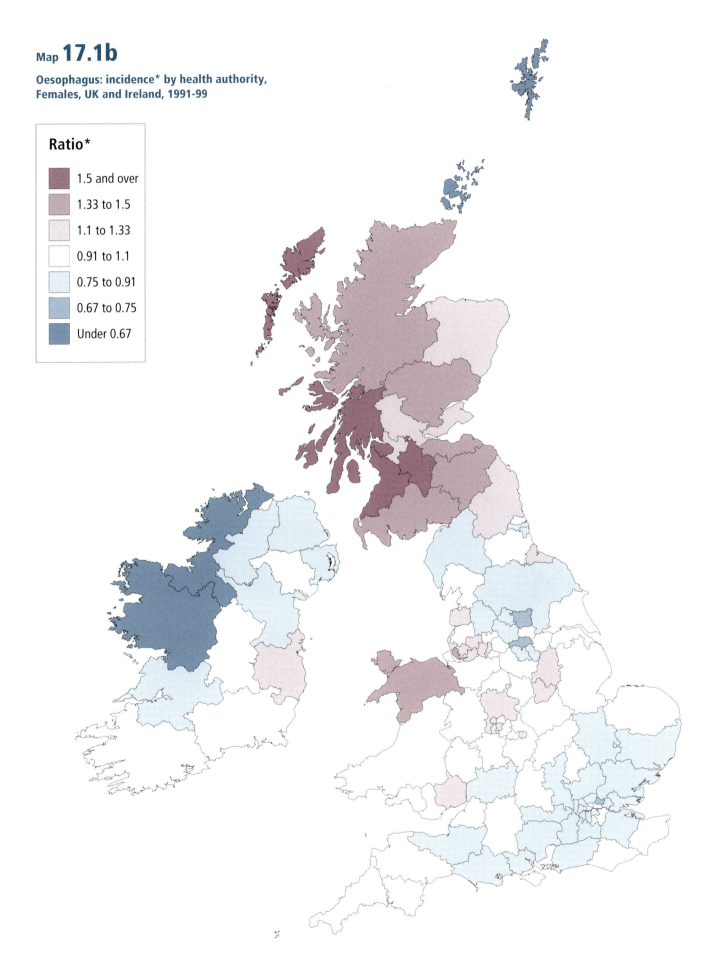

Map 17.1b

Oesophagus: incidence* by health authority, Females, UK and Ireland, 1991-99

Ratio*

- 1.5 and over
- 1.33 to 1.5
- 1.1 to 1.33
- 0.91 to 1.1
- 0.75 to 0.91
- 0.67 to 0.75
- Under 0.67

*Ratio of directly age-standardised rate in health authority to UK and Ireland average

Map **17.2a**

**Oesophagus: mortality* by health authority,
Males, UK and Ireland, 1991-2000**

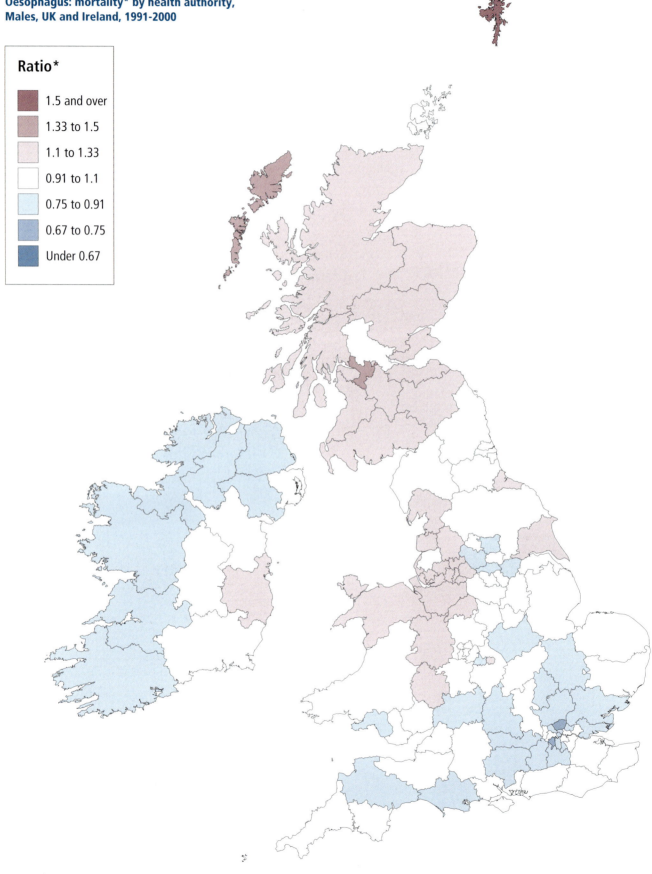

Ratio*

	1.5 and over
	1.33 to 1.5
	1.1 to 1.33
	0.91 to 1.1
	0.75 to 0.91
	0.67 to 0.75
	Under 0.67

*Ratio of directly age-standardised rate in health authority to UK and Ireland average

Map **17.2b**

**Oesophagus: mortality* by health authority,
Females, UK and Ireland, 1991-2000**

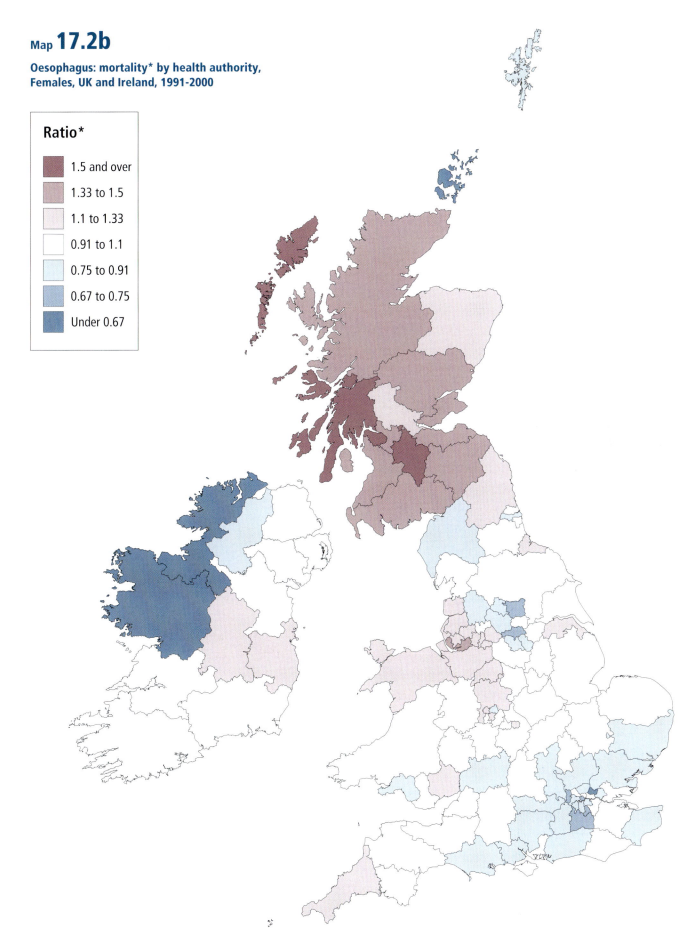

Ratio*

- 1.5 and over
- 1.33 to 1.5
- 1.1 to 1.33
- 0.91 to 1.1
- 0.75 to 0.91
- 0.67 to 0.75
- Under 0.67

*Ratio of directly age-standardised rate in health authority to UK and Ireland average

Socio-economic deprivation

In most parts of the world oesophageal cancer is a disease of the poor, and there is a clearly higher risk among people of lower socio-economic status. In England and Wales the incidence of oesophageal cancer was positively associated with deprivation, with rates around 30 per cent higher in the most deprived groups.[3] In Scotland the risk was increased two fold for males living in the most deprived areas.[15] The association between oesophageal cancer and deprivation is likely to be related to levels of smoking and alcohol consumption, and nutritional status. People in lower socio-economic classes are more likely to smoke and consume more alcohol,[16] and those in more deprived areas eat less fresh fruit and vegetables,[17] which is an indicator of poor nutrition.

Differences have been observed in the relationships between histological subtypes of oesophageal cancer and deprivation. A study of oesophageal cancer in Scotland found no association between deprivation and adenocarcinoma, but a strong positive association for other subtypes,[18] which is surprising, given that smoking is a risk factor for adenocarcinoma as well as for SCC. It has been suggested that this may be because the increased risk of oesophageal adenocarcinoma persists for up to 30 years after smoking cessation, and hence current deprivation category-specific incidence rates may reflect smoking patterns at a time when the prevalence of smoking was more evenly distributed across the social classes in the UK.[18]

The majority of the areas shown on the maps of the UK and Ireland (Maps 17.1 and 17.2) to have high rates of incidence and mortality are also areas with high deprivation scores according to the Carstairs index[19] (see Appendix F), in particular, large areas of Scotland, North Wales, and urban areas of the North West and West Midlands in England. This pattern can also be seen from the maps for cancers of the larynx (Ch 10), lip, mouth and pharynx (Ch 12), and lung (Ch 13), clearly highlighting the association between deprivation and the risk factors shared by these cancers (smoking, and in some cases, alcohol consumption).

References

1. Schouten LJ, Straatman H, Kiemeney LALM, Verbeek ALM. Cancer incidence: Life table risk versus cumulative risk. *Journal of Epidemiology and Community Health* 1994; 48: 596-600.

2. ONS. *Cancer Statistics Registrations: Registrations of cancer diagnosed in 2001, England.* Series MB1 No. 32. London: Office for National Statistics, 2004.

3. Quinn MJ, Babb PJ, Brock A, Kirby L et al. *Cancer Trends in England and Wales 1950-1999.* Studies on Medical and Population Subjects No. 66. London: The Stationery Office, 2001.

4. Newnham A, Quinn MJ, Babb P, Kang JY et al. Trends in oesophageal and gastric cancer incidence, mortality and survival in England and Wales 1971-1998/1999. *Alimentary Pharmacology and Therapeutics* 2003; 17: 655-664.

5. ONS. Cancer Survival: England and Wales, 1991-2001. March 2004. Available at *http://www.statistics.gov.uk/statbase/ ssdataset.asp?vlnk=7899.*

6. ISD Scotland. *Trends in Cancer Survival in Scotland, 1977-2001.* Edinburgh: ISD Publications, 2004.

7. Fitzpatrick D, Gavin A, Middleton R, Catney D. *Cancer in Northern Ireland 1993-2001*: A Comprehensive Report. Belfast: Northern Ireland Cancer Registry, 2004.

8. Parkin DM, Whelan SL, Ferlay J, Teppo L et al. *Cancer Incidence in Five Continents Vol. VIII.* IARC Scientific Publications No. 155. Lyons: International Agency for Research on Cancer, 2000.

9. Newnham A, Quinn MJ, Babb P, Kang JY et al. Trends in the subsite and morphology of oesophageal and gastric cancer in England and Wales 1971-1998. *Alimentary Pharmacology and Therapeutics* 2003; 17: 665-676.

10. Tuyns AJ, Pequignot G, Jensen OM. Oesophageal cancer in Ille et Villaine in relation to alcohol and tobacco consumption. Multiplicative risks (in French). *Bulletins du Cancer* 1977; 64: 45-60.

11. Munoz N, Day NE. Esophageal Cancer. In: Schottenfeld D, Fraumeni JF, Jr. (eds) *Cancer Epidemiology and Prevention, second edition.* New York: Oxford University Press, 1996.

12. Anderson LA, Murray LJ, Murphy SJ, Fitzpatrick DA et al. Mortality in Barrett's oesophagus: results from a population based study. *Gut* 2003; 52: 1081-1084.

13. Twigg L, Moon G, Walker S. *The Smoking Epidemic in England.* London: Health Development Agency, 2004.

14. Seidell JC, Flegal KM. Assessing obesity: classification and epidemiology. *British Medical Bulletin* 1997; 53: 238-252.

15. Harris V, Sandridge AL, Black RJ, Brewster DH et al. *Cancer Registration Statistics Scotland, 1986-1995.* Edinburgh: ISD Publications, 1998.

16. Rickards L, Fox K, Roberts C, Fletcher L et al. *Living in Britain: Results from the 2002 General Household Survey.* London: The Stationery Office, 2004.

17. NHS Scotland. *Health in Scotland 2001.* Edinburgh: The Stationery Office, 2002.

18. Brewster DH, Fraser LA, McKinney PA, Black RJ. Socioeconomic status and risk of adenocarcinoma of the oesophagus and cancer of the gastric cardia in Scotland. *British Journal of Cancer* 2000; 83: 387-390.

19. Carstairs V, Morris R. Deprivation and mortality: an alternative to social class? *Community Medicine* 1989; 11: 213-219.

Chapter 18

Ovary

Paul Walsh, Nicola Cooper

Summary

- In the UK and Ireland in the 1990s, ovarian cancer accounted for 1 in 20 cases of cancer and 1 in 17 deaths from cancer in females.

- There was little geographical variation in incidence and less for mortality.

- Incidence was higher than average in a few areas in the West Midlands, Wales, South East England, and in Scotland, and lower in a few areas in north east England and around London. There was a similar pattern for mortality.

- Some of the observed variations in incidence may be explained by differences in registration practices across cancer registries.

- There was no clear relationship between the geographical variations in ovarian cancer and socio-economic deprivation, or any known or suspected risk factors for the disease.

- The geographical distribution of ovarian cancer was broadly similar to that for cancers of the breast and uterus. These cancers have several risk factors in common.

Introduction

Geographic variations in ovarian cancer incidence may have arisen because of differences in diagnostic criteria or in cancer registration practice. True geographic variation in underlying risk is therefore difficult to establish. In particular, a proportion of ovarian cancers are so-called 'borderline' malignancies, of low malignant potential. Cancers of borderline malignancy are classified as malignant, according to the International Classification of Disease for Oncology, second edition (ICDO2) rules. The ICDO2 rules were introduced with the change to the International Classification of Disease, tenth revision (ICD10). Cancers of borderline malignancy were not classified as malignant prior to the introduction of ICDO2 rules. Ireland has used ICDO2 coding rules since collection of data for the entire country began in 1994. The ICDO2 classification was introduced by the cancer registries of England and Wales in 1995, Scotland in 1997 and Northern Ireland in 1996. Between

the introduction of ICDO2 and 1999, borderline cases accounted for 10 per cent of ovarian malignancies in Northern Ireland and Ireland, 14 per cent in Scotland, 8 per cent in England and 4 per cent in Wales. (For the entire period of the analysis, this might approximate to around 6 per cent of all registered invasive cases in Northern Ireland, five per cent in Scotland, 4 per cent in England and 2 per cent in Wales – compared to 10 per cent in Ireland.) The proportions of borderline malignancies in the regions of England varied from 11 per cent in the North West to 4 per cent in London. Inclusion of borderline malignancies increases the recorded incidence. Borderline malignancies are rarely fatal, and have much better prognosis than other malignant tumours arising in the ovary, thus their registration also increases the apparent survival rates for ovarian cancer.

Incidence and mortality

Ovarian cancer was the fourth commonest cancer in women, after breast, colorectal and lung and accounted for 5 per cent of all newly diagnosed cancers. About 6,700 cases of ovarian cancer were diagnosed annually in the UK and Ireland during the 1990s; the overall age-standardised rate was 18 per 100,000. Few cases occurred in pre-menopausal women; over 90 per cent of cases were in women aged 45 or over, with age-specific rates peaking in the range 70-84 years. The lifetime risk[1,2] of being diagnosed with ovarian cancer, based on data from England and Wales for 1997, was 2.1 per cent (1 in 48).[3]

During the 1990s, about 4,600 women died annually from ovarian cancer in the UK and Ireland – 6 per cent of all cancer deaths in women. Ovarian cancer was the fourth commonest cause of cancer death in women, and accounted for more deaths than all the other gynaecological cancers together. The overall age-standardised mortality rate was 11.5 per 100,000. The pattern of age-specific mortality broadly followed that of incidence, peaking in women aged 75-84.

Incidence and mortality trends

Since the early 1970s, the age-standardised incidence rate for ovarian cancer in England and Wales has gradually increased; age-specific rates increased markedly in older women,[4] whereas in women aged under 65 they were relatively stable. In Scotland, the increase in the incidence of ovarian cancer from the early 1960s to the late 1980s was attributed mainly to the increase in registered cases among women over 60 years, with no upward trend for younger women.[5] The pattern in the risk of ovarian cancer by birth cohort is similar to that for breast cancer, with rates increasing up to the cohorts born around 1930 and then declining, but appearing to increase again for women born from the early 1960s onwards.[4]

(continued on page 198)

Figure **18.1**

**Ovary: incidence by country, and region of England
UK and Ireland 1991-99[1]**

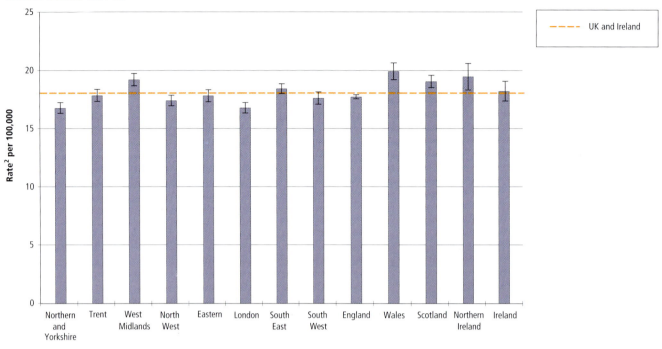

1 Northern Ireland 1993-99, Ireland 1994-99

2 Age standardised using the European standard population, with 95% confidence interval

Figure **18.2**

**Ovary: mortality by country, and region of England
UK and Ireland 1991-2000[1]**

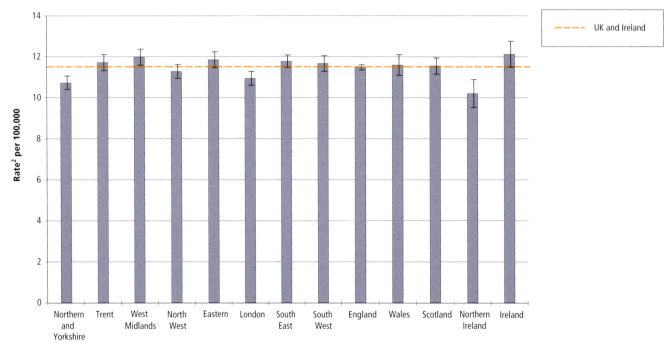

1 Scotland 1991-99, Ireland 1994-2000

2 Age standardised using the European standard population, with 95% confidence interval

Figure 18.3

**Ovary: incidence by health authority within country, and region of England
UK and Ireland 1991-99[1]**

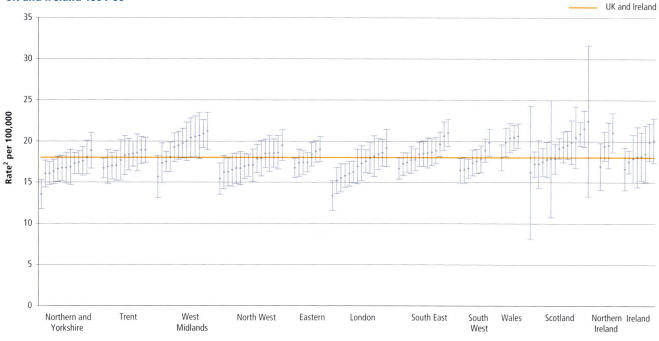

1 Northern Ireland 1993-99, Ireland 1994-99

2 Age standardised using the European standard population, with 95% confidence interval

Figure 18.4

**Ovary: mortality by health authority within country, and region of England
UK and Ireland 1991-2000[1]**

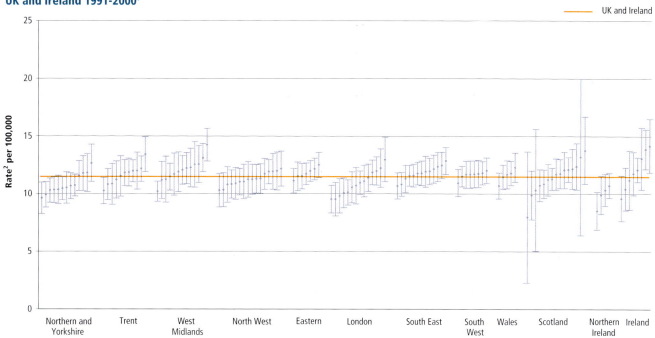

1 Scotland 1991-99, Ireland 1994-2000

2 Age standardised using the European standard population, with 95% confidence interval

Map **18.1**

Ovary: incidence* by health authority
UK and Ireland 1991-99

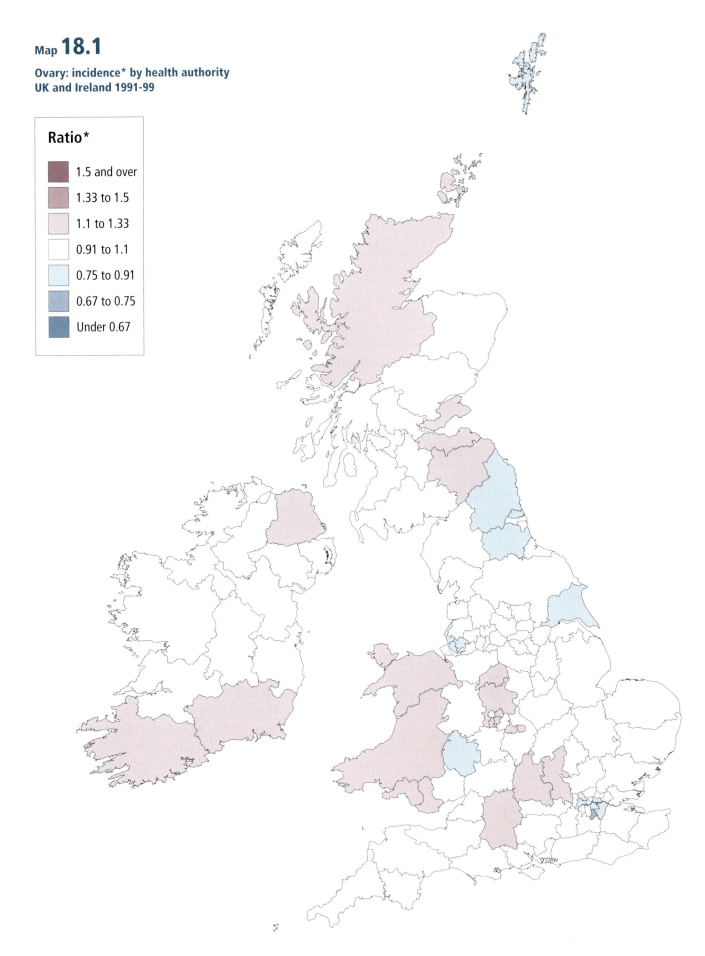

Ratio*

	1.5 and over
	1.33 to 1.5
	1.1 to 1.33
	0.91 to 1.1
	0.75 to 0.91
	0.67 to 0.75
	Under 0.67

*Ratio of directly age-standardised rate in health authority to UK and Ireland average

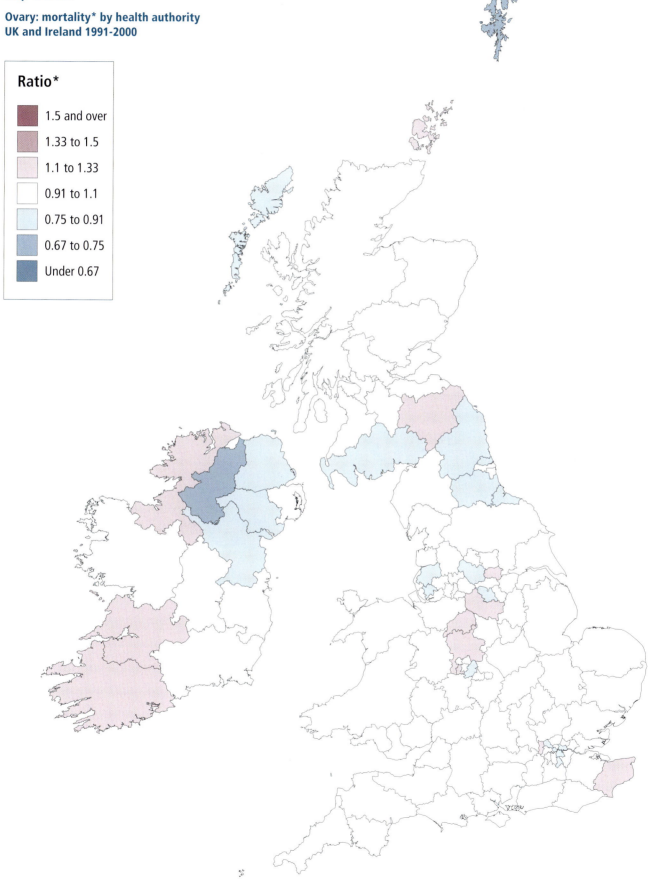

Map 18.2

**Ovary: mortality* by health authority
UK and Ireland 1991-2000**

Ratio*

	1.5 and over
	1.33 to 1.5
	1.1 to 1.33
	0.91 to 1.1
	0.75 to 0.91
	0.67 to 0.75
	Under 0.67

*Ratio of directly age-standardised rate in health authority to UK and Ireland average

In England and Wales, age-standardised mortality rates for ovarian cancer showed continuous, marked increases from early in the twentieth century until the 1980s.[4] This was attributed to a combination of more accurate diagnosis and an increase in underlying incidence. Over the period 1950-99, increases in mortality rates were greatest for older women – a doubling of rates for those aged 75 and over and a 50 per cent increase for those aged 65-74. In the age group 55-64, there was a less steep increase up to the late 1970s, then rates stabilised, and in those aged 45-54, mortality rates were stable up to the late 1970s, then declined. Mortality rates in Scotland showed no overall trend up to 1995,[5,6] but some reduction in mortality was seen among younger women.[5] In Northern Ireland, the mortality rate increased from 10.2 to 12.0 per 100,000 over the period 1993-2001.[7] In Ireland, mortality rates showed a sustained increase during the period 1956-2000, especially in older women (over 65), with no evidence for any recent downturn in rates (National Cancer Registry, in preparation).

Survival

Ovarian cancer has the lowest survival of the gynaecological cancers, largely because it is often at an advanced stage when diagnosed. Symptoms of ovarian cancer are frequently vague and are difficult to distinguish from other conditions. In England and Wales, relative survival for women diagnosed during 1996-99 was 65 per cent at one year and 36 per cent at five years after diagnosis.[8] In the early 1990s, five-year relative survival rates from ovarian cancer in England, Scotland and Wales were 5.2, 6.6 and 7.6 percentage points, respectively, below the European average.[9] Differences in the mix of histological sub-types could explain some of the variation in survival across Europe. It is difficult to compare the survival rates between the countries of the UK and Ireland because of the differences in inclusion criteria and the case mix for histological sub-types of malignant ovarian cancer.

Geographical patterns in incidence

The incidence rates for ovarian cancer show no extreme variability among the countries of the UK and Ireland (Figure 18.1). The highest incidence rates occurred in Wales, Scotland, and Northern Ireland, with incidence rates 10, 6 and 8 per cent higher, respectively, than the average. The rate for Ireland was close to the average. Within England, incidence was higher than average in the West Midlands and below in Northern and Yorkshire, and London, but the rates for all English regions were within 7 per cent of the UK and Ireland average.

There was inevitably more variation among health authorities than within countries and regions of England, but rates were all within 25 per cent of the UK and Ireland average (Figure 18.3). The map (Map 18.1) shows that the incidence of ovarian cancer was notably higher in the Highlands and also on the east coast of Scotland in Fife (19 per cent higher than average), Lothian, and Borders. There was a band of high incidence of ovarian cancer in the West Midlands covering the area from North and South Staffordshire, to the urban area surrounding Birmingham (Walsall, Dudley [18 per cent higher than average], Sandwell) and Solihull, and Coventry. A further band of high incidence is apparent in the south of England, in Buckinghamshire, Oxfordshire, and Wiltshire. In Wales, all areas with the exception of one (Gwent) had rates above the average for the UK and Ireland. The Southern and South Eastern areas in Ireland both had higher than average incidence rates, as did the Northern area in Northern Ireland.

Incidence for ovarian cancer was generally lower than average in the Northern and Yorkshire region of England, especially along the east coast with notably low rates in Northumberland; Newcastle and North Tyneside; County Durham; and East Riding (which includes Hull). In the North West region, Liverpool and its surrounding areas of St Helen's and Knowsley, and Sefton all had distinctly lower than average incidence of ovarian cancer. The north west and south east areas of London also had lower incidence of ovarian cancer, with Croydon in south London having a rate that was 25 per cent lower than the average.

As mentioned above, some of the apparent geographic variation in ovarian cancer incidence may have arisen due to differences in the classification of borderline malignancies. These differences may arise from diagnostic criteria or cancer registration practice, or both.

Geographical patterns in mortality

As for incidence, mortality rates for ovarian cancer showed relatively little variation among countries (Figure 18.2). The overall mortality rates were 5 per cent above the UK and Ireland average in Ireland, close to the overall average in England, Wales and Scotland, and relatively low in Northern Ireland (11 per cent below average). All regions in England were within 7 per cent of the average, with rates for Northern and Yorkshire, and London markedly lower than average.

The map of mortality for ovarian cancer (Map 18.2), as for that of incidence, shows markedly higher rates in Borders in Scotland, North and South Staffordshire, and Dudley in the

West Midlands, and the Southern area in Ireland. There also appeared to be isolated higher areas of ovarian cancer mortality in Wakefield, North Derbyshire, Hillingdon, and East Kent in England, and in the Mid Western and North Western areas in Ireland, although none of these areas had notably higher incidence rates.

The ovarian mortality rate in Dumfries and Galloway in Scotland was below average. Areas with low incidence rates in the Northern and Yorkshire region along the east coast (from Northumberland southwards) also tended to have lower mortality rates. South Lancashire; St Helen's and Knowsley; Calderdale and Kirklees; Sheffield; and Birmingham all had lower than average ovarian mortality. In London, three out of the six health authorities that had markedly low ovarian cancer incidence also had lower than average mortality.

Mortality rates were low, relative to incidence rates, in Scotland, Wales and, especially, Northern Ireland, but high in Ireland. These countries showed perhaps less agreement than England between the geographical patterns of mortality and incidence. It is not clear to what extent these differences reflect variations in diagnostic or registration criteria, random variation in rate estimates, or other factors.

Risk factors and aetiology

A woman's history of ovulation appears to play a role in the development of the disease. Higher rates of ovarian cancer occur among women who do not have children, and the risk decreases with the number of pregnancies (parity). The incidence is strongly correlated with breast (Chapter 5) and uterine cancer (Chapter 23); all are related to hormone levels and share many of the same risk factors. It is well established that a reduction in ovarian cancer risk is associated with an increased duration of oral contraceptive use.[10] The widespread use of the contraceptive pill is one possible explanation for the stability of rates in younger women.

Tubal ligation (a form of female sterilisation) has a protective effect on ovarian cancer with an estimated reduced risk of 39-70 per cent.[11,12] There is also evidence, although it is less conclusive, that risk is increased by hormone replacement therapy in post-menopausal women[13,14] and in women treated for infertility.[10] A small proportion of ovarian cancers are attributable to specific inherited mutations, notably those of the BRCA-1 and BRCA-2 genes.[15,16] There is limited, and inconclusive, evidence to suggest that greater height or body weight, dietary fat consumption, and talcum powder use might be associated with increased risk.[10]

It is difficult to discern any geographic patterns within the UK and Ireland that might be confidently attributed to variations in reproductive or hormonal factors. Maps of total fertility rates and live birth rates by mother's age in the United Kingdom for 1991-97[17] show no clear correlation between areas of higher fertility and those with low ovarian cancer incidence and mortality rates. In 2002-03, a quarter of women aged 16-49 in Great Britain used oral contraceptives. Women aged 18-29 were those most likely to use the contraceptive pill, with 48 per cent of women aged 20-24 using oral contraceptives.[18] However, this information is not available at a regional or lower geographical level. This makes it difficult to disentangle the influences of parity and oral contraceptive use in a given population, and it may be that their separate protective effects predominate in different regions.

Socio-economic deprivation

In England and Wales, based on data from 1993,[4] both incidence and mortality rates for ovarian cancer were marginally higher in women from more affluent areas than those from deprived areas.[4] This was similar to, but less defined than, the trend for breast cancer incidence.[4] For both breast and ovarian cancer, the trend is consistent with the known protective effect of higher parity, with total fertility rates being higher in more deprived areas.[17] However, there is no available information on oral contraceptive use by area or social class. Thus, it is not clear if the protective effect of oral contraceptive use plays any role in the relationship between ovarian cancer and deprivation. In Scottish data for 1986-95, there was no obvious relationship between ovarian cancer incidence and deprivation.[6]

For women diagnosed with ovarian cancer in 1996-99 in England and Wales, there was no difference in five-year survival between those living in the most affluent and the most deprived areas. Ovarian cancer was one of only five cancer-sex combinations in which this was the case.[19] The key issue around ovarian cancer is that diagnosis often occurs when the disease is at an advanced stage, which limits the potential for treatment to improve survival. Population screening for ovarian cancer, which has the potential to detect the disease at an earlier stage, is currently being evaluated in the United Kingdom Collaborative Trial of Ovarian Cancer Screening (UKCTOCS).[20]

References

1. Schouten LJ, Straatman H, Kiemeney LALM, Verbeek ALM. Cancer incidence: Life table risk versus cumulative risk. *Journal of Epidemiology and Community Health* 1994; 48: 596-600.

2. ONS. *Cancer Statistics Registrations: Registrations of cancer diagnosed in 2000, England.* Series MB1 No. 31. London: Office for National Statistics, 2003.

3. Quinn MJ, Babb PJ, Kirby L, Jones J. Registrations of cancer diagnosed in 1994-97, England and Wales. *Health Statistics Quarterly* 2000; 7: 71-82.

4. Quinn MJ, Babb PJ, Brock A, Kirby L et al. *Cancer Trends in England and Wales 1950-1999.* Studies on Medical and Population Subjects No. 66. London: The Stationery Office, 2001.

5. Black RJ, Macfarlane GJ, Maisonneuve P, Boyle P. *Cancer Incidence and Mortality in Scotland 1960-89.* Edinburgh: ISD Publications, 1995.

6. Harris V, Sandridge AL, Black RJ, Brewster DH et al. *Cancer Registration Statistics Scotland, 1986-1995.* Edinburgh: ISD Publications, 1998.

7. Fitzpatrick D, Gavin A, Middleton R, Catney D. *Cancer in Northern Ireland 1993-2001: A Comprehensive Report.* Belfast: Northern Ireland Cancer Registry, 2004.

8. ONS. Cancer Survival: England and Wales, 1991-2001. March 2004. Available at *http://www.statistics.gov.uk/statbase/ ssdataset.asp?vlnk=7899.*

9. Sant M, Areleid T, Berrino F, Bielska Lasota M et al. EUROCARE-3: survival of cancer patients diagnosed 1990-1994 - results and commentary. *Annals of Oncology* 2003; 14 Suppl 5: v61-v118.

10. Weiss NS, Cook LS, Farrow DC, Rosenblatt KA. Ovarian Cancer. In: Schottenfeld D, Fraumeni JF, Jr. (eds) *Cancer Epidemiology and Prevention, second edition.* New York: Oxford University Press, 1996.

11. Green A, Purdie D, Bain C, Siskind V et al. Tubal sterilisation, hysterectomy and decreased risk of ovarian cancer. Survey of Women's Health Study Group. *International Journal of Cancer* 1997; 71: 948-951.

12. Hankinson SE, Hunter DJ, Colditz GA, Willett WC et al. Tubal ligation, hysterectomy, and risk of ovarian cancer. A prospective study. *Journal of the American Medical Association* 1993; 270: 2813-2818.

13. Garg PP, Kerlikowske K, Subak L, Grady D. Hormone replacement therapy and the risk of epithelial ovarian carcinoma: a meta-analysis. *Obstetrics and Gynecology* 1998; 92: 472-479.

14. Rodriguez C, Patel AV, Calle EE, Jacob EJ et al. Estrogen replacement therapy and ovarian cancer mortality in a large prospective study of US women. *Journal of the American Medical Association* 2001; 285: 1460-1465.

15. Ford D, Easton DF, Bishop DT, Narod SA et al. Risks of cancer in BRCA1-mutation carriers. Breast Cancer Linkage Consortium. *Lancet* 1994; 343: 692-695.

16. Ford D, Easton DF, Stratton M, Narod S et al. Genetic heterogeneity and penetrance analysis of the BRCA1 and BRCA2 genes in breast cancer families. The Breast Cancer Linkage Consortium. *American Journal of Human Genetics* 1998; 62: 676-689.

17. Griffiths C, Fitzpatrick J. *Geographic Variations in Health.* Decennial Supplements No. 16. London: The Stationery Office, 2001.

18. Dawe F, Meltzer H. *Contraception and Sexual Health, 2002.* London: Office for National Statistics, 2003.

19. Coleman MP, Rachet B, Woods LM, Mitry E et al. Trends and socioeconomic inequalities in cancer survival in England and Wales up to 2001. *British Journal of Cancer* 2004; 90: 1367-1373.

20. Lewis S, Menon U. Screening for ovarian cancer. *Expert Review of Anticancer Therapy* 2003; 3: 55-62.

Chapter 19

Pancreas

Paul Walsh, Helen Wood

Summary

- In the UK and Ireland in the 1990s, pancreatic cancer accounted for around 1 in 40 cancer cases and 1 in 20 cancer deaths.

- Geographical variation in both incidence and mortality was low, with no clear patterns, and little consistency between rates for males and females.

- There was some similarity between the maps for incidence and mortality, but less than for cancers of the lung and oesophagus, which also have very low survival.

- Incidence was slightly above average in London and below average in South West England and Northern Ireland; there was little variation in mortality at the country and regional level.

- There was no obvious link between socio-economic deprivation or any known risk factor and the observed geographical distribution of pancreatic cancer.

Incidence and mortality

About 7,200 cases of pancreatic cancer were diagnosed annually in the UK and Ireland during the 1990s, with roughly equal numbers in males (3,500) and females (3,700). Pancreatic cancer was the ninth commonest cancer in males and the tenth commonest in females. Age-standardised incidence rates were 10.5 per 100,000 in males and 7.8 per 100,000 in females. The male-to-female ratio of the age-standardised rates was 1.4:1, while that for the number of cases was 0.9:1. Pancreatic cancer rates were particularly low below the age of 45 but increased strongly with age, peaking in the oldest age group (85 years and over); about 90 per cent of cases occurred in people aged 55 or older. Although rates were higher in males than females in all age groups, higher life expectancy in the female population resulted in a slight excess of female cases. The lifetime risk[1,2] of being diagnosed with pancreatic cancer, based on data from England and Wales for 1997, was 1.0 per cent for males and 1.1 per cent for females.[3]

Almost as many people died from, as were diagnosed with, pancreatic cancer annually: 7,000 deaths in total, 3,400 in males and 3,600 in females. This reflects the high fatality of this cancer, which is rarely diagnosed early enough for effective treatment. Pancreatic cancer was the seventh most common cause of cancer death in males (4 per cent of all cancer deaths) and the fifth most common in females (5 per cent), higher rankings than for incidence. Age-standardised mortality rates averaged 10.1 per 100,000 in males and 7.4 per 100,000 in females. Age-specific rates showed a similar pattern to incidence, peaking in the oldest age groups. The male-to-female ratios of both the age-standardised rates and the numbers of deaths were the same as those for the incidence rates and the numbers of cases, respectively.

Incidence and mortality trends

In England and Wales, age-standardised incidence rates of pancreatic cancer in males have fallen consistently since the early 1980s, while rates in females, which increased during the 1970s and 1980s, only began to fall in the 1990s.[4] Mortality rates in those decades paralleled the incidence rates, but had previously increased between 1950 and the late 1960s in both males and females.[4] Earlier increases may have reflected, in part, improvements in diagnosis and certification of pancreatic cancer as a cause of death. Incidence rates in Scotland declined among males, and to a lesser extent females, between 1986 and 1995.[5] Scottish mortality rates then showed relatively little change during 1970-85,[6] but some decrease thereafter, in line with incidence.[5,6] In Ireland, mortality rates have been declining among populations aged under 65 since the early 1970s in females and early 1980s in males, but with no obvious trends in older populations (National Cancer Registry of Ireland, report in preparation).

Survival

Survival rates for pancreatic cancer are lower than for any other major cancer, reflecting the advanced stage of disease and limited opportunities for effective treatment of most patients. For cases diagnosed in England and Wales during 1996-99, relative survival to one year was only 13 per cent while five-year relative survival was only 2-3 per cent.[7] For cases diagnosed in Scotland in 1992-96[8] and in Northern Ireland in 1996-99[9] figures for one- and five-year survival were similar. Survival rates in England and Wales for patients diagnosed in 1990-94 were comparable to the European average.[10]

Geographical patterns in incidence

Geographic variation in pancreatic cancer incidence was low, with all country and regional rates within 15 per cent, and most within 10 per cent, of the UK and Ireland average (Figure 19.1). Of the five countries, Wales had the highest incidence rate of pancreatic cancer for males, and Scotland for females, although these were only 6 and 5 per cent, respectively, higher than the overall average. Rates were lowest for both sexes in Northern Ireland, markedly so for females (15 per cent below average). Incidence rates in Ireland for both males and females were lower than recorded mortality rates (see next section). Within England, regional rates were high in London (11 and 6 per cent above the averages for males and females, respectively), and low in the South West region (9 per cent below average for both sexes) and for males in the Eastern region (5 per cent below average). At the country and region scale, incidence rates for males and females showed a good correlation, with low rates for both sexes in Northern Ireland and South West England and high rates in London.

At the level of health authority, there was greater variability (Figure 19.3 and Map 19.1). Incidence was relatively high among both males and females in Lambeth, Southwark and Lewisham; Manchester; North Wales; and Bexley, Bromley and Greenwich; and relatively low in both sexes in Norfolk; South Cheshire; and Avon. Otherwise, there was little correlation in rates between males and females at the health authority level. In males, rates were relatively high in Rotherham; Tees; Barking and Havering; and Merton, Sutton and Wandsworth (the very high rate in the Western Isles of Scotland should be viewed with caution as this was based on an average of only three cases per year). Relatively high rates in females were found in Dumfries and Galloway; the Western area in Ireland; Northamptonshire; Sunderland; and South Humber.

True geographic variation in incidence may be complicated by variation in diagnostic accuracy, especially among older patients. This is potentially a serious problem for pancreatic cancer, which has a higher proportion of incident cases lacking microscopic verification (diagnosis at the cellular level) than other cancers included in this atlas.[11] A high proportion of clinically diagnosed cases may lead to under-ascertainment, if such cases are more difficult for cancer registries to identify, or to over-diagnosis, if some patients in fact do not have pancreatic cancer. A related problem is that, for this highly fatal cancer, a high proportion of cases may be registered using only information from a death certificate (DCO cases), some of which may be of dubious validity.

Geographical patterns in mortality

Because pancreatic cancer is so rapidly fatal, geographic patterns for mortality can, in general, be expected to closely match those for incidence. But high fatality could also lead to a high proportion of cases being identified from death certificates, either initially or as the only source. This could exaggerate the similarity between incidence and mortality patterns. There was, in fact, a reasonably good agreement, at the level of country and English region, between incidence and mortality rates (compare Figures 19.1 and 19.2). However, mortality rates in Ireland appeared disproportionately high – 15 per cent above the UK and Ireland average for males, 11 per cent for females. Mortality rates for the other countries did not differ markedly from the average, and regional variation within England was also small, apart from slightly lower rates in the Eastern region for both males and females (both 5 per cent below average). For each sex, there was moderately good agreement at the health authority level between mapped incidence and mortality for England and Scotland (compare Maps 19.1 and 19.2). Rankings of health authorities within each country or region also showed reasonably good agreement between incidence and mortality (Table B19.1). There was, as with incidence, a fairly good correlation at the country and regional level between mortality rates for males and females, but only moderate or slight agreement at the health authority level.

In Ireland (especially in males), Northern Ireland, and South West England (males only) the mortality-to-incidence ratios were greater than one (Table B19.1), which may suggest under-registration of incident cases in these populations. Alternatively, in some instances investigation by the cancer registry prompted by the death certificate may have indicated that pancreatic cancer was not in fact involved. This could reflect regional variation in the accuracy of death certification or in the extent to which death certificate notifications are investigated further.

Risk factors and aetiology

Despite the importance of pancreatic cancer as a major cause of cancer mortality, remarkably little is known about its aetiology.[12] Tobacco smoking is the most consistently established risk factor, at least doubling a person's risk of developing pancreatic cancer, but is estimated to account for less than 30 per cent of cases.[12-14] Dietary factors are also considered likely to be involved, with the strongest evidence relating to a probable protective effect of vegetable and fruit consumption.[15]

(continued on page 210)

Figure **19.1**

**Pancreas: incidence by sex, country, and region of England
UK and Ireland 1991-99[1]**

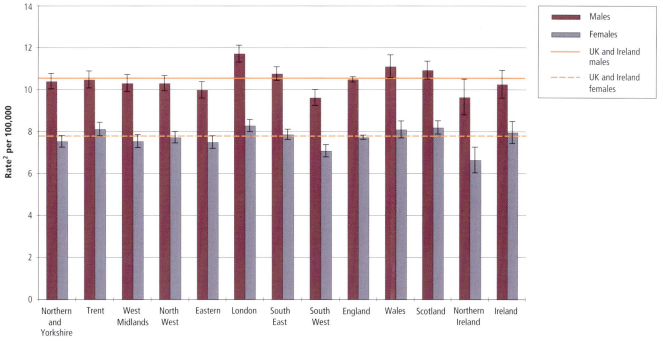

1 Northern Ireland 1993-99, Ireland 1994-99

2 Age standardised using the European standard population, with 95% confidence interval

Figure **19.2**

**Pancreas: mortality by sex, country, and region of England
UK and Ireland 1991-2000[1]**

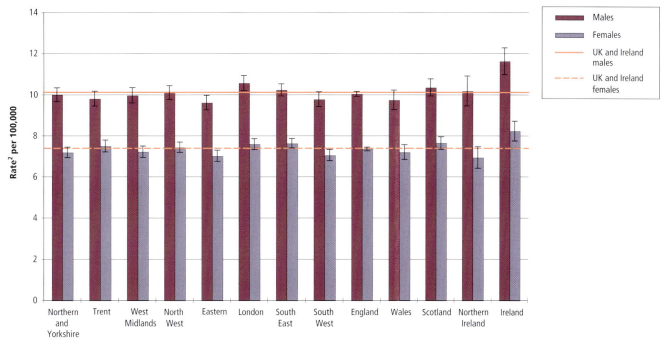

1 Scotland 1991-99, Ireland 1994-2000

2 Age standardised using the European standard population, with 95% confidence interval

Figure **19.3a**

**Pancreas: incidence by health authority within country, and region of England
Males, UK and Ireland 1991-99[1]**

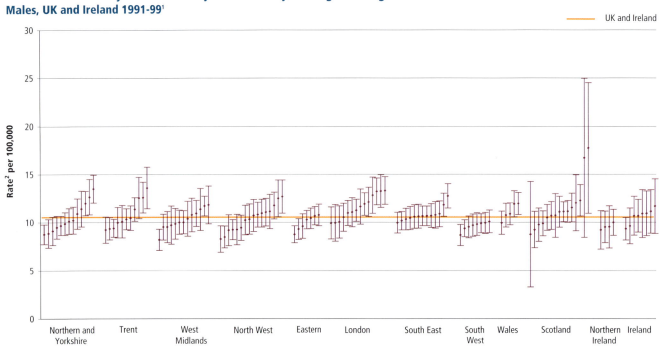

1 Northern Ireland 1993-99, Ireland 1994-99

2 Age standardised using the European standard population, with 95% confidence interval

Figure **19.3b**

**Pancreas: incidence by health authority within country, and region of England
Females, UK and Ireland 1991-99[1]**

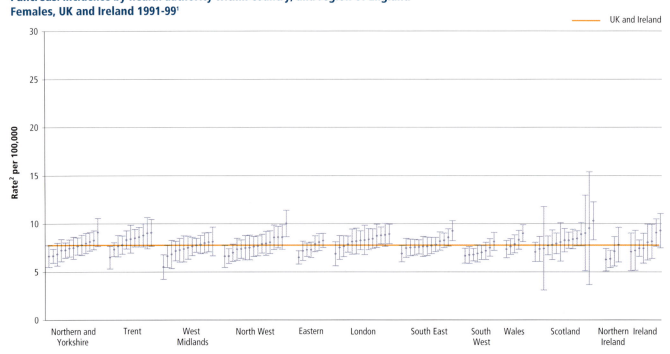

1 Northern Ireland 1993-99, Ireland 1994-99

2 Age standardised using the European standard population, with 95% confidence interval

Figure **19.4a**

**Pancreas: mortality by health authority within country, and region of England
Males, UK and Ireland 1991-2000[1]**

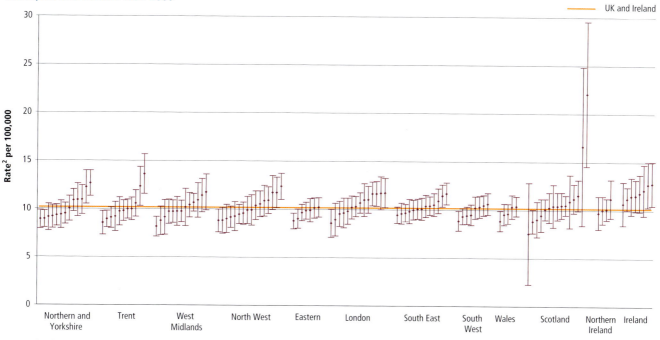

1 Scotland 1991-99, Ireland 1994-2000

2 Age standardised using the European standard population, with 95% confidence interval

Figure **19.4b**

**Pancreas: mortality by health authority within country, and region of England
Females, UK and Ireland 1991-2000[1]**

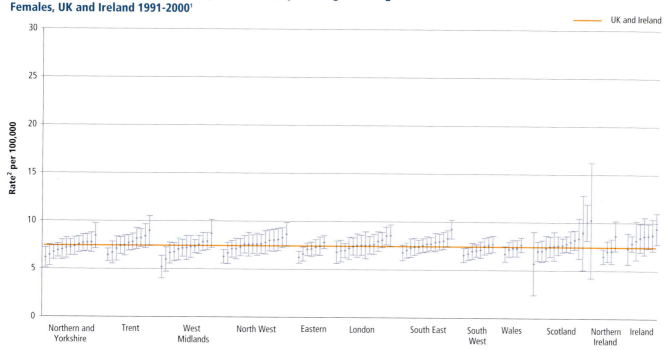

1 Scotland 1991-99, Ireland 1994-2000

2 Age standardised using the European standard population, with 95% confidence interval

Map 19.1a

**Pancreas: incidence* by health authority
Males, UK and Ireland 1991-99**

Ratio*

	1.5 and over
	1.33 to 1.5
	1.1 to 1.33
	0.91 to 1.1
	0.75 to 0.91
	0.67 to 0.75
	Under 0.67

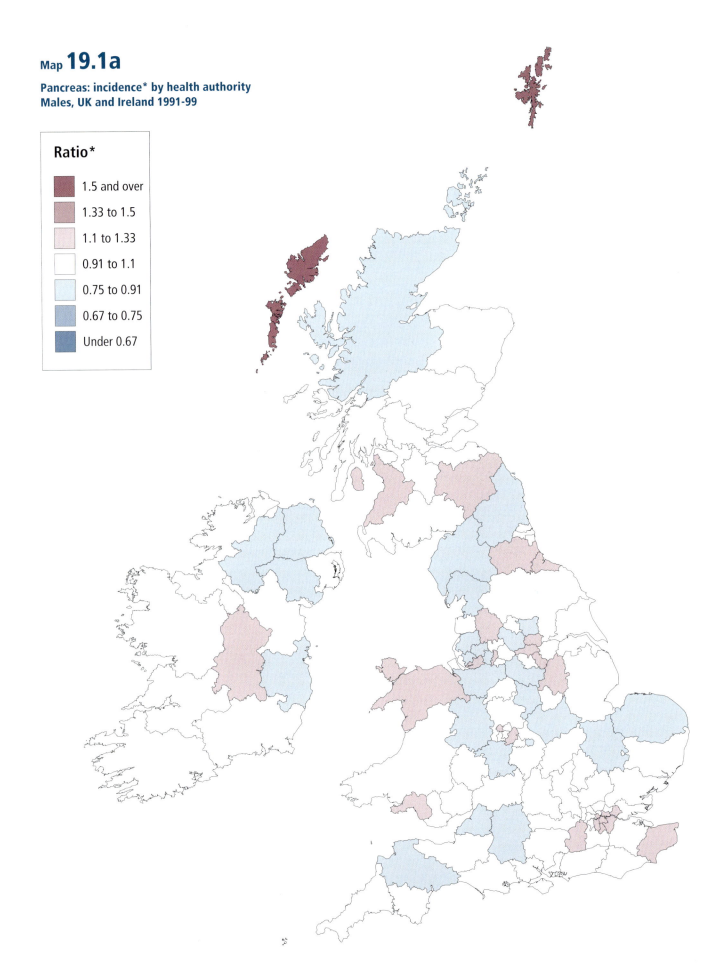

*Ratio of directly age-standardised rate in health authority to UK and Ireland average

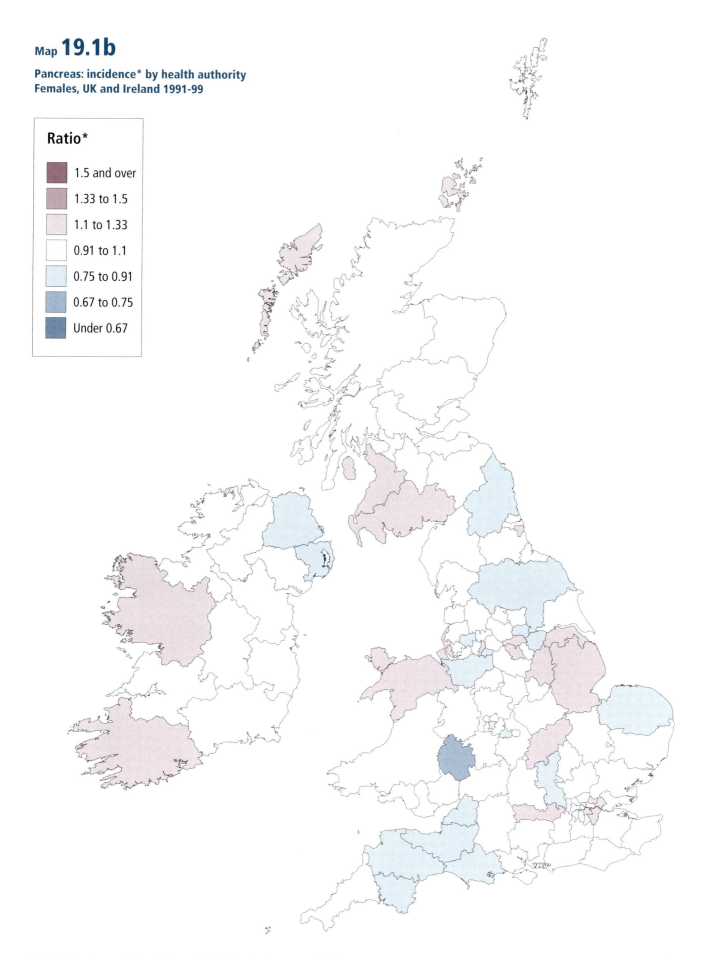

Map **19.1b**

Pancreas: incidence* by health authority
Females, UK and Ireland 1991-99

Ratio*

- 1.5 and over
- 1.33 to 1.5
- 1.1 to 1.33
- 0.91 to 1.1
- 0.75 to 0.91
- 0.67 to 0.75
- Under 0.67

Ratio of directly age-standardised rate in health authority to UK and Ireland average

Map **19.2a**

Pancreas: mortality* by health authority
Males, UK and Ireland 1991-2000

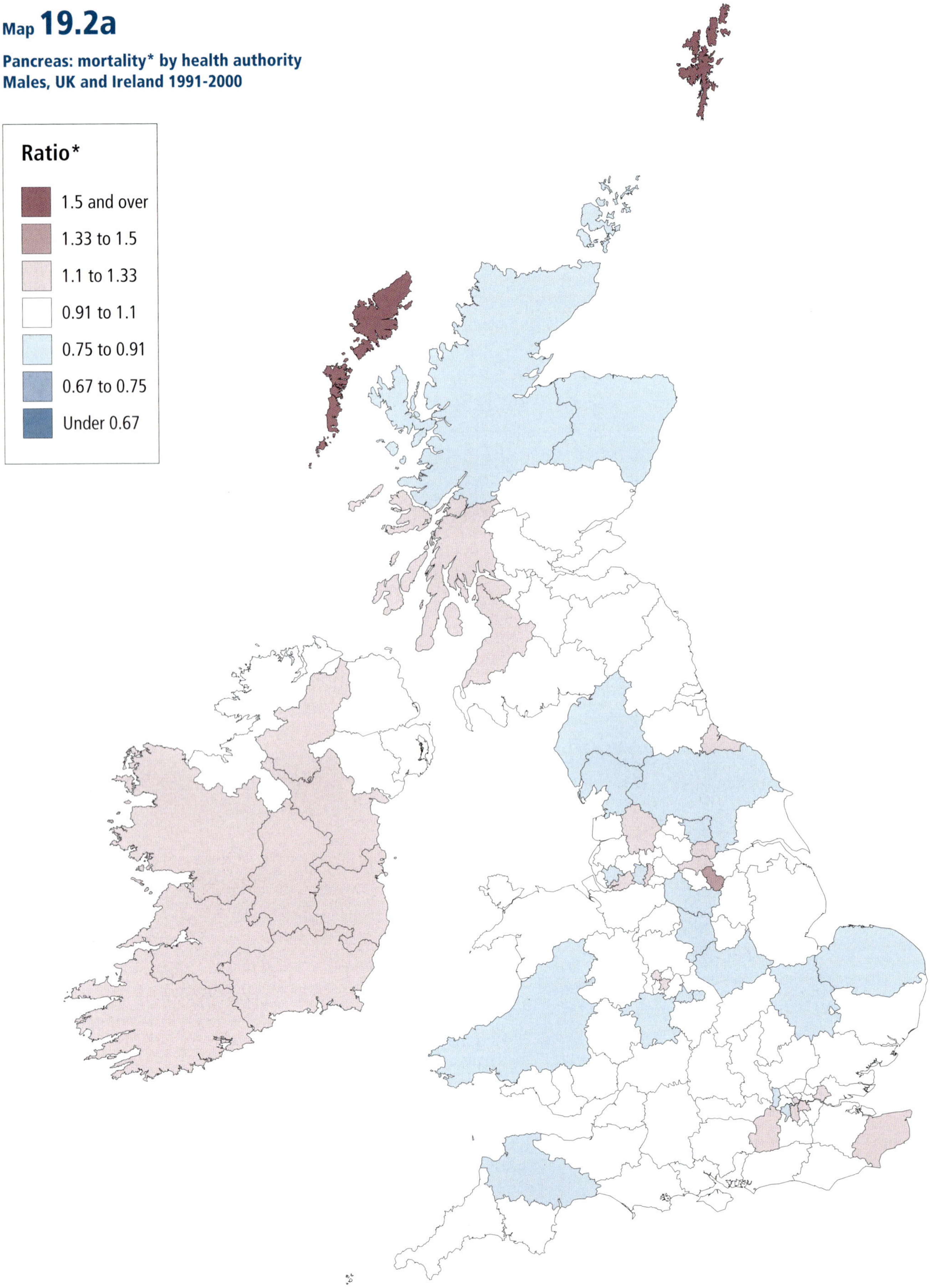

Ratio*

■	1.5 and over
■	1.33 to 1.5
■	1.1 to 1.33
□	0.91 to 1.1
■	0.75 to 0.91
■	0.67 to 0.75
■	Under 0.67

*Ratio of directly age-standardised rate in health authority to UK and Ireland average

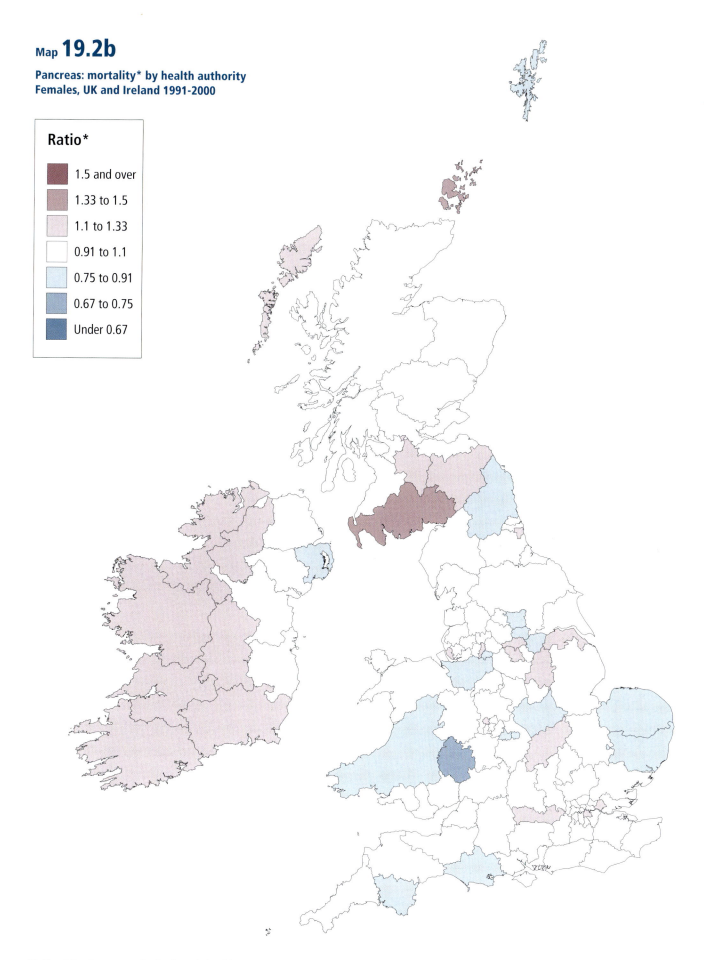

Map **19.2b**

Pancreas: mortality* by health authority
Females, UK and Ireland 1991-2000

Ratio*

	1.5 and over
	1.33 to 1.5
	1.1 to 1.33
	0.91 to 1.1
	0.75 to 0.91
	0.67 to 0.75
	Under 0.67

*Ratio of directly age-standardised rate in health authority to UK and Ireland average

For alcohol and coffee consumption, well-studied factors which some studies suggest increase risk, the consensus is that there is probably no causal, or quantitatively important, relationship.[12,15] For other dietary factors, available evidence either indicates a possible relationship to pancreatic cancer risk (for example, meat consumption, high energy intake) or is considered too inconsistent to allow judgement (for example, dietary fat).[15] Occupational variation in risk has been widely studied, but evidence from these studies, although suggestive of risk factors such as pesticides, does not support a major influence of such exposures on population rates of this cancer.[12] A small proportion (10 per cent or less) of pancreatic cancer cases reflect a family history of the disease or certain hereditary conditions, including hereditary non-polyposis colon cancer, pancreatic enzyme deficiencies and mutations in the BRCA-2 gene.[16] There is also evidence that risk is increased in patients with diabetes mellitus or chronic pancreatitis, and less consistent evidence for a number of other conditions.[12] At a geographic level, although there are no effective means of screening for pancreatic cancer, variations in diagnostic criteria between populations may obscure true variation in underlying risk.[12] This may well be the case for countries or regions within the UK and Ireland. The unquantified role of risk factors other than smoking also complicates the interpretation of mapped rates. In relation to smoking itself, geographic patterns of pancreatic cancer in the UK and Ireland appear to show only limited agreement with those for lung cancer (see Chapter 13).

Socio-economic deprivation

In England and Wales during 1992-93, incidence and mortality rates tended to be slightly higher in more deprived populations.[4] Scottish incidence data for 1986-95 also showed a fairly clear increase in rates with increases in deprivation, at least in males.[5] In both Ireland and, especially, Northern Ireland during 1989-98, pancreatic cancer mortality rates among males were significantly higher in the lowest compared with the highest occupation-defined social groups.[17] None of these trends was as clear cut as for lung cancer. Nor is it as easy to relate mapped rates of pancreatic cancer to deprived populations (see Appendix F) as it is for lung cancer (Chapter 13). There is only limited evidence that survival from pancreatic cancer, which is extremely low overall, is any worse in patients from more deprived areas (data for England and Wales).[4,18]

References

1. Schouten LJ, Straatman H, Kiemeney LALM, Verbeek ALM. Cancer incidence: Life table risk versus cumulative risk. *Journal of Epidemiology and Community Health* 1994; 48: 596-600.

2. ONS. *Cancer Statistics Registrations: Registrations of cancer diagnosed in 2001, England.* Series MB1 No. 32. London: Office for National Statistics, 2004.

3. Quinn MJ, Babb PJ, Kirby L, Jones J. Registrations of cancer diagnosed in 1994-97, England and Wales. *Health Statistics Quarterly* 2000; 7: 71-82.

4. Quinn MJ, Babb PJ, Brock A, Kirby L et al. *Cancer Trends in England and Wales 1950-1999.* Studies on Medical and Population Subjects No. 66. London: The Stationery Office, 2001.

5. Harris V, Sandridge AL, Black RJ, Brewster DH et al. *Cancer Registration Statistics Scotland, 1986-1995.* Edinburgh: ISD Publications, 1998.

6. Black RJ, Macfarlane GJ, Maisonneuve P, Boyle P. *Cancer Incidence and Mortality in Scotland 1960-89.* Edinburgh: ISD Publications, 1995.

7. ONS. Cancer Survival: England and Wales, 1991-2001. March 2004. Available at *http://www.statistics.gov.uk/statbase/ssdataset.asp?vlnk=7899.*

8. ISD Scotland. *Trends in Cancer Survival in Scotland, 1977-2001.* Edinburgh: ISD Publications, 2004.

9. Fitzpatrick D, Gavin A, Middleton R, Catney D. *Cancer in Northern Ireland 1993-2001: A Comprehensive Report.* Belfast: Northern Ireland Cancer Registry, 2004.

10. Sant M, Areleid T, Berrino F, Bielska Lasota M et al. EUROCARE-3: survival of cancer patients diagnosed 1990-1994 - results and commentary. *Annals of Oncology* 2003; 14 Suppl 5: v61-v118.

11. Parkin DM, Whelan SL, Ferlay J, Raymond L et al. *Cancer Incidence in Five Continents Vol. VII.* IARC Scientific Publications No. 143. Lyon: International Agency for Research on Cancer, 1997.

12. Anderson KE, Potter JD, Mack TM. Pancreatic Cancer. In: Schottenfeld D, Fraumeni JF, Jr. (eds) *Cancer Epidemiology and Prevention, second edition.* New York: Oxford University Press, 1996.

13. Kuper H, Boffetta P, Adami HO. Tobacco use and cancer causation: association by tumour type. *Journal of Internal Medicine* 2002; 252: 206-224.

14. Twigg L, Moon G, Walker S. *The Smoking Epidemic in England.* London: Health Development Agency, 2004.

15. Potter, J. D. *Food, Nutrition and the Prevention of Cancer: a Global Perspective.* Washington: World Cancer Research Fund in association with American Institute for Cancer Research, 1997.

16. Ghadirian P, Lynch HT, Krewski D. Epidemiology of pancreatic cancer: an overview. *Cancer Detection and Prevention* 2003; 27: 87-93.

17. Balanda, K. P., Wilde, J. *Inequalities in Mortality 1989-1998: a Report on All-Ireland Mortality Data.* Dublin: The Institute of Public Health, 2001.

18. Coleman MP, Babb P, Damiecki P, Grosclaude P et al. *Cancer Survival Trends in England and Wales, 1971-1995: Deprivation and NHS Region.* Studies on Medical and Population Subjects No. 61. London: The Stationery Office, 1999.

Chapter 20

Prostate

Henrik Møller, Mike Quinn

Summary

- In the UK and Ireland in the 1990s, prostate cancer accounted for 1 in 6 cancer cases and 1 in 9 cancer deaths in males.

- There was a band of slightly lower than average incidence across Northern Ireland and northern England, and a band of slightly higher than average incidence across the south of Ireland, Wales, London and southern England.

- A similar pattern was observed for mortality, although there was less variation than for incidence.

- Geographical variations in incidence may, to some extent, be explained by regional differences in the availability and uptake of prostate-specific antigen (PSA) testing.

- Specific causal factors for prostate cancer have not been identified; the higher incidence in London and the south of England suggests an association with affluence, although whether affluence is a marker of an as yet unidentified risk factor is not clear.

Incidence and mortality

In males in the UK and Ireland, prostate cancer accounted for 17 per cent of all cancer cases and 12 per cent of all cancer deaths in the 1990s. This made prostate cancer the second commonest malignancy in males (after lung cancer), in terms of both incidence and mortality. In the 1990s the average number of prostate cancer cases per year in the UK and Ireland was 22,500, and the annual number of deaths was about 10,000. The corresponding age-standardised incidence and mortality rates were 65 and 29 per 100,000, respectively. Prostate cancer is typically a disease of old age and incidence in men aged 85 and over was around 1,000 per 100,000 (one per cent).

Incidence and mortality trends

The age-standardised incidence of prostate cancer has doubled in the last three decades in the UK.[1] Similar increases were seen in other populations.[2-4] The recent trends are thought to be influenced mainly by changes in medical practice; a reduction in the number of cases found incidentally with transurethral resection for benign prostatic hyperplasia or urinary obstruction (leading to a decrease in recorded prostate cancer incidence) has been far outweighed by increasing use of PSA testing in the 1990s (leading to an increasing recorded incidence of localised prostate cancer).[5-9] The age-standardised mortality rates have been much more stable over time.[10,11]

Survival

In comparison with most other cancers, survival from prostate cancer is relatively high, with five-year relative survival of 65 per cent for patients diagnosed in England and Wales in 1996-99. One-year relative survival was very high at 87 per cent.[12] Survival rates were similar for patients diagnosed in the 1990s in Ireland,[13] Northern Ireland[14] and Scotland.[15] The estimated survival rates are highly sensitive to the inclusion of cases of localised cancer, detected by PSA testing. These cases tend to have a good prognosis, which has the effect of increasing the survival estimates. Using 'period' analysis, which accommodates recent changes in survival, the ten-year relative survival of prostate cancer patients in South East England has been estimated to have increased from 40 to 62 per cent from 1995 to 2000.[16]

Geographical patterns in incidence

There was remarkably little variation in prostate cancer incidence within the UK and Ireland (Map 20.1), although any real variations in incidence are likely to have been masked by the consequences of different rates of PSA testing.

Table B20.1 shows that the range of age-standardised incidence rates varied from 56 per 100,000 in Trent to 75 per 100,000 in Ireland, a difference of 34 per cent. As well as in Trent, incidence was below the UK and Ireland average in Northern and Yorkshire; Northern Ireland; and the North West. Incidence was above average in the South East, South West, Wales, Scotland and London, in addition to Ireland. Despite these apparent variations 9 of the 13 countries and regions had an incidence rate within the interval 60 to 70 per 100,000 (Figure 20.1).

Considering the variations between health authority areas within countries and regions (Figure 20.3 and Map 20.1), two outliers with relatively high values are evident: Southampton and South West Hampshire; and Dorset – on the south coast of England. Three areas had relatively low rates: Sheffield; and Rotherham in the Trent region, and Gateshead and South Tyneside in the north east of England. It is perhaps noteworthy that the two areas with high incidence are adjacent, as are the two areas in Trent with low incidence.

(continued on page 216)

Figure **20.1**

**Prostate: incidence by country, and region of England
UK and Ireland 1991-99[1]**

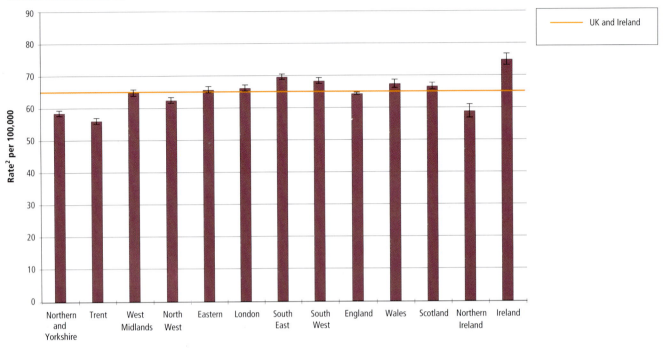

1 Northern Ireland 1993-99, Ireland 1994-99

2 Age standardised using the European standard population, with 95% confidence interval

Figure **20.2**

**Prostate: mortality by country, and region of England
UK and Ireland 1991-2000[1]**

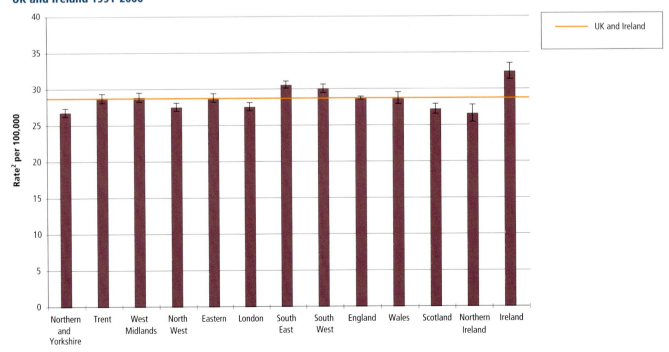

1 Scotland 1991-99, Ireland 1994-2000

2 Age standardised using the European standard population, with 95% confidence interval

Figure **20.3**

**Prostate: incidence by health authority within country, and region of England
UK and Ireland 1991-99[1]**

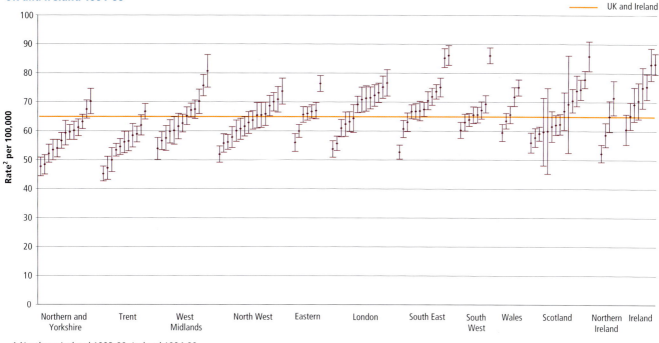

1 Northern Ireland 1993-99, Ireland 1994-99

2 Age standardised using the European standard population, with 95% confidence interval

Figure **20.4**

**Prostate: mortality by health authority within country, and region of England
UK and Ireland 1991-2000[1]**

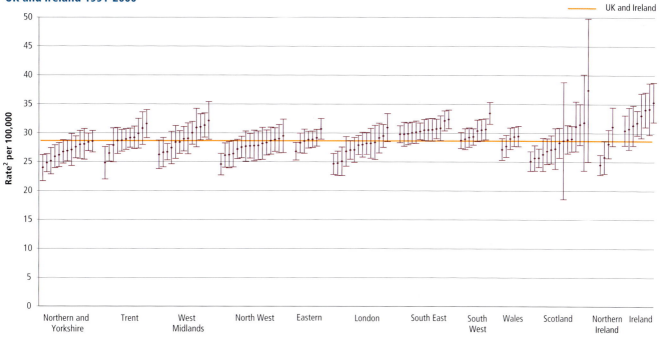

1 Scotland 1991-99, Ireland 1994-2000

2 Age standardised using the European standard population, with 95% confidence interval

Map **20.1**

Prostate: incidence* by health authority
UK and Ireland 1991-99

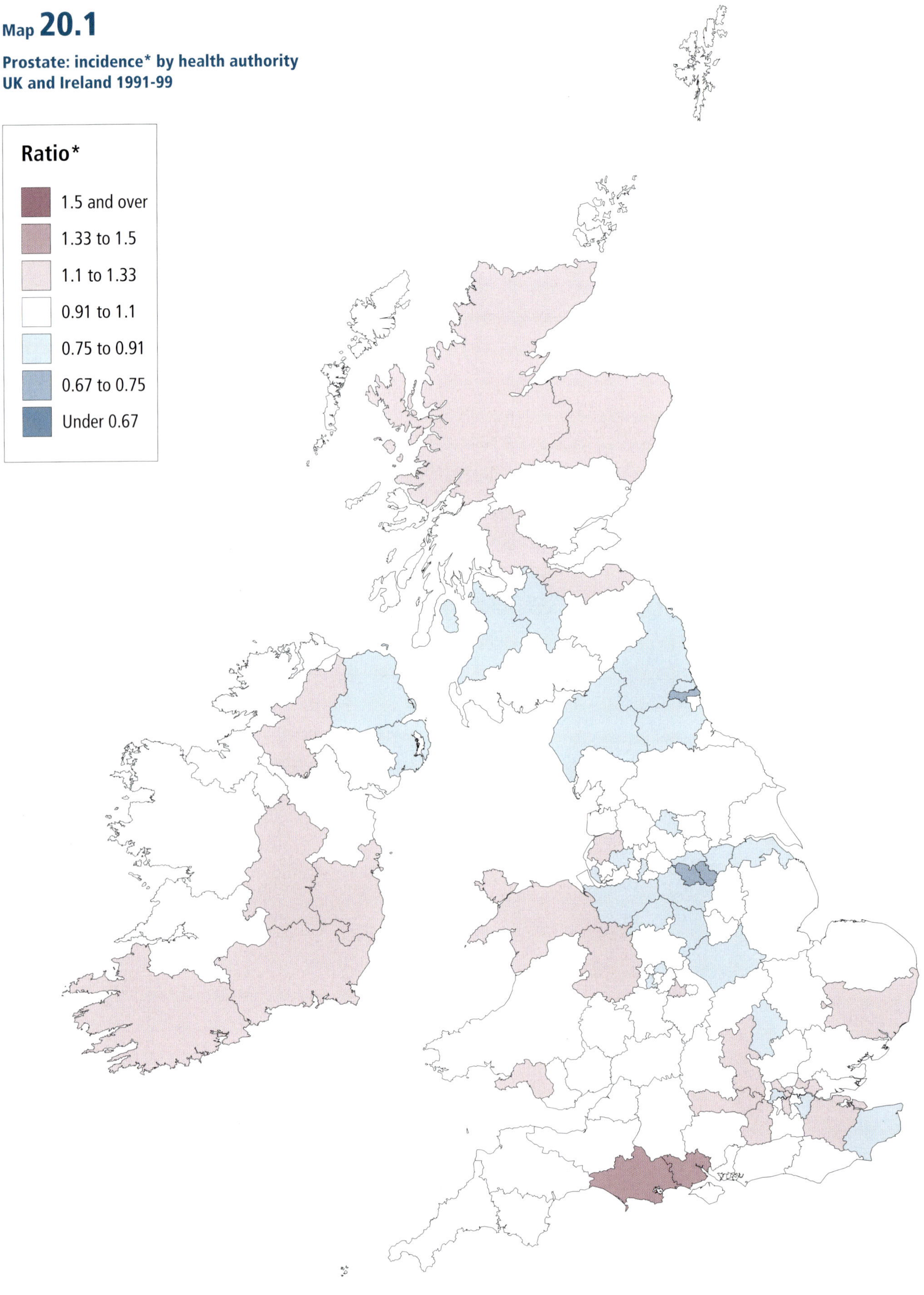

Ratio*

- 1.5 and over
- 1.33 to 1.5
- 1.1 to 1.33
- 0.91 to 1.1
- 0.75 to 0.91
- 0.67 to 0.75
- Under 0.67

*Ratio of directly age-standardised rate in health authority to UK and Ireland average

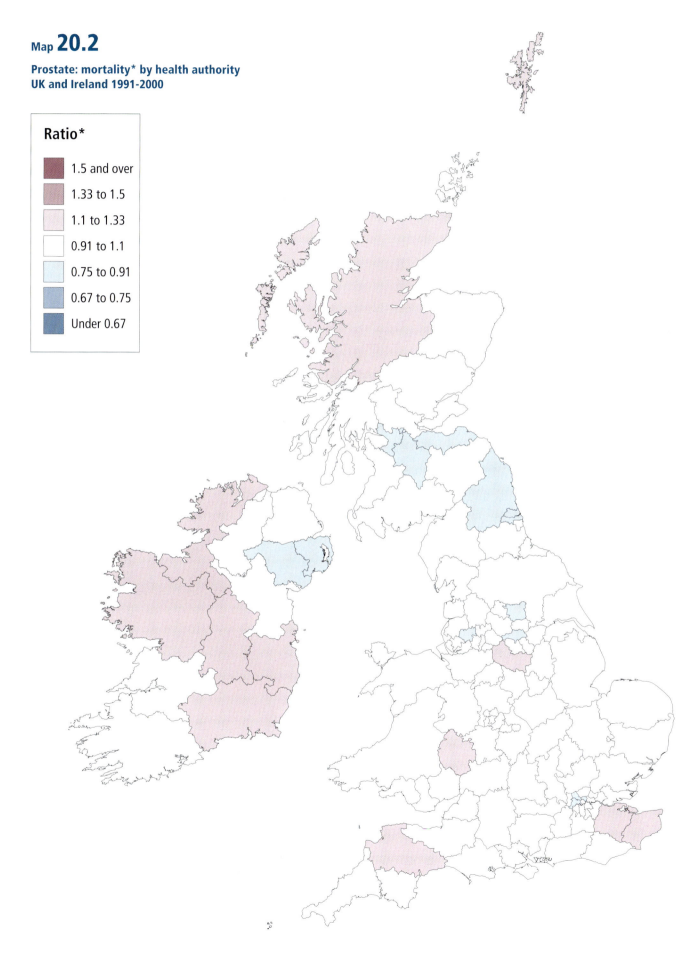

Map 20.2

**Prostate: mortality* by health authority
UK and Ireland 1991-2000**

Ratio*

- 1.5 and over
- 1.33 to 1.5
- 1.1 to 1.33
- 0.91 to 1.1
- 0.75 to 0.91
- 0.67 to 0.75
- Under 0.67

*Ratio of directly age-standardised rate in health authority to UK and Ireland average

However, a detailed analysis of the geographical epidemiology of prostate cancer in Great Britain did not find any evidence of clustering at the level of 10,510 electoral wards.[17]

Considered overall, there is a weak geographical pattern across the UK and Ireland in prostate cancer incidence of slightly lower incidence in the central area comprising Northern Ireland, the north of the Northern and Yorkshire region, and a band stretching across England from South Cheshire in the North West to South Humber in the Trent region. In particular, rates were lower than average in 10 out of 13 health authorities in Northern and Yorkshire, and in 9 of 11 health authorities in Trent. An analysis of the spread of the practice of radical prostatectomy in England, secondary to the use of testing for PSA, found that the slowest uptake of this practice was in the Trent region.[18]

Areas of higher than average incidence were more scattered, with clusters in the south east of Ireland, the north of Scotland, and London and South East England. The particularly high rates in two health authorities on the south coast of England are likely to be related to the high proportions of elderly people living in these areas.

Internationally, prostate cancer incidence is highest in North America, Europe and Australia. Ireland and the UK have intermediate levels of incidence within the European range.[19]

Geographical patterns in mortality

The variation in prostate cancer mortality in the UK and Ireland was similar to the variation in incidence, with slightly lower mortality rates in the central area comprising Northern Ireland, the far north east and the north west of England, and higher rates in Ireland (Map 20.2).

The range of variation between countries and regions was smaller for mortality than for incidence (22 per cent higher in Ireland than in Northern Ireland). Mortality was slightly below average in Northern and Yorkshire; Scotland; the North West; and London, in addition to Northern Ireland, and slightly above average in the South East and South West of England, as well as in Ireland. The ranked values in Table B20.1, and the map of mortality rates (Map 20.2) do not reveal any particular outliers or clusters. This pattern of less variation in mortality than incidence is observed across countries (Figure 20.2) as well as within countries (Figure 20.4), emphasising the likely role of PSA testing in detecting many asymptomatic cancers where it is widely used.[20]

The mortality-to-incidence ratio for prostate cancer was particularly high in Trent: 0.51 compared to 0.44 for the UK and Ireland overall (Table B20.1), which suggests lower uptake of PSA testing, some under-ascertainment of incident cases, or

lower survival of cases in the Trent region. A national analysis of cancer survival did not suggest particularly low survival rates in Trent.[21]

Risk factors and aetiology

The aetiology of prostate cancer is not well known. Despite evidence of important variations in incidence from international comparisons and studies of migrants, specific causal factors (such as environmental, life-style, diet, and occupation) have not been identified conclusively. An area of current research is diet and obesity.[22,23] In the past, many studies have explored reproductive characteristics and sexual habits, including frequency of intercourse and masturbation, but with inconsistent results.[24]

It is clear that black men have higher incidence than white men. In the SEER Program of the United States National Cancer Institute, the age-standardised incidence in black men is about 70 per cent higher than in white men.[19] In the UK, the largest concentrations of black people are in London and particularly in Lambeth, Southwark and Lewisham (19 per cent); and East London and the City (14 per cent). Neither these areas, nor London as a whole, appear to have particularly high incidence rates of prostate cancer (Map 20.2).

Consistent with the relatively high incidence in Ireland observed in this study, a study of Irish migrants found increased incidence in Irish men living in the UK.[25]

The main difficulty with aetiological research in prostate cancer is the heterogeneity of the disease, ranging from highly prevalent but clinically indolent cancers, to highly aggressive and often fatal disease.[26] Any assessment of incidence or survival will be highly sensitive to the intensity of diagnostic procedures in the community (for example, through testing for PSA) and to temporal and spatial variation in the use of such procedures. The mortality rate is much less influenced by such changes. In research into possible causes of prostate cancer it is imperative that asymptomatic, PSA-detected disease is considered separately from symptomatic disease, as the risk factors could be different. Because of these issues, it may be preferable to use prostate cancer mortality as an endpoint, even in research into possible causes.

Socio-economic deprivation

In England and Wales in the early 1990s, the incidence of prostate cancer was about 45 per cent higher in the most affluent groups compared with the most deprived; the gradient in prostate cancer mortality was of similar magnitude.[1] Survival was higher in men from affluent areas than in men from deprived areas, and the gap of about 7 percentage points in five-year relative survival between these two groups was among the largest for the major types of cancer (data for England and Wales, patients diagnosed during 1996-99).[27]

The socio-economic gradient in incidence in the early 1990s is more likely to have been due to genuine variations in the occurrence of disease rather than over-diagnosis in the most affluent group.[1] In the case of over-diagnosis we would have expected less variation in mortality than was observed, and more variation in survival. The rapid increases in the apparent incidence of prostate cancer in the late 1990s (and into the twenty-first century) may have widened the gap in rates between the affluent and the deprived.

References

1. Quinn MJ, Babb PJ, Brock A, Kirby L et al. *Cancer Trends in England and Wales 1950-1999*. Studies on Medical and Population Subjects No. 66. London: The Stationery Office, 2001.

2. Harris V, Sandridge AL, Black RJ, Brewster DH et al. *Cancer Registration Statistics Scotland, 1986-1995*. Edinburgh: ISD Publications, 1998.

3. National Cancer Registry of Ireland. *Cancer in Ireland, 1996. Incidence and Mortality*. Cork: National Cancer Registry, 1999.

4. Gavin AT, Reid J. *Cancer Incidence in Northern Ireland*, 1993-95. Belfast: The Stationery Office, 1999.

5. Oliver SE, Gunnell D, Donovan JL. Comparison of trends in prostate-cancer mortality in England and Wales and the USA. *Lancet* 2000; 355: 1788-1789.

6. Brewster DH, Fraser LA, Harris V, Black RJ. Rising incidence of prostate cancer in Scotland: increased risk or increased detection? *BJU International* 2000; 85: 463-472.

7. Majeed A, Babb P, Jones J, Quinn M. Trends in prostate cancer incidence, mortality and survival in England and Wales 1971-1998. *BJU International* 2000; 85: 1058-1062.

8. Moller H. Trends in incidence of testicular cancer and prostate cancer in Denmark. *Human Reproduction* 2001; 16: 1007-1011.

9. Evans HS, Moller H. Recent trends in prostate cancer incidence and mortality in southeast England. *European Urology* 2003; 43: 337-341.

10. Quinn M, Babb P. Patterns and trends in prostate cancer incidence, survival, prevalence and mortality. Part I: international comparisons. *BJU International* 2002; 90: 162-173.

11. Quinn M, Babb P. Patterns and trends in prostate cancer incidence, survival, prevalence and mortality. Part II: individual countries. *BJU International* 2002; 90: 174-184.

12. ONS. Cancer Survival: England and Wales, 1991-2001. March 2004. Available at *http://www.statistics.gov.uk/statbase/ssdataset.asp?vlnk=7899*.

13. National Cancer Registry of Ireland. *Cancer in Ireland, 1994 to 1998: Incidence, mortality, treatment and survival*. Cork: National Cancer Registry, 2001.

14. Fitzpatrick D, Gavin A, Middleton R, Catney D. *Cancer in Northern Ireland 1993-2001: A Comprehensive Report*. Belfast: Northern Ireland Cancer Registry, 2004.

15. Sant M, Areleid T, Berrino F, Bielska Lasota M et al. EUROCARE-3: survival of cancer patients diagnosed 1990-1994 - results and commentary. *Annals of Oncology* 2003; 14 Suppl 5: v61-v118.

16. Thames Cancer Registry. *Cancer in South East England 2000*. London: Thames Cancer Registry, 2003.

17. Jarup L, Best N, Toledano MB, Wakefield J et al. Geographical epidemiology of prostate cancer in Great Britain. *International Journal of Cancer* 2002; 97: 695-699.

18. Oliver SE, Donovan JL, Peters TJ, Frankel S et al. Recent trends in the use of radical prostatectomy in England: the epidemiology of diffusion. *BJU International* 2003; 91: 331-336.

19. Parkin DM, Whelan SL, Ferlay J, Teppo L et al. *Cancer Incidence in Five Continents Vol. VIII*. IARC Scientific Publications No. 155. Lyon: International Agency for Research on Cancer, 2000.

20. Tretli S, Engeland A, Haldorsen T, Hakulinen T et al. Prostate cancer - look to Denmark? *Journal of the National Cancer Institute* 1996; 88: 128.

21. Coleman MP, Babb P, Damiecki P, Grosclaude P et al. *Cancer Survival Trends in England and Wales, 1971-1995: Deprivation and NHS Region*. Studies on Medical and Population Subjects No. 61. London: The Stationery Office, 1999.

22. Key TJ, Allen N, Appleby P, Overvad K et al. Fruits and vegetables and prostate cancer: no association among 1104 cases in a prospective study of 130544 men in the European Prospective Investigation into Cancer and Nutrition (EPIC). *International Journal of Cancer* 2004; 109: 119-124.

23. Okasha M, McCarron P, McEwen J, Smith GD. Body mass index in young adulthood and cancer mortality: a retrospective cohort study. *Journal of Epidemiology and Community Health* 2002; 56: 780-784.

24. Leitzmann MF, Platz EA, Stampfer MJ, Willett WC et al. Ejaculation frequency and subsequent risk of prostate cancer. *Journal of the American Medical Association* 2004; 291: 1578-1586.

25. Harding S, Rosato M. Cancer incidence among first generation Scottish, Irish, West Indian and South Asian migrants living in England and Wales. *Ethnicity and Health* 1999; 4: 83-92.

26. Breslow N, Chan CW, Dhom G, Drury RA et al. Latent carcinoma of prostate at autopsy in seven areas. The International Agency for Research on Cancer, Lyon, France. *International Journal of Cancer* 1977; 20: 680-688.

27. Coleman MP, Rachet B, Woods LM, Mitry E et al. Trends and socioeconomic inequalities in cancer survival in England and Wales up to 2001. *British Journal of Cancer* 2004; 90: 1367-1373.

Chapter 21
Stomach

John Steward, Helen Wood

Summary

- In the UK and Ireland, stomach cancer accounts for about 1 in 25 diagnosed cases of cancer and about 1 in 20 deaths from cancer.

- There was a clear north-south divide in incidence across England, with higher rates in the midlands and north, and lower rates in the south and east.

- Rates were also above average in Scotland and below average in Ireland.

- The pattern of variation in mortality was very similar to that for incidence; for both, the north-south divide was more extreme for females than for males.

- Incidence was generally higher in urban areas and lower in more rural locations.

- High incidence in the industrialised areas of Scotland and the midlands and north of England confirm the known relationship with socio-economic deprivation, which is likely to be a marker for known risk factors such as smoking, poor diet, and helicobacter pylori (H pylori) infection.

- Geographical variations in stomach cancer are similar to those in lung and oesophageal cancer, although the risk factors differ somewhat.

Incidence and mortality

In the UK and Ireland in the 1990s there were around 7,000 newly diagnosed cases of stomach cancer each year in males, in whom it was the fifth most common cancer, and around 4,200 cases in females, in whom it was the sixth most common. Stomach cancer accounted for 4 per cent of all newly diagnosed cases per year; in males the figure was 5 per cent and in females was 3 per cent. The age-standardised incidence rates were 21 and 8 per 100,000 in males and females, respectively. The ratio of the number of cases in males to females was 1.7:1, but the ratio of the age-standardised rates was higher at 2.5:1.

Stomach cancer is predominantly a disease of the elderly, with age-specific rates in England and Wales increasing steeply above age 50 to peak in both males and females aged 85 years

and over.[1] Owing to demographic changes in the population at risk, the absolute number of cases peaks at 70-74 years in males and then declines rapidly whereas that in females increases with age to peak at 85 and over. The lifetime risk[2,3] of being diagnosed with stomach cancer in England and Wales was 2.3 per cent for males and 1.2 per cent for females.[1]

In the 1990s, 5,100 males and 3,300 females on average died from stomach cancer each year in the UK and Ireland. Of all deaths from cancer (excluding non-melanoma skin cancer), 5 per cent were due to stomach cancer. The figure was 6 per cent in males, in whom this was the fourth most common cause of cancer death, and 4 per cent in females, in whom it was the sixth commonest. In males, the age-standardised mortality rate was 15 per 100,000 and in females it was 6 per 100,000. The male-to-female ratio of the age-standardised mortality rates was very similar to that for incidence, 2.4:1. Following the pattern for incidence, age-specific mortality rates rose steeply with age to peak in both males and females aged 85 and over.[1]

Incidence and mortality trends

In 1980, stomach cancer was estimated to be the most common cancer in the world, but by 1990 it was the second commonest, after lung cancer.[4] The age-standardised incidence rate in England and Wales has been declining steadily since before the 1970s, at the rate of about one per cent a year, with a 40 per cent decline in males and a 50 per cent decline in females between 1971 and 1997.[1] The risk of developing stomach cancer fell substantially for females born in successive birth cohorts from the late nineteenth century up to the 1940s. The pattern for males was similar, but the decline was smaller, implying that the male-to-female ratio was lower at the beginning of the twentieth century than it is now. Mortality has been declining steadily in both sexes since before 1950 and has fallen slightly faster than incidence. In 1999, the age-standardised mortality rates in males and females were only 30 and 20 per cent, respectively, of those in 1950.

Survival

Survival from stomach cancer in England and Wales for cases diagnosed in 1996-99 was quite low – only around 33 per cent after one year and 13 per cent after five years in both men and women (relative survival).[5] The survival figures for cases diagnosed in Scotland in 1992-96,[6] in Northern Ireland in 1993-96,[7] and in Ireland in 1994-98[8] were similar. The frequently advanced stage of the disease at diagnosis, the aggressiveness of the disease, and the small number of patients who are suitable for curative surgery explain the low rates of survival from stomach cancer. There has been a small but consistent trend of improving five-year survival in England and

Wales.[9] However, in results from the EUROCARE-3 study for patients diagnosed in 1990-94, England, Scotland and Wales together with Denmark appeared to have one- and five-year survival rates that were below the European average and only slightly better than those in eastern Europe.[10]

Geographical patterns in incidence

Within the countries of the UK and Ireland, the highest incidence rates for stomach cancer in males occurred in Wales, where the incidence was 25 per cent higher. In females, the rate in both Wales and Scotland was 24 per cent higher than the average (Figure 21.1). A lower overall rate in England masks a very clear north-south divide, with incidence rates being higher than average in the North West; Northern and Yorkshire; and West Midlands, and below average in the South West; South East; and Eastern regions, and London. The notably lower overall incidence in Ireland may reflect the lower level of urbanisation and associated deprivation there compared with the UK.

There was wide variation in incidence both between health authorities within a country or region and between the constituent countries of the UK and regions of England (Figure 21.3). From the maps for incidence (Map 21.1), it can be seen that the north-south divide within England was more extreme in females than males. In Scotland, incidence was highest in the centre of the country, and in Wales rates were highest in the north and west.

Incidence was below average in all the health authorities in the South West of England and in the majority of those in the South East and Eastern regions, in both males and females. The pattern was less consistent in London, with rates close to or below average in most health authorities, but above average for both males and females in East London and the City. In Wales, incidence rates were above average in all health authorities for males and all but one for females. In the West Midlands there was a clear contrast between the high rates in the urban areas around Birmingham and in North Staffordshire, and the lower rates in the surrounding more rural areas such as Worcestershire and Herefordshire. Incidence was higher than average in most health authorities in North West; Northern and Yorkshire; and in the north western area of the Trent region, creating a band of high incidence across the country. The notable exception to this pattern was the largely rural area of North Yorkshire, where incidence was lower than in the surrounding health authorities.

Incidence rates were higher than average in both males and females in several health authorities in western and central Scotland, most notably in Greater Glasgow, and was below average in the Borders region. It should be noted that the small numbers of cases in the northern and western isles mean that the apparent rates in these health authorities should be viewed with caution (average of 0-1 case per year in each sex). In Ireland, rates were consistently lower than average in the Southern area for both sexes, and above average in the Eastern area, which includes Dublin. At the health authority level in Northern Ireland, incidence was close to or slightly above average, but there was no consistent pattern between the sexes.

Geographical patterns in mortality

Survival from stomach cancer is low, and consequently the geographic patterns in mortality (Figures 21.2 and 21.4) were similar to those in incidence. The countries and regions with the highest mortality compared to the UK and Ireland average were Wales for males and Scotland for females: 21 and 24 per cent higher, respectively.

The north-south divide across England can again be clearly seen from the maps illustrating mortality rates by health authority (Map 21.2), particularly for females. Mortality rates were uniformly low across the south and east of England, with the exception of two health authorities in London where rates were above average in both males and females (East London and the City; and Lambeth, Southwark and Lewisham).

As for incidence, there was a band of high mortality across the north of England, with again the notable exception of North Yorkshire, where mortality rates were below average. In Scotland, mortality rates tended to be higher compared to the average for females than males, but rates were particularly high for both sexes in Greater Glasgow. In Ireland, mortality rates were higher than the average in the Eastern area, but there was no overall consistent pattern between the sexes.

Risk factors and aetiology

Worldwide, the highest rates of stomach cancer are found in Japan and parts of South East Asia.[11] High rates also occur in eastern Europe, parts of the former Soviet Union and Latin America. Studies of changes in incidence in migrants from countries with higher risk to lower risk suggest that environmental factors such as diet are important.[12,13] There is strong evidence linking N-nitroso compounds to stomach cancer. Nitrates in preserved foods such as salted fish and home-cured or smoked meat are changed to nitrites in the mouth by saliva. Nitrites may react with certain foodstuffs in the stomach to form N-nitroso compounds,[14] which have been found to be carcinogenic in animals. However, a causal role in human stomach cancer is far from clear, although preserved fish appears to be a specific factor in the high incidence in Japan and South East Asia.

(continued on page 228)

Figure **21.1**

**Stomach: incidence by sex, country, and region of England
UK and Ireland 1991-99[1]**

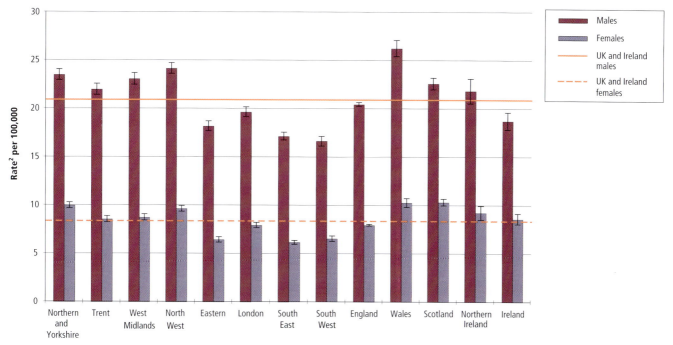

1 Northern Ireland 1993-99, Ireland 1994-99

2 Age standardised using the European standard population, with 95% confidence interval

Figure **21.2**

**Stomach: mortality by sex, country, and region of England
UK and Ireland 1991-2000[1]**

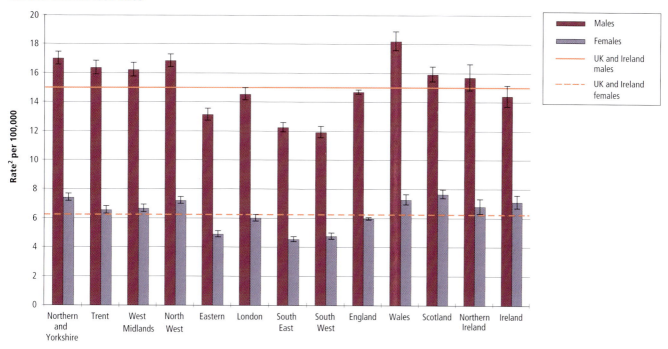

1 Scotland 1991-99, Ireland 1994-2000

2 Age standardised using the European standard population, with 95% confidence interval

Figure **21.3a**

**Stomach: incidence by health authority within country, and region of England
Males, UK and Ireland 1991-99[1]**

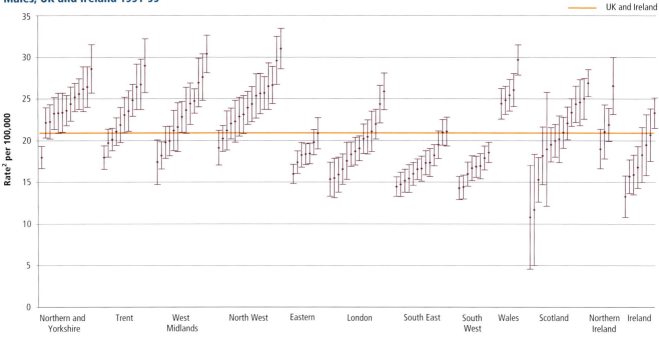

1 Northern Ireland 1993-99, Ireland 1994-99

2 Age standardised using the European standard population, with 95% confidence interval

Figure **21.3b**

**Stomach: incidence by health authority within country, and region of England
Females, UK and Ireland 1991-99[1]**

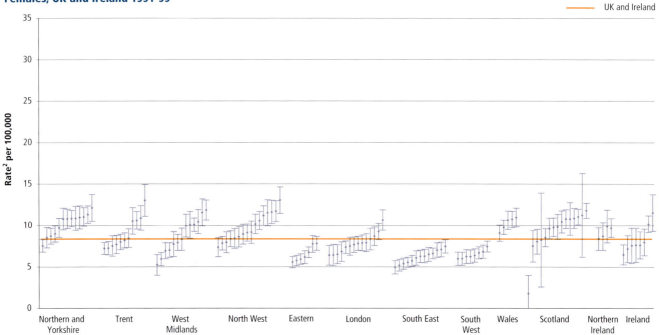

1 Northern Ireland 1991-99, Ireland 1994-99

2 Age standardised using the European standard population, with 95% confidence interval

Figure **21.4a**

**Stomach: mortality by health authority within country, and region of England
Males, UK and Ireland 1991-2000[1]**

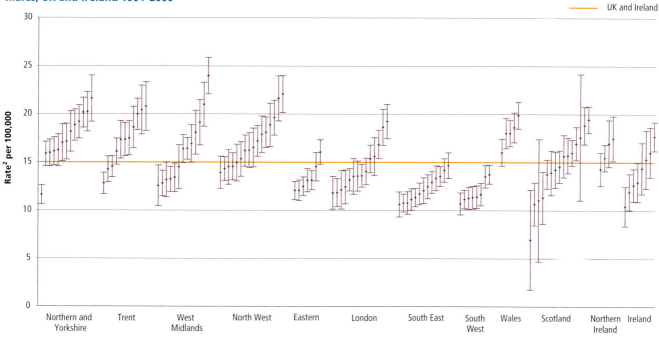

1 Scotland 1991-99, Ireland 1994-2000

2 Age standardised using the European standard population, with 95% confidence interval

Figure **21.4b**

**Stomach: mortality by health authority within country, and region of England
Females, UK and Ireland 1991-2000[1]**

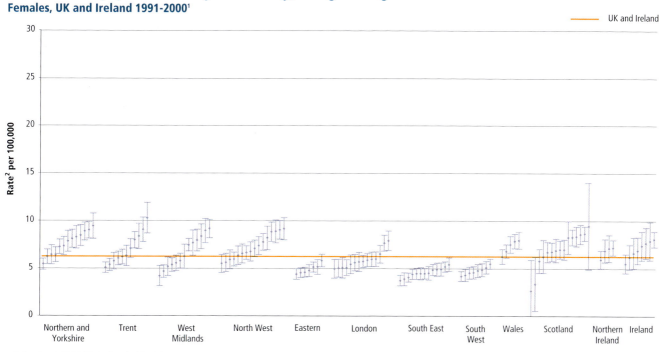

1 Scotland 1991-99, Ireland 1994-2000

2 Age standardised using the European standard population, with 95% confidence interval

Map **21.1a**

Stomach: incidence* by health authority
Males, UK and Ireland 1991-99

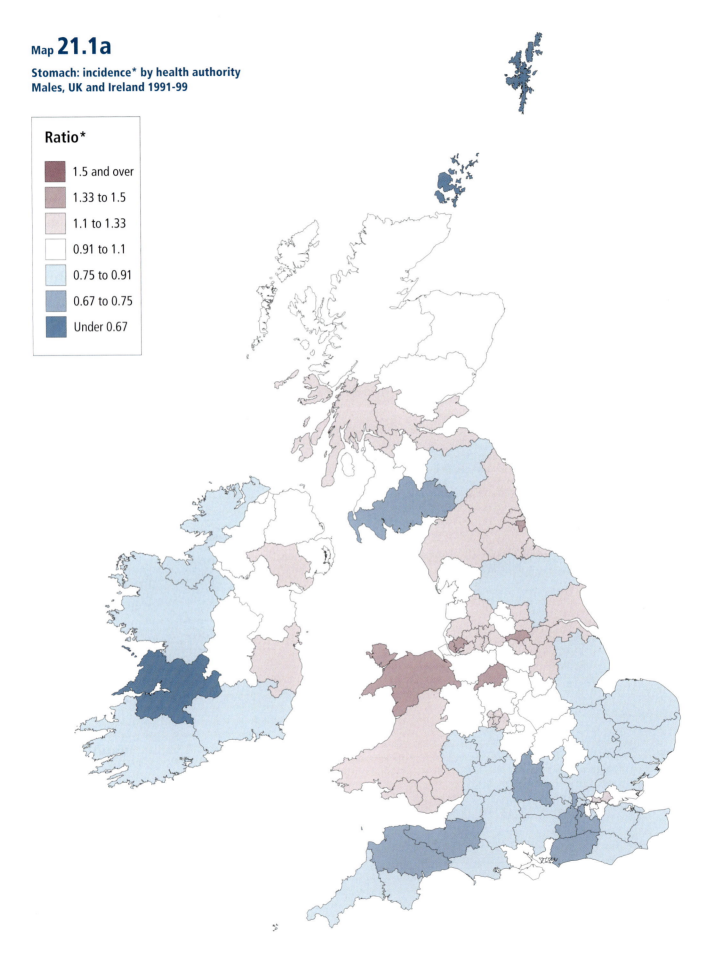

Ratio*

- 1.5 and over
- 1.33 to 1.5
- 1.1 to 1.33
- 0.91 to 1.1
- 0.75 to 0.91
- 0.67 to 0.75
- Under 0.67

*Ratio of directly age-standardised rate in health authority to UK and Ireland average

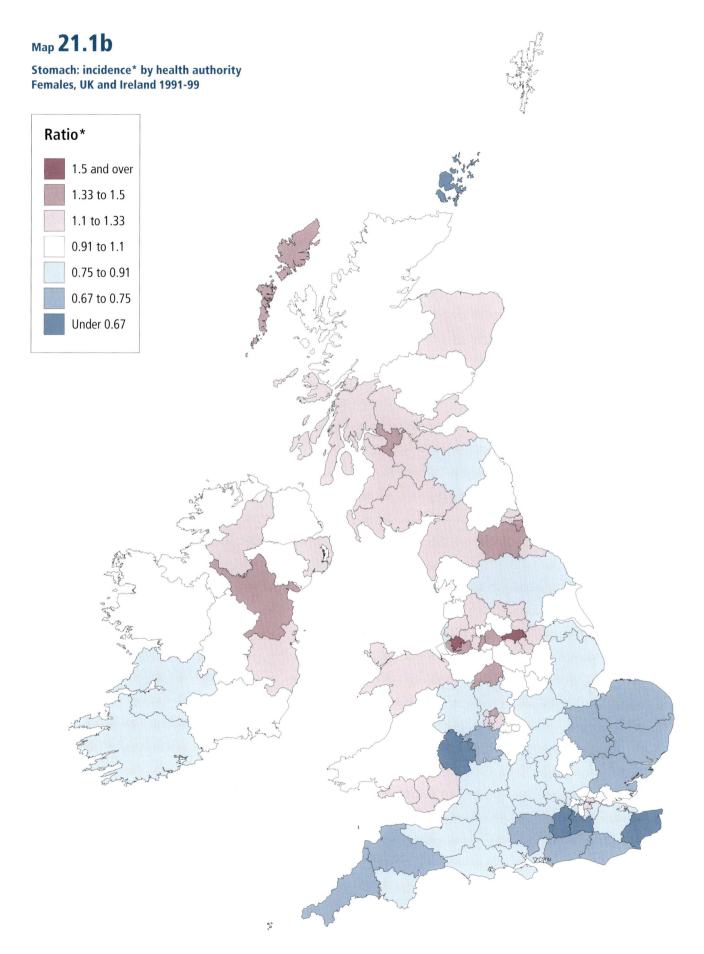

Map **21.1b**

Stomach: incidence* by health authority
Females, UK and Ireland 1991-99

Ratio*

	1.5 and over
	1.33 to 1.5
	1.1 to 1.33
	0.91 to 1.1
	0.75 to 0.91
	0.67 to 0.75
	Under 0.67

Ratio of directly age-standardised rate in health authority to UK and Ireland average

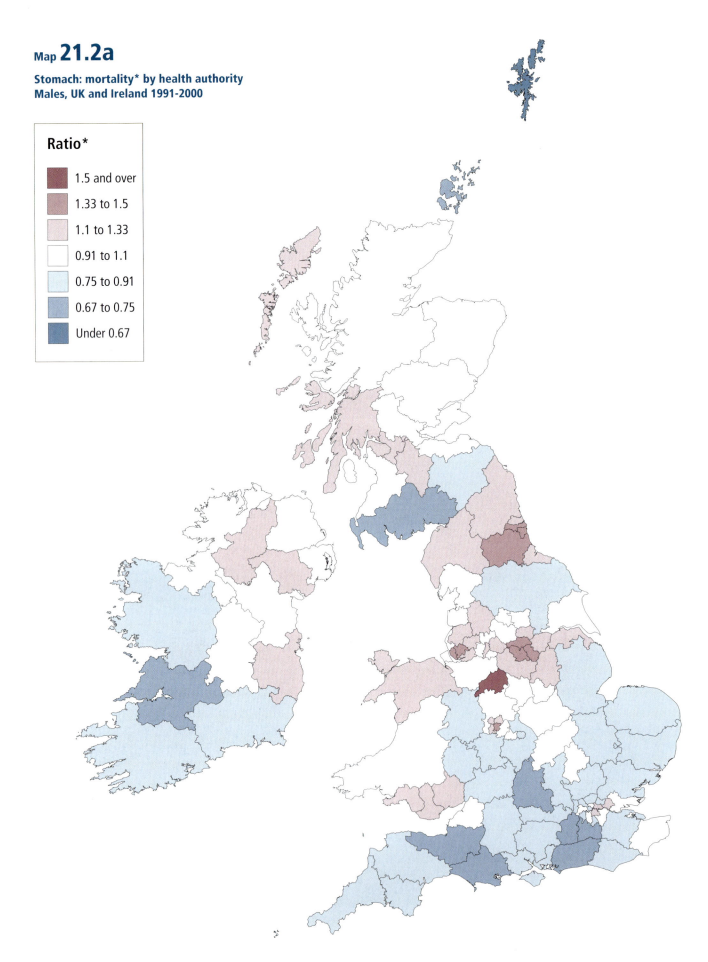

Map 21.2a

**Stomach: mortality* by health authority
Males, UK and Ireland 1991-2000**

Ratio*

- 1.5 and over
- 1.33 to 1.5
- 1.1 to 1.33
- 0.91 to 1.1
- 0.75 to 0.91
- 0.67 to 0.75
- Under 0.67

*Ratio of directly age-standardised rate in health authority to UK and Ireland average

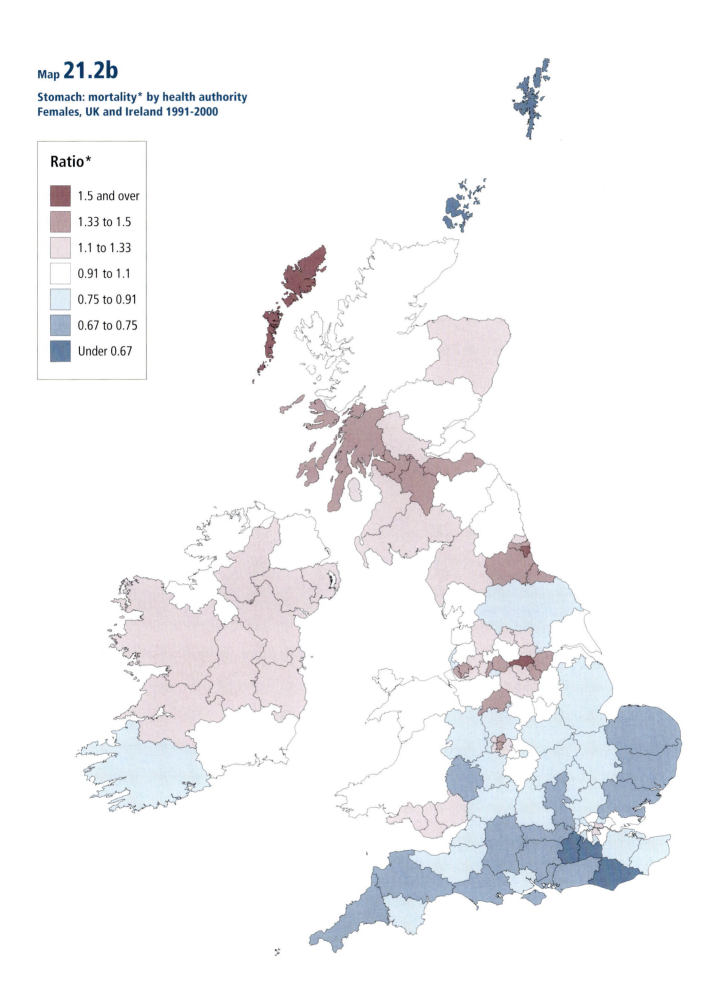

Map 21.2b

**Stomach: mortality* by health authority
Females, UK and Ireland 1991-2000**

Ratio*

- 1.5 and over
- 1.33 to 1.5
- 1.1 to 1.33
- 0.91 to 1.1
- 0.75 to 0.91
- 0.67 to 0.75
- Under 0.67

*Ratio of directly age-standardised rate in health authority to UK and Ireland average

A number of studies have shown that a high intake of fresh fruit and vegetables is protective against stomach cancer.[15] A high intake of vitamin C can prevent the production of N-nitroso compounds and vitamin A can inhibit carcinogenesis. There is also some weaker evidence that a high carbohydrate/ low protein diet may increase risk, possibly by increasing the availability of N-nitroso compounds.

Associations with smoking and alcohol intake have been reported, but dose response evidence is lacking. A recent report on the 'smoking epidemic' in England estimates that, of deaths from stomach cancer in 1998-2002, 35 per cent in males and 12 per cent in females were attributable to smoking.[16] Chronic infection with H pylori has been identified as a risk factor for cancers of the antrum.[17] Early infection leading to chronic atrophic gastritis could partly explain the link with deprivation, as H pylori infection is more common among poor populations.[18] The role of gastric ulcers in the development of stomach cancer is unclear but gastric atrophy and intestinal metaplasia are predisposing factors for the disease, and pernicious anaemia is a risk factor.

There is some evidence for familial links in stomach cancer, relatives of cases having two-to three-times the risk.[19,20] The extent to which genetic rather than shared environmental factors operate remains unclear. Associations have been found with blood group A, as well as hereditary non-polyposis colon cancer.

There are no strong occupational risk factors, although a link with poverty has been described.[21] Some weak associations have been made with coal mining, pottery, asbestos and other industries related to mineral dust. There have also been associations with the chemical and rubber industry, as well as the oil-refining industry.

The geographical variations in the incidence of stomach cancer in England, apparent from the maps shown here, suggest relationships between incidence and the levels of urbanisation and affluence. The regions with higher incidence are in the midlands and north of the country. Within these regions, the traditional industrial areas such as Birmingham, Liverpool and Sunderland have higher incidence than the more affluent rural areas. The pattern for Scotland is consistent with this, with high rates in Greater Glasgow and Lothian. In Ireland, there is a similar pattern with high rates around Dublin. In Wales, however, rates are higher in the north.

Socio-economic deprivation

Various studies both in the UK and other countries have found a clear association between stomach cancer and areas of socio-economic deprivation.[21,22] In England and Wales, there is a strongly positive gradient in the incidence of stomach cancer according to social deprivation. In 1992-93, incidence in the most deprived groups was almost twice that in the most affluent, and the variation of stomach cancer mortality with deprivation was similar to that for incidence. There was little difference in survival among deprived compared to affluent groups for patients diagnosed in England and Wales in 1996-99, the gap in five-year survival being around 2 percentage points for both men and women.[23]

Many areas with the highest level of deprivation, as measured by the Carstairs index[24] (see Appendix F), correspond to areas with high incidence and mortality on Maps 21.1 and 21.2, respectively, including Greater Glasgow; Gateshead and South Tyneside; Sunderland; Manchester; Liverpool; and East London and The City. Poor diet is associated with deprivation, as are other risk factors for stomach cancer, including H pylori infection and smoking. A large cohort study from the Netherlands suggests that this finding in ecological studies is reflected by individual-level data. A high level of educational attainment was associated with reduced risk of stomach cancer, even after adjustment for diet.[25]

References

1. Quinn MJ, Babb PJ, Brock A, Kirby L et al. *Cancer Trends in England and Wales 1950-1999*. Studies on Medical and Population Subjects No. 66. London: The Stationery Office, 2001.

2. Schouten LJ, Straatman H, Kiemeney LALM, Verbeek ALM. Cancer incidence: Life table risk versus cumulative risk. *Journal of Epidemiology and Community Health* 1994; 48: 596-600.

3. ONS. *Cancer Statistics Registrations: Registrations of cancer diagnosed in 2001, England*. Series MB1 No. 32. London: Office for National Statistics, 2004.

4. Parkin DM, Pisani P, Ferlay J. Estimates of the worldwide incidence of 25 major cancers in 1990. *International Journal of Cancer* 1999; 80: 827-841.

5. ONS. Cancer Survival: England and Wales, 1991-2001. March 2004. Available at *http://www.statistics.gov.uk/statbase/ ssdataset.asp?vlnk=7899*.

6. ISD Scotland. *Trends in Cancer Survival in Scotland, 1977-2001*. Edinburgh: ISD Publications, 2004.

7. Fitzpatrick D, Gavin A, Middleton R, Catney D. *Cancer in Northern Ireland 1993-2001: A Comprehensive Report*. Belfast: Northern Ireland Cancer Registry, 2004.

8. National Cancer Registry of Ireland. *Cancer in Ireland, 1994 to 1998: Incidence, mortality, treatment and survival*. Cork: National Cancer Registry, 2001.

9. Coleman MP, Babb P, Damiecki P, Grosclaude P et al. *Cancer Survival Trends in England and Wales, 1971-1995: Deprivation and NHS Region*. Studies on Medical and Population Subjects No. 61. London: The Stationery Office, 1999.

10. Sant M, Areleid T, Berrino F, Bielska Lasota M et al. EUROCARE-3: survival of cancer patients diagnosed 1990-1994 - results and commentary. *Annals of Oncology* 2003; 14 Suppl 5: v61-v118.

11. Parkin DM. Global cancer statistics in the year 2000. *The Lancet Oncology* 2001; 2: 533-543.

12. Gregorio DI, Flannery JT, Hansen H. Stomach cancer patterns in European immigrants to Connecticut, United States. *Cancer Causes Control* 1992; 3: 215-221.

13. Haenzel W, Kuriha M, Segi M, et al. Stomach cancer amongst Japanese in Hawaii. *Journal of the National Cancer Institute* 1972; 49: 969-988.

14. Forman D. Dietary exposure to N-nitroso compounds and the risk of human cancer. *Cancer Surveys* 1987; 6: 719-738.

15. Department of Health. *Nutritional aspects of the development of cancer. Report of the working group on diet and cancer of the Committee on Medical Aspects of Food and Nutrition Policy*. Report on Health and Social Subjects No. 48. London: HMSO, 1998.

16. Twigg L, Moon G, Walker S. The Smoking Epidemic in England. London: Health Development Agency, 2004.

17. Forman D, Newell DG, Fullerton F, Yarnell JW et al. Association between infection with Helicobacter pylori and risk of gastric cancer: evidence from a prospective investigation. *British Medical Journal* 1991; 302: 1302-1305.

18. International Agency for Research on Cancer. Liver flukes and helicobacter pylori. *IARC Monographs on the Evaluation of Carcinogenic Risks of Chemicals to Humans* 1994; 61: 177-240.

19. La Vecchia C, Negri E, Franceschi S, Gentile A. Family history and the risk of stomach and colorectal cancer. *Cancer* 1992; 70: 50-55.

20. Bakir T, Can G, Siviloglu C, Erkul S. Gastric cancer and other organ cancer history in the parents of patients with gastric cancer. *European Journal of Cancer Prevention* 2003; 12: 183-189.

21. Nomura A. Stomach Cancer. In: Schottenfeld D, Fraumeni Jnr JF (eds) *Cancer Epidemiology and Prevention, second edition*. New York: Oxford University Press, 1996.

22. Brown J, Harding S, Bethune A, Rosato M. Longitudinal study of socio-economic differences in the incidence of stomach, colorectal and pancreatic cancers. *Population Trends* 1998: 35-41.

23. Coleman MP, Rachet B, Woods LM, Mitry E et al. Trends and socioeconomic inequalities in cancer survival in England and Wales up to 2001. *British Journal of Cancer* 2004; 90: 1367-1373.

24. Carstairs V, Morris R. Deprivation and mortality: an alternative to social class? *Community Medicine* 1989; 11: 213-219.

25. van Loon AJ, Goldbohm RA, van den Brandt PA. Socioeconomic status and stomach cancer incidence in men: results from The Netherlands Cohort Study. *Journal of Epidemiology and Community Health* 1998; 52: 166-171.

Chapter 22
Testis

Henrik Møller

Summary

- In the UK and Ireland in the 1990s, testicular cancer accounted for around 1 in 75 cases of cancer, but only 1 in 800 deaths from cancer in males.

- Incidence was higher than average in Scotland and the south of England, and below average in Ireland, London, and the north of England.

- At the health authority level, there was an apparent pattern of lower incidence in more urban areas and higher incidence in more rural areas.

- This pattern may reflect the lower proportions of white males living in more urban areas, as incidence is known to be much higher in white males than in black and Asian males.

- The areas of high incidence tended to be the more affluent ones – most likely reflecting the higher proportion of white males living in these areas, rather than any causative factor inversely related to socio-economic deprivation.

- Any apparent geographical variations in mortality are not interpretable, owing to the very low numbers of deaths.

Introduction

Testicular cancer is a rare disease, accounting for 1.3 per cent of all cancers and only 0.1 per cent of all cancer deaths in males in the UK and Ireland. Unlike most other solid tumours, testicular cancer occurs most frequently in young men. Despite its low incidence overall, testicular cancer is the commonest malignancy in men in their 20s and 30s.

Incidence and mortality

The average number of testicular cancer cases per year in the UK and Ireland in the 1990s was 1,800 and the annual number of deaths was about 110. The corresponding age-standardised incidence and mortality rates were 5.9 and 0.3 per 100,000. Testicular cancer has an unusual age distribution and the highest age-specific rates were found in men in their 20s, 30s and 40s.

The majority of testicular cancers are germ cell tumours, which account for 90 per cent of cases. In younger men teratomas tend to occur, with a peak incidence in those aged 20-29, while the incidence of slower-growing seminomas is highest in those aged 30-49.[1]

Incidence and mortality trends

The incidence of testicular cancer increased in most countries in north west Europe in the last four decades of the twentieth century.[1-5]

In several north European populations, the increasing trends are strongly influenced by birth cohort effects.[6-8] In some populations there are indications that the increasing trend ceases in the most recent birth cohorts and the most recent period.[8,9]

Mortality rates from testicular cancer in England and Wales have decreased over the last three decades, by 75 per cent between 1971 and 1999, with most of this reduction in the late 1970s and early 1980s. Intensive radiotherapy for seminoma was introduced in the 1960s, but the decrease in mortality rates was particularly due to the introduction of highly effective chemotherapy for non-seminoma since the late 1970s.[1]

Survival

Survival is exceptionally high for testicular cancer. The most recent figures for patients diagnosed in England and Wales during 1996-99 indicate one-year relative survival of 98 per cent and five-year survival of 95 per cent.[10] The figures for patients diagnosed in Scotland in 1992-96 were slightly lower – 91 and 88 per cent after one year and five years, respectively.[11] For patients diagnosed in 1990-94, survival rates in England were – unusually – higher than the average for Europe.[12] Seminomas are highly sensitive to radiotherapy, and survival has been relatively good since the 1940s. In Denmark, the five-year relative survival increased from 66 per cent in 1945 to 95 per cent in 1985.[13] Non-seminoma initially had a poorer prognosis (38 per cent five-year relative survival in 1945), but the rate increased to 89 per cent by 1985. The excess mortality in testicular cancer patients is almost entirely confined to the first few years after diagnosis, and patients surviving for five years can be considered cured.

Geographical patterns in incidence

There was considerable variation in testicular cancer incidence between countries and regions in the UK and Ireland (Figure 22.1).

(continued on page 236)

Figure 22.1

Testis: incidence by country, and region of England
UK and Ireland 1991-99[1]

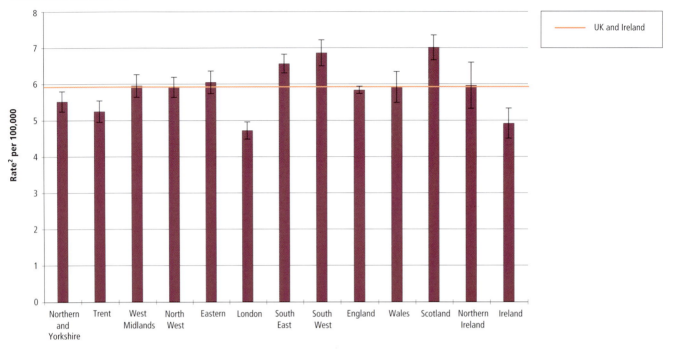

1 Northern Ireland 1993-99, Ireland 1994-99

2 Age standardised using the European standard population, with 95% confidence interval

Figure 22.2

Testis: mortality by country, and region of England
UK and Ireland 1991-2000[1]

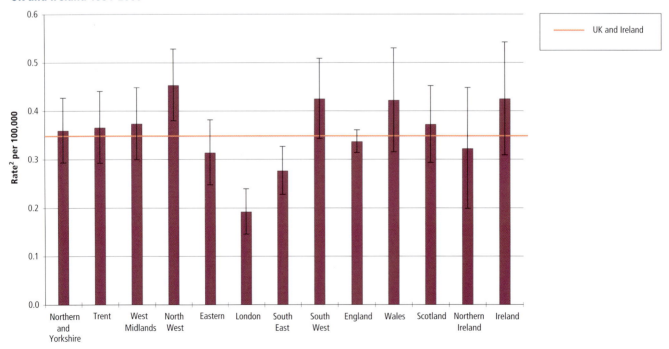

1 Scotland 1991-99, Ireland 1994-2000

2 Age standardised using the European standard population, with 95% confidence interval

Figure **22.3**

**Testis: incidence by health authority within country, and region of England
UK and Ireland 1991-99[1]**

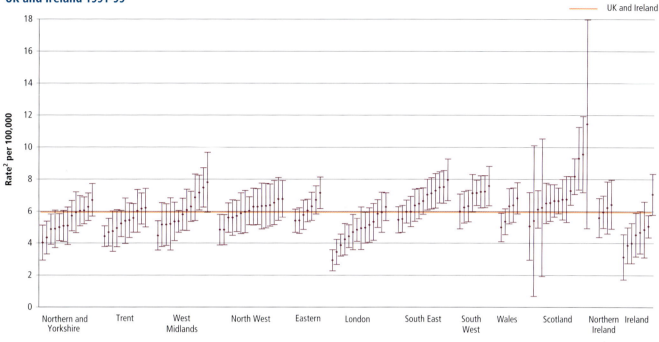

1 Northern Ireland 1993-99, Ireland 1994-99

2 Age standardised using the European standard population, with 95% confidence interval

Figure **22.4**

**Testis: mortality by health authority within country, and region of England
UK and Ireland 1991-2000[1]**

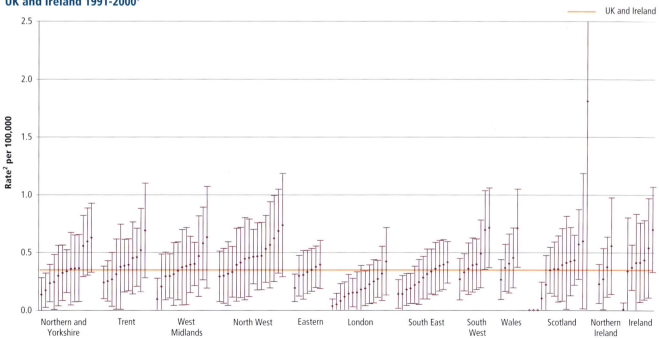

1 Scotland 1991-99, Ireland 1994-2000

2 Age standardised using the European standard population, with 95% confidence interval

Map 22.1

**Testis: incidence* by health authority
UK and Ireland 1991-99**

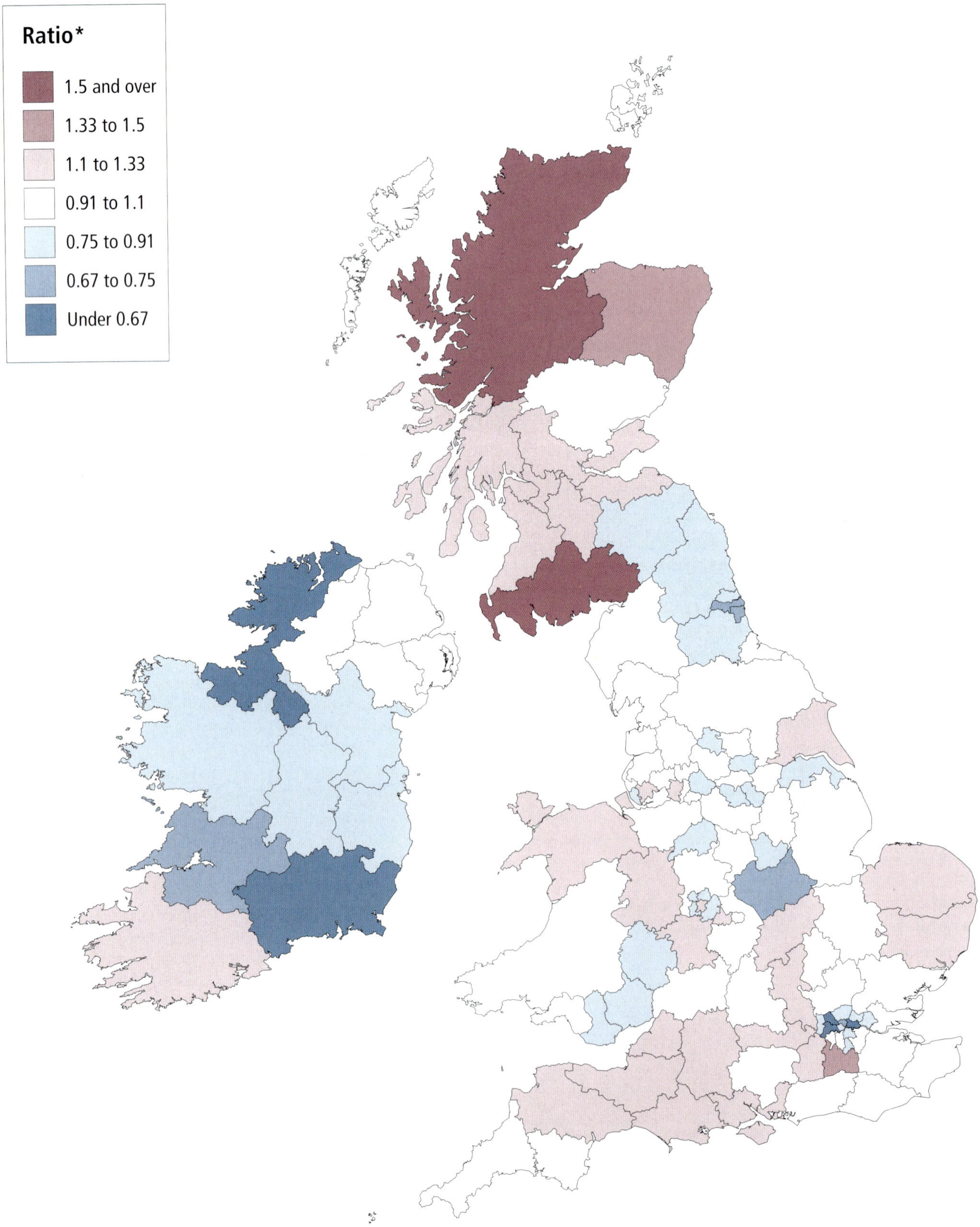

Ratio*

■	1.5 and over
■	1.33 to 1.5
■	1.1 to 1.33
□	0.91 to 1.1
■	0.75 to 0.91
■	0.67 to 0.75
■	Under 0.67

Ratio of directly age-standardised rate in health authority to UK and Ireland average

Map **22.2**

**Testis: mortality* by health authority
UK and Ireland 1991-2000**

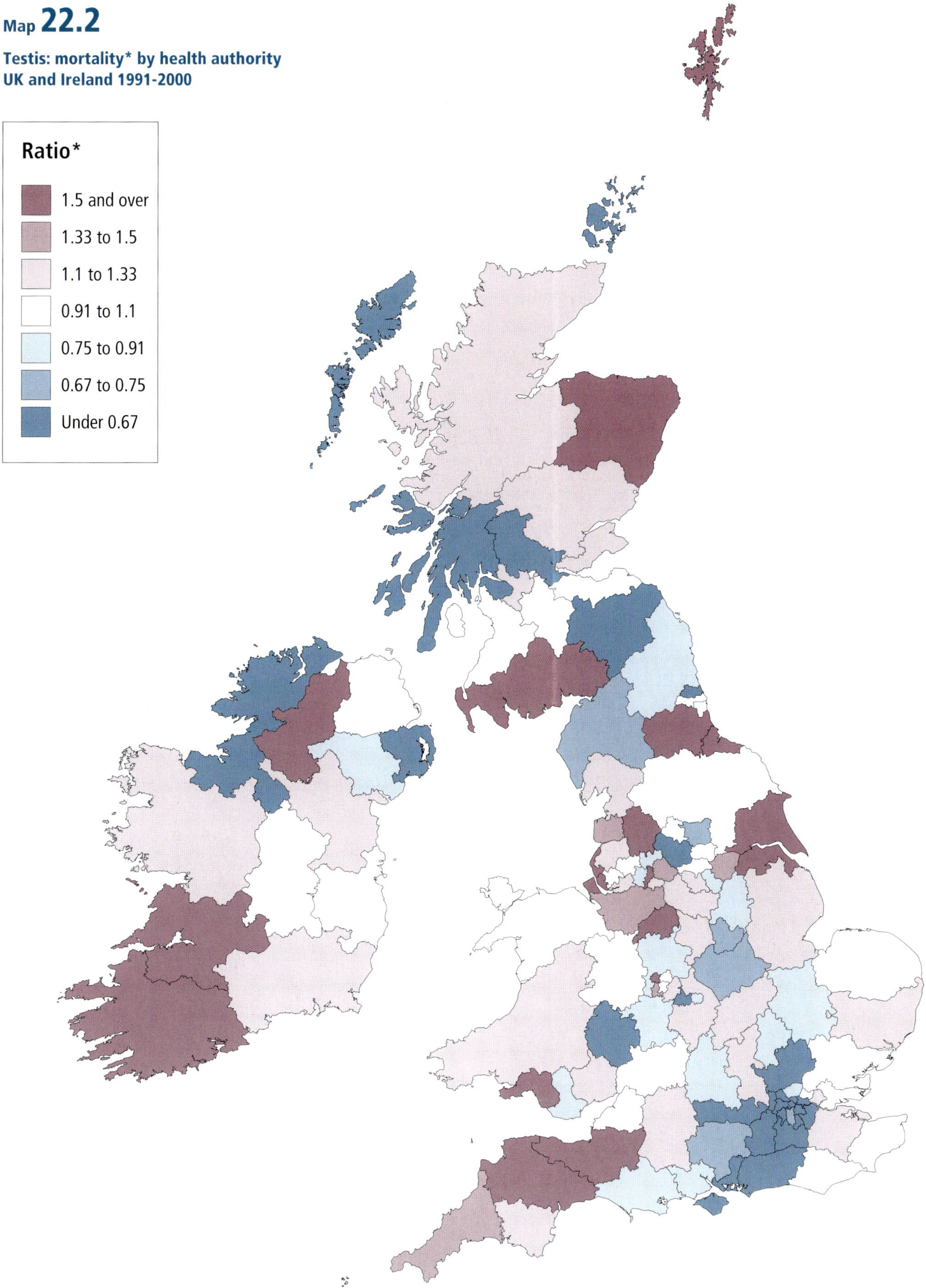

Ratio*

	1.5 and over
	1.33 to 1.5
	1.1 to 1.33
	0.91 to 1.1
	0.75 to 0.91
	0.67 to 0.75
	Under 0.67

*Ratio of directly age-standardised rate in health authority to UK and Ireland average

Age-standardised rates were low in London; Ireland; Trent; and Northern and Yorkshire; and high in Scotland, and South West and South East England. Incidence rates in London were about 65 per cent of those in Scotland.

In the Nordic countries, where rates are slightly higher than in Great Britain, there is remarkable variation along the national boundaries.[14] A similar phenomenon appears in the map of the UK and Ireland, where rates were quite consistently low in Ireland, intermediate in England and Wales and high in Scotland (Map 22.1).

Considering the variations between health authority areas within countries and regions (Figure 22.3), there were no particular outliers, with the possible exception of the relatively high rate in the Southern area in Ireland.

In England there was a suggestion of a pattern with the lowest rates being found in the smaller areas – urban areas with high population density – and the highest rates were in the larger areas – more rural areas with lower population density (Map 22.1). There are important variations in testicular cancer incidence between different ethnic groups (see below) and there is a much higher incidence in white males than in black or Asian males. The variation with population density, urbanisation and size of health authority area could possibly reflect the higher proportion of males from ethnic minorities in more urban areas.

A study of geographical variation across 10,530 wards in Great Britain reported significant variation between cancer registry areas (which could be due to genuine regional differences and/ or to differences in case ascertainment), but there was no clear evidence for any local spatial variation at the small-area level.[15]

A Danish study found a distinct geographical pattern of incidence with region of birth, but no such pattern with respect to region of residence at the time of diagnosis.[16] A study of Finnish emigrants in Sweden showed that Finnish men retained their typically low level of incidence compared with Swedish men. Their reduced risk, compared to Swedish men, was independent of age at emigration or duration of stay in Sweden.[17] The latter observations are consistent with the hypothesis of a prenatal or perinatal aetiology of testicular cancer (see below). Hence, spatial variations at the small-area level would be expected to be found in association with the place of birth rather than with the place of residence at the time of cancer diagnosis.

Geographical patterns in mortality

The geographical pattern of testicular cancer mortality rates was hardly interpretable (Map 22.2). Due to the low number of events, the mortality rates were unstable and many areas in the map appeared to have high or low mortality simply due to chance variations. Table B22.2 shows that the average count of testicular cancer deaths per year in health authority areas was typically 0, 1 or 2.

Risk factors and aetiology

There is a considerable body of evidence, both indirect and direct, which indicates that the rate-limiting steps in the development of testicular cancer occur very early in life, most probably early in gestation around the time of differentiation of the genital organs.[5,18-21] The susceptible cell population from which testicular germ cell cancers arise are most likely the primordial germ cells, which in normal development differentiate into spermatogonia in males or oocytes in females. Occasionally, primordial germ cells persist in the undifferentiated state and give rise to a pre-malignant condition known as testicular carcinoma *in situ*, which in turn has a very high probability of progression to invasive cancer after the onset of puberty.[22]

Consistent risk factors for testicular cancer are congenital malformations of the genital organs (particularly cryptorchidism – failure of the testicles to descend into the scrotum), low birth weight and intrauterine growth retardation, low maternal parity, and sub-fertility.[23-25] Other postulated risk factors include maternal exposure to exogenous oestrogens during the first trimester of pregnancy, and a history of trauma to the affected testicle.[26] It has been suggested that heritability plays a role in the aetiology of testicular cancer, with a family history of testicular cancer among first degree relatives being a postulated risk factor.[26] It has been estimated that the risk of testicular cancer in the brothers of cases is 2.2 per cent.[27]

Rates of testicular cancer vary between different ethnic groups and different European populations, with higher rates in males of European origin than Asian and black males.[28] The incidence in England and Wales is amongst the highest in the world.[1] In the SEER cancer registries of the United States National Cancer Institute, the age-standardised incidence rate is five times higher in white males than in black males. In Los Angeles, where incidence rates are reported separately by ethnic group, the rates are: non-Hispanic white 5.7 per 100,000; Hispanic white 3.1; Japanese 2.3; black 1.4; Chinese 1.0; Filipinos 0.8; and Koreans 0.6. Rates around 1.0 per 100,000 or lower are reported from China, India and Pakistan.

Socio-economic deprivation

In England and Wales there is an inverse socio-economic gradient in testicular cancer incidence, such that slightly higher rates are found in the more affluent groups.[1] For patients diagnosed in England and Wales during 1986-90, five-year relative survival was 6 percentage points lower in males living in the most deprived areas (87 per cent), compared to those living in the most affluent areas (93 per cent).[29] However, the survival gap between these groups has subsequently narrowed and for those patients diagnosed in 1996-99, five-year survival was just over one percentage point lower in the most deprived than the most affluent males.[30]

References

1. Quinn MJ, Babb PJ, Brock A, Kirby L et al. *Cancer Trends in England and Wales 1950-1999.* Studies on Medical and Population Subjects No. 66. London: The Stationery Office, 2001.

2. Harris V, Sandridge AL, Black RJ, Brewster DH et al. *Cancer Registration Statistics Scotland, 1986-1995.* Edinburgh: ISD Publications, 1998.

3. National Cancer Registry of Ireland. *Cancer in Ireland, 1996. Incidence and Mortality.* Cork: National Cancer Registry, 1999.

4. Gavin AT, Reid J. *Cancer Incidence in Northern Ireland, 1993-95.* Belfast: The Stationery Office, 1999.

5. Power DA, Brown RS, Brock CS, Payne HA et al. Trends in testicular carcinoma in England and Wales, 1971-99. *BJU International* 2001; 87: 361-365.

6. Moller H. Clues to the aetiology of testicular germ cell tumours from descriptive epidemiology. *European Urology* 1993; 23: 8-13.

7. Bergstrom R, Adami HO, Mohner M, Zatonski W et al. Increase in testicular cancer incidence in six European countries: a birth cohort phenomenon. *Journal of the National Cancer Institute* 1996; 88: 727-733.

8. Moller H. Trends in incidence of testicular cancer and prostate cancer in Denmark. *Human Reproduction* 2001; 16: 1007-1011.

9. McGlynn KA, Devesa SS, Sigurdson AJ, Brown LM et al. Trends in the incidence of testicular germ cell tumors in the United States. *Cancer* 2003; 97: 63-70.

10. ONS. Cancer Survival: England and Wales, 1991-2001. March 2004. Available at *http://www.statistics.gov.uk/statbase/ ssdataset.asp?vlnk=7899.*

11. ISD Scotland. *Trends in Cancer Survival in Scotland, 1977-2001.* Edinburgh: ISD Publications, 2004.

12. Sant M, Areleid T, Berrino F, Bielska Lasota M et al. EUROCARE-3: survival of cancer patients diagnosed 1990-1994 - results and commentary. *Annals of Oncology* 2003; 14 Suppl 5: v61-v118.

13. Moller H, Friis S, Kjaer SK. Survival of Danish cancer patients 1943-1987. Male genital organs. *Acta Pathologica, Microbiologica et Immunologica Scandinavica Supplement* 1993; 33: 122-136.

14. Cartensen B, Jensen OM. *Atlas of Cancer Incidence in Denmark* 1970-79. Copenhagen: Danish Cancer Society, 1986.

15. Toledano M, Järup L, Best N, Wakefield J et al. Spatial variation and temporal trends of testicular cancer in Great Britain. *British Journal of Cancer* 2001; 84: 1482-1487.

16. Moller H. Work in agriculture, childhood residence, nitrate exposure, and testicular cancer risk: a case-control study in Denmark. *Cancer Epidemiology Biomarkers and Prevention* 1997; 6: 141-144.

17. Ekbom A, Richiardi L, Akre O, Montgomery SM et al. Age at immigration and duration of stay in relation to risk for testicular cancer among Finnish immigrants in Sweden. *Journal of the National Cancer Institute* 2003; 95: 1238-1240.

18. Cartwright RA, Elwood PC, Birch J, Tyrell C et al. Aetiology of testicular cancer: association with congenital abnormalities, age at puberty, infertility, and exercise. *British Medical Journal* 1994; 308: 1393-1399.

19. Swerdlow AJ, De Stavola BL, Swanwick MA, Mangtani P et al. Risk factors for testicular cancer: a case-control study in twins. *British Journal of Cancer* 1999; 80: 1098-1102.

20. Swerdlow AJ, Stavola B, Swanwick M, Mavconochie N. Risks of breast and testicular cancers in young adult twins in England and Wales: evidence on prenatal and genetic aetiology. *Lancet* 1997; 350: 1723-1728.

21. Moller H, Evans H. Epidemiology of gonadal germ cell cancer in males and females. *Acta Pathologica, Microbiologica et Immunologica Scandinavica Supplement* 2003; 111: 43-46.

22. Skakkebaek NE, Berthelsen JG, Giwercman A, Muller J. Carcinoma-in-situ of the testis: possible origin from gonocytes and precursor of all types of germ cell tumours except spermatocytoma. *International Journal of Andrology* 1987; 10: 19-28.

23. Moller H, Skakkebaek NE. Testicular cancer and cryptorchidism in relation to prenatal factors: case-control studies in Denmark. *Cancer Causes and Control* 1997; 8: 904-912.

24. Richiardi L, Akre O, Bellocco R, Ekbom A. Perinatal determinants of germ-cell testicular cancer in relation to histological subtypes. *British Journal of Cancer* 2002; 87: 545-550.

25. Jacobsen R, Bostofte E, Engholm G, Hansen J et al. Risk of testicular cancer in men with abnormal semen characteristics: cohort study. *British Medical Journal* 2000; 321: 789-792.

26. Schottenfeld D. Testicular cancer. In: Schottenfeld D, Fraumeni JF, Jr. (eds) *Cancer Epidemiology and Prevention.* New York: Oxford University Press, 0 AD.

27. Forman D, Oliver RT, Brett AR, Marsh SG et al. Familial testicular cancer: a report of the UK family register, estimation of risk and an HLA class 1 sib-pair analysis. *British Journal of Cancer* 1992; 65: 255-262.

28. Parkin DM, Whelan SL, Ferlay J, Teppo L et al. *Cancer Incidence in Five Continents Vol. VIII.* IARC Scientific Publications No. 155. Lyon: International Agency for Research on Cancer, 2000.

29. Coleman MP, Babb P, Damiecki P, Grosclaude P et al. *Cancer Survival Trends in England and Wales, 1971-1995: Deprivation and NHS Region.* Studies on Medical and Population Subjects No. 61. London: The Stationery Office, 1999.

30. Coleman MP, Rachet B, Woods LM, Mitry E et al. Trends and socioeconomic inequalities in cancer survival in England and Wales up to 2001. *British Journal of Cancer* 2004; 90: 1367-1373.

Chapter 23

Uterus

Nicola Cooper

Summary

- Among women in the UK and Ireland, cancer of the body of the uterus accounted for around 1 in 30 cancer cases and 1 in 50 cancer deaths in the 1990s.

- There was some suggestion of a north-south divide in incidence rates across Great Britain – incidence was slightly higher than average in the midlands, south and east of England, and slightly lower than average in Scotland and the north of England.

- The areas of slightly higher incidence mostly fell into two bands across England – one across the West Midlands and part of the Trent region, and the other across the South West, South East and Eastern regions from Cornwall to Norfolk.

- The geographic variations in cancer of the uterus were similar to those for breast and ovarian cancers – all three share the same main risk factors.

- There was no obvious link between the observed variations in incidence and any known risk factors for cancer of the uterus, although incidence tended to be lower in more deprived areas suggesting that the protective effect of higher parity is more important than the risk conferred by obesity.

Introduction

The incidence data included in the analysis described below do not include those cancers assigned to the non-specific code for uterus (179 in ICD9). Cancers of the uterus should not normally be registered without sufficient information being available to classify them to the cervix or the body of the uterus. In the mid-1990s, the proportion of newly diagnosed cases of uterine cancer assigned to the non-specific code for uterus was 10 per cent of cases in Scotland,[1] around 15 per cent in Northern Ireland[2] and Ireland,[3] 19 per cent in Wales[4] and 7 per cent in England.[4] The proportions varied widely between the regions of England from around one per cent in London, the South East and Eastern regions to around 12-15 per cent in the South West, and Northern and Yorkshire,[4] reflecting the registration

practice of the regional cancer registries. The mortality data, however, does include those registrations assigned to the non-specific code for uterus because around one third of deaths from cancer of the uterus are coded to the non-specific code for uterus in England (with little variation between the regions of England),[5] Wales,[5] and Ireland,[3] and more than half in Northern Ireland[2] (although only 16 per cent in Scotland).[1]

Geographic variations in the incidence of cancer of the body of the uterus are also potentially subject to the prevalence of hysterectomy in the population, which should be considered in interpreting rates. Since women who have had their uterus removed are not at risk from the disease, their inclusion in the calculation of population-based cancer rates artificially lowers estimates of the disease occurrence. By 1995, 2.3 million women in England and Wales had had a hysterectomy, with a peak prevalence of 21 per cent in the age group 55-59.[6] There is a deprivation gradient in hysterectomy – women with less education were more likely to have had a hysterectomy by the age of 52.[7]

Incidence and mortality

Cancer of the body of the uterus was the fifth most common cancer in women and accounted for around 4 per cent of all newly diagnosed cancers. About 4,900 cases of cancer of the uterus were diagnosed annually in the UK and Ireland during the 1990s; the age-standardised incidence rate was 13 per 100,000. Few cases occurred in pre-menopausal women and over 90 per cent of cases were in women aged 50 or over. Age-specific rates peaked in the range 60-74 years, accounting for almost 45 per cent of the total cases. The lifetime risk[8,9] of being diagnosed with cancer of the uterus, based on England and Wales data for 1997, was 1.4 per cent (1 in 73).[10]

During the 1990s, about 1,500 women died annually from cancer of the uterus in the UK and Ireland, 2 per cent of all cancer deaths in women. This was the twelfth most common cause of cancer death in women. The age-standardised mortality rate was 3.3 per 100,000.

Incidence and mortality trends

Trends in incidence show that the age-standardised rate for cancer of the uterus gradually increased during the 1970s and 1980s by around 10 per cent, but the rate of increase doubled after 1990. Age-specific rates increased especially in older women,[11] whereas in women aged under 65, they remained relatively stable. The pattern in the risk of cancer of the uterus by birth cohort is similar to that for breast cancer, with rates increasing up to the cohorts born around 1930 and then declining.[11]

(continued on page 244)

Figure **23.1**

**Uterus: incidence by country, and region of England
UK and Ireland 1991-99[1]**

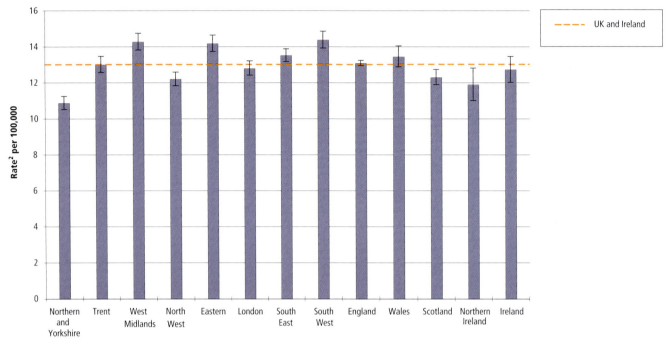

1 Northern Ireland 1993-99, Ireland 1994-99

2 Age standardised using the European standard population, with 95% confidence interval

Figure **23.2**

**Uterus: mortality by country, and region of England
UK and Ireland 1991-2000[1]**

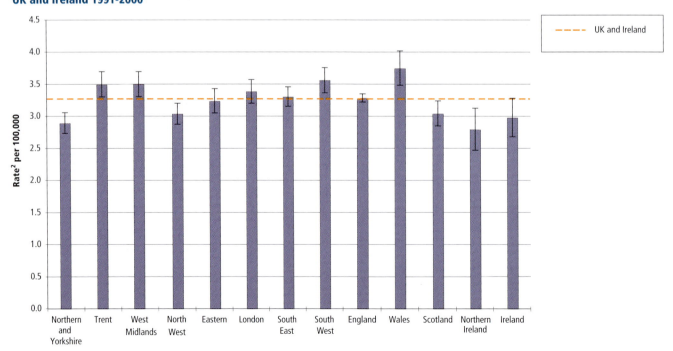

1 Scotland 1991-99, Ireland 1994-2000

2 Age standardised using the European standard population, with 95% confidence interval

Figure 23.3

**Uterus: incidence by health authority within country, and region of England
UK and Ireland 1991-99[1]**

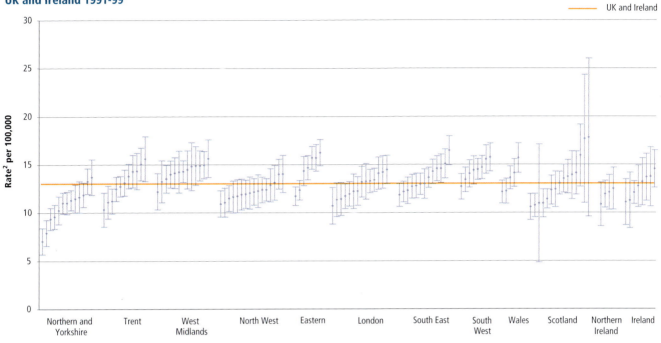

1 Northern Ireland 1993-99, Ireland 1994-99

2 Age standardised using the European standard population, with 95% confidence interval

Figure 23.4

**Uterus: mortality by health authority within country, and region of England
UK and Ireland 1991-2000[1]**

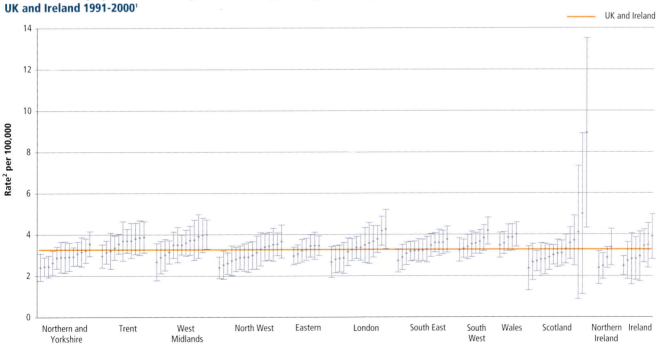

1 Scotland 1991-99, Ireland 1994-2000

2 Age standardised using the European standard population, with 95% confidence interval

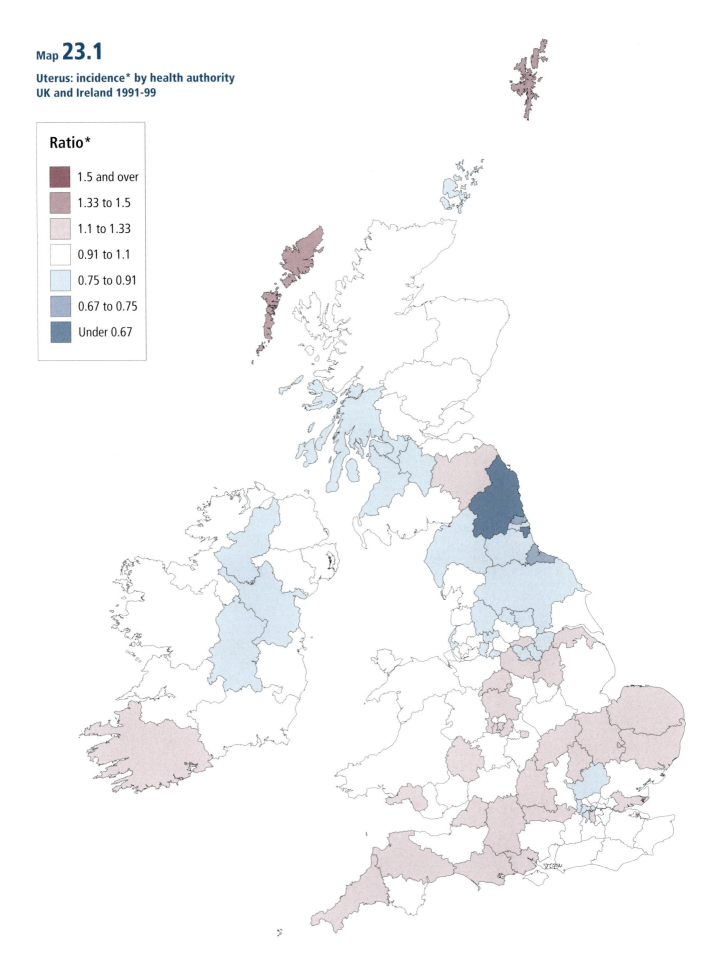

Map **23.1**

**Uterus: incidence* by health authority
UK and Ireland 1991-99**

Ratio*

- 1.5 and over
- 1.33 to 1.5
- 1.1 to 1.33
- 0.91 to 1.1
- 0.75 to 0.91
- 0.67 to 0.75
- Under 0.67

*Ratio of directly age-standardised rate in health authority to UK and Ireland average

Map **23.2**

**Uterus: mortality* by health authority
UK and Ireland 1991-2000**

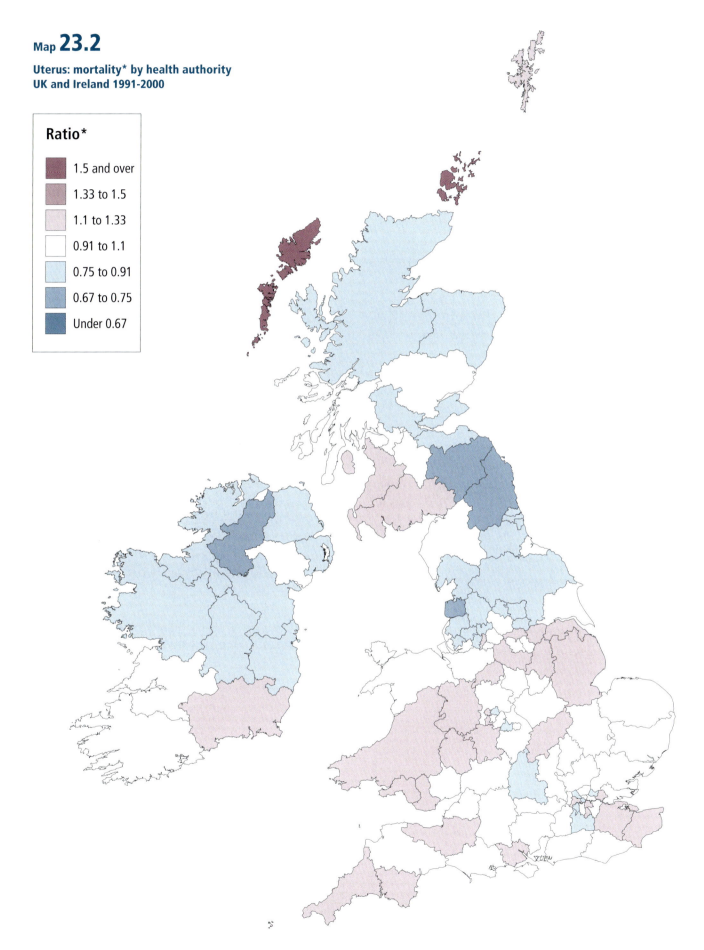

Ratio*

- 1.5 and over
- 1.33 to 1.5
- 1.1 to 1.33
- 0.91 to 1.1
- 0.75 to 0.91
- 0.67 to 0.75
- Under 0.67

*Ratio of directly age-standardised rate in health authority to UK and Ireland average

In contrast to the increase in incidence, there have been long-term declines in mortality from cancer of the uterus in all age groups except the very elderly. The age-standardised rate halved from 6 per 100,000 in 1950 to 3 per 100,000 in 1999.[11] Cohort mortality reached a maximum for those born up to the 1920s and decreased for those born subsequently.[12]

Survival

Cancer of the uterus has a higher survival rate than for many other cancers, largely because it is often at an early stage when diagnosed. Early diagnosis of cancer of the uterus is aided by its early presentation with post-menopausal bleeding or particularly irregular or heavy bleeding around the time of the menopause. If detected at an early stage, both cancer of the uterus and pre-cancerous changes that carry a high risk of progressing to cancer are curable in most cases. In England and Wales, age-standardised relative survival for women diagnosed during 1996-99 was 87 per cent at one year and 73 per cent at five years after diagnosis.[13] One- and five-year relative survival both rose significantly by around 2.5 per cent every five years between 1986-90 and 1996-99. The increase in five-year survival was particularly marked between the early and late 1990s.[14] Survival from cancer of the uterus varied widely across Europe in the early 1990s, with five-year survival rates for England, Scotland and Wales being 2.3, 4.2 and 5.3 percentage points, respectively, below the European average.[15]

Geographical patterns in incidence

The incidence rates for cancer of the uterus showed relatively little variability among the countries of the UK and Ireland (Figure 23.1). Incidence in Wales was around 4 per cent higher than the average for the UK and Ireland. Rates for Northern Ireland, Scotland and Ireland were 8.5, 5.4 and 2.3 per cent, respectively, below the average. There was more variation in incidence rates between the regions of England than between the countries of the UK and Ireland. Age-standardised incidence rates were around 10 per cent higher than the average in the South West, West Midlands and Eastern regions; and rates were noticeably below average in the Northern and Yorkshire (16 per cent) and North West (6 per cent) regions.

Within countries and the regions of England, there was inevitably wider variation among health authorities, with incidence rates ranging from 7 per 100,000 in areas of the Northern and Yorkshire region to 16 per 100,000 in areas of the South East and Eastern regions (Figure 23.3). The map (Map 23.1) shows that most of the higher than average incidence of uterine cancer occurred in two bands across England. In addition, there were three areas of noticeably higher than average incidence, one in the Border area in

Scotland, another in Morgannwg in Wales and one in the Southern area of Ireland. In England, the first band of incidence covers the area from the South Humber, through North Nottingham and North Derbyshire to Staffordshire, Birmingham and the surrounding area. The second band is apparent in the south of England covering the Eastern, South East and South West regions from Norfolk through Northamptonshire (26 per cent above the average), Oxfordshire and Dorset to Cornwall.

The incidence of cancer of the uterus was lower than average on the west coast of Scotland around Glasgow, the Western area of Northern Ireland, and the Northern Eastern and Midland areas of Ireland. No area in the Northern and Yorkshire or the North West regions of England had rates significantly above the average. On the east coast of England, in Northumberland and Sunderland, incidence rates were around 40 per cent lower than average. In Trent, low rates occurred in the Doncaster, Rotherham and Sheffield areas, that were sandwiched between areas of higher than average incidence. In the south of England, the only areas with apparently lower than average rates were Hertfordshire and four areas in west London: Hillingdon; Ealing, Hammersmith and Hounslow; Kingston and Richmond; and Kensington, Chelsea and Westminster.

Any of the apparent geographical patterns among the regions and health authorities may, however, have been affected by the varying proportions of cases assigned to the non-specific code for uterus by the regional cancer registries (as these were excluded from the analysis of cancer incidence). In addition, the apparent incidence rates exclude any real differences resulting from variation in the prevalence of women who have had a hysterectomy.

Geographical patterns in mortality

Mortality rates for cancer of the uterus showed more variation among countries than did incidence rates (Figure 23.3). Wales, the country with the highest incidence also had the highest mortality rates (5 per cent above the UK and Ireland average). Northern Ireland had the lowest mortality rate (16 per cent below the average), and Ireland and Scotland both had rates 10 per cent below the average. The regions in England also showed some variation, with two of the three regions, the South West and West Midlands (but not Eastern), that had high incidence rates also having high mortality rates, 9 per cent and 6 per cent above the average, respectively. Trent, with an incidence rate close to the average, had a mortality rate that was 6 per cent above it. The regions of Northern and Yorkshire and the North West, with lower than average incidence, also had mortality rates that were 13 per cent and 10 per cent below the UK and Ireland average, respectively.

The map of mortality from cancer of the uterus (Map 23.2) was broadly similar to that for incidence. Slightly higher than average mortality rates were apparent in Ayrshire and Arran, and Dumfries and Galloway in Scotland, and the South Eastern area of Ireland. There were areas with higher than average rates in England from the South Humber through North Nottinghamshire and North Derby to Staffordshire, a band incorporating Doncaster – an area of lower than average incidence with higher than average mortality. The neighbouring areas of Wolverhampton, Dudley, Worcestershire, Shropshire and Herefordshire in the West Midlands, and of Dyfed Powys, Morgannwg and Bro Taf in Wales, all had higher than average mortality rates, as did Northamptonshire, Kent, Southampton and South West Hampshire, Somerset, and Cornwall. In London, higher mortality rates were apparent in Barking and Havering; Lambeth, Southwark and Lewisham; Croydon; and Ealing, Hammersmith and Hounslow – the last of these had a noticeably lower than average incidence rate.

There were slightly lower than average mortality rates for cancer of the uterus for most of the areas in Scotland, Northern Ireland, Ireland and the regions of Northern and Yorkshire and North West in England. The following areas all had mortality rates that were more than a third lower than the UK and Ireland average: the Western area of Ireland, Northumberland and North West Lancashire in England, and Borders in Scotland. The incidence map for cancer of the uterus showed the latter area had a rate above the UK and Ireland average. The maps of incidence and mortality, although broadly similar, showed less agreement between the geographical areas than might have been expected. The differences may to some extent reflect variation among the regional cancer registries in the classification of newly diagnosed cases to the code for 'uterus unspecified', or geographical variation in the prevalence of women who have had a hysterectomy.

Risk factors and aetiology

About 90 per cent of cancers of the body of the uterus occur in the inner lining of the womb (endometrium). The pattern of geographic variation and the main risk factors for endometrial cancer are broadly similar to those for cancers of the breast (Chapter 5) and ovary (Chapter 18);[12,16,17] the main risk factors being early age at menarche, low parity and late age at menopause. These risk factors are all related to hormone levels, and result in either prolonged or increased amounts of oestrogen to which the uterus is exposed. Another source of oestrogen is hormone treatments that contain only oestrogen. Oestrogen-only hormone replacement therapy (HRT) and unopposed oestrogen therapy used for alleviating the

symptoms and harmful effects of the menopause increase the risk of endometrial cancer (among women who have not had a hysterectomy).[18] HRT formulations that include progestin appear to reduce the risk.[19] Sequential oral contraceptives (oestrogen followed by progesterone) increase the risk of cancer of the uterus.[17] However, the combined oral contraceptives that contain both the hormones have a long-lasting protective effect.[20] There is also a slight increased risk of endometrial cancer in women treated with tamoxifen for breast cancer.[21]

Excess body weight and physical inactivity account for approximately a quarter to one third of cases of cancer of the endometruim.[22] Hormones in the body are affected by obesity. Fatty tissue contains important enzymes used in the production of oestrogen-like compounds. The more fat there is in a body, the more oestrogen it can make and the greater the risk of endometrial cancer. Excess body weight is also associated with high blood pressure and diabetes; this association increases the likelihood that those with such conditions may develop endometrial cancer.

Changes in the prevalence of these aetiological factors over time may be responsible for much of the observed change in incidence of cancer of the body of the uterus. It is difficult to discern any geographic patterns in these aetiological factors within the UK and Ireland because variations in hysterectomy status and reproductive or hormonal factors are not well documented. Maps of total fertility rates and live birth rates by mother's age in the UK for 1991-97[23] show no clear inverse correlation between areas of higher fertility and those with low incidence and mortality rates for cancer of the uterus. In 2002-03, a quarter of women aged 16-49 in Great Britain used oral contraceptives. Women aged 18-29 were those most likely to use the contraceptive pill, with 48 per cent of women aged 20-24 using oral contraceptives.[24] However, this information is not available at a regional or lower geographical level. This makes it difficult to disentangle the influences of higher parity and oral contraceptive use in a given population, and it may be that their separate protective effects predominate in different regions.

The proportions of women who are obese or overweight rose markedly in the 1980s and 1990s. In 2001, over 20 per cent of women (aged 16 and over) in England were classified as obese and a further third as overweight.[25,26] Further information on the prevalence of obesity at a regional level in England is not available. In Scotland, 17 per cent of women were classified as obese and a further 30 per cent were classified as overweight. The prevalence of age-adjusted obesity among women varied significantly between regions. Women in Greater Glasgow were significantly less likely to be obese than Scottish women in general.[27] In Ireland, 10 per cent of people reported height and weight consistent with being obese and 32 per cent reported being overweight; there was significant geographic variation in the percentage of respondents who were obese, the highest rate of obesity among females occurred in the Northern area of Ireland (15 per cent).[28,29] In Ireland, 19 per cent of respondents were obese and a further 37 per cent were overweight.[30] The available information on obesity at the regional level for the UK and Ireland does not explain the geographical variation observed for cancer of the uterus.

Socio-economic deprivation

Incidence rates for cancer of the body of the uterus were marginally higher in women from more affluent areas than more deprived areas.[11,31] This was similar to, but less well defined than, the trend for breast and ovarian cancer incidence.[11] For cancer of the uterus, as for both breast and ovarian cancer, this trend is consistent with the known protective effect of higher parity, as total fertility rates are higher in more deprived areas.[23] Obesity is a known risk factor for cancer of the uterus and is linked to social class. Obesity is more common among those in the routine or semi-routine occupational groups than the managerial and professional groups. This link is stronger for women than men. In 2001, 30 per cent of women in routine occupations were classified as obese compared with 16 per cent in higher managerial and professional occupations.[25] This upward trend in the prevalence of obesity over the last seven years was more marked in manual social classes than in non-manual classes.[32] In Scotland, the age-standardised prevalence of obesity among women was lower in the non-manual than the manual social classes in the mid-1990s.[27] The relationship between obesity and deprivation seems to conflict with the observed pattern in deprivation for cancer of the uterus and implies that the protective effect of higher parity is more important than the risk conferred by obesity. There is no variation in mortality from cancer of the uterus with deprivation.[11]

Survival for cancer of the uterus has been consistently lower for patients in deprived areas. Although survival has improved in all socio-economic groups, the deprivation gap in survival between the affluent and deprived, which was small but statistically significant during the late 1980s, widened in the late 1990s. There was a statistically significant deprivation gap of 4.2 percentage points for five-year relative survival between the most affluent and the most deprived groups in women diagnosed during 1996-99 in England and Wales.[14] Social class and deprivation can explain a small part of the geographical variation in the incidence of, and mortality from, cancer of the

uterus in the UK and Ireland. It is not possible at present to identify whether later stage at presentation might contribute to socio-economic differences in mortality from cancer of the uterus; but a higher proportion of cancer of the cervix (labelled as cancer of the uterus unspecified) may also contribute to the higher mortality in more deprived groups.

References

1. NHS Scotland. Cancer Data. August 2004. Available at *http://www.isdscotland.org/isd/info3.jsp?pContentID=402&p_applic=CCC&p_service=Content.show&*.

2. Gavin AT, Reid J. *Cancer Incidence in Northern Ireland, 1993-95.* Belfast: The Stationery Office, 1999.

3. National Cancer Registry of Ireland. *Cancer in Ireland, 1994: Incidence and Mortality.* Cork: National Cancer Registry Board, 1997.

4. ONS. *Cancer Statistics Registrations: Registrations of cancer diagnosed in 1994-97, England.* Series MB1 No. 28. London: The Stationery Office, 2000.

5. ONS. *Mortality Statistics: cause 1993 (revised) and 1994.* Series DH2 No. 21. London: HMSO, 1996.

6. Redburn JC, Murphy MFG. Hysterectomy prevalence and adjusted cervical and uterine cancer rates in England and Wales. *British Journal of Obstetrics and Gynaecology* 2001; 108: 388-395.

7. Marshall SF, Hardy RJ, Kuh D. Socioeconomic variation in hysterectomy up to age 52: national, population based, prospective cohort study. *British Medical Journal* 2000; 320: 1579.

8. Schouten LJ, Straatman H, Kiemeney LALM, Verbeek ALM. Cancer incidence: Life table risk versus cumulative risk. *Journal of Epidemiology and Community Health* 1994; 48: 596-600.

9. ONS. *Cancer Statistics Registrations: Registrations of cancer diagnosed in 2001, England.* Series MB1 No. 32. London: Office for National Statistics, 2004.

10. Quinn MJ, Babb PJ, Kirby L, Jones J. Registrations of cancer diagnosed in 1994-97, England and Wales. *Health Statistics Quarterly* 2000; 7: 71-82.

11. Quinn MJ, Babb PJ, Brock A, Kirby L et al. *Cancer Trends in England and Wales 1950-1999.* Studies on Medical and Population Subjects No. 66. London: The Stationery Office, 2001.

12. dos Santos-Silva, I, Swerdlow AJ. Recent trends in incidence of and mortality from breast, ovarian and endometrial cancers in England and Wales and their relation to changing fertility and oral contraceptive use. *British Journal of Cancer* 1995; 72: 485-492.

13. ONS. Cancer Survival: England and Wales, 1991-2001. March 2004. Available at *http://www.statistics.gov.uk/statbase/ssdataset.asp?vlnk=7899*.

14. Coleman MP, Rachet B, Woods LM, Mitry E et al. Trends and socioeconomic inequalities in cancer survival in England and Wales up to 2001. *British Journal of Cancer* 2004; 90: 1367-1373.

15. Sant M, Areleid T, Berrino F, Bielska Lasota M et al. EUROCARE-3: survival of cancer patients diagnosed 1990-1994 - results and commentary. *Annals of Oncology* 2003; 14 Suppl 5: v61-v118.

16. Elwood JM, Cole P, Rothman KJ, Kaplan SD. Epidemiology of endometrial cancer. *Journal of the National Cancer Institute* 1977; 59: 1055-1060.

17. Henderson BE, Casagrande JT, Pike MC, Mack T et al. The epidemiology of endometrial cancer in young women. *British Journal of Cancer* 1983; 47: 749-756.

18. Oestrogen replacement and endometrial cancer. A statement by the British Gynaecological Cancer Group. *Lancet* 1981; 1: 1359-1360.

19. Weiderpass E, Adami HO, Baron JA, Magnusson C et al. Risk of endometrial cancer following estrogen replacement with and without progestins. *Journal of the National Cancer Institute* 1999; 91: 1131-1137.

20. Endometrial cancer and combined oral contraceptives. The Who Collaborative Study of Neoplasia and Steroid Contraceptives. *International Journal of Epidemiology* 1988; 17: 263-269.

21. IARC Working Group. *Some Pharmaceutical Drugs.* IARC Monographs on the Evaluation of Carcinogenic Risk to Humans, No. 66. Lyon: IARC, 1996.

22. Vainio H, Kaaks R, Bianchini F. Weight control and physical activity in cancer prevention: international evaluation of the evidence. *European Journal of Cancer Prevention* 2002; 11 Suppl 2: S94-100.

23. Griffiths C, Fitzpatrick J. *Geographic Variations in Health.* Decennial Supplements No. 16. London: The Stationery Office, 2001.

24. Dawe F, Meltzer H. *Contraception and Sexual Health, 2002.* London: Office for National Statistics, 2003.

25. Department of Health. *Health Survey for England, 2001.* London: The Stationery Office, 2003.

26. ONS. *Social Trends 34.* London: The Stationery Office, 2004.

27. The Scottish Office Department of Health. *Scotland's Health: Scottish Health Survey 1995.* London: The Stationery Office, 1997.

28. Department of Health and Children. *Quality and Fairness. A Health System for You: Health Strategy.* Dublin: The Stationery Office, 2001.

29. Department of Health and Children. *The National Health and Lifestyle Surveys 2003. Regional results of the National Health and Lifestyle Surveys SLAN (Survey of Lifestyle, Attitudes and Nutrition) and HBSC (Health Behaviour in School-Aged Children).* Dublin: The Stationery Office, 2003.

30. Department of Health, Social Services and Public Safety. *Health and Lifestyle Report. A Report from the Health and Social Wellbeing Survey, 1997.* Belfast: Department of Health, Social Services and Public Safety, 2001.

31. Harris V, Sandridge AL, Black RJ, Brewster DH et al. *Cancer Registration Statistics Scotland, 1986-1995.* Edinburgh: ISD Publications, 1998.

32. Department of Health. *Health Survey for England, 2002: The Health of Children and Young People.* London: The Stationery Office, 2003.

Appendix A
Key maps

This appendix gives 'key' or reference maps to assist in the interpretation of information given in Chapters 3-23.

Map A1 is a key map of the UK and Ireland, by health authority. The health regional offices (in England) and all health authorities are labelled. The term 'health authorities' has been used throughout this atlas, although these areas are actually called 'health boards' in Scotland and Ireland, and 'health and social services boards' in Northern Ireland (see Appendix G for details).

Map A2 shows the cancer registry boundaries and health regional office boundaries (within England), to illustrate where the areas differ. Details of the United Kingdom Association of Cancer Registries and contact details for the individual cancer registries are given in Appendix J.

Map A1

United Kingdom and Ireland: Health geography, 2001

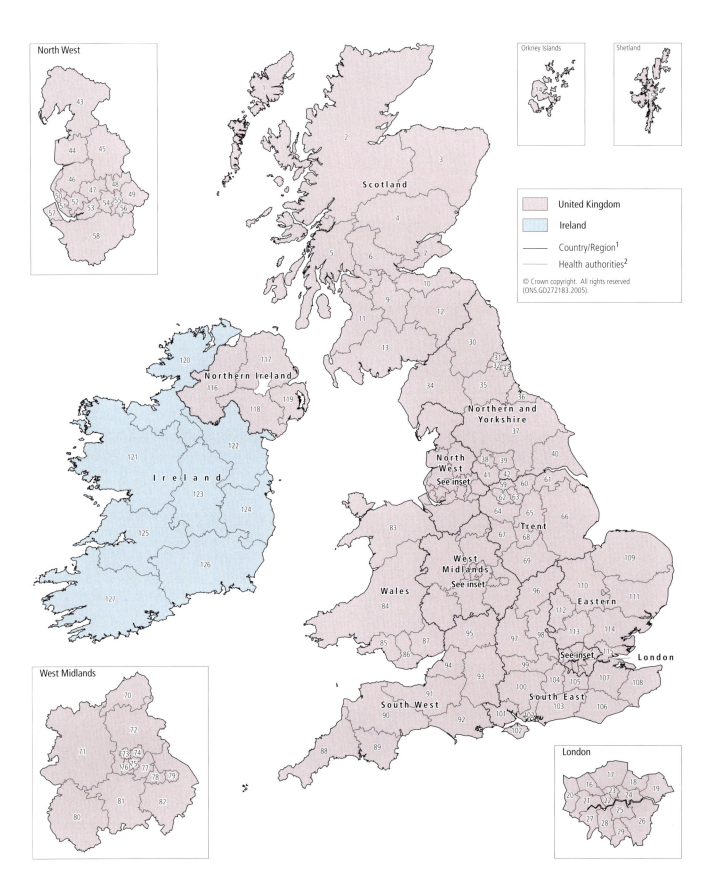

Index of health authorities[2] within the United Kingdom and Ireland, 2001

Northern and Yorkshire

Bradford	38
Calderdale and Kirklees	41
County Durham	35
East Riding	40
Gateshead and South Tyneside	32
Leeds	39
Newcastle and North Tyneside	31
North Cumbria	34
Northumberland	30
North Yorkshire	37
Sunderland	33
Tees	36
Wakefield	42

Trent

Barnsley	59
Doncaster	60
Leicestershire	69
Lincolnshire	66
North Derbyshire	64
North Nottinghamshire	65
Nottingham	68
Rotherham	63
Sheffield	62
South Humber	61
Southern Derbyshire	67

West Midlands

Birmingham	77
Coventry	79
Dudley	76
Herefordshire	80
North Staffordshire	70
Sandwell	75
Shropshire	71
Solihull	78
South Staffordshire	72
Walsall	74
Warwickshire	82
Wolverhampton	73
Worcestershire	81

North West

Bury and Rochdale	48
East Lancashire	45
Liverpool	51
Manchester	55
Morecambe Bay	43
North Cheshire	53
North West Lancashire	44
St Helen's and Knowsley	52
Salford and Trafford	54
Sefton	50
South Cheshire	58
South Lancashire	46
Stockport	56
West Pennine	49
Wigan and Bolton	47
Wirral	57

Eastern

Bedfordshire	112
Cambridgeshire	110
Hertfordshire	113
Norfolk	109
North Essex	114
South Essex	115
Suffolk	111

London

Bexley, Bromley and Greenwich	26
Brent and Harrow	16
Barking and Havering	19
Barnet, Enfield and Haringey	17
Camden and Islington	23
Croydon	29
Ealing, Hammersmith and Hounslow	21
East London and The City	24
Hillingdon	20
Kensington and Chelsea and Westminster	22
Kingston and Richmond	27
Lambeth, Southwark and Lewisham	25
Merton, Sutton and Wandsworth	28
Redbridge and Waltham Forest	18

South East

Berkshire	99
Buckinghamshire	98
East Kent	108
East Surrey	105
East Sussex, Brighton and Hove	106
Isle of Wight, Portsmouth and South East Hampshire	102
Northamptonshire	96
North and Mid Hampshire	100
Oxfordshire	97
Southampton and South West Hampshire	101
West Kent	107
West Surrey	104
West Sussex	103

South West

Avon	94
Cornwall and Isles of Scilly	88
Dorset	92
Gloucestershire	95
North and East Devon	90
Somerset	91
South and West Devon	89
Wiltshire	93

Wales

Bro Taf	86
Dyfed Powys	84
Gwent	87
Morgannwg	85
North Wales	83

Scotland

Argyll and Clyde	5
Ayrshire and Arran	11
Borders	12
Dumfries and Galloway	13
Fife	7
Forth Valley	6
Grampian	3
Greater Glasgow	8
Highland	2
Lanarkshire	9
Lothian	10
Orkney	14
Shetland	15
Tayside	4
Western Isles	1

Northern Ireland

Eastern	119
Northern	117
Southern	118
Western	116

Ireland

Eastern	124
Midland	123
Mid Western	125
North East	122
North Western	120
South Eastern	126
Southern	127
Western	121

1   *Health regional offices in England.*

2   *Health authorities (England and Wales); health boards (Scotland); health and social services boards (Northern Ireland); 7 regional health boards and 1 regional health authority (Ireland; the Eastern regional health authority is divided into 3 smaller area health boards, which are not shown).*

Map A2

Cancer registries in the UK and Ireland and health regional offices within England

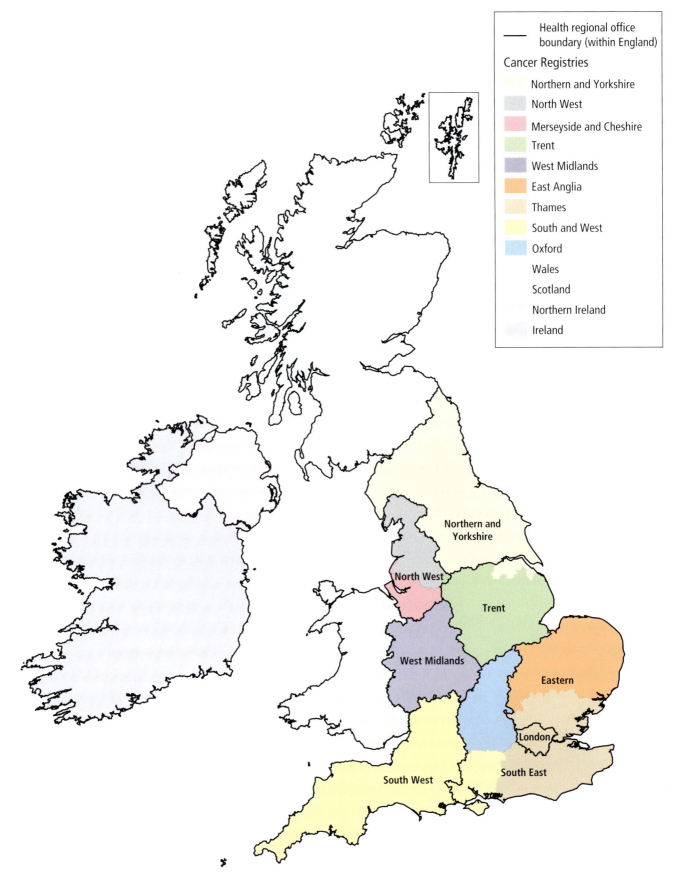

Health regional office boundary (within England)

Cancer Registries

Northern and Yorkshire
North West
Merseyside and Cheshire
Trent
West Midlands
East Anglia
Thames
South and West
Oxford
Wales
Scotland
Northern Ireland
Ireland

Appendix B
Data tables

This appendix gives tables of the incidence and mortality data by sex, country, health regional office (within England), and health authority for the UK and Ireland, for all cancers combined (excluding non-melanoma skin cancer) and for each of the 21 individual cancers discussed in detail in this volume.

For ease of reference the tables are numbered according to the relevant chapter, thus Table B2.1 gives data for all cancers (Chapter 2), Table B3.1 gives data for bladder cancer (Chapter 3 – Bladder), and so on (there is no Table B1.1).

Table Bx.1 gives the age-standardised mortality and incidence rates, the mortality-to-incidence ratio (of the age-standardised rates), a ranking of health authorities within region (of England) or country, and an indication of whether the rate is significantly higher or lower than the average for the UK and Ireland. In the rankings, 1 indicates the lowest rate. The rankings can be used to identify individual health authorities on Figures x.3 and x.4 in Chapters 3-23, as on these charts the health authorities are ordered from lowest to highest rate. In the table, a rate significantly higher than the average is indicated by '*', and '~' indicates a rate significantly lower than the average (based on non-overlapping 95 per cent confidence intervals – see Appendix H for details).

Table Bx.2 gives the average number of registrations and deaths per year, rounded where necessary to three meaningful digits.

Table **B2.1**

All cancers[1]: annual average age-standardised incidence and mortality by health authority within country and region of England, UK and Ireland, 1991–2000

Country/ region/ health authority	Incidence[2]						Mortality[3]						M:I Ratio[4]	
		Males			Females			Males			Females		Males	Females
	Rank[5]	Rate[6]	Signif[7]	Rank	Rate	Signif	Rank	Rate	Signif	Rank	Rate	Signif		
UK and Ireland		404.7			336.7			253.0			172.6		0.63	0.51
England		396.3	~		331.0	~		249.6	~		170.3	~	0.63	0.51
Northern and Yorkshire		411.0	*		332.4	~		271.4	*		180.7	*	0.66	0.54
Bradford	4	396.2		7	336.0		6	264.2	*	5	175.6		0.67	0.52
Calderdale and Kirklees	5	401.1		2	316.2	~	2	253.4		2	166.8	~	0.63	0.53
County Durham	8	425.7	*	8	336.1		9	284.5	*	9	185.3	*	0.67	0.55
East Riding	6	403.0		6	335.3		7	265.0	*	7	179.1	*	0.66	0.53
Gateshead and South Tyneside	12	446.3	*	11	345.0		13	320.0	*	11	198.7	*	0.72	0.58
Leeds	7	421.6	*	13	353.2	*	4	259.2	*	6	176.2		0.61	0.50
Newcastle and North Tyneside	13	446.8	*	9	336.9		12	311.8	*	13	204.8	*	0.70	0.61
North Cumbria	3	383.2	~	4	319.8	~	3	256.7		4	175.4		0.67	0.55
North Yorkshire	1	369.0	~	3	319.0	~	1	224.1	~	1	156.7	~	0.61	0.49
Northumberland	2	376.6	~	1	297.6	~	5	263.1	*	3	175.0		0.70	0.59
Sunderland	11	442.7	*	10	338.4		11	306.9	*	12	203.3	*	0.69	0.60
Tees	10	434.2	*	5	335.3		10	304.0	*	10	195.5	*	0.70	0.58
Wakefield	9	426.4	*	12	348.9	*	8	270.2	*	8	184.9	*	0.63	0.53
Trent		391.2	~		323.5	~		254.0			172.2		0.65	0.53
Barnsley	10	424.9	*	11	363.0	*	10	276.3	*	11	192.5	*	0.65	0.53
Doncaster	11	428.8	*	10	339.4		9	272.1	*	10	185.4	*	0.63	0.55
Leicestershire	1	351.9	~	1	306.7	~	1	224.6	~	1	158.2	~	0.64	0.52
Lincolnshire	3	388.5	~	5	327.4	~	2	237.0	~	3	165.4	~	0.61	0.51
North Derbyshire	6	397.3		6	327.4		5	260.2		7	179.2	*	0.65	0.55
North Nottinghamshire	7	407.4		4	324.4	~	6	260.4	*	4	172.5		0.64	0.53
Nottingham	8	410.9		9	334.2		4	258.2		5	174.3		0.63	0.52
Rotherham	9	411.2		7	327.9		7	269.0	*	9	181.2	*	0.65	0.55
Sheffield	4	390.1	~	2	308.8	~	11	280.7	*	6	178.1	*	0.72	0.58
South Humber	5	392.1	~	8	328.1		8	271.1	*	8	180.1	*	0.69	0.55
Southern Derbyshire	2	378.0	~	3	314.6	~	3	248.6		2	165.2	~	0.66	0.53
West Midlands		406.3			329.4	~		255.6	*		170.6	~	0.63	0.52
Birmingham	12	430.8	*	12	334.6		9	264.9	*	8	172.2		0.61	0.51
Coventry	8	411.0		11	333.8		6	249.7		10	173.0		0.61	0.52
Dudley	5	395.4		4	325.6		8	260.2		6	168.1		0.66	0.52
Herefordshire	1	353.3	~	2	323.3	~	1	226.2	~	1	157.6	~	0.64	0.49
North Staffordshire	9	415.2	*	7	328.8	~	12	280.4	*	13	186.6	*	0.68	0.57
Sandwell	13	438.9	*	13	337.8		13	292.5	*	12	183.4	*	0.67	0.54
Shropshire	7	408.1		9	331.6		5	246.1		4	166.2		0.60	0.50
Solihull	6	399.5		10	333.2		3	233.7	~	5	167.2		0.59	0.50
South Staffordshire	3	394.3	~	8	328.8	~	7	252.0		7	169.6		0.64	0.52
Walsall	11	417.2	*	6	328.4		10	270.8	*	9	172.6		0.65	0.53
Warwickshire	4	394.5	~	5	326.6	~	4	241.4	~	3	165.5	~	0.61	0.51
Wolverhampton	10	415.6	*	1	318.3	~	11	273.2	*	11	174.8		0.66	0.55
Worcestershire	2	378.7	~	3	323.4	~	2	229.1	~	2	158.8	~	0.61	0.49
North West		423.2	*		342.6	*		275.0	*		184.4	*	0.65	0.54
Bury and Rochdale	7	417.3	*	12	348.4	*	8	270.1	*	10	186.5	*	0.65	0.54
East Lancashire	3	395.7	~	3	331.7		6	262.7	*	6	174.5	*	0.66	0.53
Liverpool	15	491.3	*	15	380.4	*	15	331.0	*	16	220.6	*	0.67	0.58
Manchester	16	505.9	*	16	391.6	*	16	336.5	*	15	214.2	*	0.67	0.55
Morecambe Bay	2	387.3	~	2	322.7	~	3	249.0		2	166.6	~	0.64	0.52
North Cheshire	11	433.1	*	7	337.5		12	285.4	*	11	188.7	*	0.66	0.56
North West Lancashire	10	422.6	*	10	338.9		5	256.4		5	173.8		0.61	0.51
Salford and Trafford	13	443.9	*	14	362.0	*	13	287.7	*	13	194.9	*	0.65	0.54
Sefton	12	437.0	*	5	336.2		10	280.4	*	8	185.7	*	0.64	0.55
South Cheshire	1	368.3	~	1	309.6	~	1	242.8	~	1	166.3	~	0.66	0.54
South Lancashire	4	406.1		8	337.6		2	247.7		4	170.5		0.61	0.50
St Helen's and Knowsley	14	459.5	*	13	353.3	*	14	303.6	*	14	196.1	*	0.66	0.56
Stockport	6	412.1		4	335.1		4	252.1		3	170.2		0.61	0.51
West Pennine	8	419.0	*	11	345.0	*	11	283.5	*	9	185.8	*	0.68	0.54
Wigan and Bolton	5	411.2		9	338.7		7	267.8	*	7	181.4	*	0.65	0.54
Wirral	9	421.3	*	6	337.1		9	278.2	*	12	189.2	*	0.66	0.56
Eastern		362.8	~		319.0	~		229.1	~		161.6	~	0.63	0.51
Bedfordshire	3	358.7	~	3	320.8	~	6	233.6	~	5	161.6	~	0.65	0.50
Cambridgeshire	4	362.5	~	7	328.4	~	2	223.1	~	3	159.8	~	0.62	0.49
Hertfordshire	5	364.3	~	6	324.8	~	4	226.6	~	6	163.6	~	0.62	0.50
Norfolk	2	357.1	~	2	318.5	~	3	224.4	~	1	157.0	~	0.63	0.49
North Essex	1	346.5	~	1	302.3	~	5	229.9	~	4	161.1	~	0.66	0.53
South Essex	7	381.8	~	4	321.1	~	7	249.2	~	7	170.9	~	0.65	0.53
Suffolk	6	375.9	~	5	322.1	~	1	221.7	~	2	157.2	~	0.59	0.49

1 All malignancies excluding non-melanoma skin cancer (ICD9 codes 140-208 excluding 173; ICD10 codes C00-C97 excluding C44).
2 England, Wales and Scotland 1991-99; Northern Ireland 1993-99; Ireland 1994-99.
3 England, Wales and Northern Ireland 1991-2000; Scotland 1991-99; Ireland 1994-2000.
4 Mortality to incidence ratio; age-standardised mortality rate divided by age-standardised incidence rate.
5 Rank within region or country; 1 is lowest.
6 Directly age standardised using the European standard population; per 100,000 population.
7 Rate significantly different from that for UK and Ireland; * significantly higher; ~ significantly lower. See Appendix H.

Country/ region/ health authority	Incidence[2]						Mortality[3]						M:I Ratio[4]	
	Rank[5]	Males Rate[6]	Signif[7]	Rank	Females Rate	Signif	Rank	Males Rate	Signif	Rank	Females Rate	Signif	Males	Females
London		**394.8**	~		**325.8**	~		**249.9**	~		**171.2**		**0.63**	**0.53**
Barking and Havering	11	421.7	*	12	345.6	*	12	275.0	*	12	182.8	*	0.65	0.53
Barnet, Enfield and Haringey	4	373.2	~	4	318.2	~	3	227.6	~	3	160.8	~	0.61	0.51
Bexley, Bromley and Greenwich	7	376.7	~	2	313.4	~	9	249.0		7	171.5		0.66	0.55
Brent and Harrow	3	368.3	~	3	313.6	~	1	216.0	~	1	155.3	~	0.59	0.50
Camden and Islington	14	452.7	*	14	348.7	*	11	273.2	*	13	183.1	*	0.60	0.53
Croydon	2	365.8	~	8	326.5	~	4	232.0	~	6	166.6		0.63	0.51
Ealing, Hammersmith and Hounslow	1	358.4	~	1	310.8	~	5	236.7	~	8	171.6		0.66	0.55
East London and The City	12	427.8	*	7	325.6	~	14	294.7	*	14	190.0	*	0.69	0.58
Hillingdon	5	375.2	~	5	320.1	~	8	243.5	~	10	173.1		0.65	0.54
Kensington and Chelsea, and Westminster	9	414.2		11	337.1		7	241.4	~	5	166.6		0.58	0.49
Kingston and Richmond	6	376.2	~	10	332.3		2	226.1	~	4	166.0	~	0.60	0.50
Lambeth, Southwark and Lewisham	10	418.0	*	6	322.9	~	13	282.6	*	11	181.3	*	0.68	0.56
Merton, Sutton and Wandsworth	13	430.7	*	13	346.6	*	10	254.7		9	172.5		0.59	0.50
Redbridge and Waltham Forest	8	385.6	~	9	327.7	~	6	236.8	~	2	159.3	~	0.61	0.49
South East		**387.6**	~		**334.7**			**234.2**	~		**162.6**	~	**0.60**	**0.49**
Berkshire	11	406.8		8	338.5		6	227.8	~	6	161.4	~	0.56	0.48
Buckinghamshire	9	403.4		10	356.9	*	5	227.1	~	10	164.3	~	0.56	0.46
East Kent	2	363.9	~	5	322.2	~	13	255.9		12	172.7		0.70	0.54
East Surrey	5	375.6	~	6	323.7	~	1	218.8	~	1	150.3	~	0.58	0.46
East Sussex, Brighton and Hove	3	365.5	~	1	313.9	~	8	231.9	~	8	162.9	~	0.63	0.52
Isle of Wight, Portsmouth and South East Hampshire	12	424.0	*	13	365.5	*	10	242.8	~	9	163.0	~	0.57	0.45
North and Mid-Hampshire	6	376.7	~	7	336.9		7	229.3	~	5	160.1	~	0.61	0.48
Northamptonshire	10	403.7		9	349.9	*	12	248.9		13	173.6		0.62	0.50
Oxfordshire	8	385.8	~	12	359.7	*	4	225.5	~	7	161.9	~	0.58	0.45
Southampton and South West Hampshire	13	440.3	*	11	359.6	*	9	238.8	~	4	158.0	~	0.54	0.44
West Kent	7	380.1	~	3	316.6	~	11	247.8	~	11	170.2	~	0.65	0.54
West Surrey	4	375.2	~	4	319.9	~	2	224.0	~	2	155.7	~	0.60	0.49
West Sussex	1	363.1	~	2	315.6	~	3	224.3	~	3	156.9	~	0.62	0.50
South West		**389.7**	~		**336.8**			**227.9**	~		**158.6**	~	**0.58**	**0.47**
Avon	4	385.0	~	2	326.9	~	7	232.9	~	5	159.6	~	0.60	0.49
Cornwall and Isles of Scilly	5	388.2	~	6	334.0	~	6	230.0	~	8	165.7	~	0.59	0.50
Dorset	8	429.0	*	8	368.9	*	1	220.8	~	1	152.8	~	0.51	0.41
Gloucestershire	2	371.2	~	1	311.9	~	2	222.7	~	4	156.9	~	0.60	0.50
North and East Devon	1	362.0	~	4	328.5	~	3	223.3	~	3	156.1	~	0.62	0.48
Somerset	3	377.0	~	5	329.5		5	226.2	~	2	154.7	~	0.60	0.47
South and West Devon	6	391.9	~	3	327.6	~	8	241.2	~	7	163.8	~	0.62	0.50
Wiltshire	7	397.2	~	7	362.3	*	4	225.2	~	6	160.6	~	0.57	0.44
Wales		**436.6**	*		**356.7**	*		**255.8**			**174.9**	*	**0.59**	**0.49**
Bro Taf	3	440.6	*	3	352.6	*	5	266.9	*	5	179.7	*	0.61	0.51
Dyfed Powys	1	401.0		2	345.3	*	1	233.6	*	1	165.8	~	0.58	0.48
Gwent	2	425.0	*	1	343.7	*	4	261.1	*	4	177.5	*	0.61	0.52
Morgannwg	4	450.5	*	4	358.3	*	3	260.6	*	2	174.2		0.58	0.49
North Wales	5	458.9	*	5	378.0	*	2	255.3		3	175.4		0.56	0.46
Scotland		**471.1**	*		**381.1**	*		**290.0**	*		**195.4**	*	**0.62**	**0.51**
Argyll and Clyde	13	484.0	*	12	385.3	*	14	310.4	*	14	201.9	*	0.64	0.52
Ayrshire and Arran	4	437.4	*	9	375.2	*	8	275.1	*	11	192.3	*	0.63	0.51
Borders	2	420.8		10	379.3	*	3	262.0		4	179.8		0.62	0.47
Dumfries and Galloway	3	423.7	*	7	372.4	*	2	249.1		6	185.3	*	0.59	0.50
Fife	10	456.6	*	5	370.5	*	7	272.2	*	7	185.4	*	0.60	0.50
Forth Valley	11	465.8	*	8	373.7	*	9	276.4	*	9	188.4	*	0.59	0.50
Grampian	8	455.0	*	4	369.1	*	4	263.8	*	5	180.0	*	0.58	0.49
Greater Glasgow	15	531.0	*	15	405.5	*	15	345.6	*	15	220.6	*	0.65	0.54
Highland	9	455.9	*	11	380.4	*	5	265.2	*	3	179.0		0.58	0.47
Lanarkshire	7	448.5	*	3	361.5	*	11	290.0	*	13	197.3	*	0.65	0.55
Lothian	14	489.2	*	14	390.9	*	10	285.7	*	12	193.3	*	0.58	0.49
Orkney	1	350.9	~	1	316.3		1	213.8	~	1	159.3	~	0.61	0.50
Shetland	6	448.2	*	2	344.8		12	291.6	*	2	165.0		0.65	0.48
Tayside	5	445.7	*	6	371.9	*	6	271.5	*	10	189.1	*	0.61	0.51
Western Isles	12	470.7	*	13	385.8	*	13	306.3	*	8	185.8		0.65	0.48
Northern Ireland		**405.1**			**345.4**	*		**245.0**	~		**167.0**	~	**0.60**	**0.48**
Eastern	4	419.3	*	2	347.2	*	4	257.8	*	4	169.7		0.61	0.49
Northern	1	379.4	~	1	333.7		2	230.9	~	1	163.5	~	0.61	0.49
Southern	2	405.3		4	352.6	*	1	230.8	~	2	164.5	~	0.57	0.47
Western	3	409.3		3	350.7	*	3	249.4		3	167.7		0.61	0.48
Ireland		**394.6**	~		**320.9**	~		**252.7**			**174.3**		**0.64**	**0.54**
Eastern	8	452.4	*	8	344.7	*	8	284.5	*	8	184.6	*	0.63	0.54
Mid-Western	1	342.5	~	2	294.2	~	3	239.6	~	4	171.7		0.70	0.58
Midland	4	374.4	~	7	333.2		7	247.3		6	173.3		0.66	0.52
North Eastern	6	379.8	~	3	305.1	~	2	236.3	~	2	166.1	~	0.62	0.54
North Western	5	375.6	~	5	312.6	~	4	240.0	~	3	166.9	~	0.64	0.53
South Eastern	3	373.3	~	4	305.8	~	5	243.1	~	7	174.0		0.65	0.57
Southern	7	395.2	~	6	325.5	~	6	247.1	~	5	172.5	~	0.63	0.53
Western	2	351.0	~	1	286.8	~	1	228.3	~	1	159.5	~	0.65	0.56

Table **B2.2**

All cancers[1]: annual average numbers of registrations and deaths by health authority within country and region of England, UK and Ireland, 1991–2000

Country/ region/ health authority	Number of registrations[2]		Number of deaths[3]	
	Males	Females	Males	Females
UK and Ireland	**134,000**	**136,000**	**85,100**	**78,500**
England	**105,000**	**107,000**	**67,300**	**61,900**
Northern and Yorkshire	**14,000**	**14,000**	**9,330**	**8,490**
Bradford	908	982	609	576
Calderdale and Kirklees	1,190	1,190	759	708
County Durham	1,410	1,370	952	836
East Riding	1,280	1,280	853	772
Gateshead and South Tyneside	892	863	645	552
Leeds	1,580	1,640	986	924
Newcastle and North Tyneside	1,130	1,100	797	745
North Cumbria	707	713	482	436
North Yorkshire	1,620	1,690	1,010	964
Northumberland	682	652	484	426
Sunderland	657	626	456	408
Tees	1,210	1,150	851	732
Wakefield	697	697	445	407
Trent	**11,100**	**10,900**	**7,320**	**6,480**
Barnsley	521	532	342	314
Doncaster	678	636	435	384
Leicestershire	1,690	1,760	1,100	1,010
Lincolnshire	1,530	1,450	960	831
North Derbyshire	858	839	571	513
North Nottinghamshire	897	842	583	491
Nottingham	1,390	1,350	887	789
Rotherham	542	531	359	321
Sheffield	1,150	1,150	838	745
South Humber	682	671	479	408
Southern Derbyshire	1,160	1,140	774	678
West Midlands	**11,600**	**11,300**	**7,410**	**6,530**
Birmingham	2,180	2,050	1,350	1,180
Coventry	674	641	413	374
Dudley	670	672	447	389
Herefordshire	381	400	254	227
North Staffordshire	1,070	1,040	730	649
Sandwell	698	659	468	400
Shropshire	958	918	589	518
Solihull	442	446	261	247
South Staffordshire	1,180	1,180	759	673
Walsall	576	546	377	318
Warwickshire	1,100	1,070	688	608
Wolverhampton	563	508	376	310
Worcestershire	1,120	1,130	697	635
North West	**14,700**	**15,000**	**9,640**	**8,990**
Bury and Rochdale	805	851	523	505
East Lancashire	1,050	1,100	703	648
Liverpool	1,140	1,150	771	732
Manchester	974	952	648	571
Morecambe Bay	728	751	479	445
North Cheshire	647	629	426	383
North West Lancashire	1,200	1,180	749	700
Salford and Trafford	1,060	1,080	693	647
Sefton	736	743	480	466
South Cheshire	1,370	1,410	917	844
South Lancashire	646	665	398	370
St Helen's and Knowsley	749	734	496	445
Stockport	643	658	399	383
West Pennine	989	1,039	674	618
Wigan and Bolton	1,170	1,220	765	716
Wirral	777	808	523	512
Eastern	**10,800**	**11,100**	**6,960**	**6,370**
Bedfordshire	938	984	615	543
Cambridgeshire	1,300	1,350	813	743
Hertfordshire	1,940	2,080	1,220	1,170
Norfolk	1,850	1,830	1,210	1,070
North Essex	1,770	1,840	1,210	1,110
South Essex	1,450	1,510	962	897
Suffolk	1,520	1,490	924	840

1 All malignancies excluding non-melanoma skin cancer (ICD9 codes 140-208 excluding 173; ICD10 codes C00-C97 excluding C44).
2 England, Wales and Scotland 1991-99; Northern Ireland 1993-99; Ireland 1994-99.
3 England, Wales and Northern Ireland 1991-2000; Scotland 1991-99; Ireland 1994-2000.

Country/ region/ health authority	Number of registrations[2]		Number of deaths[3]	
	Males	Females	Males	Females
London	**12,900**	**13,500**	**8,220**	**7,770**
Barking and Havering	895	915	588	535
Barnet, Enfield and Haringey	1,390	1,520	862	857
Bexley, Bromley and Greenwich	1,430	1,530	964	931
Brent and Harrow	792	841	469	457
Camden and Islington	710	689	428	391
Croydon	558	633	357	355
Ealing, Hammersmith and Hounslow	1,040	1,120	688	668
East London and The City	1,030	950	704	586
Hillingdon	450	473	296	282
Kensington and Chelsea, and Westminster	646	662	382	363
Kingston and Richmond	602	683	370	383
Lambeth, Southwark and Lewisham	1,290	1,260	865	759
Merton, Sutton and Wandsworth	1,220	1,300	730	717
Redbridge and Waltham Forest	843	905	520	489
South East	**18,200**	**19,100**	**11,300**	**10,600**
Berkshire	1,470	1,480	824	773
Buckinghamshire	1,270	1,360	722	689
East Kent	1,360	1,450	994	915
East Surrey	922	964	553	516
East Sussex, Brighton and Hove	1,800	1,960	1,200	1,210
Isle of Wight, Portsmouth and South East Hampshire	1,640	1,710	970	885
North and Mid-Hampshire	991	1,100	608	579
Northamptonshire	1,260	1,290	787	711
Oxfordshire	1,160	1,260	693	648
Southampton and South West Hampshire	1,360	1,310	757	664
West Kent	1,890	1,960	1,250	1,170
West Surrey	1,300	1,350	790	740
West Sussex	1,750	1,890	1,130	1,120
South West	**11,700**	**12,000**	**7,100**	**6,610**
Avon	2,090	2,170	1,290	1,210
Cornwall and Isles of Scilly	1,250	1,260	774	735
Dorset	2,090	2,100	1,130	1,050
Gloucestershire	1,210	1,200	741	692
North and East Devon	1,160	1,240	759	710
Somerset	1,170	1,190	732	660
South and West Devon	1,450	1,490	929	870
Wiltshire	1,270	1,360	736	686
Wales	**7,340**	**7,330**	**4,390**	**4,050**
Bro Taf	1,660	1,660	1,020	941
Dyfed Powys	1,250	1,240	747	683
Gwent	1,290	1,270	807	725
Morgannwg	1,310	1,280	775	701
North Wales	1,840	1,890	1,040	1,000
Scotland	**12,400**	**13,000**	**7,640**	**7,380**
Argyll and Clyde	1,060	1,130	682	648
Ayrshire and Arran	884	974	558	548
Borders	280	310	178	171
Dumfries and Galloway	384	411	230	227
Fife	840	873	504	481
Forth Valley	655	676	390	377
Grampian	1,190	1,220	691	662
Greater Glasgow	2,370	2,500	1,550	1,490
Highland	511	525	299	277
Lanarkshire	1,170	1,240	751	720
Lothian	1,840	1,950	1,080	1,070
Orkney	40	41	25	24
Shetland	49	46	32	25
Tayside	1,000	1,070	617	612
Western Isles	83	80	55	46
Northern Ireland	**3,050**	**3,260**	**1,840**	**1,730**
Eastern	1,330	1,450	824	789
Northern	727	782	437	416
Southern	529	560	297	284
Western	464	465	278	236
Ireland	**6,210**	**5,840**	**3,990**	**3,450**
Eastern	2,080	2,150	1,280	1,220
Mid-Western	494	467	347	299
Midland	370	335	246	191
North Eastern	516	459	321	267
North Western	428	361	283	219
South Eastern	669	605	435	374
Southern	993	928	625	540
Western	660	534	446	340

Table B3.1

Bladder: annual average age-standardised incidence and mortality by health authority within country and region of England, UK and Ireland, 1991–2000

Country/ region/ health authority	Incidence[1]						Mortality[2]						M:I Ratio[3]	
		Males			Females			Males			Females		Males	Females
	Rank[4]	Rate[5]	Signif[6]	Rank	Rate	Signif	Rank	Rate	Signif	Rank	Rate	Signif		
UK and Ireland		29.2			8.3			10.5			3.3		0.36	0.40
England		29.1			8.1	~		10.7			3.3		0.37	0.40
Northern and Yorkshire		25.4	~		7.4	~		11.4	*		3.8	*	0.45	0.51
Bradford	3	22.8	~	7	7.4		9	12.0		11	4.2	*	0.52	0.56
Calderdale and Kirklees	10	27.4		8	7.4		8	11.8		9	4.1	*	0.43	0.55
County Durham	13	29.0		9	7.9		7	11.5		4	3.5		0.40	0.44
East Riding	8	26.4	~	4	6.8	~	11	12.3	*	6	3.7		0.47	0.55
Gateshead and South Tyneside	12	28.9		13	8.7		10	12.3	*	10	4.1	*	0.43	0.47
Leeds	7	25.0	~	10	7.9		5	10.8		8	3.9	*	0.43	0.49
Newcastle and North Tyneside	11	27.9		11	8.1		6	10.9		13	4.4	*	0.39	0.54
North Cumbria	6	24.0	~	5	7.0	~	3	10.3		2	3.3		0.43	0.47
North Yorkshire	2	22.4	~	3	6.7	~	1	9.9		1	3.3		0.44	0.49
Northumberland	1	20.9	~	6	7.2		2	10.2		5	3.6		0.49	0.50
Sunderland	9	27.2		12	8.5		4	10.5		3	3.5		0.38	0.41
Tees	4	23.5	~	1	6.1	~	13	13.0	*	7	3.8		0.55	0.63
Wakefield	5	23.6	~	2	6.7	~	12	13.0	*	12	4.3	*	0.55	0.64
Trent		32.2	*		9.5	*		11.4	*		3.5		0.36	0.37
Barnsley	9	35.9	*	8	10.3	*	8	11.5		5	3.4		0.32	0.33
Doncaster	8	35.4	*	10	10.7	*	5	11.3		10	4.1		0.32	0.38
Leicestershire	3	29.8		2	8.8		3	10.6		2	3.1		0.36	0.35
Lincolnshire	5	32.4	*	4	9.3		9	11.8	*	3	3.3		0.36	0.36
North Derbyshire	7	33.5	*	6	9.8	*	2	10.6		6	3.6		0.32	0.37
North Nottinghamshire	6	33.3	*	7	10.0	*	6	11.4		8	3.8		0.34	0.38
Nottingham	4	31.8	*	3	9.0		4	11.3		4	3.4		0.36	0.38
Rotherham	11	36.8	*	11	11.4	*	7	11.5		9	4.0		0.31	0.35
Sheffield	10	36.4	*	9	10.6	*	10	12.9	*	7	3.6		0.36	0.34
South Humber	1	25.9	~	1	7.2		11	13.6	*	11	4.4	*	0.52	0.61
Southern Derbyshire	2	29.2		5	9.5		1	10.4		1	3.1		0.35	0.33
West Midlands		32.7	*		9.0	*		9.9	~		3.3		0.30	0.37
Birmingham	11	33.8	*	12	9.8	*	5	10.1		7	3.4		0.30	0.34
Coventry	9	33.0	*	11	9.6		12	10.9		12	3.8		0.33	0.39
Dudley	5	32.5	*	13	10.3	*	10	10.7		13	3.8		0.33	0.37
Herefordshire	1	29.5		2	8.0		1	8.3	~	2	2.5		0.28	0.32
North Staffordshire	7	32.9	*	5	8.5		9	10.6		10	3.7		0.32	0.44
Sandwell	10	33.2	*	10	9.3		8	10.3		8	3.4		0.31	0.36
Shropshire	6	32.8	*	4	8.5		3	8.4	~	6	3.3		0.26	0.39
Solihull	12	33.8	*	6	8.5		4	9.3		3	2.8		0.28	0.33
South Staffordshire	3	31.3		7	8.6		13	11.2		11	3.8		0.36	0.44
Walsall	2	29.6		8	8.8		11	10.7		9	3.5		0.36	0.39
Warwickshire	8	33.0	*	9	9.1		6	10.1		5	3.1		0.31	0.34
Wolverhampton	13	36.3	*	1	7.7		7	10.2		4	3.1		0.28	0.40
Worcestershire	4	32.1	*	3	8.3		2	8.3	~	1	2.5	~	0.26	0.30
North West		30.3	*		9.0	*		11.2	*		3.7	*	0.37	0.41
Bury and Rochdale	5	27.5		10	8.9		12	12.4	*	16	4.4	*	0.45	0.50
East Lancashire	8	29.0		12	9.2		15	13.2	*	12	4.3	*	0.45	0.46
Liverpool	12	35.1	*	15	12.2	*	6	10.5		15	4.4	*	0.30	0.36
Manchester	10	29.6		14	9.7	*	14	13.0	*	11	4.2	*	0.44	0.43
Morecambe Bay	1	23.6	~	1	7.0	~	3	9.8		3	3.0		0.42	0.43
North Cheshire	13	35.6	*	13	9.3		11	12.3		6	3.4		0.35	0.37
North West Lancashire	9	29.5		5	8.4		4	10.2		7	3.5		0.35	0.41
Salford and Trafford	3	25.8	~	8	8.8		10	11.6		10	4.1	*	0.45	0.47
Sefton	14	36.7	*	3	8.1		13	12.4	*	8	3.8		0.34	0.47
South Cheshire	11	32.5	*	7	8.8		1	9.4	~	2	2.8		0.29	0.32
South Lancashire	6	28.3		11	9.0		2	9.8		5	3.3		0.35	0.37
St Helen's and Knowsley	16	38.2	*	16	12.6	*	16	13.7	*	13	4.3	*	0.36	0.34
Stockport	7	28.9		4	8.2		7	10.9		1	2.7		0.38	0.33
West Pennine	4	27.5		9	8.9		8	11.1		14	4.3	*	0.40	0.49
Wigan and Bolton	2	25.4	~	2	8.0		5	10.3		9	4.0	*	0.41	0.50
Wirral	15	36.7	*	6	8.4		9	11.1		4	3.1		0.30	0.37
Eastern		23.7	~		6.4	~		10.7			3.1		0.45	0.48
Bedfordshire	6	26.3	~	5	6.7	~	2	10.4		4	3.2		0.39	0.48
Cambridgeshire	1	18.4	~	1	4.9	~	1	10.0		1	2.4	~	0.54	0.48
Hertfordshire	5	26.3	~	6	7.3	~	3	10.4		5	3.2		0.39	0.44
Norfolk	2	19.4	~	3	5.3	~	6	10.7		3	3.0		0.55	0.56
North Essex	4	25.0	~	4	6.6	~	4	10.5		6	3.2		0.42	0.49
South Essex	7	30.9		7	8.6		7	12.6	*	7	3.6		0.41	0.42
Suffolk	3	20.5	~	2	5.1	~	5	10.6		2	2.9		0.52	0.57

1 England, Wales and Scotland 1991-99; Northern Ireland 1993-99; Ireland 1994-99.
2 England, Wales and Northern Ireland 1991-2000; Scotland 1991-99; Ireland 1994-2000.
3 Mortality to incidence ratio; age-standardised mortality rate divided by age-standardised incidence rate.
4 Rank within region or country; 1 is lowest.
5 Directly age standardised using the European standard population; per 100,000 population.
6 Rate significantly different from that for UK and Ireland; * significantly higher; ~ significantly lower. See Appendix H.

Country/ region/ health authority	Incidence[1]						Mortality[2]						M:I Ratio[3]	
	Males			Females			Males			Females			Males	Females
	Rank[4]	Rate[5]	Signif[6]	Rank	Rate	Signif	Rank	Rate	Signif	Rank	Rate	Signif		
London		**26.0**	~		**7.3**	~		**10.7**			**3.0**	~	**0.41**	**0.42**
Barking and Havering	14	31.0		12	8.2		14	12.7	*	14	3.7		0.41	0.45
Barnet, Enfield and Haringey	6	25.0	~	8	7.2	~	7	10.3		5	2.8		0.41	0.39
Bexley, Bromley and Greenwich	4	24.2	~	6	7.0	~	13	12.1	*	4	2.8	~	0.50	0.40
Brent and Harrow	8	25.6	~	9	7.6		1	8.3	~	9	3.0		0.32	0.40
Camden and Islington	13	30.2		10	7.8		11	11.7		12	3.6		0.39	0.46
Croydon	5	24.8	~	5	7.0	~	6	10.0		2	2.6	~	0.40	0.37
Ealing, Hammersmith and Hounslow	7	25.6	~	3	6.5	~	10	11.1		3	2.7	~	0.44	0.42
East London and The City	1	22.3		2	6.5	~	4	9.9		13	3.6		0.44	0.56
Hillingdon	10	27.1		14	8.6		9	11.1		10	3.2		0.41	0.37
Kensington and Chelsea, and Westminster	2	22.4	~	1	5.7	~	2	8.5	~	8	3.0		0.38	0.53
Kingston and Richmond	9	26.9		11	8.2		8	10.5		6	2.9		0.39	0.35
Lambeth, Southwark and Lewisham	3	22.9	~	7	7.1	~	12	11.9		11	3.5		0.52	0.49
Merton, Sutton and Wandsworth	12	29.5		13	8.6		5	10.0		7	3.0		0.34	0.35
Redbridge and Waltham Forest	11	29.4		4	6.9	~	3	9.8		1	2.5	~	0.33	0.36
South East		**29.6**			**7.8**	~		**10.5**			**3.0**	~	**0.36**	**0.39**
Berkshire	10	34.1	*	8	8.4		8	10.7		5	2.8	~	0.31	0.33
Buckinghamshire	11	35.0	*	10	9.3		6	10.3		3	2.7	~	0.29	0.29
East Kent	2	24.6	~	1	6.1	~	13	12.5	*	13	3.6		0.51	0.59
East Surrey	6	26.4	~	5	6.6	~	1	9.4		2	2.6	~	0.35	0.40
East Sussex, Brighton and Hove	1	23.9	~	2	6.3	~	11	11.0		7	3.0		0.46	0.48
Isle of Wight, Portsmouth and South East Hampshire	13	37.5	*	13	10.8	*	10	11.0		11	3.3		0.29	0.30
North and Mid-Hampshire	7	29.6		9	8.9		3	9.5		10	3.1		0.32	0.35
Northamptonshire	8	31.9	*	7	8.1		9	10.9		4	2.7	~	0.34	0.34
Oxfordshire	9	32.8	*	11	9.5	*	7	10.3		8	3.1		0.31	0.32
Southampton and South West Hampshire	12	36.3	*	12	10.3	*	5	10.0		9	3.1		0.27	0.30
West Kent	3	25.3	~	4	6.5	~	12	11.4		12	3.6		0.45	0.55
West Surrey	4	26.2	~	6	6.7	~	4	9.9		1	2.6	~	0.38	0.38
West Sussex	5	26.3	~	3	6.4	~	2	9.4	~	6	2.8	~	0.36	0.44
South West		**33.4**	*		**8.6**			**9.9**	~		**2.8**	~	**0.30**	**0.32**
Avon	5	33.7	*	5	8.7		8	11.0		8	3.2		0.33	0.37
Cornwall and Isles of Scilly	6	34.0	*	6	8.8		1	9.2	~	5	2.7	~	0.27	0.31
Dorset	8	37.4	*	8	10.3	*	3	9.5	~	6	2.8	~	0.26	0.27
Gloucestershire	3	31.9	*	3	8.2		2	9.4		4	2.6	~	0.29	0.32
North and East Devon	1	29.2		1	7.2	~	4	9.7		2	2.5	~	0.33	0.35
Somerset	2	31.7	*	4	8.5		7	10.5		3	2.5	~	0.33	0.29
South and West Devon	4	32.1	*	2	7.4		6	10.0		1	2.5	~	0.31	0.34
Wiltshire	7	35.4	*	7	9.4		5	9.8		7	3.0		0.28	0.32
Wales		**34.0**	*		**9.8**	*		**9.3**	~		**3.1**		**0.27**	**0.32**
Bro Taf	3	32.9	*	2	9.5	*	4	9.4		2	3.1		0.29	0.32
Dyfed Powys	1	31.6	*	1	9.1		2	8.9	~	1	2.8		0.28	0.31
Gwent	2	32.7	*	4	9.9	*	5	10.3		5	3.6		0.31	0.37
Morgannwg	4	34.2	*	3	9.7	*	1	8.6	~	3	3.1		0.25	0.32
North Wales	5	37.5	*	5	10.4	*	3	9.2	~	4	3.1		0.24	0.30
Scotland		**32.1**	*		**10.3**	*		**11.5**	*		**4.1**	*	**0.36**	**0.40**
Argyll and Clyde	9	31.8		14	11.1	*	14	12.8	*	12	4.5	*	0.40	0.41
Ayrshire and Arran	12	32.8	*	13	11.0	*	8	11.6		6	3.9		0.35	0.35
Borders	4	28.5		10	10.1		7	11.4		14	4.9	*	0.40	0.49
Dumfries and Galloway	15	36.4	*	6	9.3		2	9.2		4	3.1		0.25	0.33
Fife	11	32.7	*	9	9.6		6	11.4		15	4.9	*	0.35	0.51
Forth Valley	5	29.8		8	9.4		10	12.0		10	4.3	*	0.40	0.46
Grampian	7	31.3		7	9.3		9	11.6		11	4.5	*	0.37	0.48
Greater Glasgow	13	33.3	*	15	11.6	*	12	12.4	*	13	4.7	*	0.37	0.41
Highland	6	30.0		5	9.2		3	9.6		3	2.9		0.32	0.31
Lanarkshire	8	31.8	*	4	9.2		13	12.4	*	7	3.9		0.39	0.43
Lothian	10	32.3	*	11	10.5	*	5	10.5		5	3.4		0.33	0.33
Orkney	1	17.5	~	3	5.9		1	4.2	~	9	4.0		0.24	0.69
Shetland	2	22.7		2	4.7		15	14.1		1	0.4	~	0.62	0.09
Tayside	14	33.8	*	12	10.8	*	4	10.5		8	3.9		0.31	0.36
Western Isles	3	27.5		1	4.5	~	11	12.1		2	2.0		0.44	0.46
Northern Ireland		**20.3**	~		**6.1**	~		**7.5**	~		**2.7**	~	**0.37**	**0.45**
Eastern	4	22.8	~	4	6.3	~	4	8.2	~	3	2.8		0.36	0.44
Northern	1	17.7	~	3	6.2	~	3	7.1	~	2	2.6	~	0.40	0.42
Southern	3	19.2	~	1	5.6	~	1	6.4	~	4	3.1		0.33	0.55
Western	2	18.8	~	2	6.0	~	2	7.1	~	1	2.3		0.38	0.38
Ireland		**21.0**	~		**6.6**	~		**6.9**	~		**2.3**	~	**0.33**	**0.36**
Eastern	8	25.4	~	8	7.7	~	8	7.7	~	4	2.3	~	0.30	0.30
Mid-Western	3	18.3	~	6	7.1	~	4	6.4	~	6	2.4	~	0.35	0.34
Midland	1	16.4	~	7	7.6	~	1	5.7	~	8	3.3		0.35	0.43
North Eastern	5	19.6	~	2	5.5	~	2	5.9	~	3	2.3	~	0.30	0.41
North Western	6	20.9	~	5	6.5	~	7	7.3	~	2	2.1	~	0.35	0.32
South Eastern	4	19.4	~	4	6.3	~	5	7.1	~	5	2.3	~	0.37	0.36
Southern	7	21.1	~	3	6.2	~	6	7.2	~	7	2.6		0.34	0.41
Western	2	17.2	~	1	3.9	~	3	6.1	~	1	1.9	~	0.35	0.49

259

Table **B3.2**

Bladder: annual average numbers of registrations and deaths by health authority within country and region of England, UK and Ireland, 1991–2000

Country/ region/ health authority	Number of registrations[1]		Number of deaths[2]	
	Males	Females	Males	Females
UK and Ireland	**9,700**	**3,910**	**3,590**	**1,800**
England	**7,790**	**3,090**	**2,960**	**1,450**
Northern and Yorkshire	**871**	**374**	**392**	**215**
Bradford	53	27	28	16
Calderdale and Kirklees	82	35	36	22
County Durham	97	37	38	19
East Riding	85	32	40	21
Gateshead and South Tyneside	58	27	25	14
Leeds	95	46	42	25
Newcastle and North Tyneside	71	30	28	18
North Cumbria	45	19	19	10
North Yorkshire	102	45	45	25
Northumberland	38	19	19	10
Sunderland	40	17	16	8
Tees	67	25	36	17
Wakefield	38	16	21	11
Trent	**925**	**368**	**336**	**158**
Barnsley	45	16	15	7
Doncaster	57	22	19	9
Leicestershire	144	59	52	25
Lincolnshire	131	48	49	20
North Derbyshire	73	29	23	12
North Nottinghamshire	74	30	26	13
Nottingham	108	41	40	18
Rotherham	49	20	16	8
Sheffield	109	46	40	20
South Humber	45	18	24	11
Southern Derbyshire	90	40	33	15
West Midlands	**940**	**353**	**290**	**149**
Birmingham	171	70	52	28
Coventry	55	22	19	10
Dudley	55	24	18	11
Herefordshire	31	11	10	4
North Staffordshire	85	30	27	15
Sandwell	53	22	17	9
Shropshire	78	27	20	12
Solihull	37	14	10	5
South Staffordshire	94	34	33	17
Walsall	41	16	15	7
Warwickshire	94	34	29	13
Wolverhampton	49	15	14	7
Worcestershire	96	34	26	12
North West	**1,060**	**457**	**399**	**217**
Bury and Rochdale	54	26	24	15
East Lancashire	78	36	36	19
Liverpool	82	41	25	17
Manchester	58	30	25	14
Morecambe Bay	45	20	20	11
North Cheshire	53	19	18	8
North West Lancashire	87	36	32	16
Salford and Trafford	62	31	28	16
Sefton	63	21	22	12
South Cheshire	123	46	36	17
South Lancashire	45	20	16	9
St Helen's and Knowsley	62	28	23	11
Stockport	46	18	17	7
West Pennine	65	30	27	17
Wigan and Bolton	72	33	30	18
Wirral	69	24	22	10
Eastern	**715**	**268**	**335**	**148**
Bedfordshire	68	23	27	12
Cambridgeshire	67	24	38	13
Hertfordshire	140	55	56	28
Norfolk	106	39	61	24
North Essex	129	48	57	26
South Essex	119	48	50	24
Suffolk	85	31	46	20

1 *England, Wales and Scotland 1991-99; Northern Ireland 1993-99; Ireland 1994-99.*
2 *England, Wales and Northern Ireland 1991-2000; Scotland 1991-99; Ireland 1994-2000.*

Country/ region/ health authority	Number of registrations[1]		Number of deaths[2]	
	Males	Females	Males	Females
London	**852**	**361**	**357**	**171**
Barking and Havering	66	26	28	13
Barnet, Enfield and Haringey	94	44	40	19
Bexley, Bromley and Greenwich	93	40	47	19
Brent and Harrow	56	23	18	11
Camden and Islington	47	18	19	10
Croydon	38	16	16	7
Ealing, Hammersmith and Hounslow	73	30	33	14
East London and The City	53	22	24	13
Hillingdon	33	15	14	6
Kensington and Chelsea, and Westminster	35	14	14	8
Kingston and Richmond	43	20	18	9
Lambeth, Southwark and Lewisham	71	33	37	17
Merton, Sutton and Wandsworth	84	38	29	16
Redbridge and Waltham Forest	65	23	23	10
South East	**1,410**	**540**	**525**	**248**
Berkshire	122	40	39	16
Buckinghamshire	110	39	33	13
East Kent	95	36	51	24
East Surrey	65	25	25	12
East Sussex, Brighton and Hove	125	53	63	31
Isle of Wight, Portsmouth and South East Hampshire	148	61	45	23
North and Mid-Hampshire	78	33	25	13
Northamptonshire	100	35	35	14
Oxfordshire	99	38	32	15
Southampton and South West Hampshire	113	45	33	16
West Kent	126	48	57	30
West Surrey	92	36	36	16
West Sussex	132	51	51	26
South West	**1,020**	**371**	**323**	**144**
Avon	185	68	63	29
Cornwall and Isles of Scilly	111	39	33	15
Dorset	186	71	52	25
Gloucestershire	106	37	32	14
North and East Devon	95	35	35	15
Somerset	99	37	35	13
South and West Devon	122	42	41	18
Wiltshire	114	42	33	16
Wales	**582**	**233**	**164**	**90**
Bro Taf	126	52	37	19
Dyfed Powys	100	37	30	14
Gwent	100	42	32	18
Morgannwg	103	41	27	16
North Wales	153	62	39	23
Scotland	**849**	**392**	**305**	**179**
Argyll and Clyde	70	37	28	17
Ayrshire and Arran	67	31	24	12
Borders	19	9	8	5
Dumfries and Galloway	33	12	8	4
Fife	61	26	22	14
Forth Valley	42	19	17	9
Grampian	81	34	31	19
Greater Glasgow	150	81	56	38
Highland	33	13	11	5
Lanarkshire	82	33	31	16
Lothian	122	58	40	22
Orkney	2	1	1	1
Shetland	2	1	1	0
Tayside	77	35	24	14
Western Isles	5	2	2	1
Northern Ireland	**153**	**67**	**56**	**33**
Eastern	74	31	27	16
Northern	34	16	14	7
Southern	24	10	8	6
Western	21	9	8	3
Ireland	**327**	**127**	**110**	**53**
Eastern	115	50	33	17
Mid-Western	26	12	9	5
Midland	16	8	6	4
North Eastern	26	9	8	4
North Western	25	8	9	4
South Eastern	35	14	13	6
Southern	52	20	18	9
Western	33	8	13	5

Table **B4.1**

Brain: annual average age-standardised incidence and mortality by health authority within country and region of England, UK and Ireland, 1991–2000

Country/ region/ health authority	Incidence[1]						Mortality[2]						M:I Ratio[3]	
	Males			Females			Males			Females			Males	Females
	Rank[4]	Rate[5]	Signif[6]	Rank	Rate	Signif	Rank	Rate	Signif	Rank	Rate	Signif		
UK and Ireland		7.9			5.3			6.1			3.9		0.77	0.74
England		7.8			5.2			6.0			3.8		0.77	0.74
Northern and Yorkshire		7.6			5.1			5.9			3.9		0.78	0.76
Bradford	7	7.5		3	4.4		7	5.8		1	3.3		0.77	0.76
Calderdale and Kirklees	2	6.6	~	2	4.3	~	2	5.3		2	3.4		0.80	0.79
County Durham	8	7.6		7	5.2		3	5.4		8	3.9		0.71	0.76
East Riding	5	7.3		5	5.1		11	6.5		12	4.3		0.89	0.85
Gateshead and South Tyneside	3	7.2		4	5.0		12	6.6		3	3.4		0.92	0.69
Leeds	9	7.7		6	5.2		9	6.1		11	4.2		0.80	0.82
Newcastle and North Tyneside	11	8.1		13	5.8		10	6.1		13	4.5		0.76	0.78
North Cumbria	4	7.2		10	5.4		4	5.6		9	4.0		0.78	0.74
North Yorkshire	10	7.7		12	5.5		8	6.1		7	3.8		0.79	0.70
Northumberland	13	9.2		8	5.2		13	6.9		10	4.0		0.75	0.78
Sunderland	1	6.1	~	11	5.5		1	5.1		5	3.6		0.84	0.66
Tees	12	8.3		9	5.3		6	5.7		4	3.5		0.69	0.66
Wakefield	6	7.4		1	4.1	~	5	5.7		6	3.8		0.78	0.94
Trent		8.2			5.4			5.8			3.8		0.71	0.70
Barnsley	2	7.6		2	4.9		1	4.8		1	2.7	~	0.63	0.55
Doncaster	7	8.4		10	5.9		2	5.4		3	3.6		0.64	0.60
Leicestershire	11	8.8		6	5.3		9	6.2		6	3.7		0.70	0.71
Lincolnshire	4	8.2		8	5.8		4	5.5		2	3.4		0.67	0.59
North Derbyshire	9	8.8		9	5.9		6	5.9		4	3.6		0.67	0.61
North Nottinghamshire	3	7.7		7	5.3		5	5.6		10	4.1		0.72	0.77
Nottingham	6	8.4		5	5.2		10	6.3		7	3.9		0.75	0.75
Rotherham	8	8.8		11	6.3		8	6.0		8	4.0		0.69	0.63
Sheffield	1	6.7	~	4	5.1		3	5.4		5	3.7		0.80	0.73
South Humber	10	8.8		1	4.7		11	6.6		9	4.0		0.75	0.86
Southern Derbyshire	5	8.3		3	5.0		7	6.0		11	4.3		0.72	0.87
West Midlands		7.4	~		4.7	~		5.8			3.7		0.78	0.78
Birmingham	6	7.4		9	4.8		3	5.3	~	7	3.7		0.71	0.76
Coventry	1	6.6		4	4.5		1	4.5	~	2	3.3		0.69	0.74
Dudley	12	8.2		1	4.1	~	12	6.5		8	3.7		0.80	0.91
Herefordshire	8	7.5		13	5.3		10	6.2		4	3.5		0.83	0.66
North Staffordshire	9	7.5		3	4.5		2	5.2		13	4.0		0.69	0.89
Sandwell	13	8.4		2	4.4		4	5.6		1	3.2		0.66	0.72
Shropshire	11	8.0		7	4.8		13	6.9		11	3.8		0.86	0.79
Solihull	4	6.9		11	5.0		5	5.6		10	3.8		0.81	0.75
South Staffordshire	3	6.8		5	4.6		8	6.1		12	4.0		0.89	0.87
Walsall	5	7.1		12	5.1		7	5.8		6	3.7		0.82	0.72
Warwickshire	7	7.4		10	5.0		6	5.7		5	3.6		0.77	0.72
Wolverhampton	2	6.8		8	4.8		11	6.3		3	3.5		0.92	0.72
Worcestershire	10	7.7		6	4.8		9	6.1		9	3.7		0.79	0.79
North West		7.8			5.1			5.6	~		3.6	~	0.73	0.70
Bury and Rochdale	4	7.0		15	5.9		3	4.9	~	14	4.0		0.70	0.68
East Lancashire	5	7.0		2	4.5		10	5.8		6	3.4		0.82	0.77
Liverpool	9	8.1		6	5.0		2	4.9	~	1	3.1		0.61	0.63
Manchester	10	8.2		3	4.6		11	5.9		7	3.4		0.72	0.75
Morecambe Bay	12	8.7		8	5.1		16	6.9		16	4.2		0.80	0.82
North Cheshire	16	9.4		14	5.9		12	6.1		3	3.3		0.65	0.55
North West Lancashire	7	7.4		7	5.1		9	5.7		10	3.7		0.77	0.73
Salford and Trafford	2	6.7		12	5.5		5	5.4		9	3.6		0.81	0.65
Sefton	13	8.7		13	5.8		4	5.3		12	3.8		0.61	0.66
South Cheshire	8	7.7		10	5.4		8	5.7		11	3.8		0.75	0.71
South Lancashire	11	8.5		4	4.6		15	6.7		13	3.9		0.78	0.85
St Helen's and Knowsley	15	8.9		11	5.5		7	5.4		2	3.2		0.61	0.59
Stockport	6	7.4		9	5.2		14	6.2		8	3.5		0.85	0.67
West Pennine	3	7.0		5	4.9		6	5.4		4	3.3		0.77	0.69
Wigan and Bolton	1	6.6	~	1	4.4		1	4.6	~	5	3.4		0.69	0.76
Wirral	14	8.7		16	6.0		13	6.2		15	4.0		0.71	0.67
Eastern		7.4	~		5.2			6.0			4.0		0.81	0.77
Bedfordshire	2	6.7	~	1	4.4		2	5.8		1	3.6		0.87	0.82
Cambridgeshire	3	7.4		4	5.3		4	6.1		5	4.0		0.82	0.76
Hertfordshire	1	6.1	~	2	4.9		1	5.1	~	2	3.9		0.83	0.78
Norfolk	7	8.4		6	5.7		7	6.5		4	4.0		0.78	0.70
North Essex	4	7.5		3	4.9		6	6.3		3	3.9		0.85	0.79
South Essex	5	7.6		7	5.9		3	6.0		7	4.6		0.79	0.78
Suffolk	6	8.3		5	5.5		5	6.3		6	4.2		0.76	0.77

1 England, Wales and Scotland 1991-99; Northern Ireland 1993-99; Ireland 1994-99.
2 England, Wales and Northern Ireland 1991-2000; Scotland 1991-99; Ireland 1994-2000.
3 Mortality to incidence ratio; age-standardised mortality rate divided by age-standardised incidence rate.
4 Rank within region or country; 1 is lowest.
5 Directly age standardised using the European standard population; per 100,000 population.
6 Rate significantly different from that for UK and Ireland; * significantly higher; ~ significantly lower. See Appendix H.

Country/ region/ health authority	Incidence[1] Males			Incidence[1] Females			Mortality[2] Males			Mortality[2] Females			M:I Ratio[3]	
	Rank[4]	Rate[5]	Signif[6]	Rank	Rate	Signif	Rank	Rate	Signif	Rank	Rate	Signif	Males	Females
London		7.2	~		4.8	~		5.8			3.5	~	0.79	0.74
Barking and Havering	12	8.4		14	6.3		14	6.7		14	4.7		0.80	0.76
Barnet, Enfield and Haringey	8	7.2		8	4.8		10	6.0		10	3.7		0.83	0.75
Bexley, Bromley and Greenwich	7	7.2		1	4.0	~	12	6.2		6	3.4		0.87	0.86
Brent and Harrow	1	6.3	~	3	4.3		4	5.2		1	2.9	~	0.82	0.67
Camden and Islington	4	6.7		9	5.0		3	5.1		7	3.4		0.77	0.68
Croydon	11	8.0		13	5.5		8	5.7		9	3.7		0.71	0.66
Ealing, Hammersmith and Hounslow	5	6.8	~	6	4.6		6	5.6		5	3.3		0.82	0.72
East London and The City	3	6.4	~	5	4.6		2	5.0	~	3	3.2		0.79	0.69
Hillingdon	9	7.5		10	5.2		9	5.8		12	3.9		0.77	0.76
Kensington and Chelsea, and Westminster	10	7.9		7	4.8		1	4.9		2	2.9	~	0.62	0.60
Kingston and Richmond	13	8.5		11	5.3		13	6.5		11	3.8		0.77	0.71
Lambeth, Southwark and Lewisham	6	6.9		2	4.1	~	7	5.7		4	3.2		0.83	0.80
Merton, Sutton and Wandsworth	14	8.7		12	5.4		11	6.2		13	4.0		0.71	0.75
Redbridge and Waltham Forest	2	6.3	~	4	4.5		5	5.6		8	3.5		0.88	0.78
South East		8.3	*		5.6	*		6.5	*		4.2	*	0.78	0.74
Berkshire	8	8.4		3	5.3		6	6.2		3	3.7		0.74	0.70
Buckinghamshire	10	8.9		13	6.7	*	9	6.6		6	4.3		0.74	0.63
East Kent	1	6.8	~	8	5.8		1	5.7		11	4.6		0.85	0.80
East Surrey	13	10.4	*	12	6.5	*	13	8.4	*	13	4.9	*	0.81	0.75
East Sussex, Brighton and Hove	11	9.0		1	5.0		11	6.8		2	3.6		0.75	0.72
Isle of Wight, Portsmouth and South East Hampshire	9	8.5		9	5.8		2	6.0		4	4.0		0.70	0.68
North and Mid-Hampshire	7	8.4		7	5.8		4	6.0		7	4.3		0.72	0.74
Northamptonshire	5	8.1		4	5.3		8	6.5		1	3.4		0.81	0.63
Oxfordshire	4	8.0		6	5.6		5	6.2		9	4.4		0.77	0.79
Southampton and South West Hampshire	3	7.9		10	5.9		3	6.0		8	4.3		0.76	0.73
West Kent	2	7.4		2	5.2		7	6.2		5	4.2		0.84	0.81
West Surrey	12	9.2	*	11	6.0		12	7.4	*	12	4.7		0.81	0.78
West Sussex	6	8.2		5	5.4		10	6.6		10	4.6		0.81	0.86
South West		8.5	*		5.4			6.4			3.9		0.75	0.72
Avon	6	8.7		6	5.4		5	6.6		3	3.8		0.76	0.69
Cornwall and Isles of Scilly	1	8.0		2	5.3		1	5.7		5	4.0		0.71	0.76
Dorset	5	8.6		8	6.0		4	6.5		6	4.1		0.75	0.67
Gloucestershire	4	8.3		4	5.3		7	6.8		8	4.3		0.82	0.82
North and East Devon	3	8.2		3	5.3		2	5.9		1	3.4		0.72	0.64
Somerset	7	8.7		5	5.4		3	5.9		7	4.3		0.68	0.80
South and West Devon	2	8.2		1	5.2		8	6.9		4	3.9		0.84	0.76
Wiltshire	8	9.1		7	5.4		6	6.7		2	3.6		0.74	0.67
Wales		9.0	*		6.2	*		5.7			3.7		0.64	0.59
Bro Taf	3	8.7		1	5.9		1	5.5		1	3.3		0.63	0.56
Dyfed Powys	2	8.4		4	6.4	*	2	5.5		4	3.9		0.66	0.60
Gwent	1	8.3		2	6.1		3	5.7		3	3.8		0.68	0.62
Morgannwg	4	8.8		5	6.5	*	4	5.7		5	4.3		0.65	0.66
North Wales	5	10.4	*	3	6.3	*	5	6.2		2	3.4		0.60	0.54
Scotland		7.7			5.2			6.4			4.2		0.84	0.80
Argyll and Clyde	8	8.0		9	5.5		11	7.1		11	4.4		0.88	0.79
Ayrshire and Arran	13	8.6		1	4.6		10	6.9		5	3.7		0.81	0.82
Borders	14	9.1		12	5.8		14	9.0	*	7	4.0		0.99	0.68
Dumfries and Galloway	7	7.5		13	5.9		2	5.4		14	4.4		0.72	0.75
Fife	6	7.4		5	4.9		8	6.2		4	3.7		0.84	0.74
Forth Valley	2	6.4	~	3	4.6		5	5.7		8	4.3		0.90	0.94
Grampian	12	8.4		2	4.6		9	6.7		6	3.8		0.80	0.84
Greater Glasgow	5	7.3		11	5.6		7	6.1		15	4.5		0.83	0.82
Highland	11	8.4		10	5.6		4	5.6		10	4.4		0.67	0.79
Lanarkshire	3	6.7		4	4.7		3	5.5		2	3.6		0.81	0.76
Lothian	9	8.2		8	5.4		12	7.4	*	12	4.4		0.89	0.82
Orkney	1	4.2		7	5.2		1	4.1		3	3.6		0.97	0.70
Shetland	10	8.3		14	6.9		15	9.4		1	3.1		1.13	0.46
Tayside	4	7.2		6	5.2		6	5.9		9	4.3		0.81	0.84
Western Isles	15	9.6		15	7.7		13	8.6		13	4.4		0.90	0.58
Northern Ireland		7.9			5.3			5.9			3.8		0.74	0.71
Eastern	3	8.3		3	5.2		3	6.3		4	4.0		0.76	0.76
Northern	2	7.3		1	4.9		2	5.5		2	3.8		0.74	0.77
Southern	1	6.6		4	6.1		1	4.9		3	3.9		0.75	0.65
Western	4	9.3		2	5.2		4	6.3		1	3.0		0.68	0.57
Ireland		8.8	*		5.7			7.5	*		5.0	*	0.85	0.88
Eastern	7	9.3	*	8	6.1		6	8.3	*	7	5.4	*	0.89	0.88
Mid-Western	3	7.6		2	5.2		4	6.8		1	3.7		0.89	0.71
Midland	1	6.7		4	5.5		3	6.7		2	4.4		1.00	0.80
North Eastern	6	9.3		5	5.7		5	7.5		5	5.1		0.81	0.89
North Western	4	8.4		1	4.3		2	6.2		3	4.6		0.74	1.07
South Eastern	2	7.5		7	5.9		1	5.1		8	5.4	*	0.67	0.91
Southern	8	10.3	*	6	5.8		8	8.9	*	4	4.7		0.86	0.82
Western	5	9.1		3	5.4		7	8.3	*	6	5.3	*	0.91	0.98

Table **B4.2**

Brain: annual average numbers of registrations and deaths by health authority within country and region of England, UK and Ireland, 1991–2000

Country/ region/ health authority	Number of registrations[1]		Number of deaths[2]	
	Males	Females	Males	Females
UK and Ireland	**2,440**	**1,850**	**1,890**	**1,410**
England	**1,910**	**1,440**	**1,470**	**1,090**
Northern and Yorkshire	**239**	**184**	**189**	**143**
Bradford	16	10	13	8
Calderdale and Kirklees	19	13	15	11
County Durham	23	18	17	14
East Riding	21	17	19	15
Gateshead and South Tyneside	14	11	13	8
Leeds	27	21	21	18
Newcastle and North Tyneside	19	17	14	13
North Cumbria	12	11	10	9
North Yorkshire	30	23	24	17
Northumberland	15	10	12	8
Sunderland	9	9	7	6
Tees	22	16	15	11
Wakefield	12	7	9	7
Trent	**216**	**159**	**154**	**114**
Barnsley	9	6	5	3
Doncaster	13	10	8	6
Leicestershire	39	26	28	19
Lincolnshire	28	21	19	13
North Derbyshire	17	14	12	9
North Nottinghamshire	16	12	11	9
Nottingham	27	20	21	15
Rotherham	11	9	8	6
Sheffield	18	17	15	12
South Humber	14	8	11	7
Southern Derbyshire	24	16	17	14
West Midlands	**198**	**141**	**156**	**112**
Birmingham	35	26	25	19
Coventry	10	8	7	6
Dudley	13	7	11	7
Herefordshire	7	5	6	4
North Staffordshire	18	12	13	11
Sandwell	12	7	8	6
Shropshire	17	12	15	10
Solihull	7	6	6	5
South Staffordshire	20	15	18	13
Walsall	9	8	8	6
Warwickshire	19	14	15	11
Wolverhampton	9	6	8	5
Worcestershire	21	15	17	12
North West	**253**	**195**	**185**	**140**
Bury and Rochdale	13	12	9	9
East Lancashire	18	13	15	10
Liverpool	18	13	11	8
Manchester	15	10	11	8
Morecambe Bay	14	9	11	8
North Cheshire	14	10	9	5
North West Lancashire	18	15	14	11
Salford and Trafford	15	14	12	10
Sefton	13	10	8	7
South Cheshire	27	22	20	16
South Lancashire	13	8	10	7
St Helen's and Knowsley	15	10	9	6
Stockport	11	9	9	6
West Pennine	16	13	12	10
Wigan and Bolton	18	14	13	11
Wirral	14	12	11	8
Eastern	**202**	**158**	**165**	**124**
Bedfordshire	18	13	15	10
Cambridgeshire	25	20	21	15
Hertfordshire	31	27	26	22
Norfolk	37	28	30	21
North Essex	34	25	30	20
South Essex	27	23	21	18
Suffolk	30	23	23	18

1 *England, Wales and Scotland 1991-99; Northern Ireland 1993-99; Ireland 1994-99.*
2 *England, Wales and Northern Ireland 1991-2000; Scotland 1991-99; Ireland 1994-2000.*

Country/ region/ health authority	Number of registrations[1]		Number of deaths[2]	
	Males	Females	Males	Females
London	**228**	**177**	**179**	**133**
Barking and Havering	16	15	13	11
Barnet, Enfield and Haringey	25	21	21	16
Bexley, Bromley and Greenwich	25	17	22	15
Brent and Harrow	13	10	11	7
Camden and Islington	10	9	8	7
Croydon	12	10	8	7
Ealing, Hammersmith and Hounslow	19	15	16	11
East London and The City	16	14	12	9
Hillingdon	8	7	7	5
Kensington and Chelsea, and Westminster	12	8	8	5
Kingston and Richmond	12	10	10	8
Lambeth, Southwark and Lewisham	21	14	17	11
Merton, Sutton and Wandsworth	24	17	16	13
Redbridge and Waltham Forest	13	11	11	9
South East	**354**	**270**	**279**	**206**
Berkshire	31	21	23	15
Buckinghamshire	28	24	21	15
East Kent	21	20	18	17
East Surrey	22	16	19	13
East Sussex, Brighton and Hove	36	24	27	18
Isle of Wight, Portsmouth and South East Hampshire	29	23	21	16
North and Mid-Hampshire	22	16	16	12
Northamptonshire	24	18	20	11
Oxfordshire	23	18	18	14
Southampton and South West Hampshire	22	17	17	13
West Kent	35	27	30	23
West Surrey	29	21	24	17
West Sussex	33	24	28	22
South West	**217**	**157**	**168**	**119**
Avon	43	31	33	22
Cornwall and Isles of Scilly	22	16	16	12
Dorset	33	25	25	18
Gloucestershire	24	18	20	15
North and East Devon	21	15	16	11
Somerset	22	16	16	14
South and West Devon	25	19	22	15
Wiltshire	27	18	21	13
Wales	**139**	**109**	**90**	**65**
Bro Taf	31	25	19	14
Dyfed Powys	23	19	16	12
Gwent	24	19	17	12
Morgannwg	24	20	16	13
North Wales	37	26	22	15
Scotland	**194**	**155**	**162**	**128**
Argyll and Clyde	17	15	15	12
Ayrshire and Arran	16	11	13	9
Borders	5	4	5	4
Dumfries and Galloway	6	5	5	4
Fife	13	10	11	7
Forth Valley	9	7	8	7
Grampian	22	14	17	12
Greater Glasgow	32	29	26	24
Highland	9	7	6	6
Lanarkshire	18	15	14	11
Lothian	30	24	26	20
Orkney	0	1	0	0
Shetland	1	1	1	0
Tayside	15	12	12	11
Western Isles	2	1	1	1
Northern Ireland	**60**	**46**	**43**	**34**
Eastern	25	19	19	16
Northern	14	11	10	9
Southern	9	9	6	6
Western	11	7	7	4
Ireland	**144**	**103**	**119**	**91**
Eastern	49	39	41	33
Mid-Western	11	8	10	6
Midland	7	6	6	5
North Eastern	13	9	11	8
North Western	9	5	6	5
South Eastern	14	11	9	11
Southern	26	16	22	14
Western	16	10	14	10

Table B5.1

Breast: annual average age-standardised incidence and mortality by health authority within country and region of England, UK and Ireland, 1991–2000

Country/ region/ health authority	Incidence[1] Males			Incidence[1] Females			Mortality[2] Males			Mortality[2] Females			M:I Ratio[3]	
	Rank[4]	Rate[5]	Signif[6]	Rank	Rate	Signif	Rank	Rate	Signif	Rank	Rate	Signif	Males	Females
UK and Ireland					108.2						35.3			0.33
England					108.4						35.3			0.33
Northern and Yorkshire					99.7	~					33.4	~		0.33
Bradford				8	99.1	~				3	31.0	~		0.31
Calderdale and Kirklees				7	98.6	~				1	30.2	~		0.31
County Durham				9	100.9	~				5	33.3			0.33
East Riding				10	101.2	~				11	34.7			0.34
Gateshead and South Tyneside				5	96.8	~				4	33.1			0.34
Leeds				13	107.1					2	30.5	~		0.28
Newcastle and North Tyneside				2	91.0	~				13	38.0	*		0.42
North Cumbria				3	95.3	~				10	34.7			0.36
North Yorkshire				12	106.8					8	33.9			0.32
Northumberland				1	86.2	~				7	33.6			0.39
Sunderland				4	96.5	~				12	36.1			0.37
Tees				6	97.5	~				9	34.6			0.36
Wakefield				11	106.6					6	33.4			0.31
Trent					102.8	~					35.9			0.35
Barnsley				10	108.3					3	35.3			0.33
Doncaster				9	108.3					9	37.0			0.34
Leicestershire				3	99.5	~				1	33.6			0.34
Lincolnshire				11	109.0					7	36.4			0.33
North Derbyshire				8	107.9					11	38.4	*		0.36
North Nottinghamshire				6	102.9	~				5	35.9			0.35
Nottingham				5	102.0	~				10	37.4			0.37
Rotherham				2	98.0	~				4	35.4			0.36
Sheffield				1	92.5	~				2	35.1			0.38
South Humber				4	101.6	~				8	36.4			0.36
Southern Derbyshire				7	105.3					6	36.3			0.34
West Midlands					106.4	~					36.5	*		0.34
Birmingham				6	104.6	~				3	35.4			0.34
Coventry				5	104.3					1	33.4			0.32
Dudley				8	105.5					7	36.6			0.35
Herefordshire				11	112.8					2	35.0			0.31
North Staffordshire				2	98.4	~				11	37.4			0.38
Sandwell				4	104.2					12	37.7			0.36
Shropshire				13	113.6	*				10	37.3			0.33
Solihull				12	113.3					13	39.0	*		0.34
South Staffordshire				7	105.5					6	36.6			0.35
Walsall				3	103.3					9	37.2			0.36
Warwickshire				9	108.4					5	36.1			0.33
Wolverhampton				1	97.8	~				4	36.0			0.37
Worcestershire				10	112.3					8	36.7			0.33
North West					104.4	~					34.6			0.33
Bury and Rochdale				12	106.5					10	35.1			0.33
East Lancashire				8	103.5					2	32.1	~		0.31
Liverpool				2	99.4	~				11	35.3			0.35
Manchester				14	108.1					8	34.8			0.32
Morecambe Bay				4	101.1	~				1	31.6	~		0.31
North Cheshire				3	100.4	~				6	34.2			0.34
North West Lancashire				13	107.3					7	34.5			0.32
Salford and Trafford				16	112.0					15	36.5			0.33
Sefton				6	103.3					9	35.0			0.34
South Cheshire				7	103.4	~				5	34.1			0.33
South Lancashire				10	104.5					14	35.8			0.34
St Helen's and Knowsley				1	98.3	~				3	33.5			0.34
Stockport				15	110.1					12	35.4			0.32
West Pennine				11	106.1					16	36.9			0.35
Wigan and Bolton				5	102.4	~				4	34.0			0.33
Wirral				9	104.3					13	35.4			0.34
Eastern					113.0	*					36.4	*		0.32
Bedfordshire				5	113.5	*				1	34.3			0.30
Cambridgeshire				6	116.4	*				6	37.0			0.32
Hertfordshire				7	120.9	*				5	36.7			0.30
Norfolk				4	111.5					2	35.5			0.32
North Essex				1	107.1					3	36.2			0.34
South Essex				3	111.2					7	37.7	*		0.34
Suffolk				2	109.5					4	36.3			0.33

1 England, Wales and Scotland 1991-99; Northern Ireland 1993-99; Ireland 1994-99.
2 England, Wales and Northern Ireland 1991-2000; Scotland 1991-99; Ireland 1994-2000.
3 Mortality to incidence ratio; age-standardised mortality rate divided by age-standardised incidence rate.
4 Rank within region or country; 1 is lowest.
5 Directly age standardised using the European standard population; per 100,000 population.
6 Rate significantly different from that for UK and Ireland; * significantly higher; ~ significantly lower. See Appendix H.

Country/ region/ health authority	Incidence[1]						Mortality[2]						M:I Ratio[3]	
	Males			Females			Males			Females			Males	Females
	Rank[4]	Rate[5]	Signif[6]	Rank	Rate	Signif	Rank	Rate	Signif	Rank	Rate	Signif		
London					**108.7**						**35.6**			**0.33**
Barking and Havering				12	115.9	*				12	36.5			0.32
Barnet, Enfield and Haringey				10	113.9	*				11	36.2			0.32
Bexley, Bromley and Greenwich				4	104.1	~				8	35.7			0.34
Brent and Harrow				6	110.0					3	34.7			0.32
Camden and Islington				7	110.7					9	35.9			0.32
Croydon				11	114.0					10	36.0			0.32
Ealing, Hammersmith and Hounslow				3	103.2	~				2	33.9			0.33
East London and The City				1	96.1	~				6	35.1			0.36
Hillingdon				5	109.5					7	35.5			0.32
Kensington and Chelsea, and Westminster				8	110.9					1	32.3	~		0.29
Kingston and Richmond				14	118.9	*				14	40.4	*		0.34
Lambeth, Southwark and Lewisham				2	97.5	~				4	34.8			0.36
Merton, Sutton and Wandsworth				9	112.7	*				13	37.0			0.33
Redbridge and Waltham Forest				13	117.2	*				5	34.9			0.30
South East					**115.7**	*					**35.6**			**0.31**
Berkshire				8	117.2	*				10	36.5			0.31
Buckinghamshire				12	124.2	*				11	36.9			0.30
East Kent				4	110.7					5	35.2			0.32
East Surrey				6	116.6	*				1	33.4			0.29
East Sussex, Brighton and Hove				3	110.7					7	35.4			0.32
Isle of Wight, Portsmouth and South East Hampshire				11	120.3	*				8	35.5			0.29
North and Mid-Hampshire				10	118.0	*				9	36.5			0.31
Northamptonshire				7	117.2	*				13	37.4			0.32
Oxfordshire				13	129.1	*				4	34.8			0.27
Southampton and South West Hampshire				9	117.5	*				6	35.3			0.30
West Kent				2	109.6					12	37.0			0.34
West Surrey				5	113.7	*				3	34.3			0.30
West Sussex				1	108.4					2	34.3			0.32
South West					**114.6**	*					**35.2**			**0.31**
Avon				2	107.8					5	35.1			0.33
Cornwall and Isles of Scilly				3	110.1					4	34.8			0.32
Dorset				8	130.0	*				2	34.1			0.26
Gloucestershire				5	112.1					8	37.0			0.33
North and East Devon				4	111.7					1	33.7			0.30
Somerset				6	114.0	*				3	34.6			0.30
South and West Devon				1	106.9					7	36.5			0.34
Wiltshire				7	123.4	*				6	36.3			0.29
Wales					**110.9**	*					**35.9**			**0.32**
Bro Taf				1	103.4	~				2	35.5			0.34
Dyfed Powys				4	111.6					3	35.7			0.32
Gwent				2	104.3					4	36.5			0.35
Morgannwg				3	109.2					1	34.1			0.31
North Wales				5	124.4	*				5	37.2			0.30
Scotland					**110.2**	*					**35.5**			**0.32**
Argyll and Clyde				11	111.1					10	35.9			0.32
Ayrshire and Arran				13	112.2					12	37.5			0.33
Borders				3	107.6					2	31.6			0.29
Dumfries and Galloway				1	102.0					4	33.2			0.33
Fife				5	108.1					6	33.8			0.31
Forth Valley				10	110.5					5	33.5			0.30
Grampian				4	107.6					3	32.9	~		0.31
Greater Glasgow				6	109.1					13	37.9	*		0.35
Highland				14	121.1	*				7	34.0			0.28
Lanarkshire				8	109.4					11	37.0			0.34
Lothian				12	111.8					8	34.5			0.31
Orkney				2	103.5					1	29.0			0.28
Shetland				7	109.3					15	46.3			0.42
Tayside				9	109.6					9	35.7			0.33
Western Isles				15	123.8					14	44.4			0.36
Northern Ireland					**100.5**	~					**33.3**	~		**0.33**
Eastern				1	94.9	~				1	32.6	~		0.34
Northern				2	99.5	~				3	33.9			0.34
Southern				4	109.3					4	34.7			0.32
Western				3	108.0					2	32.6			0.30
Ireland					**96.7**	~					**35.3**			**0.37**
Eastern				8	103.8	~				3	34.8			0.34
Mid-Western				3	90.6	~				5	35.6			0.39
Midland				7	101.4					7	37.8			0.37
North Eastern				1	85.6	~				6	36.3			0.42
North Western				4	91.9	~				1	31.7			0.34
South Eastern				2	89.1	~				2	34.3			0.38
Southern				6	98.1	~				4	35.5			0.36
Western				5	92.9	~				8	37.9			0.41

Table **B5.2**

Breast: annual average numbers of registrations and deaths by health authority within country and region of England, UK and Ireland, 1991–2000

Country/ region/ health authority	Number of registrations[1]		Number of deaths[2]	
	Males	Females	Males	Females
UK and Ireland		**38,900**		**14,600**
England		**31,000**		**11,700**
Northern and Yorkshire		**3,650**		**1,420**
Bradford		252		93
Calderdale and Kirklees		326		120
County Durham		365		137
East Riding		340		137
Gateshead and South Tyneside		207		82
Leeds		432		143
Newcastle and North Tyneside		245		121
North Cumbria		186		77
North Yorkshire		494		183
Northumberland		164		74
Sunderland		158		67
Tees		298		119
Wakefield		189		67
Trent		**3,090**		**1,250**
Barnsley		144		55
Doncaster		182		70
Leicestershire		510		198
Lincolnshire		434		168
North Derbyshire		245		99
North Nottinghamshire		243		96
Nottingham		367		157
Rotherham		143		60
Sheffield		297		131
South Humber		189		77
Southern Derbyshire		338		137
West Midlands		**3,270**		**1,280**
Birmingham		565		223
Coventry		178		67
Dudley		196		78
Herefordshire		123		45
North Staffordshire		279		119
Sandwell		177		76
Shropshire		283		105
Solihull		137		53
South Staffordshire		350		137
Walsall		157		65
Warwickshire		323		123
Wolverhampton		138		59
Worcestershire		358		134
North West		**4,030**		**1,530**
Bury and Rochdale		233		87
East Lancashire		301		108
Liverpool		264		108
Manchester		228		85
Morecambe Bay		203		74
North Cheshire		168		64
North West Lancashire		318		123
Salford and Trafford		291		107
Sefton		198		80
South Cheshire		421		159
South Lancashire		189		72
St Helen's and Knowsley		185		71
Stockport		191		73
West Pennine		287		112
Wigan and Bolton		331		124
Wirral		218		87
Eastern		**3,510**		**1,310**
Bedfordshire		323		110
Cambridgeshire		435		158
Hertfordshire		702		241
Norfolk		559		219
North Essex		581		230
South Essex		468		181
Suffolk		444		173

1 England, Wales and Scotland 1991-99; Northern Ireland 1993-99; Ireland 1994-99.
2 England, Wales and Northern Ireland 1991-2000; Scotland 1991-99; Ireland 1994-2000.

Country/ region/ health authority	Number of registrations[1]		Number of deaths[2]	
	Males	Females	Males	Females
London		**4,010**		**1,480**
Barking and Havering		273		99
Barnet, Enfield and Haringey		488		176
Bexley, Bromley and Greenwich		443		172
Brent and Harrow		267		93
Camden and Islington		195		71
Croydon		199		72
Ealing, Hammersmith and Hounslow		334		121
East London and The City		254		101
Hillingdon		144		52
Kensington and Chelsea, and Westminster		196		65
Kingston and Richmond		215		82
Lambeth, Southwark and Lewisham		343		134
Merton, Sutton and Wandsworth		371		143
Redbridge and Waltham Forest		290		97
South East		**5,850**		**2,100**
Berkshire		473		163
Buckinghamshire		438		145
East Kent		429		168
East Surrey		306		103
East Sussex, Brighton and Hove		561		227
Isle of Wight, Portsmouth and South East Hampshire		501		173
North and Mid-Hampshire		354		123
Northamptonshire		398		143
Oxfordshire		413		128
Southampton and South West Hampshire		384		132
West Kent		616		236
West Surrey		428		150
West Sussex		551		212
South West		**3,580**		**1,310**
Avon		627		239
Cornwall and Isles of Scilly		365		140
Dorset		638		203
Gloucestershire		382		147
North and East Devon		362		135
Somerset		360		134
South and West Devon		423		173
Wiltshire		417		141
Wales		**2,030**		**756**
Bro Taf		436		171
Dyfed Powys		354		133
Gwent		347		137
Morgannwg		341		123
North Wales		546		192
Scotland		**3,360**		**1,220**
Argyll and Clyde		290		106
Ayrshire and Arran		261		97
Borders		76		27
Dumfries and Galloway		101		36
Fife		225		78
Forth Valley		181		62
Grampian		319		110
Greater Glasgow		590		232
Highland		153		49
Lanarkshire		344		126
Lothian		498		171
Orkney		12		4
Shetland		13		6
Tayside		273		100
Western Isles		23		9
Northern Ireland		**853**		**309**
Eastern		350		134
Northern		212		77
Southern		157		55
Western		133		43
Ireland		**1,620**		**638**
Eastern		607		214
Mid-Western		133		57
Midland		94		37
North Eastern		119		53
North Western		97		37
South Eastern		164		67
Southern		257		103
Western		154		69

Table B6.1

Cervix: annual average age-standardised incidence and mortality by health authority within country and region of England, UK and Ireland, 1991–2000

Country/ region/ health authority	Incidence[1]						Mortality[2]						M:I Ratio[3]	
	Males			Females			Males			Females			Males	Females
	Rank[4]	Rate[5]	Signif[6]	Rank	Rate	Signif	Rank	Rate	Signif	Rank	Rate	Signif		
UK and Ireland					10.9						4.2			**0.39**
England					10.6	~					4.1			**0.39**
Northern and Yorkshire					12.8	*					4.7	*		**0.37**
Bradford				12	15.7	*				11	5.6	*		0.36
Calderdale and Kirklees				10	14.6	*				10	5.4	*		0.37
County Durham				3	11.0					3	3.9			0.35
East Riding				13	16.9	*				5	4.6			0.27
Gateshead and South Tyneside				11	15.2	*				7	4.9			0.32
Leeds				8	13.2	*				6	4.6			0.35
Newcastle and North Tyneside				7	12.5					8	5.0			0.40
North Cumbria				4	11.2					4	4.6			0.41
North Yorkshire				2	9.6					2	3.3	~		0.34
Northumberland				1	9.3					1	3.0	~		0.32
Sunderland				5	11.3					12	5.7	*		0.51
Tees				6	12.2					9	5.2	*		0.43
Wakefield				9	13.9					13	6.4	*		0.46
Trent					10.7						4.4			**0.41**
Barnsley				11	16.5	*				11	7.6	*		0.46
Doncaster				9	13.0					9	5.5	*		0.42
Leicestershire				3	9.3					2	3.7			0.40
Lincolnshire				4	9.8					8	4.5			0.46
North Derbyshire				1	8.7					3	3.8			0.44
North Nottinghamshire				7	10.6					4	4.2			0.39
Nottingham				8	11.5					6	4.4			0.38
Rotherham				6	10.5					7	4.4			0.42
Sheffield				5	9.9					5	4.3			0.44
South Humber				10	15.1	*				10	6.2	*		0.41
Southern Derbyshire				2	9.0					1	3.4	~		0.37
West Midlands					12.0	*					4.1			**0.34**
Birmingham				9	12.5					9	4.3			0.34
Coventry				12	14.6	*				12	4.8			0.33
Dudley				10	12.8					10	4.3			0.34
Herefordshire				2	10.0					7	4.2			0.42
North Staffordshire				8	12.0					11	4.6			0.38
Sandwell				13	15.7	*				13	5.3			0.34
Shropshire				7	11.7					4	3.8			0.32
Solihull				1	9.4					6	4.0			0.42
South Staffordshire				6	11.7					8	4.2			0.36
Walsall				5	11.4					5	3.9			0.34
Warwickshire				3	10.9					2	3.4	~		0.31
Wolverhampton				11	12.8					3	3.7			0.29
Worcestershire				4	11.0					1	3.2	~		0.29
North West					13.7	*					5.3	*		**0.39**
Bury and Rochdale				12	15.1	*				10	5.8	*		0.39
East Lancashire				14	16.3	*				13	6.5	*		0.40
Liverpool				15	17.8	*				16	7.5	*		0.42
Manchester				16	19.6	*				15	6.9	*		0.35
Morecambe Bay				5	12.2					5	4.2			0.35
North Cheshire				4	11.5					11	5.9	*		0.51
North West Lancashire				6	12.9	*				8	5.1	*		0.40
Salford and Trafford				9	14.1	*				9	5.8	*		0.41
Sefton				10	14.2	*				12	6.4	*		0.45
South Cheshire				1	8.2	~				3	3.9			0.47
South Lancashire				7	13.2	*				4	4.2			0.32
St Helen's and Knowsley				13	16.1	*				14	6.9	*		0.43
Stockport				3	10.9					1	3.4			0.31
West Pennine				8	13.7	*				2	3.9			0.28
Wigan and Bolton				11	14.9	*				7	4.7			0.32
Wirral				2	10.6					6	4.3			0.41
Eastern					8.2	~					3.3	~		**0.40**
Bedfordshire				6	9.5	~				5	3.6			0.38
Cambridgeshire				4	8.1	~				4	3.6			0.44
Hertfordshire				3	8.1	~				2	3.0	~		0.38
Norfolk				7	9.6	~				7	3.7			0.38
North Essex				1	6.9	~				1	2.7	~		0.40
South Essex				2	7.3	~				6	3.7			0.50
Suffolk				5	8.8	~				3	3.2	~		0.37

1 England, Wales and Scotland 1991-99; Northern Ireland 1993-99; Ireland 1994-99.
2 England, Wales and Northern Ireland 1991-2000; Scotland 1991-99; Ireland 1994-2000.
3 Mortality to incidence ratio; age-standardised mortality rate divided by age-standardised incidence rate.
4 Rank within region or country; 1 is lowest.
5 Directly age standardised using the European standard population; per 100,000 population.
6 Rate significantly different from that for UK and Ireland; * significantly higher; ~ significantly lower. See Appendix H.

Country/ region/ health authority	Incidence[1]						Mortality[2]						M:I Ratio[3]	
	Males			Females			Males			Females			Males	Females
	Rank[4]	Rate[5]	Signif[6]	Rank	Rate	Signif	Rank	Rate	Signif	Rank	Rate	Signif		
London					9.4	~					4.0			0.42
Barking and Havering				12	11.1					11	4.4			0.39
Barnet, Enfield and Haringey				4	8.3	~				3	3.3	~		0.39
Bexley, Bromley and Greenwich				3	8.0	~				4	3.7			0.46
Brent and Harrow				2	7.5	~				1	2.8	~		0.37
Camden and Islington				7	9.2	~				10	4.2			0.46
Croydon				6	8.9	~				12	4.6			0.52
Ealing, Hammersmith and Hounslow				9	9.4	~				8	3.9			0.42
East London and The City				11	10.9					14	5.5	*		0.51
Hillingdon				10	9.9					6	3.8			0.39
Kensington and Chelsea, and Westminster				13	11.4					9	4.1			0.35
Kingston and Richmond				1	7.4					2	3.2	~		0.43
Lambeth, Southwark and Lewisham				14	13.2	*				13	5.2	*		0.40
Merton, Sutton and Wandsworth				8	9.2	~				7	3.9			0.42
Redbridge and Waltham Forest				5	8.6	~				5	3.8			0.44
South East					8.7	~					3.4	~		0.38
Berkshire				3	7.8	~				5	3.0	~		0.38
Buckinghamshire				4	8.0	~				4	2.8	~		0.36
East Kent				9	8.9	~				13	4.5			0.51
East Surrey				2	7.6	~				3	2.8	~		0.37
East Sussex, Brighton and Hove				6	8.2	~				10	3.8			0.47
Isle of Wight, Portsmouth and South East Hampshire				13	11.5					12	4.1			0.36
North and Mid-Hampshire				5	8.0	~				2	2.5	~		0.32
Northamptonshire				11	9.5	~				11	4.0			0.42
Oxfordshire				7	8.3	~				1	2.4	~		0.28
Southampton and South West Hampshire				12	10.6					8	3.5			0.33
West Kent				10	9.0	~				9	3.8			0.42
West Surrey				1	7.6	~				6	3.0	~		0.39
West Sussex				8	8.8	~				7	3.2	~		0.36
South West					10.1	~					3.8	~		0.38
Avon				6	10.9					5	4.0			0.36
Cornwall and Isles of Scilly				2	9.0	~				8	4.9			0.55
Dorset				5	10.7					3	3.7			0.35
Gloucestershire				1	6.9	~				1	2.8	~		0.41
North and East Devon				3	9.0	~				7	4.1			0.46
Somerset				4	9.9					4	3.9			0.40
South and West Devon				7	10.9					6	4.0			0.37
Wiltshire				8	12.0					2	3.2	~		0.27
Wales					12.1	*					4.9	*		0.40
Bro Taf				4	13.0	*				5	5.4	*		0.42
Dyfed Powys				2	11.2					3	4.8			0.43
Gwent				3	11.3					1	4.4			0.39
Morgannwg				1	11.1					4	4.9			0.44
North Wales				5	13.4	*				2	4.7			0.35
Scotland					13.0	*					4.8	*		0.37
Argyll and Clyde				10	13.7	*				9	4.7			0.34
Ayrshire and Arran				9	13.3	*				14	5.7	*		0.42
Borders				12	14.0					1	2.8			0.20
Dumfries and Galloway				8	12.2					11	5.0			0.41
Fife				6	11.7					7	4.5			0.38
Forth Valley				15	15.9	*				15	5.8	*		0 36
Grampian				4	11.4					2	3.8			0.34
Greater Glasgow				13	14.6	*				13	5.4	*		0.37
Highland				14	15.7	*				12	5.3			0.34
Lanarkshire				7	11.9					10	5.0			0.42
Lothian				11	13.8	*				8	4.5			0.33
Orkney				2	7.8					6	4.3			0.56
Shetland				1	7.4					4	3.9			0.52
Tayside				3	9.9					3	3.9			0.39
Western Isles				5	11.5					5	4.1			0.35
Northern Ireland					9.9	~					3.8			0.38
Eastern				4	10.9					3	4.0			0.37
Northern				1	8.7	~				2	3.7			0.43
Southern				2	9.1					1	2.9	~		0.31
Western				3	10.2					4	4.7			0.46
Ireland					10.2						4.3			0.42
Eastern				7	10.7					2	3.8			0.36
Mid-Western				3	8.9					5	4.6			0.51
Midland				8	14.4	*				7	5.2			0.36
North Eastern				2	8.9					1	3.1			0.35
North Western				4	9.6					8	5.6			0.59
South Eastern				6	10.2					6	5.2			0.51
Southern				5	9.9					3	4.1			0.41
Western				1	8.8	~				4	4.5			0.52

Table **B6.2**

Cervix: annual average numbers of registrations and deaths by health authority within country and region of England, UK and Ireland, 1991–2000

Country/ region/ health authority	Number of registrations[1]		Number of deaths[2]	
	Males	Females	Males	Females
UK and Ireland		3,730		1,600
England		2,900		1,250
Northern and Yorkshire		450		184
Bradford		39		16
Calderdale and Kirklees		46		19
County Durham		39		16
East Riding		51		16
Gateshead and South Tyneside		31		12
Leeds		51		20
Newcastle and North Tyneside		34		15
North Cumbria		21		9
North Yorkshire		42		17
Northumberland		17		6
Sunderland		18		10
Tees		37		17
Wakefield		25		12
Trent		302		139
Barnsley		21		10
Doncaster		21		10
Leicestershire		46		20
Lincolnshire		35		18
North Derbyshire		18		9
North Nottinghamshire		23		10
Nottingham		40		16
Rotherham		15		7
Sheffield		28		14
South Humber		26		12
Southern Derbyshire		28		11
West Midlands		342		133
Birmingham		64		24
Coventry		23		9
Dudley		22		9
Herefordshire		9		5
North Staffordshire		30		14
Sandwell		24		9
Shropshire		27		10
Solihull		10		5
South Staffordshire		38		16
Walsall		16		6
Warwickshire		29		10
Wolverhampton		17		6
Worcestershire		32		11
North West		501		214
Bury and Rochdale		32		13
East Lancashire		46		20
Liverpool		44		20
Manchester		40		15
Morecambe Bay		21		9
North Cheshire		20		11
North West Lancashire		34		16
Salford and Trafford		35		16
Sefton		24		12
South Cheshire		32		17
South Lancashire		23		9
St Helen's and Knowsley		30		14
Stockport		18		6
West Pennine		35		11
Wigan and Bolton		46		16
Wirral		21		10
Eastern		246		112
Bedfordshire		28		11
Cambridgeshire		30		14
Hertfordshire		46		19
Norfolk		44		21
North Essex		35		16
South Essex		30		15
Suffolk		34		15

1 *England, Wales and Scotland 1991-99; Northern Ireland 1993-99; Ireland 1994-99.*
2 *England, Wales and Northern Ireland 1991-2000; Scotland 1991-99; Ireland 1994-2000.*

Country/ region/ health authority	Number of registrations[1]		Number of deaths[2]	
	Males	Females	Males	Females
London		351		156
Barking and Havering		24		10
Barnet, Enfield and Haringey		35		15
Bexley, Bromley and Greenwich		34		17
Brent and Harrow		18		7
Camden and Islington		18		9
Croydon		16		9
Ealing, Hammersmith and Hounslow		31		14
East London and The City		29		15
Hillingdon		12		5
Kensington and Chelsea, and Westminster		20		8
Kingston and Richmond		14		6
Lambeth, Southwark and Lewisham		47		19
Merton, Sutton and Wandsworth		32		14
Redbridge and Waltham Forest		21		10
South East		422		183
Berkshire		33		13
Buckinghamshire		29		11
East Kent		31		18
East Surrey		19		8
East Sussex, Brighton and Hove		37		20
Isle of Wight, Portsmouth and South East Hampshire		43		18
North and Mid-Hampshire		24		8
Northamptonshire		33		15
Oxfordshire		26		8
Southampton and South West Hampshire		32		12
West Kent		49		22
West Surrey		28		12
West Sussex		38		17
South West		284		127
Avon		59		24
Cornwall and Isles of Scilly		28		18
Dorset		44		19
Gloucestershire		23		11
North and East Devon		26		14
Somerset		28		13
South and West Devon		37		17
Wiltshire		39		12
Wales		203		90
Bro Taf		51		22
Dyfed Powys		33		16
Gwent		35		15
Morgannwg		33		17
North Wales		51		20
Scotland		377		153
Argyll and Clyde		34		13
Ayrshire and Arran		29		13
Borders		8		2
Dumfries and Galloway		10		5
Fife		25		11
Forth Valley		25		10
Grampian		33		13
Greater Glasgow		73		30
Highland		18		7
Lanarkshire		36		16
Lothian		59		21
Orkney		1		1
Shetland		1		1
Tayside		23		10
Western Isles		2		1
Northern Ireland		82		33
Eastern		38		15
Northern		18		8
Southern		13		4
Western		13		6
Ireland		173		74
Eastern		67		23
Mid-Western		13		7
Midland		13		5
North Eastern		13		5
North Western		9		6
South Eastern		18		9
Southern		25		11
Western		14		7

Table B7.1

Colorectal: annual average age-standardised incidence and mortality by health authority within country and region of England, UK and Ireland, 1991–2000

Country/ region/ health authority	Incidence[1]						Mortality[2]						M:I Ratio[3]	
	Males			Females			Males			Females			Males	Females
	Rank[4]	Rate[5]	Signif[6]	Rank	Rate	Signif	Rank	Rate	Signif	Rank	Rate	Signif		
UK and Ireland		52.9			34.9			28.1			18.0		0.53	0.52
England		50.9	~		33.8	~		27.3	~		17.6	~	0.54	0.52
Northern and Yorkshire		54.1	*		33.1	~		29.3	*		17.4	~	0.54	0.52
Bradford	6	53.2		4	33.0		3	26.4		3	16.5		0.50	0.50
Calderdale and Kirklees	1	48.4	~	1	28.5	~	2	26.1		2	16.3	~	0.54	0.57
County Durham	12	59.7	*	9	33.9		10	32.4	*	10	18.5		0.54	0.55
East Riding	3	51.0		5	33.3		5	28.0		5	16.7		0.55	0.50
Gateshead and South Tyneside	7	54.4		8	33.6		11	32.6	*	12	18.9		0.60	0.56
Leeds	8	54.6		11	34.3		1	25.4	~	1	15.5	~	0.47	0.45
Newcastle and North Tyneside	5	52.9		3	31.7		9	32.1	*	9	17.7		0.61	0.56
North Cumbria	9	55.5		13	35.3		8	29.9		8	17.6		0.54	0.50
North Yorkshire	4	52.6		10	34.0		4	26.5		4	16.5	~	0.50	0.49
Northumberland	2	49.4		2	30.1	~	7	29.5		6	17.0		0.60	0.56
Sunderland	11	56.1		6	33.6		12	32.6	*	11	18.6		0.58	0.56
Tees	13	60.4	*	12	34.8		13	35.7	*	13	20.2	*	0.59	0.58
Wakefield	10	56.0		7	33.6		6	28.2		7	17.5		0.50	0.52
Trent		50.5	~		32.6	~		28.2			17.6		0.56	0.54
Barnsley	11	59.3	*	11	38.2		11	32.7	*	11	20.4		0.55	0.53
Doncaster	10	53.6		4	31.9	~	5	28.8		8	19.5		0.54	0.61
Leicestershire	1	47.7	~	7	33.0	~	1	24.5	~	2	16.6	~	0.51	0.50
Lincolnshire	5	51.2		8	33.5		4	28.2		5	17.3		0.55	0.52
North Derbyshire	8	51.8		3	31.2	~	10	31.8	*	6	18.0		0.61	0.58
North Nottinghamshire	7	51.8		10	34.0		7	29.3		7	18.6		0.57	0.55
Nottingham	4	49.9	~	9	34.0		2	25.9	~	1	16.5	~	0.52	0.48
Rotherham	9	52.2		5	32.8		9	31.4	*	9	19.6		0.60	0.60
Sheffield	3	48.6	~	2	30.4	~	6	29.0		4	16.7		0.60	0.55
South Humber	6	51.3		6	32.8		8	30.9		10	19.8		0.60	0.60
Southern Derbyshire	2	48.1	~	1	29.0	~	3	27.9		3	16.6		0.58	0.57
West Midlands		56.7	*		35.6	*		30.0	*		18.4	*	0.53	0.52
Birmingham	6	55.8	*	5	34.4		6	29.7		3	17.6		0.53	0.51
Coventry	4	54.1		2	33.6		2	27.5		6	18.1		0.51	0.54
Dudley	10	58.8	*	8	35.7		9	31.0	*	10	19.0		0.53	0.53
Herefordshire	2	53.8		13	39.8	*	1	26.7		12	20.0		0.50	0.50
North Staffordshire	5	55.7		6	35.0		11	31.9	*	13	20.5	*	0.57	0.58
Sandwell	3	53.9		3	34.1		13	32.8	*	5	18.1		0.61	0.53
Shropshire	12	59.2	*	12	38.8	*	4	28.9		9	18.9		0.49	0.49
Solihull	1	50.8		4	34.2		3	28.1		7	18.6		0.55	0.54
South Staffordshire	8	57.3	*	10	36.4		8	30.2		8	18.6		0.53	0.51
Walsall	11	59.1	*	11	37.6		10	31.7	*	11	19.8		0.54	0.53
Warwickshire	7	56.4	*	9	35.8		7	30.1		1	17.4		0.53	0.49
Wolverhampton	9	57.5		1	33.6		12	32.0	*	4	17.8		0.56	0.53
Worcestershire	13	59.3	*	7	35.5		5	29.5		2	17.5		0.50	0.49
North West		54.9	*		34.6			30.6	*		18.8	*	0.56	0.54
Bury and Rochdale	9	53.9		8	34.4		8	30.2		12	19.4		0.56	0.56
East Lancashire	2	49.8		1	32.4	~	6	29.8		5	18.2		0.60	0.56
Liverpool	16	63.2	*	13	35.4		16	36.1	*	6	18.3		0.57	0.52
Manchester	15	62.9	*	16	39.3	*	15	34.5	*	16	19.9	*	0.55	0.51
Morecambe Bay	8	53.2		6	34.0		4	28.3		4	17.9		0.53	0.53
North Cheshire	4	51.6		5	33.9		7	29.8		3	17.9		0.58	0.53
North West Lancashire	10	54.6		11	34.8		5	28.5		8	18.8		0.52	0.54
Salford and Trafford	11	57.4	*	10	34.5		11	32.2	*	13	19.6		0.56	0.57
Sefton	14	60.6	*	15	37.9	*	14	33.1	*	11	19.3		0.55	0.51
South Cheshire	1	49.7	~	9	34.5		3	28.2		7	18.7		0.57	0.54
South Lancashire	3	50.2		14	36.7		2	28.2		14	19.6		0.56	0.54
St Helen's and Knowsley	13	59.0	*	12	35.0		10	30.3		10	19.1		0.51	0.55
Stockport	5	52.4		4	33.5		1	27.7		1	17.6		0.53	0.53
West Pennine	7	53.0		2	32.7		13	32.7	*	2	17.7		0.62	0.54
Wigan and Bolton	12	58.2	*	7	34.4		9	30.2		15	19.8	*	0.52	0.58
Wirral	6	52.6		3	32.8		12	32.5	*	9	18.9		0.62	0.58
Eastern		47.3	~		33.7	~		24.8	~		17.2	~	0.53	0.51
Bedfordshire	3	45.7	~	3	31.7	~	4	25.0	~	5	17.6		0.55	0.56
Cambridgeshire	7	52.0		6	36.8		6	25.2	~	3	17.1		0.48	0.47
Hertfordshire	4	47.6	~	4	33.1	~	7	25.4	~	1	16.1	~	0.53	0.49
Norfolk	5	50.0	~	5	34.6		1	24.0	~	2	16.4	~	0.48	0.47
North Essex	1	41.3	~	1	30.8	~	2	24.6	~	6	18.0		0.60	0.58
South Essex	2	44.8	~	2	31.3	~	3	24.9	~	4	17.3		0.56	0.55
Suffolk	6	50.4		7	38.3	*	5	25.1	~	7	18.3		0.50	0.48

1 England, Wales and Scotland 1991-99; Northern Ireland 1993-99; Ireland 1994-99.
2 England, Wales and Northern Ireland 1991-2000; Scotland 1991-99; Ireland 1994-2000.
3 Mortality to incidence ratio; age-standardised mortality rate divided by age-standardised incidence rate.
4 Rank within region or country; 1 is lowest.
5 Directly age standardised using the European standard population; per 100,000 population.
6 Rate significantly different from that for UK and Ireland; * significantly higher; ~ significantly lower. See Appendix H.

Country/ region/ health authority	Incidence[1]						Mortality[2]						M:I Ratio[3]	
	Males			Females			Males			Females			Males	Females
	Rank[4]	Rate[5]	Signif[6]	Rank	Rate	Signif	Rank	Rate	Signif	Rank	Rate	Signif		
London		**43.1**	~		**30.1**	~		**23.9**	~		**16.2**	~	**0.56**	**0.54**
Barking and Havering	10	44.7	~	11	32.3	~	8	24.6	~	14	18.6		0.55	0.58
Barnet, Enfield and Haringey	2	39.8	~	4	28.5	~	2	22.2	~	2	15.0	~	0.56	0.53
Bexley, Bromley and Greenwich	6	42.5	~	6	29.3	~	11	24.8	~	12	17.6		0.58	0.60
Brent and Harrow	3	41.0	~	10	30.5	~	1	20.6	~	3	15.1	~	0.50	0.49
Camden and Islington	12	44.9	~	8	29.8	~	14	26.0		9	16.5		0.58	0.55
Croydon	11	44.7	~	13	33.7		9	24.6	~	13	18.2		0.55	0.54
Ealing, Hammersmith and Hounslow	4	41.0	~	3	28.2	~	4	23.8	~	7	15.8	~	0.58	0.56
East London and The City	1	38.1	~	1	26.8	~	5	23.9	~	6	15.7	~	0.63	0.58
Hillingdon	9	44.6	~	9	30.2	~	6	24.0	~	5	15.5	~	0.54	0.51
Kensington and Chelsea, and Westminster	7	43.4	~	2	27.2	~	12	24.9	~	4	15.5	~	0.57	0.57
Kingston and Richmond	14	49.9		12	33.6		10	24.7	~	11	17.3		0.49	0.51
Lambeth, Southwark and Lewisham	5	42.2	~	5	28.7	~	7	24.4	~	8	16.1	~	0.58	0.56
Merton, Sutton and Wandsworth	13	49.1	~	14	35.2		13	25.7	~	10	16.7		0.52	0.47
Redbridge and Waltham Forest	8	43.7	~	7	29.7	~	3	22.2	~	1	13.3	~	0.51	0.45
South East		**49.4**	~		**34.8**			**25.8**	~		**17.5**	~	**0.52**	**0.50**
Berkshire	5	49.7		9	36.3		5	25.2	~	11	18.1		0.51	0.50
Buckinghamshire	10	51.6		11	37.3	*	2	24.9	~	5	17.2		0.48	0.46
East Kent	1	42.8	~	1	29.3	~	13	27.7		13	19.1		0.65	0.65
East Surrey	8	50.5		5	34.6		8	25.7		8	17.7		0.51	0.51
East Sussex, Brighton and Hove	3	44.7	~	3	32.8	~	7	25.5	~	6	17.2		0.57	0.53
Isle of Wight, Portsmouth and South East Hampshire	11	54.6		12	38.3	*	3	25.0	~	1	16.0	~	0.46	0.42
North and Mid-Hampshire	9	51.0		10	36.4		1	24.0	~	3	17.0		0.47	0.47
Northamptonshire	6	49.7	~	8	36.1		10	26.4		12	18.3		0.53	0.51
Oxfordshire	12	55.6		6	34.9		9	26.0	~	4	17.1		0.47	0.49
Southampton and South West Hampshire	13	55.9		13	39.0	*	11	26.9		2	16.5	~	0.48	0.42
West Kent	2	44.3	~	2	31.1	~	12	27.2		9	17.8		0.61	0.57
West Surrey	4	47.9	~	4	34.6		4	25.0	~	7	17.5		0.52	0.51
West Sussex	7	50.3		7	35.7		6	25.4		10	18.0		0.51	0.50
South West		**51.7**	~		**35.7**			**26.0**	~		**17.4**	~	**0.50**	**0.49**
Avon	3	48.7	~	1	32.3	~	3	25.2	~	1	16.6	~	0.52	0.51
Cornwall and Isles of Scilly	2	48.5	~	4	34.9		2	25.2	~	7	18.7		0.52	0.54
Dorset	8	57.2	*	8	40.3	*	1	24.5	~	4	17.1		0.43	0.42
Gloucestershire	6	53.9		3	34.3		7	27.7		5	17.5		0.51	0.51
North and East Devon	4	49.7	~	5	35.4		6	26.8		3	17.1		0.54	0.48
Somerset	5	52.6		6	36.1		8	27.9		6	17.5		0.53	0.49
South and West Devon	1	48.0	~	2	34.0		4	25.6	~	2	16.8		0.53	0.49
Wiltshire	7	55.1		7	38.8	*	5	26.5		8	18.8		0.48	0.48
Wales		**56.3**	*		**35.7**			**30.8**	*		**19.0**	*	**0.55**	**0.53**
Bro Taf	2	56.6	*	2	34.4		5	32.7	*	3	18.5		0.58	0.54
Dyfed Powys	1	50.3		3	35.0		1	27.2		2	18.3		0.54	0.52
Gwent	4	57.7	*	1	34.1		3	31.8	*	1	17.7		0.55	0.52
Morgannwg	5	58.9	*	4	37.3	*	4	32.0	*	4	18.9		0.54	0.51
North Wales	3	57.6	*	5	37.5	*	2	30.4	*	5	20.9	*	0.53	0.56
Scotland		**64.3**	*		**41.5**	*		**32.7**	*		**20.6**	*	**0.51**	**0.50**
Argyll and Clyde	8	63.3	*	9	41.1	*	5	30.9	*	9	21.4	*	0.49	0.52
Ayrshire and Arran	4	59.8	*	3	39.7	*	2	30.0		6	20.4		0.50	0.51
Borders	6	62.1	*	14	47.6	*	14	36.0	*	14	24.2	*	0.58	0.51
Dumfries and Galloway	5	61.8	*	5	40.7	*	9	34.1	*	1	19.3		0.55	0.47
Fife	7	63.0	*	8	41.0	*	4	30.5		7	20.6	*	0.48	0.50
Forth Valley	3	57.6	*	4	40.0	*	1	28.9		12	21.8		0.50	0.54
Grampian	14	69.3	*	10	45.0	*	11	34.9	*	8	21.1	*	0.50	0.47
Greater Glasgow	12	68.2	*	7	40.8	*	13	35.7	*	3	19.6	*	0.52	0.48
Highland	13	69.2	*	13	46.6	*	8	33.7	*	10	21.6	*	0.49	0.46
Lanarkshire	2	55.9		1	35.8		3	30.1		5	19.9	*	0.54	0.55
Lothian	9	65.4	*	6	40.8	*	7	30.9	*	2	19.5	*	0.47	0.48
Orkney	1	53.6		2	38.0		6	30.9		4	19.6		0.58	0.52
Shetland	15	78.7	*	12	45.6		15	43.2	*	11	21.7		0.55	0.47
Tayside	10	67.0	*	11	45.2	*	12	35.3	*	13	22.2	*	0.53	0.49
Western Isles	11	67.8	*	15	52.0	*	10	34.2		15	24.2		0.50	0.47
Northern Ireland		**62.6**	*		**43.0**	*		**28.9**			**19.5**	*	**0.46**	**0.45**
Eastern	3	63.7	*	1	42.0	*	4	29.6		1	18.1		0.46	0.43
Northern	1	60.5	*	3	43.4	*	1	27.8		4	20.8	*	0.46	0.48
Southern	2	62.0	*	4	44.6	*	3	29.1		2	20.3	*	0.47	0.45
Western	4	63.8	*	2	43.4	*	2	28.7		3	20.3		0.45	0.47
Ireland		**63.7**	*		**38.8**	*		**32.8**	*		**19.4**	*	**0.51**	**0.50**
Eastern	8	70.5	*	5	39.8	*	8	35.0	*	6	20.3	*	0.50	0.51
Mid-Western	1	55.2		1	30.3	~	4	31.1		2	17.9		0.56	0.59
Midland	2	55.6		7	43.0	*	1	29.3		5	19.1		0.53	0.45
North Eastern	6	63.1	*	4	38.3		3	30.6		4	18.6		0.49	0.48
North Western	4	61.1	*	6	42.5	*	2	29.4		7	20.5		0.48	0.48
South Eastern	3	59.4	*	3	36.7		6	33.3	*	3	18.5		0.56	0.50
Southern	7	66.7	*	8	44.0	*	7	34.2	*	8	20.5	*	0.51	0.47
Western	5	61.2	*	2	34.0		5	32.1	*	1	16.9		0.52	0.50

Table **B7.2**

Colorectal: annual average numbers of registrations and deaths by health authority within country and region of England, UK and Ireland, 1991–2000

Country/ region/ health authority	Number of registrations[1]		Number of deaths[2]	
	Males	Females	Males	Females
UK and Ireland	**17,600**	**16,300**	**9,450**	**9,050**
England	**13,500**	**12,600**	**7,330**	**7,090**
Northern and Yorkshire	**1,840**	**1,610**	**1,000**	**908**
Bradford	122	114	61	61
Calderdale and Kirklees	144	129	79	77
County Durham	198	159	108	92
East Riding	162	147	89	81
Gateshead and South Tyneside	108	97	66	57
Leeds	205	188	96	92
Newcastle and North Tyneside	135	124	82	76
North Cumbria	103	90	56	49
North Yorkshire	232	210	119	114
Northumberland	89	76	54	46
Sunderland	84	68	48	40
Tees	169	133	101	81
Wakefield	91	76	47	43
Trent	**1,440**	**1,270**	**811**	**733**
Barnsley	73	63	41	36
Doncaster	85	68	46	45
Leicestershire	229	215	118	115
Lincolnshire	204	174	114	97
North Derbyshire	114	91	70	58
North Nottinghamshire	114	100	66	57
Nottingham	169	158	88	83
Rotherham	70	61	43	38
Sheffield	143	135	86	79
South Humber	89	78	55	51
Southern Derbyshire	147	123	87	74
West Midlands	**1,620**	**1,390**	**872**	**771**
Birmingham	285	243	152	133
Coventry	88	75	46	43
Dudley	99	83	54	48
Herefordshire	59	58	31	32
North Staffordshire	143	127	83	77
Sandwell	87	76	53	43
Shropshire	138	123	69	65
Solihull	57	52	32	30
South Staffordshire	172	144	91	79
Walsall	81	72	44	39
Warwickshire	158	133	86	70
Wolverhampton	78	62	44	35
Worcestershire	176	144	90	78
North West	**1,910**	**1,750**	**1,070**	**1,020**
Bury and Rochdale	103	95	59	58
East Lancashire	131	123	79	76
Liverpool	147	125	83	69
Manchester	122	113	67	61
Morecambe Bay	101	92	55	53
North Cheshire	77	70	44	39
North West Lancashire	157	145	83	86
Salford and Trafford	137	121	77	71
Sefton	102	98	56	54
South Cheshire	186	176	107	103
South Lancashire	80	81	45	47
St Helen's and Knowsley	95	80	49	47
Stockport	82	75	44	44
West Pennine	126	113	78	66
Wigan and Bolton	164	141	86	85
Wirral	98	96	60	59
Eastern	**1,400**	**1,340**	**747**	**746**
Bedfordshire	119	107	65	64
Cambridgeshire	184	170	91	86
Hertfordshire	253	236	135	125
Norfolk	259	237	129	126
North Essex	212	217	128	138
South Essex	169	170	95	100
Suffolk	203	207	104	109

1 England, Wales and Scotland 1991-99; Northern Ireland 1993-99; Ireland 1994-99.
2 England, Wales and Northern Ireland 1991-2000; Scotland 1991-99; Ireland 1994-2000.

Country/ region/ health authority	Incidence[1]						Mortality[2]						M:I Ratio[3]	
	Males			Females			Males			Females			Males	Females
	Rank[4]	Rate[5]	Signif[6]	Rank	Rate	Signif	Rank	Rate	Signif	Rank	Rate	Signif		
London		**9.6**			**4.4**			**4.4**			**1.8**	*	**0.46**	**0.41**
Barking and Havering	1	5.6	~	3	3.8		1	2.7	~	2	1.4		0.48	0.36
Barnet, Enfield and Haringey	8	9.2		4	4.2		5	3.5	~	3	1.4		0.38	0.33
Bexley, Bromley and Greenwich	4	7.3	~	2	3.6	~	3	3.1	~	4	1.4		0.43	0.40
Brent and Harrow	7	8.8		7	4.4		8	4.5		9	2.1		0.51	0.48
Camden and Islington	13	14.7	*	14	6.1	*	14	6.9	*	14	2.8	*	0.47	0.46
Croydon	6	8.1	~	11	4.7		7	3.9		6	1.8		0.48	0.38
Ealing, Hammersmith and Hounslow	10	11.4	*	12	5.0		10	5.4	*	10	2.1		0.47	0.43
East London and The City	12	12.5	*	8	4.4		13	6.3	*	8	2.0		0.51	0.45
Hillingdon	3	7.1	~	10	4.7		6	3.7		12	2.3		0.53	0.49
Kensington and Chelsea, and Westminster	14	15.3	*	13	6.0	*	11	5.9	*	13	2.5	*	0.38	0.43
Kingston and Richmond	5	7.7	~	9	4.4		4	3.5		11	2.2		0.45	0.50
Lambeth, Southwark and Lewisham	11	11.8	*	6	4.4		12	6.0	*	7	1.9		0.51	0.44
Merton, Sutton and Wandsworth	9	10.3		5	4.3		9	4.7		1	1.3		0.45	0.31
Redbridge and Waltham Forest	2	6.1	~	1	3.3	~	2	2.8	~	5	1.7		0.46	0.51
South East		**8.1**	~		**4.0**			**3.4**	~		**1.4**	~	**0.42**	**0.36**
Berkshire	8	8.5	~	7	4.0		9	3.4		8	1.5		0.40	0.37
Buckinghamshire	5	7.5	~	10	4.4		1	2.7	~	1	1.1	~	0.36	0.26
East Kent	9	8.5	~	6	3.9		12	4.2		5	1.4		0.49	0.35
East Surrey	4	7.0	~	5	3.9		7	3.3	~	6	1.4		0.47	0.36
East Sussex, Brighton and Hove	10	8.5	~	3	3.4	~	11	3.9		4	1.3		0.46	0.39
Isle of Wight, Portsmouth and South East Hampshire	12	10.2		11	4.5		13	4.2		3	1.3		0.41	0.29
North and Mid-Hampshire	3	7.0	~	2	3.3	~	2	2.8	~	2	1.2		0.40	0.38
Northamptonshire	7	8.3	~	9	4.3		10	3.5	~	13	1.8		0.42	0.41
Oxfordshire	11	8.9		12	4.7		6	3.2	~	9	1.5		0.36	0.32
Southampton and South West Hampshire	13	10.4		13	5.3	*	3	3.1	~	11	1.5		0.30	0.29
West Kent	1	6.4	~	1	3.3	~	8	3.3	~	10	1.5		0.52	0.46
West Surrey	6	8.1	~	8	4.2		4	3.2	~	12	1.7		0.39	0.40
West Sussex	2	7.0	~	4	3.7		5	3.2	~	7	1.4		0.46	0.38
South West		**7.7**	~		**3.8**	~		**3.1**	~		**1.4**	~	**0.40**	**0.38**
Avon	3	7.2	~	5	3.7		1	2.7	~	4	1.4		0.37	0.37
Cornwall and Isles of Scilly	1	6.1	~	4	3.4	~	2	2.8	~	6	1.6		0.47	0.46
Dorset	7	9.3		8	4.5		3	3.0	~	5	1.4		0.32	0.31
Gloucestershire	2	6.7	~	2	3.3	~	4	3.1	~	1	1.2	~	0.47	0.36
North and East Devon	4	7.3	~	3	3.4	~	6	3.3	~	2	1.3		0.45	0.38
Somerset	5	7.8	~	1	3.2	~	7	3.3	~	3	1.3		0.43	0.41
South and West Devon	6	7.9	~	6	3.9		8	3.6		8	1.6		0.46	0.42
Wiltshire	8	9.3		7	4.5		5	3.2		7	1.6		0.34	0.35
Wales		**12.3**	*		**4.9**	*		**4.5**			**1.6**		**0.37**	**0.33**
Bro Taf	4	13.7	*	2	4.5		5	5.5	*	1	1.3	~	0.40	0.28
Dyfed Powys	3	11.9	*	3	4.6		1	3.8		3	1.7		0.32	0.36
Gwent	1	9.6		1	3.9		2	3.9		2	1.5		0.40	0.37
Morgannwg	5	14.8	*	5	5.9	*	4	4.6		4	1.8		0.31	0.31
North Wales	2	11.4	*	4	5.7	*	3	4.4		5	2.0		0.38	0.34
Scotland		**15.6**	*		**6.1**	*		**6.5**	*		**2.3**	*	**0.42**	**0.38**
Argyll and Clyde	12	18.6	*	13	7.1	*	12	7.4	*	13	2.6	*	0.40	0.37
Ayrshire and Arran	3	13.0	*	10	6.0	*	5	5.3		11	2.4	*	0.41	0.40
Borders	1	9.9		2	4.5		4	5.2		8	2.2		0.53	0.48
Dumfries and Galloway	2	10.7		3	4.9		1	4.3		5	2.0		0.41	0.42
Fife	5	13.5	*	6	5.4		3	4.8		4	1.9		0.36	0.36
Forth Valley	11	15.8	*	11	6.2	*	10	6.9	*	2	1.7		0.44	0.27
Grampian	9	14.9	*	7	5.6	*	2	4.5		3	1.8		0.31	0.32
Greater Glasgow	14	20.9	*	14	7.4	*	13	9.3	*	14	2.9	*	0.45	0.39
Highland	10	15.3	*	5	5.2		7	5.9	*	1	1.6		0.38	0.31
Lanarkshire	8	14.6	*	8	5.8	*	11	7.0	*	12	2.5	*	0.48	0.43
Lothian	6	14.2	*	9	5.9	*	8	6.3	*	10	2.3	*	0.44	0.40
Orkney	7	14.5		1	4.2		9	6.7		7	2.1		0.47	0.50
Shetland	15	25.6	*	15	8.6		15	10.7		15	3.2		0.42	0.37
Tayside	4	13.3	*	4	5.0		6	5.7	*	6	2.0		0.43	0.41
Western Isles	13	19.6	*	12	6.3		14	9.8	*	9	2.2		0.50	0.36
Northern Ireland		**13.5**	*		**5.1**	*		**4.4**			**1.2**	~	**0.33**	**0.24**
Eastern	3	13.6	*	3	5.2		3	4.8		3	1.4		0.35	0.27
Northern	2	12.4	*	2	4.9		1	3.6		2	1.0	~	0.29	0.20
Southern	1	12.3		1	4.6		2	3.9		1	0.8	~	0.32	0.18
Western	4	16.4	*	4	5.7		4	5.4		4	1.5		0.33	0.27
Ireland		**13.5**	*		**3.8**	~		**6.4**	*		**1.9**		**0.47**	**0.50**
Eastern	6	14.4	*	6	4.1		7	6.9	*	6	2.0		0.48	0.48
Mid-Western	3	11.3		1	2.6	~	8	6.9	*	1	1.4		0.61	0.56
Midland	5	13.0	*	7	4.4		3	5.9		3	1.6		0.45	0.36
North Eastern	2	10.2		4	3.7		1	4.3		2	1.6		0.42	0.44
North Western	7	16.9	*	8	5.0		4	6.3	*	4	1.8		0.37	0.35
South Eastern	1	10.1		3	3.5		2	5.4		8	2.2		0.54	0.64
Southern	4	11.6		2	3.4		6	6.8	*	7	2.1		0.59	0.61
Western	8	19.7	*	5	3.8		5	6.7	*	5	1.8		0.34	0.47

Table **B12.2**

Lip, mouth and pharynx: annual average numbers of registrations and deaths by health authority within country and region of England, UK and Ireland, 1991–2000

Country/ region/ health authority	Number of registrations[1]		Number of deaths[2]	
	Males	Females	Males	Females
UK and Ireland	**3,100**	**1,690**	**1,360**	**736**
England	**2,210**	**1,270**	**995**	**564**
Northern and Yorkshire	**337**	**173**	**156**	**78**
Bradford	22	13	11	6
Calderdale and Kirklees	27	15	12	7
County Durham	34	15	16	8
East Riding	27	16	12	8
Gateshead and South Tyneside	25	9	13	4
Leeds	37	21	16	8
Newcastle and North Tyneside	32	14	14	7
North Cumbria	17	8	10	4
North Yorkshire	35	23	14	9
Northumberland	16	8	8	5
Sunderland	16	9	7	3
Tees	35	15	17	6
Wakefield	17	9	8	4
Trent	**225**	**120**	**101**	**56**
Barnsley	9	6	3	3
Doncaster	16	6	7	3
Leicestershire	34	23	15	11
Lincolnshire	27	13	11	7
North Derbyshire	15	9	7	4
North Nottinghamshire	17	9	7	4
Nottingham	30	16	15	7
Rotherham	11	6	5	3
Sheffield	24	10	13	6
South Humber	17	7	8	3
Southern Derbyshire	25	14	10	6
West Midlands	**223**	**128**	**109**	**60**
Birmingham	47	29	22	12
Coventry	15	8	8	4
Dudley	14	7	6	3
Herefordshire	5	4	3	2
North Staffordshire	20	11	10	6
Sandwell	13	8	7	4
Shropshire	17	10	9	4
Solihull	7	5	3	2
South Staffordshire	21	12	10	6
Walsall	13	7	6	3
Warwickshire	19	11	9	5
Wolverhampton	13	5	6	3
Worcestershire	19	12	10	5
North West	**366**	**188**	**171**	**90**
Bury and Rochdale	21	11	9	5
East Lancashire	24	16	12	7
Liverpool	38	14	18	6
Manchester	32	13	15	6
Morecambe Bay	16	11	9	6
North Cheshire	15	6	7	3
North West Lancashire	30	18	14	9
Salford and Trafford	27	14	12	6
Sefton	15	9	6	4
South Cheshire	26	15	13	8
South Lancashire	14	8	6	4
St Helen's and Knowsley	19	8	8	4
Stockport	16	8	6	4
West Pennine	27	13	13	6
Wigan and Bolton	28	14	12	6
Wirral	19	11	11	5
Eastern	**195**	**121**	**78**	**49**
Bedfordshire	17	11	8	3
Cambridgeshire	27	17	8	7
Hertfordshire	32	23	15	10
Norfolk	36	21	13	9
North Essex	29	18	14	9
South Essex	19	14	9	6
Suffolk	34	18	11	6

1 England, Wales and Scotland 1991-99; Northern Ireland 1993-99; Ireland 1994-99.
2 England, Wales and Northern Ireland 1991-2000; Scotland 1991-99; Ireland 1994-2000.

Country/ region/ health authority	Number of registrations[1]		Number of deaths[2]	
	Males	Females	Males	Females
London	**296**	**170**	**137**	**77**
Barking and Havering	11	10	5	4
Barnet, Enfield and Haringey	32	19	12	7
Bexley, Bromley and Greenwich	26	17	11	8
Brent and Harrow	18	11	9	6
Camden and Islington	21	11	10	5
Croydon	12	9	6	4
Ealing, Hammersmith and Hounslow	31	16	15	8
East London and The City	29	12	14	6
Hillingdon	8	6	4	3
Kensington and Chelsea, and Westminster	23	11	9	5
Kingston and Richmond	11	8	5	4
Lambeth, Southwark and Lewisham	34	16	17	7
Merton, Sutton and Wandsworth	27	16	13	5
Redbridge and Waltham Forest	12	9	6	5
South East	**356**	**228**	**154**	**95**
Berkshire	30	18	12	7
Buckinghamshire	24	17	8	5
East Kent	29	17	15	8
East Surrey	16	11	8	5
East Sussex, Brighton and Hove	36	22	18	11
Isle of Wight, Portsmouth and South East Hampshire	36	22	15	7
North and Mid-Hampshire	18	11	8	4
Northamptonshire	25	17	11	8
Oxfordshire	26	16	10	6
Southampton and South West Hampshire	29	19	9	6
West Kent	31	21	16	11
West Surrey	27	17	11	8
West Sussex	30	21	14	10
South West	**212**	**137**	**89**	**58**
Avon	37	24	14	10
Cornwall and Isles of Scilly	19	13	9	7
Dorset	38	25	14	9
Gloucestershire	20	13	10	6
North and East Devon	21	13	11	6
Somerset	23	13	10	6
South and West Devon	26	17	12	8
Wiltshire	28	18	10	7
Wales	**190**	**101**	**72**	**36**
Bro Taf	48	21	20	7
Dyfed Powys	34	17	11	7
Gwent	28	14	12	6
Morgannwg	39	21	13	7
North Wales	41	28	17	10
Scotland	**389**	**205**	**165**	**84**
Argyll and Clyde	39	20	16	8
Ayrshire and Arran	25	16	10	6
Borders	6	4	3	2
Dumfries and Galloway	9	6	4	2
Fife	23	12	9	5
Forth Valley	21	11	9	3
Grampian	38	19	11	6
Greater Glasgow	87	45	39	19
Highland	17	7	6	3
Lanarkshire	38	19	18	9
Lothian	51	29	23	13
Orkney	2	1	1	0
Shetland	3	1	1	1
Tayside	28	14	12	7
Western Isles	3	1	2	0
Northern Ireland	**98**	**50**	**32**	**13**
Eastern	41	22	15	7
Northern	23	12	7	3
Southern	16	7	5	2
Western	18	8	6	2
Ireland	**208**	**71**	**99**	**38**
Eastern	67	27	32	13
Mid-Western	16	4	10	3
Midland	13	5	6	2
North Eastern	14	6	6	3
North Western	18	6	7	3
South Eastern	17	7	9	5
Southern	28	10	17	6
Western	35	7	12	4

Table B13.1

Lung: annual average age-standardised incidence and mortality by health authority within country and region of England, UK and Ireland, 1991–2000

Country/ region/ health authority	Incidence[1]						Mortality[2]						M:I Ratio[3]	
	Rank[4]	Males Rate[5]	Signif[6]	Rank	Females Rate	Signif	Rank	Males Rate	Signif	Rank	Females Rate	Signif	Males	Females
UK and Ireland		81.0			34.9			71.7			30.2		0.89	0.87
England		79.0	~		33.4	~		70.3	~		29.2	~	0.89	0.87
Northern and Yorkshire		94.1	*		43.5	*		84.3	*		38.5	*	0.90	0.89
Bradford	5	91.3	*	5	41.7	*	6	82.8	*	5	35.9	*	0.91	0.86
Calderdale and Kirklees	3	84.8		3	36.2		3	76.8	*	3	32.5	*	0.91	0.90
County Durham	9	101.0	*	9	46.5	*	9	89.3	*	8	40.0	*	0.88	0.86
East Riding	4	88.8	*	4	40.0	*	4	78.5	*	4	35.3	*	0.88	0.88
Gateshead and South Tyneside	13	125.3	*	13	60.6	*	13	114.0	*	13	53.6	*	0.91	0.89
Leeds	8	94.6	*	8	45.9	*	5	82.1	*	9	40.2	*	0.87	0.88
Newcastle and North Tyneside	12	119.5	*	12	56.6	*	12	106.6	*	12	49.7	*	0.89	0.88
North Cumbria	2	76.3	~	2	33.8		2	71.6		2	32.3		0.94	0.96
North Yorkshire	1	64.8	~	1	27.5	~	1	57.9	~	1	24.5	~	0.89	0.89
Northumberland	6	92.4	*	6	41.9	*	7	83.7	*	6	36.3	*	0.91	0.86
Sunderland	11	115.0	*	11	54.4	*	11	102.4	*	11	49.7	*	0.89	0.91
Tees	10	105.9	*	10	51.5	*	10	96.3	*	10	45.7	*	0.91	0.89
Wakefield	7	93.1	*	7	42.3	*	8	83.8	*	7	38.7	*	0.90	0.92
Trent		82.5	*		32.3	~		72.4			28.1	~	0.88	0.87
Barnsley	11	103.7	*	10	40.7	*	11	89.4	*	10	35.5	*	0.86	0.87
Doncaster	10	98.1	*	11	42.4	*	9	84.7	*	11	36.2	*	0.86	0.85
Leicestershire	1	66.2	~	1	25.9	~	1	57.5	~	1	22.3	~	0.87	0.86
Lincolnshire	2	72.0	~	2	27.7	~	2	62.8	~	2	24.2	~	0.87	0.87
North Derbyshire	5	82.7		4	31.4	~	4	72.7		4	27.5		0.88	0.88
North Nottinghamshire	6	85.8	*	6	32.5		6	75.8	*	5	28.0		0.88	0.86
Nottingham	7	89.4	*	7	35.0		7	77.2	*	7	29.7		0.86	0.85
Rotherham	9	97.1	*	8	36.7		8	83.5	*	8	33.3	*	0.86	0.91
Sheffield	8	94.4	*	9	38.8	*	10	85.2	*	9	34.5	*	0.90	0.89
South Humber	4	79.3		5	32.2		5	73.6		6	28.7		0.93	0.89
Southern Derbyshire	3	78.8		3	27.8	~	3	69.9		3	24.6	~	0.89	0.88
West Midlands		81.2			30.1	~		72.5			26.6	~	0.89	0.88
Birmingham	10	93.0	*	11	35.1		9	80.1	*	11	31.1		0.86	0.89
Coventry	7	81.6		13	36.1		7	73.2		12	31.5		0.90	0.87
Dudley	8	83.3		3	26.1	~	8	74.8		3	23.1	~	0.90	0.89
Herefordshire	1	54.1	~	1	21.9	~	1	50.2	~	1	20.6	~	0.93	0.94
North Staffordshire	11	93.1	*	10	32.6		12	84.4	*	10	28.8		0.91	0.88
Sandwell	13	101.8	*	12	35.8		13	92.0	*	13	33.5	*	0.90	0.94
Shropshire	6	74.2	~	6	27.9	~	6	67.8	~	4	24.2	~	0.91	0.87
Solihull	3	70.1	~	9	30.3	~	3	60.3	~	8	25.6	~	0.86	0.84
South Staffordshire	5	73.8	~	4	27.7	~	5	66.6	~	5	24.5	~	0.90	0.88
Walsall	12	94.6	*	8	30.1		10	81.7	*	9	27.2	~	0.86	0.90
Warwickshire	4	71.8	~	5	27.8	~	4	64.1	~	6	24.9	~	0.89	0.90
Wolverhampton	9	90.0	*	7	28.7	~	11	83.0	*	7	25.5	~	0.92	0.89
Worcestershire	2	64.8	~	2	24.6	~	2	59.4	~	2	21.0	~	0.92	0.85
North West		95.1	*		42.8	*		84.4	*		37.4	*	0.89	0.87
Bury and Rochdale	7	92.3	*	9	43.1	*	7	80.3	*	8	37.1	*	0.87	0.86
East Lancashire	4	82.4		6	37.0		6	75.0		6	33.1		0.91	0.89
Liverpool	15	131.8	*	16	69.1	*	15	117.1	*	16	59.7	*	0.89	0.86
Manchester	16	139.0	*	15	60.8	*	16	121.6	*	15	53.9	*	0.88	0.89
Morecambe Bay	2	74.5	~	2	32.3		2	66.5	~	4	29.5		0.89	0.91
North Cheshire	12	98.7	*	11	43.6	*	10	86.8	*	9	37.2	*	0.88	0.85
North West Lancashire	6	85.0	*	5	35.6		5	74.4		5	31.2		0.88	0.88
Salford and Trafford	13	103.1	*	13	50.1	*	13	92.7	*	13	42.9	*	0.90	0.86
Sefton	11	98.5	*	10	43.3	*	11	87.4	*	11	38.3	*	0.89	0.88
South Cheshire	1	72.5	~	1	28.8	~	1	65.1	~	1	25.3	~	0.90	0.88
South Lancashire	5	84.4		4	35.1		4	73.0		3	29.3		0.87	0.84
St Helen's and Knowsley	14	114.4	*	14	54.1	*	14	101.8	*	14	46.4	*	0.89	0.86
Stockport	3	81.5		3	33.3		3	73.0		2	29.0		0.89	0.87
West Pennine	9	96.0	*	8	42.5	*	12	88.7	*	10	38.2	*	0.92	0.90
Wigan and Bolton	8	93.3	*	7	38.7	*	8	81.8	*	7	34.6	*	0.88	0.89
Wirral	10	97.2	*	12	48.3	*	9	85.6	*	12	41.2	*	0.88	0.85
Eastern		66.5	~		27.3	~		59.7	~		24.0	~	0.90	0.88
Bedfordshire	6	71.6	~	6	29.5	~	6	64.0	~	5	25.3	~	0.89	0.86
Cambridgeshire	4	68.0	~	4	27.3	~	4	59.7	~	2	22.5	~	0.88	0.82
Hertfordshire	5	68.9	~	5	28.9	~	5	59.8	~	6	25.4	~	0.87	0.88
Norfolk	1	60.9	~	1	23.7	~	1	55.1	~	1	21.4	~	0.91	0.90
North Essex	3	64.5	~	3	27.3	~	3	59.7	~	4	24.1	~	0.93	0.88
South Essex	7	74.3	~	7	29.8	~	7	67.5	~	7	27.1	~	0.91	0.91
Suffolk	2	61.4	~	2	25.7	~	2	55.6	~	3	22.6	~	0.91	0.88

1 England, Wales and Scotland 1991-99; Northern Ireland 1993-99; Ireland 1994-99.
2 England, Wales and Northern Ireland 1991-2000; Scotland 1991-99; Ireland 1994-2000.
3 Mortality to incidence ratio; age-standardised mortality rate divided by age-standardised incidence rate.
4 Rank within region or country; 1 is lowest.
5 Directly age standardised using the European standard population; per 100,000 population.
6 Rate significantly different from that for UK and Ireland; * significantly higher; ~ significantly lower. See Appendix H.

Country/ region/ health authority	Incidence[1]						Mortality[2]						M:I Ratio[3]	
	Males			Females			Males			Females			Males	Females
	Rank[4]	Rate[5]	Signif[6]	Rank	Rate	Signif	Rank	Rate	Signif	Rank	Rate	Signif		
London		**82.4**	*		**36.6**	*		**73.1**	*		**31.7**	*	**0.89**	**0.87**
Barking and Havering	11	90.0	*	10	38.7	*	11	80.2	*	10	33.7	*	0.89	0.87
Barnet, Enfield and Haringey	4	70.3	~	2	29.8	~	2	60.6	~	3	25.7	~	0.86	0.86
Bexley, Bromley and Greenwich	9	80.0		7	35.4		9	72.0		7	31.5		0.90	0.89
Brent and Harrow	1	67.4	~	1	28.6	~	1	59.0	~	1	23.3	~	0.88	0.81
Camden and Islington	12	96.8	*	13	44.5	*	12	85.3	*	12	37.8	*	0.88	0.85
Croydon	3	68.7	~	4	33.0		4	62.0	~	5	28.6		0.90	0.87
Ealing, Hammersmith and Hounslow	6	73.5	~	8	36.7		6	66.9	~	9	33.2	*	0.91	0.90
East London and The City	14	109.8	*	14	45.5	*	14	97.5	*	14	39.9	*	0.89	0.88
Hillingdon	5	73.2	~	6	33.9		7	68.2		6	31.0		0.93	0.91
Kensington and Chelsea, and Westminster	8	78.8		11	42.0	*	5	66.1	~	11	35.6	*	0.84	0.85
Kingston and Richmond	2	68.2	~	3	31.4	~	3	60.9	~	2	25.5	~	0.89	0.81
Lambeth, Southwark and Lewisham	13	102.6	*	12	43.1	*	13	90.3	*	13	38.2	*	0.88	0.88
Merton, Sutton and Wandsworth	10	86.9	*	9	38.0	*	10	76.7	*	8	32.4	*	0.88	0.85
Redbridge and Waltham Forest	7	78.1		5	33.3		8	71.0		4	28.1		0.91	0.84
South East		**67.7**	~		**27.9**	~		**59.7**	~		**24.2**	~	**0.88**	**0.87**
Berkshire	7	67.6	~	5	27.3	~	6	58.4	~	4	22.8	~	0.86	0.84
Buckinghamshire	8	68.9	~	12	30.4	~	8	58.9	~	10	25.8	~	0.85	0.85
East Kent	12	75.1	~	11	30.4	~	13	69.9		13	28.5		0.93	0.94
East Surrey	3	62.3	~	2	24.5	~	3	54.0	~	2	21.4	~	0.87	0.87
East Sussex, Brighton and Hove	4	63.9	~	6	27.4	~	5	57.0	~	8	24.0	~	0.89	0.88
Isle of Wight, Portsmouth and South East Hampshire	11	72.9	~	10	30.0	~	9	61.2	~	9	25.2	~	0.84	0.84
North and Mid-Hampshire	5	65.2	~	4	26.8	~	7	58.6	~	5	23.2	~	0.90	0.87
Northamptonshire	13	75.3	~	13	31.8	~	11	65.9	~	12	27.0	~	0.88	0.85
Oxfordshire	6	65.8	~	7	27.8	~	4	56.1	~	7	23.6	~	0.85	0.85
Southampton and South West Hampshire	9	72.1	~	8	28.3	~	10	62.4	~	6	23.3	~	0.87	0.83
West Kent	10	72.4	~	9	28.3	~	12	66.9	~	11	26.4	~	0.92	0.93
West Surrey	2	61.8	~	3	26.0	~	2	54.0	~	3	21.8	~	0.87	0.84
West Sussex	1	59.0	~	1	24.4	~	1	53.2	~	1	21.0	~	0.90	0.86
South West		**62.7**	~		**24.9**	~		**56.7**	~		**21.7**	~	**0.90**	**0.87**
Avon	8	67.3	~	7	26.7	~	8	60.8	~	7	23.4	~	0.90	0.88
Cornwall and Isles of Scilly	7	65.8	~	4	24.7	~	6	58.0	~	6	22.3	~	0.88	0.90
Dorset	4	62.7	~	5	24.9	~	4	55.3	~	3	20.3	~	0.88	0.82
Gloucestershire	3	58.8	~	2	22.8	~	3	55.2	~	4	20.9	~	0.94	0.92
North and East Devon	1	55.8	~	3	22.9	~	1	50.1	~	2	19.6	~	0.90	0.86
Somerset	2	58.7	~	1	21.3	~	2	54.0	~	1	19.4	~	0.92	0.91
South and West Devon	6	65.7	~	8	28.0	~	7	60.7	~	8	24.9	~	0.92	0.89
Wiltshire	5	63.5	~	6	26.4	~	5	56.8	~	5	22.1	~	0.89	0.84
Wales		**83.1**	*		**34.4**			**70.0**	~		**28.7**	~	**0.84**	**0.83**
Bro Taf	3	84.7	*	5	38.0	*	4	73.6		5	32.5	*	0.87	0.86
Dyfed Powys	1	69.9	~	1	27.2	~	1	58.9	~	1	23.0	~	0.84	0.85
Gwent	4	85.4	*	3	35.3		5	74.4		4	31.1		0.87	0.88
Morgannwg	2	83.9		2	33.8		3	71.9		3	28.1	~	0.86	0.83
North Wales	5	89.6	*	4	36.2		2	70.5		2	28.0	~	0.79	0.77
Scotland		**108.2**	*		**51.8**	*		**95.6**	*		**43.5**	*	**0.88**	**0.84**
Argyll and Clyde	14	118.5	*	14	54.7	*	14	108.0	*	14	47.1	*	0.91	0.86
Ayrshire and Arran	11	104.0	*	11	48.4	*	11	93.0	*	11	41.7	*	0.89	0.86
Borders	4	81.3		4	36.7		3	68.2		4	29.7		0.84	0.81
Dumfries and Galloway	5	84.2		7	44.4	*	5	74.2		7	36.8	*	0.88	0.83
Fife	9	97.6	*	8	45.0	*	9	87.0	*	9	38.3	*	0.89	0.85
Forth Valley	10	102.2	*	10	47.2	*	10	89.6	*	8	37.9	*	0.88	0.80
Grampian	6	86.7	*	6	43.0	*	7	77.7	*	6	36.2	*	0.90	0.84
Greater Glasgow	15	146.6	*	15	71.7	*	15	130.1	*	15	60.2	*	0.89	0.84
Highland	3	81.0		5	37.1		4	71.3		5	32.4		0.88	0.87
Lanarkshire	13	111.3	*	12	51.2	*	13	99.8	*	13	43.4	*	0.90	0.85
Lothian	12	108.8	*	13	52.4	*	12	93.8	*	12	42.7	*	0.86	0.82
Orkney	1	64.5	~	2	30.9		1	56.8	~	2	24.0		0.88	0.78
Shetland	2	79.6		3	31.4		2	67.2		3	25.6		0.84	0.82
Tayside	8	94.3	*	9	45.3	*	8	80.5	*	10	39.0	*	0.85	0.86
Western Isles	7	91.3		1	28.0		6	76.9		1	19.5		0.84	0.69
Northern Ireland		**74.3**	~		**32.7**	~		**67.2**	~		**27.4**	~	**0.90**	**0.84**
Eastern	4	86.5	*	4	40.6	*	4	77.8	*	4	33.0	*	0.90	0.81
Northern	2	64.0	~	2	26.5	~	2	59.0	~	2	23.2	~	0.92	0.87
Southern	1	62.8	~	1	23.1	~	1	56.6	~	1	21.0	~	0.90	0.91
Western	3	70.6	~	3	30.8	~	3	63.9	~	3	24.9	~	0.90	0.81
Ireland		**62.1**	~		**27.1**	~		**60.9**	~		**26.8**	~	**0.98**	**0.99**
Eastern	8	84.3		8	36.8		8	80.4	*	8	35.6	*	0.95	0.97
Mid-Western	2	51.4	~	2	21.2	~	3	53.4	~	5	24.1	~	1.04	1.14
Midland	3	52.7	~	3	22.0	~	2	52.9	~	2	21.6	~	1.00	0.98
North Eastern	5	54.9	~	7	24.8	~	6	56.0	~	7	24.9	~	1.02	1.00
North Western	7	57.6	~	4	22.7	~	5	55.7	~	4	23.5	~	0.97	1.03
South Eastern	4	54.8	~	5	22.9	~	7	56.2	~	6	24.4	~	1.03	1.06
Southern	6	55.1	~	6	23.3	~	4	54.2	~	3	21.6	~	0.98	0.93
Western	1	46.5	~	1	18.5	~	1	45.9	~	1	17.2	~	0.99	0.93

Table **B13.2**

Lung: annual average numbers of registrations and deaths by health authority within country and region of England, UK and Ireland, 1991–2000

Country/ region/ health authority	Number of registrations[1]		Number of deaths[2]	
	Males	Females	Males	Females
UK and Ireland	**27,000**	**15,000**	**24,300**	**13,400**
England	**21,200**	**11,500**	**19,100**	**10,300**
Northern and Yorkshire	**3,240**	**1,940**	**2,930**	**1,750**
Bradford	210	126	191	112
Calderdale and Kirklees	252	141	230	129
County Durham	343	199	306	177
East Riding	285	165	257	149
Gateshead and South Tyneside	258	163	236	147
Leeds	358	224	314	201
Newcastle and North Tyneside	307	197	275	177
North Cumbria	143	81	136	78
North Yorkshire	290	161	264	148
Northumberland	172	98	157	87
Sunderland	174	105	155	98
Tees	299	187	274	169
Wakefield	154	89	141	83
Trent	**2,380**	**1,150**	**2,120**	**1,030**
Barnsley	129	62	113	57
Doncaster	159	85	138	75
Leicestershire	323	157	284	138
Lincolnshire	290	135	258	121
North Derbyshire	183	85	163	76
North Nottinghamshire	194	87	173	77
Nottingham	308	151	268	130
Rotherham	130	62	113	57
Sheffield	281	151	256	138
South Humber	141	71	132	64
Southern Derbyshire	243	107	218	98
West Midlands	**2,360**	**1,100**	**2,130**	**1,000**
Birmingham	474	229	411	207
Coventry	135	76	121	67
Dudley	143	56	130	51
Herefordshire	60	30	57	28
North Staffordshire	245	108	223	97
Sandwell	165	74	150	70
Shropshire	177	84	164	76
Solihull	79	44	69	38
South Staffordshire	223	105	204	95
Walsall	133	52	116	49
Warwickshire	205	98	184	91
Wolverhampton	126	50	117	45
Worcestershire	197	93	183	83
North West	**3,330**	**1,970**	**2,990**	**1,760**
Bury and Rochdale	179	110	157	97
East Lancashire	220	129	200	118
Liverpool	308	220	275	193
Manchester	267	150	235	134
Morecambe Bay	143	81	129	75
North Cheshire	148	85	131	74
North West Lancashire	246	134	218	121
Salford and Trafford	248	156	225	137
Sefton	168	104	152	94
South Cheshire	273	140	248	126
South Lancashire	135	71	118	61
St Helen's and Knowsley	190	118	171	104
Stockport	129	72	116	64
West Pennine	227	133	212	122
Wigan and Bolton	270	146	237	133
Wirral	180	121	161	106
Eastern	**2,010**	**1,030**	**1,830**	**933**
Bedfordshire	188	95	170	84
Cambridgeshire	245	120	218	103
Hertfordshire	373	199	327	179
Norfolk	327	156	302	144
North Essex	338	180	318	162
South Essex	288	151	265	141
Suffolk	253	133	234	120

1 England, Wales and Scotland 1991-99; Northern Ireland 1993-99; Ireland 1994-99.
2 England, Wales and Northern Ireland 1991-2000; Scotland 1991-99; Ireland 1994-2000.

Country/ region/ health authority	Number of registrations[1]		Number of deaths[2]	
	Males	Females	Males	Females
London	**2,700**	**1,590**	**2,410**	**1,410**
Barking and Havering	196	113	177	100
Barnet, Enfield and Haringey	263	151	228	134
Bexley, Bromley and Greenwich	311	185	282	168
Brent and Harrow	145	81	128	68
Camden and Islington	151	91	133	79
Croydon	105	68	96	60
Ealing, Hammersmith and Hounslow	213	138	194	128
East London and The City	264	134	233	118
Hillingdon	89	53	83	50
Kensington and Chelsea, and Westminster	124	90	104	78
Kingston and Richmond	110	69	100	58
Lambeth, Southwark and Lewisham	315	174	277	155
Merton, Sutton and Wandsworth	247	149	220	130
Redbridge and Waltham Forest	170	96	157	83
South East	**3,220**	**1,740**	**2,880**	**1,560**
Berkshire	243	127	212	109
Buckinghamshire	218	122	189	106
East Kent	287	156	271	151
East Surrey	156	81	137	72
East Sussex, Brighton and Hove	324	197	294	178
Isle of Wight, Portsmouth and South East Hampshire	286	152	244	133
North and Mid-Hampshire	171	93	156	83
Northamptonshire	236	120	209	105
Oxfordshire	200	108	173	94
Southampton and South West Hampshire	226	114	199	98
West Kent	364	186	340	176
West Surrey	218	120	192	104
West Sussex	292	168	268	153
South West	**1,930**	**987**	**1,770**	**894**
Avon	371	192	340	172
Cornwall and Isles of Scilly	218	105	196	99
Dorset	316	163	283	139
Gloucestershire	195	97	186	94
North and East Devon	185	97	170	87
Somerset	188	86	175	82
South and West Devon	248	139	234	128
Wiltshire	205	109	187	94
Wales	**1,430**	**751**	**1,220**	**644**
Bro Taf	324	191	285	165
Dyfed Powys	224	106	191	93
Gwent	266	137	234	124
Morgannwg	250	128	217	109
North Wales	363	189	290	153
Scotland	**2,870**	**1,870**	**2,550**	**1,600**
Argyll and Clyde	262	170	240	147
Ayrshire and Arran	213	131	191	115
Borders	56	32	47	27
Dumfries and Galloway	79	54	70	44
Fife	183	112	164	98
Forth Valley	145	91	128	75
Grampian	228	149	205	128
Greater Glasgow	663	465	591	398
Highland	92	53	82	48
Lanarkshire	292	185	262	159
Lothian	411	277	356	230
Orkney	7	4	7	3
Shetland	9	4	7	3
Tayside	216	139	187	123
Western Isles	16	6	14	5
Northern Ireland	**560**	**319**	**510**	**276**
Eastern	276	176	249	149
Northern	123	64	112	58
Southern	82	38	74	35
Western	79	41	71	34
Ireland	**971**	**518**	**964**	**530**
Eastern	382	234	365	234
Mid-Western	74	36	78	41
Midland	53	24	54	24
North Eastern	74	38	77	40
North Western	65	29	64	31
South Eastern	99	48	102	52
Southern	137	71	137	68
Western	87	40	87	40

Table B14.1

Melanoma of skin: annual average age-standardised incidence and mortality by health authority within country and region of England, UK and Ireland, 1991–2000

Country/ region/ health authority	Incidence[1]						Mortality[2]						M:I Ratio[3]	
		Males			Females			Males			Females		Males	Females
	Rank[4]	Rate[5]	Signif[6]	Rank	Rate	Signif	Rank	Rate	Signif	Rank	Rate	Signif		
UK and Ireland		7.7			9.7			2.5			2.0		0.32	0.20
England		7.5			9.4	~		2.5			2.0		0.33	0.21
Northern and Yorkshire		6.4	~		8.3	~		2.2	~		1.7	~	0.34	0.21
Bradford	8	6.5	~	11	9.6		5	1.9		9	1.9		0.30	0.20
Calderdale and Kirklees	6	5.8	~	5	7.5	~	3	1.9	~	10	2.0		0.32	0.26
County Durham	7	6.4	~	9	8.3	~	8	2.3		7	1.8		0.36	0.21
East Riding	10	6.8		8	8.1	~	7	2.1		3	1.6		0.32	0.19
Gateshead and South Tyneside	4	4.6	~	1	5.6	~	1	1.4	~	1	1.1	~	0.30	0.20
Leeds	11	6.9		10	9.4		12	2.4		6	1.7		0.35	0.18
Newcastle and North Tyneside	5	4.8	~	4	7.0	~	6	2.1		2	1.3	~	0.43	0.18
North Cumbria	13	10.5	*	12	10.8		13	3.1		12	2.1		0.29	0.19
North Yorkshire	12	9.2	*	13	11.4	*	10	2.4		5	1.6		0.26	0.14
Northumberland	2	4.4	~	7	8.0	~	2	1.8		11	2.0		0.42	0.25
Sunderland	3	4.5	~	3	6.6	~	11	2.4		4	1.6		0.53	0.25
Tees	1	3.8	~	2	5.9	~	4	1.9	~	8	1.8		0.49	0.30
Wakefield	9	6.7		6	7.8	~	9	2.3		13	2.1		0.35	0.27
Trent		6.3	~		8.0	~		2.3			1.9		0.36	0.24
Barnsley	3	5.2	~	5	7.6	~	5	2.1		2	1.6		0.40	0.21
Doncaster	1	4.4	~	1	5.8	~	1	1.7		5	1.7		0.40	0.30
Leicestershire	4	5.6	~	4	7.1	~	6	2.2		3	1.7		0.39	0.24
Lincolnshire	10	7.5		7	8.3	~	10	2.7		8	2.1		0.36	0.25
North Derbyshire	9	7.0		3	6.9	~	11	2.7		4	1.7		0.38	0.24
North Nottinghamshire	7	6.6		6	8.1	~	3	2.0		9	2.2		0.30	0.27
Nottingham	11	7.5		11	10.9		7	2.2		6	1.8		0.29	0.16
Rotherham	8	6.8		8	8.6		2	2.0		7	1.8		0.29	0.21
Sheffield	6	6.2	~	2	6.9	~	9	2.6		1	1.5		0.42	0.23
South Humber	2	4.9	~	10	8.7		8	2.2		11	2.7		0.46	0.31
Southern Derbyshire	5	5.9	~	9	8.6		4	2.0		10	2.4		0.34	0.28
West Midlands		6.6	~		8.5	~		2.2	~		1.8		0.33	0.21
Birmingham	4	5.0		3	6.3	~	2	1.6	~	1	1.2	~	0.32	0.20
Coventry	8	7.1		7	8.1		8	2.3		10	2.2		0.32	0.27
Dudley	1	4.0	~	6	7.7	~	1	1.5	~	6	1.7		0.38	0.22
Herefordshire	12	8.7		13	11.5		13	2.8		8	1.8		0.32	0.15
North Staffordshire	7	6.5	~	5	7.2	~	5	1.9		2	1.4	~	0.29	0.19
Sandwell	2	4.4	~	1	4.6	~	4	1.7	~	4	1.6		0.39	0.34
Shropshire	11	8.4		11	10.9		11	2.7		13	2.6		0.32	0.24
Solihull	6	6.2		8	9.7		9	2.5		11	2.3		0.41	0.24
South Staffordshire	10	7.7		9	10.2		10	2.6		3	1.5		0.34	0.15
Walsall	5	5.4	~	4	6.5	~	7	2.2		7	1.7		0.41	0.27
Warwickshire	9	7.4		12	11.1		12	2.8		12	2.6		0.38	0.23
Wolverhampton	3	4.8	~	2	6.2	~	3	1.6	~	5	1.6		0.34	0.26
Worcestershire	13	9.1	*	10	10.6		6	2.2		9	1.9		0.24	0.18
North West		7.1	~		9.5			2.1	~		1.7	~	0.29	0.18
Bury and Rochdale	7	6.8		8	9.0		4	1.7		14	2.1		0.25	0.24
East Lancashire	4	6.4	~	7	8.9		1	1.5	~	7	1.6		0.23	0.18
Liverpool	5	6.6		6	8.7		7	1.9		9	1.8		0.29	0.21
Manchester	2	5.4	~	1	7.4	~	3	1.7	~	1	1.2	~	0.32	0.16
Morecambe Bay	15	9.3		14	11.5	*	16	3.1		13	2.1		0.33	0.18
North Cheshire	14	8.0		10	9.9		10	2.1		3	1.3	~	0.26	0.13
North West Lancashire	8	6.9		13	11.3	*	8	2.0		16	2.2		0.29	0.19
Salford and Trafford	11	7.4		9	9.5		9	2.0		8	1.7		0.27	0.18
Sefton	10	7.4		12	11.2		14	2.6		12	1.9		0.34	0.17
South Cheshire	13	7.8		11	10.0		11	2.3		10	1.9		0.29	0.19
South Lancashire	12	7.7		16	13.8	*	6	1.9		2	1.3	~	0.24	0.09
St Helen's and Knowsley	6	6.8		4	7.9	~	12	2.4		11	1.9		0.35	0.24
Stockport	16	9.3		15	11.6		15	2.9		15	2.1		0.31	0.19
West Pennine	3	6.3	~	3	7.9	~	2	1.7	~	5	1.5		0.27	0.19
Wigan and Bolton	1	5.2	~	2	7.5	~	5	1.8	~	4	1.3		0.35	0.18
Wirral	9	7.1		5	8.0	~	13	2.5		6	1.5		0.36	0.19
Eastern		8.2			9.9			2.6			2.2	*	0.32	0.22
Bedfordshire	3	7.6		4	10.4		1	1.9		1	2.0		0.25	0.19
Cambridgeshire	5	9.2	*	5	11.0	*	3	2.5		4	2.1		0.27	0.19
Hertfordshire	1	6.6	~	1	7.6	~	5	2.8		7	2.5	*	0.43	0.33
Norfolk	6	9.3	*	6	11.1	*	4	2.5		5	2.3		0.27	0.20
North Essex	4	7.6		3	9.6		7	3.0		3	2.0		0.40	0.21
South Essex	2	7.6		2	8.1	~	6	2.9		6	2.4		0.38	0.30
Suffolk	7	9.7	*	7	13.0	*	2	2.5		2	2.0		0.25	0.16

1 England, Wales and Scotland 1991-99; Northern Ireland 1993-99; Ireland 1994-99.
2 England, Wales and Northern Ireland 1991-2000; Scotland 1991-99; Ireland 1994-2000.
3 Mortality to incidence ratio; age-standardised mortality rate divided by age-standardised incidence rate.
4 Rank within region or country; 1 is lowest.
5 Directly age standardised using the European standard population; per 100,000 population.
6 Rate significantly different from that for UK and Ireland; * significantly higher; ~ significantly lower. See Appendix H.

Country/ region/ health authority	Incidence[1]						Mortality[2]						M:I Ratio[3]	
	Rank[4]	Males Rate[5]	Signif[6]	Rank	Females Rate	Signif	Rank	Males Rate	Signif	Rank	Females Rate	Signif	Males	Females
London		**5.7**	~		**6.3**	~		**2.4**			**1.8**		**0.42**	**0.29**
Barking and Havering	10	6.1	~	11	7.9	~	7	2.3		8	1.9		0.38	0.24
Barnet, Enfield and Haringey	7	5.7	~	7	5.5	~	5	2.3		3	1.5	~	0.41	0.26
Bexley, Bromley and Greenwich	12	7.4		12	8.6		10	2.7		12	2.2		0.36	0.25
Brent and Harrow	2	4.2	~	4	5.3	~	8	2.4		10	2.0		0.57	0.39
Camden and Islington	13	7.5		8	6.2	~	9	2.5		9	2.0		0.33	0.33
Croydon	8	6.1	~	10	7.1	~	12	3.1		13	2.3		0.50	0.33
Ealing, Hammersmith and Hounslow	4	4.6	~	2	4.2	~	4	2.1		6	1.7		0.46	0.41
East London and The City	1	2.9	~	1	2.8	~	1	1.7	~	1	1.4	~	0.58	0.49
Hillingdon	3	4.2	~	3	4.6	~	14	3.2		7	1.8		0.76	0.39
Kensington and Chelsea, and Westminster	9	6.1	~	9	6.2	~	13	3.2		11	2.1		0.52	0.34
Kingston and Richmond	14	8.0		14	9.1		11	2.8		4	1.5		0.34	0.16
Lambeth, Southwark and Lewisham	6	5.0	~	6	5.5	~	3	2.0		2	1.4	~	0.41	0.26
Merton, Sutton and Wandsworth	11	6.7		13	8.9		6	2.3		14	2.4		0.35	0.27
Redbridge and Waltham Forest	5	4.9	~	5	5.3	~	2	1.8		5	1.6		0.38	0.29
South East		**9.0**	*		**11.0**	*		**3.1**	*		**2.3**	*	**0.34**	**0.21**
Berkshire	9	10.0	*	9	12.3	*	7	3.2	*	5	2.1		0.32	0.17
Buckinghamshire	10	10.5	*	11	13.3	*	5	3.0		7	2.2		0.28	0.17
East Kent	1	5.6	~	1	6.7	~	2	2.3		10	2.5		0.42	0.37
East Surrey	3	7.8		6	10.8		4	2.9		4	2.1		0.38	0.19
East Sussex, Brighton and Hove	6	8.1		5	10.1		10	3.3	*	13	2.9	*	0.41	0.28
Isle of Wight, Portsmouth and South East Hampshire	12	12.2	*	12	14.0	*	12	3.6	*	12	2.6	*	0.30	0.18
North and Mid-Hampshire	8	9.0		8	12.1	*	3	2.5		2	2.0		0.28	0.16
Northamptonshire	4	7.9		7	11.4	*	1	2.0		1	1.8		0.25	0.16
Oxfordshire	11	10.6	*	10	12.9	*	8	3.2	*	8	2.4		0.30	0.18
Southampton and South West Hampshire	13	12.8	*	13	15.6	*	13	3.8	*	3	2.0		0.30	0.13
West Kent	2	7.4		2	7.4	~	6	3.1	*	6	2.1		0.42	0.29
West Surrey	7	8.4		3	9.3		11	3.5	*	9	2.5		0.42	0.27
West Sussex	5	8.0		4	9.8		9	3.3	*	11	2.5	*	0.41	0.26
South West		**11.1**	*		**13.8**	*		**3.3**	*		**2.6**	*	**0.29**	**0.19**
Avon	1	8.9	*	2	12.3	*	3	2.9		3	2.4		0.33	0.19
Cornwall and Isles of Scilly	6	12.8	*	8	16.7	*	6	3.7	*	7	3.1	*	0.28	0.18
Dorset	8	13.1	*	7	15.6	*	7	3.7	*	5	2.7	*	0.28	0.17
Gloucestershire	2	8.9	*	1	9.2		1	2.8		1	2.2		0.31	0.24
North and East Devon	4	11.2	*	6	14.9	*	4	3.1		8	3.2	*	0.28	0.22
Somerset	3	10.1	*	5	14.6	*	2	2.9		4	2.4		0.29	0.17
South and West Devon	7	12.9	*	4	14.4	*	8	3.8	*	6	2.9	*	0.29	0.20
Wiltshire	5	11.5	*	3	13.8	*	5	3.1		2	2.3		0.27	0.17
Wales		**6.3**	~		**7.0**	~		**2.7**			**1.8**		**0.42**	**0.26**
Bro Taf	5	7.2		4	7.3	~	3	2.7		4	1.9		0.38	0.26
Dyfed Powys	1	5.5	~	2	6.2	~	4	2.7		2	1.8		0.50	0.29
Gwent	2	5.8	~	1	5.7	~	2	2.5		1	1.5		0.44	0.27
Morgannwg	4	7.0		3	7.3	~	5	3.1		5	2.0		0.45	0.28
North Wales	3	6.1	~	5	8.2	~	1	2.3		3	1.8		0.37	0.22
Scotland		**9.6**	*		**11.9**	*		**2.4**			**1.6**	~	**0.25**	**0.13**
Argyll and Clyde	9	9.4	*	9	13.0	*	10	2.4		11	1.7		0.25	0.13
Ayrshire and Arran	10	9.7	*	11	13.3	*	9	2.4		12	1.8		0.25	0.13
Borders	15	14.0	*	15	16.7	*	14	3.0		10	1.7		0.22	0.10
Dumfries and Galloway	14	13.4	*	14	16.1	*	13	2.9		9	1.6		0.22	0.10
Fife	8	9.0		12	13.3	*	3	1.9		7	1.4		0.21	0.10
Forth Valley	7	8.9		7	10.8		8	2.4		15	2.1		0.27	0.19
Grampian	11	9.7	*	8	11.6	*	11	2.6		5	1.3	~	0.27	0.11
Greater Glasgow	6	8.8	*	5	9.6		6	2.3		14	1.9		0.27	0.20
Highland	3	8.0		6	9.9		4	2.1		4	1.0	~	0.26	0.10
Lanarkshire	2	7.8		4	9.2		7	2.4		8	1.6		0.30	0.17
Lothian	13	10.8	*	13	14.0	*	5	2.2		6	1.4	~	0.21	0.10
Orkney	5	8.7		2	7.6		1	0.9		1	0.0	~	0.10	0.00
Shetland	4	8.2		1	6.9		2	1.9		2	0.0	~	0.24	0.00
Tayside	12	10.2	*	10	13.2	*	12	2.7		13	1.9		0.27	0.15
Western Isles	1	6.9		3	8.0		15	3.8		3	0.8		0.55	0.10
Northern Ireland		**8.6**			**12.0**	*		**1.7**	~		**1.5**	~	**0.19**	**0.13**
Eastern	3	9.2	*	3	12.1	*	2	1.5	~	1	1.2	~	0.17	0.10
Northern	2	8.3		2	11.0	*	1	1.5		3	1.8		0.19	0.17
Southern	4	9.2		4	16.4	*	3	1.7	~	4	1.9		0.19	0.12
Western	1	6.4		1	8.9		4	2.2		2	1.5		0.34	0.17
Ireland		**9.3**	*		**13.3**	*		**2.0**	~		**1.8**		**0.21**	**0.13**
Eastern	8	11.1	*	6	14.0	*	8	2.7		8	2.3		0.24	0.16
Mid-Western	4	8.3		2	11.6		3	1.6	~	5	1.5		0.19	0.13
Midland	6	9.2		4	12.5		5	1.8		1	1.1	~	0.20	0.09
North Eastern	5	8.9		5	13.3	*	1	1.4	~	3	1.5		0.16	0.11
North Western	2	7.2		1	10.3		6	1.9		2	1.2		0.27	0.12
South Eastern	3	7.5		7	14.2	*	4	1.6		6	1.6		0.21	0.11
Southern	7	10.3	*	8	14.7	*	2	1.5	~	4	1.5		0.14	0.10
Western	1	7.1		3	11.8		7	2.0		7	1.7		0.29	0.15

Table **B14.2**

Melanoma of skin: annual average numbers of registrations and deaths by health authority within country and region of England, UK and Ireland, 1991–2000

Country/ region/ health authority	Number of registrations[1]		Number of deaths[2]	
	Males	Females	Males	Females
UK and Ireland	**2,410**	**3,490**	**779**	**795**
England	**1,860**	**2,660**	**634**	**653**
Northern and Yorkshire	**203**	**305**	**69**	**72**
Bradford	14	25	4	5
Calderdale and Kirklees	16	25	5	7
County Durham	20	29	7	7
East Riding	20	27	7	6
Gateshead and South Tyneside	8	12	3	3
Leeds	24	38	8	8
Newcastle and North Tyneside	11	20	5	4
North Cumbria	18	21	5	5
North Yorkshire	36	51	10	9
Northumberland	7	16	3	4
Sunderland	7	11	3	3
Tees	10	18	5	6
Wakefield	10	14	4	4
Trent	**165**	**235**	**60**	**66**
Barnsley	6	10	2	2
Doncaster	6	9	2	3
Leicestershire	26	38	10	10
Lincolnshire	25	31	10	9
North Derbyshire	14	15	6	5
North Nottinghamshire	13	18	4	6
Nottingham	24	38	7	8
Rotherham	8	12	2	3
Sheffield	17	21	7	6
South Humber	8	15	4	5
Southern Derbyshire	17	27	6	9
West Midlands	**176**	**256**	**59**	**60**
Birmingham	24	34	8	8
Coventry	11	13	3	4
Dudley	6	14	3	3
Herefordshire	8	12	3	2
North Staffordshire	15	19	5	4
Sandwell	7	8	3	3
Shropshire	18	27	6	7
Solihull	7	12	3	3
South Staffordshire	23	34	8	5
Walsall	7	9	3	3
Warwickshire	19	32	8	8
Wolverhampton	6	9	2	3
Worcestershire	25	33	6	7
North West	**231**	**356**	**69**	**74**
Bury and Rochdale	13	19	3	5
East Lancashire	16	27	4	6
Liverpool	15	24	4	5
Manchester	10	16	3	3
Morecambe Bay	15	22	5	5
North Cheshire	12	17	3	3
North West Lancashire	17	31	5	7
Salford and Trafford	16	25	5	5
Sefton	11	20	4	4
South Cheshire	27	39	8	8
South Lancashire	12	24	3	3
St Helen's and Knowsley	11	14	4	4
Stockport	14	19	4	4
West Pennine	14	21	4	4
Wigan and Bolton	14	23	5	5
Wirral	12	15	4	4
Eastern	**223**	**298**	**74**	**76**
Bedfordshire	20	29	5	6
Cambridgeshire	32	42	9	9
Hertfordshire	34	44	15	15
Norfolk	42	52	12	14
North Essex	35	49	14	12
South Essex	27	33	11	11
Suffolk	33	50	9	9

1 England, Wales and Scotland 1991-99; Northern Ireland 1993-99; Ireland 1994-99.
2 England, Wales and Northern Ireland 1991-2000; Scotland 1991-99; Ireland 1994-2000.

Country/ region/ health authority	Number of registrations[1]		Number of deaths[2]	
	Males	Females	Males	Females
London	**182**	**250**	**77**	**77**
Barking and Havering	12	18	5	5
Barnet, Enfield and Haringey	21	26	9	8
Bexley, Bromley and Greenwich	27	37	10	11
Brent and Harrow	9	14	5	5
Camden and Islington	12	13	4	4
Croydon	9	13	5	5
Ealing, Hammersmith and Hounslow	14	16	6	7
East London and The City	7	9	4	4
Hillingdon	5	6	4	3
Kensington and Chelsea, and Westminster	10	12	5	4
Kingston and Richmond	12	17	4	3
Lambeth, Southwark and Lewisham	16	23	7	6
Merton, Sutton and Wandsworth	19	32	7	9
Redbridge and Waltham Forest	10	14	4	4
South East	**392**	**546**	**137**	**134**
Berkshire	37	50	12	9
Buckinghamshire	34	47	10	9
East Kent	18	25	8	11
East Surrey	18	28	7	7
East Sussex, Brighton and Hove	33	51	15	19
Isle of Wight, Portsmouth and South East Hampshire	42	57	13	12
North and Mid-Hampshire	24	37	7	7
Northamptonshire	23	38	6	7
Oxfordshire	31	41	10	8
Southampton and South West Hampshire	36	49	11	8
West Kent	36	41	15	13
West Surrey	27	35	12	10
West Sussex	33	49	14	16
South West	**290**	**410**	**89**	**95**
Avon	44	71	15	16
Cornwall and Isles of Scilly	36	52	11	12
Dorset	52	73	16	16
Gloucestershire	26	29	8	9
North and East Devon	30	44	9	12
Somerset	28	42	9	9
South and West Devon	41	53	13	13
Wiltshire	34	46	10	9
Wales	**98**	**124**	**42**	**37**
Bro Taf	25	31	10	10
Dyfed Powys	15	19	8	6
Gwent	17	19	7	6
Morgannwg	18	22	9	7
North Wales	22	33	8	9
Scotland	**241**	**362**	**60**	**56**
Argyll and Clyde	20	33	5	5
Ayrshire and Arran	19	31	5	5
Borders	8	11	2	1
Dumfries and Galloway	11	15	2	2
Fife	16	27	3	3
Forth Valley	12	17	3	4
Grampian	25	35	7	5
Greater Glasgow	38	54	10	12
Highland	9	12	2	1
Lanarkshire	20	30	6	5
Lothian	40	61	8	7
Orkney	1	1	0	0
Shetland	1	1	0	0
Tayside	21	32	6	5
Western Isles	1	1	1	0
Northern Ireland	**64**	**107**	**12**	**15**
Eastern	29	46	5	6
Northern	16	25	3	4
Southern	12	24	2	3
Western	7	12	2	2
Ireland	**146**	**239**	**31**	**33**
Eastern	55	88	13	15
Mid-Western	12	18	2	3
Midland	9	13	2	1
North Eastern	12	20	2	2
North Western	7	11	2	1
South Eastern	13	27	3	3
Southern	26	41	4	5
Western	12	20	4	3

Table B15.1

Multiple myeloma: annual average age-standardised incidence and mortality by health authority within country and region of England, UK and Ireland, 1991–2000

Country/ region/ health authority	Incidence[1]						Mortality[2]						M:I Ratio[3]	
		Males			Females			Males			Females		Males	Females
	Rank[4]	Rate[5]	Signif[6]	Rank	Rate	Signif	Rank	Rate	Signif	Rank	Rate	Signif		
UK and Ireland		5.5			3.7			3.7			2.6		0.68	0.70
England		5.4			3.6			3.7			2.6		0.68	0.71
Northern and Yorkshire		5.2			3.5			3.6			2.5		0.69	0.70
Bradford	5	5.0		9	3.5		3	3.3		9	2.6		0.65	0.73
Calderdale and Kirklees	3	4.8		2	2.9	~	5	3.4		3	2.2		0.72	0.76
County Durham	4	4.9		8	3.5		10	3.8		8	2.5		0.77	0.73
East Riding	10	5.5		13	4.6	*	12	3.9		12	2.9		0.72	0.62
Gateshead and South Tyneside	7	5.2		4	3.1		4	3.4		7	2.5		0.65	0.82
Leeds	13	6.1		11	3.8		8	3.6		5	2.4		0.58	0.63
Newcastle and North Tyneside	8	5.3		10	3.5		7	3.6		11	2.7		0.68	0.78
North Cumbria	9	5.5		6	3.3		13	4.3		13	3.0		0.78	0.90
North Yorkshire	6	5.0		12	4.1		9	3.6		6	2.5		0.71	0.61
Northumberland	2	4.7		1	2.7	~	1	3.1		4	2.3		0.67	0.84
Sunderland	11	5.6		3	3.0		6	3.4		10	2.6		0.61	0.86
Tees	1	4.3	~	5	3.2		2	3.2		2	2.1		0.74	0.67
Wakefield	12	6.1		7	3.4		11	3.8		1	2.0		0.63	0.59
Trent		5.5			3.7			3.8			2.7		0.68	0.73
Barnsley	9	6.4		10	4.1		11	4.5		10	3.0		0.71	0.72
Doncaster	11	6.6		7	3.9		9	4.0		9	3.0		0.61	0.77
Leicestershire	6	5.4		5	3.6		6	3.7		2	2.5		0.68	0.68
Lincolnshire	8	5.5		1	3.1		5	3.6		1	2.4		0.67	0.78
North Derbyshire	1	4.7		8	3.9		2	3.5		8	2.8		0.74	0.73
North Nottinghamshire	7	5.4		6	3.6		3	3.5		4	2.6		0.65	0.73
Nottingham	10	6.4		11	4.6	*	4	3.5		11	3.1		0.55	0.66
Rotherham	3	5.1		4	3.6		1	2.7		3	2.6		0.54	0.71
Sheffield	5	5.3		2	3.5		10	4.4		7	2.8		0.83	0.80
South Humber	2	4.9		9	4.1		7	3.8		6	2.7		0.79	0.67
Southern Derbyshire	4	5.3		3	3.5		8	4.0		5	2.7		0.75	0.77
West Midlands		5.4			3.6			3.9			2.7		0.72	0.73
Birmingham	10	5.6		9	3.8		11	4.2		7	2.7		0.75	0.71
Coventry	3	4.7		2	2.5	~	9	4.1		4	2.3		0.86	0.94
Dudley	7	5.2		8	3.8		8	4.0		9	2.7		0.78	0.71
Herefordshire	2	4.6		5	3.4		6	3.7		11	2.7		0.81	0.82
North Staffordshire	4	4.7		6	3.5		2	3.4		10	2.7		0.71	0.78
Sandwell	12	6.3		13	4.4		13	4.9		6	2.7		0.77	0.61
Shropshire	13	6.5		10	4.0		10	4.2		3	2.3		0.65	0.58
Solihull	6	5.0		4	3.3		7	3.8		2	2.3		0.76	0.70
South Staffordshire	9	5.6		11	4.1		5	3.7		12	2.8		0.66	0.69
Walsall	5	4.9		1	2.4	~	3	3.4		1	2.0		0.71	0.85
Warwickshire	11	6.3		12	4.2		12	4.4		13	3.4	*	0.69	0.81
Wolverhampton	8	5.4		3	3.2		4	3.5		5	2.6		0.66	0.80
Worcestershire	1	4.3	~	7	3.5		1	3.1		8	2.7		0.71	0.75
North West		4.9	~		3.3	~		3.4	~		2.4		0.69	0.73
Bury and Rochdale	8	4.7		7	3.1		3	3.0		7	2.3		0.65	0.74
East Lancashire	16	6.1		13	3.7		14	3.7		13	2.6		0.61	0.71
Liverpool	7	4.6		9	3.6		11	3.6		16	3.1		0.78	0.87
Manchester	15	6.0		6	3.0		16	4.1		4	2.1		0.68	0.68
Morecambe Bay	12	5.5		11	3.6		10	3.5		11	2.6		0.62	0.71
North Cheshire	5	4.3		8	3.2		12	3.6		15	3.0		0.83	0.92
North West Lancashire	10	5.2		16	3.9		15	3.8		14	2.7		0.72	0.68
Salford and Trafford	9	5.1		12	3.7		4	3.1		6	2.3		0.61	0.62
Sefton	6	4.4		3	2.8	~	9	3.4		3	2.0		0.77	0.71
South Cheshire	3	4.2	~	4	2.9	~	7	3.2		12	2.6		0.77	0.91
South Lancashire	11	5.4		5	3.0		6	3.2		5	2.3		0.59	0.76
St Helen's and Knowsley	2	3.9	~	2	2.4	~	2	2.9		1	1.4	~	0.74	0.59
Stockport	13	5.6		15	3.8		8	3.3		8	2.4		0.59	0.64
West Pennine	4	4.3	~	10	3.6		5	3.1		9	2.5		0.73	0.70
Wigan and Bolton	14	5.9		14	3.7		13	3.6		10	2.5		0.62	0.68
Wirral	1	3.1	~	1	1.9	~	1	2.9	~	2	1.7	~	0.92	0.88
Eastern		5.2			3.5			3.8			2.6		0.73	0.74
Bedfordshire	4	5.3		7	3.8		7	4.4		7	2.9		0.83	0.78
Cambridgeshire	7	5.9		1	3.2		6	4.2		2	2.3		0.72	0.69
Hertfordshire	2	4.9		6	3.7		5	3.8		4	2.6		0.79	0.70
Norfolk	5	5.4		5	3.6		4	3.7		6	2.8		0.69	0.79
North Essex	1	4.4	~	4	3.5		1	3.5		5	2.8		0.78	0.82
South Essex	3	5.1		2	3.4		2	3.5		1	2.2		0.68	0.64
Suffolk	6	5.5		3	3.4		3	3.5		3	2.5		0.64	0.73

1 England, Wales and Scotland 1991-99; Northern Ireland 1993-99; Ireland 1994-99.
2 England, Wales and Northern Ireland 1991-2000; Scotland 1991-99; Ireland 1994-2000.
3 Mortality to incidence ratio; age-standardised mortality rate divided by age-standardised incidence rate.
4 Rank within region or country; 1 is lowest.
5 Directly age standardised using the European standard population; per 100,000 population.
6 Rate significantly different from that for UK and Ireland; * significantly higher; ~ significantly lower. See Appendix H.

Country/ region/ health authority	Incidence[1]						Mortality[2]						M:I Ratio[3]	
	Males			Females			Males			Females			Males	Females
	Rank[4]	Rate[5]	Signif[6]	Rank	Rate	Signif	Rank	Rate	Signif	Rank	Rate	Signif		
London		**5.4**			**3.7**			**3.7**			**2.6**		**0.68**	**0.71**
Barking and Havering	11	6.0		11	3.9		13	4.6		9	2.7		0.78	0.70
Barnet, Enfield and Haringey	6	5.1		5	3.5		3	3.4		3	2.4		0.67	0.68
Bexley, Bromley and Greenwich	5	5.1		3	3.4		1	3.1	~	7	2.6		0.61	0.76
Brent and Harrow	7	5.4		13	4.1		9	3.7		14	3.0		0.68	0.72
Camden and Islington	9	5.5		10	3.8		6	3.6		8	2.7		0.65	0.71
Croydon	13	6.3		6	3.5		14	4.8		4	2.4		0.76	0.69
Ealing, Hammersmith and Hounslow	2	4.4	~	4	3.4		2	3.2		6	2.5		0.74	0.74
East London and The City	8	5.5		14	4.3		10	3.8		13	3.0		0.69	0.70
Hillingdon	4	4.9		2	3.4		5	3.4		11	2.9		0.70	0.84
Kensington and Chelsea, and Westminster	3	4.9		7	3.6		7	3.6		10	2.7		0.75	0.77
Kingston and Richmond	1	4.3		1	2.9		4	3.4		2	2.2		0.79	0.78
Lambeth, Southwark and Lewisham	12	6.2		12	3.9		11	3.9		1	2.2		0.63	0.57
Merton, Sutton and Wandsworth	14	6.5		8	3.6		12	4.3		12	2.9		0.66	0.79
Redbridge and Waltham Forest	10	5.7		9	3.8		8	3.6		5	2.5		0.64	0.66
South East		**5.5**			**3.7**			**3.8**			**2.6**		**0.68**	**0.69**
Berkshire	6	5.1		9	4.2		5	3.5		11	2.8		0.68	0.68
Buckinghamshire	13	7.6	*	10	4.3		12	4.3		2	2.3		0.56	0.54
East Kent	1	4.8		3	3.3		2	3.3		6	2.5		0.69	0.76
East Surrey	8	5.5		7	3.7		1	3.0		4	2.5		0.55	0.66
East Sussex, Brighton and Hove	5	5.1		4	3.3		6	3.7		8	2.6		0.74	0.79
Isle of Wight, Portsmouth and South East Hampshire	7	5.2		1	3.0	~	8	3.9		1	2.2		0.75	0.73
North and Mid-Hampshire	11	6.4		6	3.6		13	4.7	*	5	2.5		0.73	0.69
Northamptonshire	10	6.3		11	4.3		11	4.2		12	2.9		0.67	0.66
Oxfordshire	9	6.1		13	4.6	*	10	4.1		9	2.7		0.68	0.59
Southampton and South West Hampshire	12	6.7	*	12	4.5		9	3.9		13	3.0		0.59	0.68
West Kent	2	4.8		2	3.2		3	3.5		3	2.4		0.72	0.74
West Surrey	4	5.0		5	3.5		7	3.8		10	2.8		0.76	0.82
West Sussex	3	4.9		8	3.8		4	3.5		7	2.6		0.70	0.68
South West		**5.8**			**4.0**	*		**3.6**			**2.6**		**0.62**	**0.65**
Avon	7	6.1		5	4.2		5	3.6		4	2.6		0.59	0.61
Cornwall and Isles of Scilly	1	5.3		8	5.0	*	1	3.0	~	8	2.9		0.57	0.58
Dorset	6	5.9		6	4.2		3	3.5		2	2.4		0.60	0.58
Gloucestershire	2	5.5		4	3.8		4	3.6		5	2.6		0.65	0.70
North and East Devon	3	5.8		7	4.3		6	3.6		7	2.8		0.63	0.66
Somerset	8	6.2		2	3.7		8	4.2		3	2.6		0.69	0.70
South and West Devon	5	5.8		1	3.4		7	4.0		1	2.4		0.69	0.72
Wiltshire	4	5.8		3	3.7		2	3.3		6	2.7		0.57	0.72
Wales		**5.6**			**3.9**			**3.7**			**2.5**		**0.65**	**0.65**
Bro Taf	5	6.5	*	5	4.4		5	4.4		4	2.6		0.69	0.60
Dyfed Powys	2	5.1		3	3.7		3	3.5		2	2.4		0.68	0.64
Gwent	1	5.0		1	3.7		1	3.2		5	2.9		0.66	0.77
Morgannwg	4	6.0		2	3.7		4	3.8		3	2.5		0.63	0.68
North Wales	3	5.5		4	4.0		2	3.4		1	2.4		0.62	0.59
Scotland		**5.5**			**4.1**	*		**3.5**			**2.5**		**0.64**	**0.62**
Argyll and Clyde	11	6.6		10	4.3		11	4.0		10	2.7		0.61	0.63
Ayrshire and Arran	3	4.2	~	12	4.6		1	2.3	~	7	2.5		0.55	0.54
Borders	2	4.1		7	3.9		4	3.3		13	3.0		0.79	0.77
Dumfries and Galloway	13	6.7		15	5.4	*	9	3.8		3	2.1		0.56	0.39
Fife	6	5.1		3	3.2		13	4.2		4	2.3		0.82	0.72
Forth Valley	8	5.5		5	3.5		2	3.1		5	2.4		0.56	0.68
Grampian	10	6.3		14	5.4	*	10	4.0		14	3.0		0.63	0.56
Greater Glasgow	7	5.4		4	3.4		6	3.4		2	2.0	~	0.62	0.59
Highland	12	6.7		2	2.8		12	4.2		12	3.0		0.63	1.07
Lanarkshire	9	6.1		11	4.4		8	3.7		11	2.8		0.60	0.64
Lothian	5	5.1		8	4.2		5	3.3		9	2.6		0.64	0.62
Orkney	1	3.2		13	5.1		3	3.2		15	4.6		1.00	0.90
Shetland	15	8.1		1	2.5		15	7.0		1	0.7	~	0.86	0.29
Tayside	4	4.8		6	3.6		7	3.6		8	2.5		0.75	0.68
Western Isles	14	7.1		9	4.3		14	5.6		6	2.4		0.80	0.56
Northern Ireland		**6.6**	*		**4.1**			**4.3**	*		**2.7**		**0.66**	**0.65**
Eastern	2	6.2		3	4.2		1	4.2		1	2.3		0.67	0.54
Northern	3	6.8		2	4.0		2	4.3		4	3.2		0.63	0.79
Southern	4	7.7	*	4	4.7		4	4.6		2	2.8		0.60	0.59
Western	1	6.2		1	3.4		3	4.6		3	3.0		0.75	0.88
Ireland		**6.2**	*		**3.9**			**5.0**	*		**3.2**	*	**0.80**	**0.82**
Eastern	3	6.1		3	3.4		3	4.5		5	3.1		0.74	0.90
Mid-Western	4	6.3		4	3.8		2	4.4		6	3.3		0.69	0.88
Midland	5	6.6		6	4.2		8	6.2		7	3.4		0.95	0.81
North Eastern	8	7.4		5	4.0		5	5.4	*	2	2.9		0.73	0.74
North Western	6	6.8		7	4.4		7	5.8	*	4	3.1		0.86	0.69
South Eastern	1	4.5		1	3.1		4	5.0		3	3.0		1.13	0.96
Southern	7	7.4	*	8	5.9	*	6	5.7	*	8	4.1	*	0.77	0.70
Western	2	5.7		2	3.2		1	4.2		1	2.6		0.74	0.80

Table B15.2

Multiple myeloma: annual average numbers of registrations and deaths by health authority within country and region of England, UK and Ireland, 1991–2000

Country/ region/ health authority	Number of registrations[1]		Number of deaths[2]	
	Males	Females	Males	Females
UK and Ireland	**1,810**	**1,700**	**1,270**	**1,260**
England	**1,420**	**1,330**	**1,000**	**1,000**
Northern and Yorkshire	**177**	**169**	**124**	**126**
Bradford	11	12	7	9
Calderdale and Kirklees	14	13	10	10
County Durham	16	17	13	13
East Riding	17	20	13	13
Gateshead and South Tyneside	11	9	7	8
Leeds	23	20	13	14
Newcastle and North Tyneside	14	13	9	11
North Cumbria	10	8	8	8
North Yorkshire	22	25	16	17
Northumberland	9	7	6	6
Sunderland	9	7	5	6
Tees	12	12	9	8
Wakefield	10	7	6	5
Trent	**156**	**142**	**109**	**107**
Barnsley	8	7	6	5
Doncaster	10	7	6	6
Leicestershire	26	23	18	17
Lincolnshire	21	16	15	13
North Derbyshire	10	11	8	9
North Nottinghamshire	12	11	8	8
Nottingham	21	20	13	14
Rotherham	7	7	4	5
Sheffield	15	14	13	12
South Humber	8	9	7	7
Southern Derbyshire	16	15	13	12
West Midlands	**154**	**138**	**113**	**108**
Birmingham	28	25	22	19
Coventry	8	6	7	5
Dudley	9	8	7	6
Herefordshire	5	5	4	5
North Staffordshire	12	12	9	10
Sandwell	10	10	8	7
Shropshire	15	12	10	7
Solihull	6	5	4	3
South Staffordshire	17	16	11	12
Walsall	6	4	5	4
Warwickshire	17	15	12	13
Wolverhampton	7	6	5	5
Worcestershire	13	14	9	12
North West	**170**	**163**	**120**	**126**
Bury and Rochdale	9	9	6	7
East Lancashire	16	14	10	10
Liverpool	10	13	8	11
Manchester	12	8	8	6
Morecambe Bay	11	9	7	7
North Cheshire	6	7	5	6
North West Lancashire	15	16	11	12
Salford and Trafford	12	13	8	8
Sefton	8	8	6	6
South Cheshire	16	14	12	14
South Lancashire	8	6	5	5
St Helen's and Knowsley	6	5	5	3
Stockport	9	8	5	6
West Pennine	10	12	8	9
Wigan and Bolton	16	16	10	11
Wirral	6	6	6	5
Eastern	**152**	**140**	**114**	**111**
Bedfordshire	13	13	12	10
Cambridgeshire	21	15	16	11
Hertfordshire	26	26	21	20
Norfolk	28	25	20	21
North Essex	23	24	18	21
South Essex	19	19	13	13
Suffolk	22	19	15	14

1 *Englandg, Wales and Scotland 1991-99; Northern Ireland 1993-99; Ireland 1994-99.*
2 *England, Wales and Northern Ireland 1991-2000; Scotland 1991-99; Ireland 1994-2000.*

Country/ region/ health authority	Number of registrations[1]		Number of deaths[2]	
	Males	Females	Males	Females
London	**175**	**163**	**122**	**122**
Barking and Havering	12	11	10	8
Barnet, Enfield and Haringey	19	18	13	13
Bexley, Bromley and Greenwich	19	18	12	15
Brent and Harrow	12	12	8	9
Camden and Islington	9	8	6	6
Croydon	9	7	7	5
Ealing, Hammersmith and Hounslow	13	13	10	9
East London and The City	13	13	9	10
Hillingdon	6	6	4	5
Kensington and Chelsea, and Westminster	8	7	6	6
Kingston and Richmond	7	7	6	6
Lambeth, Southwark and Lewisham	19	16	12	10
Merton, Sutton and Wandsworth	18	15	12	12
Redbridge and Waltham Forest	12	12	8	8
South East	**259**	**245**	**181**	**183**
Berkshire	19	19	13	14
Buckinghamshire	24	18	14	11
East Kent	17	19	13	15
East Surrey	14	12	8	9
East Sussex, Brighton and Hove	25	25	20	21
Isle of Wight, Portsmouth and South East Hampshire	20	17	15	14
North and Mid-Hampshire	17	14	12	10
Northamptonshire	19	17	13	12
Oxfordshire	18	19	13	12
Southampton and South West Hampshire	20	19	13	14
West Kent	24	22	18	17
West Surrey	17	17	13	14
West Sussex	24	27	18	21
South West	**176**	**168**	**114**	**118**
Avon	33	32	20	21
Cornwall and Isles of Scilly	17	21	10	13
Dorset	29	27	19	18
Gloucestershire	18	17	12	13
North and East Devon	19	21	13	15
Somerset	19	16	14	12
South and West Devon	22	19	16	15
Wiltshire	18	16	11	12
Wales	**94**	**92**	**64**	**64**
Bro Taf	24	23	17	15
Dyfed Powys	16	15	11	11
Gwent	15	16	10	13
Morgannwg	18	15	12	11
North Wales	21	22	14	14
Scotland	**145**	**156**	**95**	**103**
Argyll and Clyde	14	14	9	9
Ayrshire and Arran	8	13	5	8
Borders	3	4	2	3
Dumfries and Galloway	6	7	3	3
Fife	10	9	8	6
Forth Valley	8	7	4	5
Grampian	17	19	11	12
Greater Glasgow	24	25	16	15
Highland	7	5	5	5
Lanarkshire	16	16	10	11
Lothian	19	23	13	14
Orkney	0	1	0	1
Shetland	1	0	1	0
Tayside	11	13	8	9
Western Isles	1	1	1	1
Northern Ireland	**50**	**43**	**33**	**29**
Eastern	20	20	14	12
Northern	13	10	8	8
Southern	10	8	6	5
Western	7	5	5	4
Ireland	**98**	**78**	**79**	**67**
Eastern	28	22	20	20
Mid-Western	9	6	6	6
Midland	7	5	6	4
North Eastern	10	7	7	5
North Western	8	6	7	5
South Eastern	8	7	9	7
Southern	19	18	15	13
Western	11	8	9	6

Table **B16.1**

Non-Hodgkin's lymphoma: annual average age-standardised incidence and mortality by health authority within country and region of England, UK and Ireland, 1991–2000

Country/ region/ health authority	Incidence[1]						Mortality[2]						M:I Ratio[3]	
	Males			Females			Males			Females			Males	Females
	Rank[4]	Rate[5]	Signif[6]	Rank	Rate	Signif	Rank	Rate	Signif	Rank	Rate	Signif		
UK and Ireland		14.2			9.9			7.3			4.8		0.51	0.49
England		14.1			9.7	~		7.3			4.8		0.52	0.49
Northern and Yorkshire		12.2	~		8.8	~		6.7	~		4.6		0.55	0.52
Bradford	11	13.1		9	9.0		4	6.3		5	4.5		0.48	0.50
Calderdale and Kirklees	3	11.4	~	1	7.2	~	2	6.0	~	1	3.3	~	0.53	0.46
County Durham	13	13.4		8	8.9		3	6.2	~	7	4.6		0.47	0.52
East Riding	8	12.2	~	11	9.5		9	7.0		9	4.8		0.57	0.50
Gateshead and South Tyneside	2	11.0	~	2	7.7	~	5	6.4		3	4.4		0.59	0.57
Leeds	10	12.9		13	10.0		6	6.5		6	4.6		0.50	0.46
Newcastle and North Tyneside	7	11.9	~	12	9.6		13	8.0		12	5.3		0.67	0.55
North Cumbria	6	11.8	~	4	8.3	~	11	7.5		10	4.8		0.64	0.58
North Yorkshire	12	13.2		7	8.9	~	7	6.6		8	4.7		0.50	0.53
Northumberland	9	12.5		5	8.5		10	7.4		11	5.1		0.60	0.60
Sunderland	4	11.4	~	3	8.3	~	12	7.6		13	5.7		0.67	0.69
Tees	1	10.6	~	10	9.2		1	6.0	~	4	4.4		0.57	0.48
Wakefield	5	11.6	~	6	8.5		8	6.8		2	4.1		0.59	0.48
Trent		13.1	~		9.0	~		6.8	~		4.5	~	0.52	0.50
Barnsley	3	12.3		10	10.1		3	6.4		1	4.0		0.52	0.40
Doncaster	9	14.2		9	10.0		5	6.6		3	4.3		0.46	0.43
Leicestershire	2	11.5	~	5	8.9	~	6	6.6		7	4.6		0.57	0.52
Lincolnshire	6	13.2		2	8.5	~	4	6.5		4	4.4		0.49	0.51
North Derbyshire	7	13.6		11	10.2		1	6.1	~	6	4.6		0.45	0.45
North Nottinghamshire	8	13.6		4	8.8		11	8.8	*	8	4.6		0.65	0.53
Nottingham	11	15.7		8	9.8		10	7.7		5	4.5		0.49	0.46
Rotherham	10	14.6		3	8.7		7	6.7		11	4.9		0.46	0.57
Sheffield	1	11.2	~	1	7.1	~	2	6.3		2	4.3		0.57	0.60
South Humber	5	12.9		6	9.1		9	6.9		9	4.7		0.53	0.51
Southern Derbyshire	4	12.6	~	7	9.4		8	6.8		10	4.8		0.54	0.51
West Midlands		12.5	~		8.4	~		7.0			4.5		0.56	0.54
Birmingham	11	13.5		9	8.8	~	6	6.7		4	4.4		0.50	0.50
Coventry	12	13.5		3	7.7	~	1	5.9	~	8	4.7		0.44	0.61
Dudley	4	11.2	~	5	8.1	~	11	7.5		2	4.0		0.67	0.50
Herefordshire	10	13.3		13	9.5		13	8.2		6	4.6		0.62	0.49
North Staffordshire	3	11.1	~	4	7.8	~	7	7.3		13	4.9		0.66	0.63
Sandwell	13	14.3		2	7.6	~	12	8.1		7	4.6		0.57	0.61
Shropshire	1	10.5	~	1	7.5	~	2	6.2		9	4.7		0.59	0.62
Solihull	9	13.1		10	8.9		10	7.5		3	4.2		0.57	0.48
South Staffordshire	5	12.6	~	7	8.5	~	5	6.7		5	4.5		0.53	0.53
Walsall	6	12.7		8	8.5		4	6.5		1	3.8	~	0.51	0.44
Warwickshire	8	12.7		11	8.9		9	7.4		10	4.7		0.58	0.53
Wolverhampton	2	11.1	~	6	8.3		3	6.4		11	4.8		0.58	0.58
Worcestershire	7	12.7		12	9.3		8	7.4		12	4.8		0.58	0.52
North West		13.2	~		9.0	~		6.5	~		4.5	~	0.49	0.50
Bury and Rochdale	10	13.7		7	8.7		10	6.8		3	4.1		0.50	0.47
East Lancashire	11	14.0		8	9.2		4	6.2	~	5	4.2		0.44	0.46
Liverpool	14	14.3		16	10.8		16	7.2		15	5.1		0.51	0.48
Manchester	13	14.3		14	10.0		12	7.0		16	5.4		0.49	0.54
Morecambe Bay	8	13.6		10	9.3		6	6.4		2	4.0		0.47	0.42
North Cheshire	5	12.4		2	8.0	~	11	7.0		4	4.1		0.56	0.51
North West Lancashire	15	15.1		15	10.0		15	7.2		9	4.5		0.48	0.45
Salford and Trafford	16	15.1		6	8.6	~	8	6.7		6	4.2		0.44	0.49
Sefton	4	12.0	~	1	7.4	~	5	6.2		10	4.6		0.52	0.62
South Cheshire	6	12.7	~	3	8.2	~	13	7.0		12	4.7		0.55	0.58
South Lancashire	12	14.2		12	9.4		14	7.1		8	4.3		0.50	0.46
St Helen's and Knowsley	2	11.3	~	11	9.4		2	5.2	~	14	5.0		0.46	0.54
Stockport	9	13.7		13	9.7		7	6.6		11	4.7		0.48	0.48
West Pennine	1	10.2	~	4	8.2	~	3	6.1	~	1	3.5	~	0.60	0.43
Wigan and Bolton	3	11.3	~	9	9.2		1	5.1	~	7	4.2		0.45	0.46
Wirral	7	13.4		5	8.6		9	6.7		13	5.0		0.50	0.59
Eastern		13.9			9.7			7.5			4.9		0.54	0.51
Bedfordshire	5	14.5		7	10.5		7	8.0		7	5.3		0.55	0.51
Cambridgeshire	6	14.5		6	10.1		1	7.0		3	4.8		0.48	0.48
Hertfordshire	2	13.6		4	9.9		3	7.2		6	5.1		0.53	0.52
Norfolk	4	14.3		3	9.8		6	8.0		4	4.9		0.56	0.50
North Essex	1	12.6	~	2	9.4		4	7.5		2	4.7		0.59	0.50
South Essex	3	13.9		1	8.4	~	5	7.8		5	5.0		0.56	0.60
Suffolk	7	14.5		5	9.9		2	7.1		1	4.6		0.49	0.46

1 England, Wales and Scotland 1991-99; Northern Ireland 1993-99; Ireland 1994-99.
2 England, Wales and Northern Ireland 1991-2000; Scotland 1991-99; Ireland 1994-2000.
3 Mortality to incidence ratio; age-standardised mortality rate divided by age-standardised incidence rate.
4 Rank within region or country; 1 is lowest.
5 Directly age standardised using the European standard population; per 100,000 population.
6 Rate significantly different from that for UK and Ireland; * significantly higher; ~ significantly lower. See Appendix H.

Country/ region/ health authority	Incidence[1]						Mortality[2]						M:I Ratio[3]	
	Males			Females			Males			Females			Males	Females
	Rank[4]	Rate[5]	Signif[6]	Rank	Rate	Signif	Rank	Rate	Signif	Rank	Rate	Signif		
London		**16.0**	*		**10.4**	*		**8.2**	*		**5.1**		**0.51**	**0.49**
Barking and Havering	2	13.8		5	9.9		2	6.8		6	4.9		0.50	0.49
Barnet, Enfield and Haringey	9	16.5	*	10	11.2	*	8	8.0		10	5.3		0.48	0.48
Bexley, Bromley and Greenwich	5	14.9		4	9.6		7	7.9		5	4.8		0.53	0.50
Brent and Harrow	7	15.2		11	11.3		5	7.5		11	5.4		0.50	0.48
Camden and Islington	14	18.9	*	14	12.1	*	12	9.3	*	7	5.1		0.49	0.42
Croydon	4	14.3		8	10.8		10	8.5		13	5.5		0.60	0.51
Ealing, Hammersmith and Hounslow	6	15.0		3	8.9		9	8.1		9	5.3		0.54	0.59
East London and The City	10	17.0	*	12	11.3		11	9.1	*	12	5.4		0.53	0.48
Hillingdon	1	13.3		1	8.3		1	6.4		3	4.7		0.48	0.57
Kensington and Chelsea, and Westminster	13	18.6	*	13	11.8	*	14	9.9	*	14	5.9		0.53	0.50
Kingston and Richmond	3	14.2		2	8.9		3	7.5		1	4.3		0.53	0.49
Lambeth, Southwark and Lewisham	11	17.1	*	7	10.7		13	9.8	*	8	5.2		0.57	0.49
Merton, Sutton and Wandsworth	12	17.8	*	6	10.4		6	7.8		4	4.8		0.44	0.46
Redbridge and Waltham Forest	8	15.3		9	11.0		4	7.5		2	4.6		0.49	0.42
South East		**14.9**	*		**10.6**	*		**7.5**			**5.0**		**0.50**	**0.48**
Berkshire	5	14.7		9	10.9		8	7.6		8	5.3		0.52	0.48
Buckinghamshire	9	15.1		10	11.2	*	2	6.6		3	4.8		0.44	0.43
East Kent	2	14.1		2	9.5		12	8.5	*	12	5.4		0.61	0.57
East Surrey	13	18.1	*	4	10.0		9	7.7		7	5.1		0.43	0.51
East Sussex, Brighton and Hove	8	15.0		5	10.0		13	8.6	*	6	5.0		0.57	0.50
Isle of Wight, Portsmouth and South East Hampshire	11	15.5		11	11.5	*	5	7.2		4	4.8		0.46	0.42
North and Mid-Hampshire	6	14.7		3	9.7		10	7.7		1	4.0	~	0.52	0.41
Northamptonshire	4	14.6		6	10.3		6	7.4		9	5.3		0.51	0.52
Oxfordshire	7	15.0		13	12.0	*	4	7.1		5	5.0		0.47	0.41
Southampton and South West Hampshire	12	15.7		12	11.9	*	1	6.4		10	5.4		0.41	0.45
West Kent	1	13.4		1	9.5		3	6.7		2	4.6		0.50	0.49
West Surrey	10	15.2		8	10.6		11	8.2		13	5.5		0.54	0.52
West Sussex	3	14.3		7	10.6		7	7.5		11	5.4		0.52	0.50
South West		**16.5**	*		**11.0**	*		**7.6**			**4.8**		**0.46**	**0.43**
Avon	3	15.5	*	6	11.2	*	4	7.5		7	5.4		0.48	0.48
Cornwall and Isles of Scilly	5	17.0	*	5	11.1		5	7.7		2	4.5		0.45	0.40
Dorset	8	18.8	*	7	11.8	*	6	7.7		5	4.6		0.41	0.39
Gloucestershire	2	15.3		2	10.3		2	7.0		8	5.4		0.46	0.53
North and East Devon	7	17.6	*	4	10.5		3	7.5		4	4.5		0.43	0.43
Somerset	4	16.2	*	3	10.4		8	8.2		1	4.2		0.51	0.40
South and West Devon	6	17.6	*	1	10.1		7	8.1		3	4.5		0.46	0.45
Wiltshire	1	13.7		8	12.1	*	1	6.7		6	4.8		0.49	0.39
Wales		**14.4**			**9.8**			**7.0**			**4.5**	~	**0.49**	**0.45**
Bro Taf	4	14.7		1	9.0		4	7.8		5	4.7		0.53	0.52
Dyfed Powys	3	14.4		4	10.2		2	6.4		4	4.7		0.45	0.46
Gwent	5	15.9	*	5	10.4		5	7.9		3	4.4		0.50	0.43
Morgannwg	2	13.7		2	9.5		3	6.8		1	4.1	~	0.49	0.43
North Wales	1	13.5		3	10.1		1	6.1	~	2	4.4		0.45	0.44
Scotland		**14.9**	*		**11.6**	*		**7.3**			**5.4**	*	**0.49**	**0.47**
Argyll and Clyde	10	15.0		13	12.2	*	7	7.2		9	5.4		0.48	0.45
Ayrshire and Arran	3	13.6		8	11.9	*	3	6.6		5	5.2		0.49	0.43
Borders	12	16.3		15	13.9	*	12	8.3		12	5.6		0.51	0.41
Dumfries and Galloway	11	15.2		7	11.4		2	6.6		4	5.0		0.43	0.44
Fife	9	14.8		10	11.9	*	10	7.6		14	6.1	*	0.51	0.51
Forth Valley	2	13.4		12	12.0	*	4	6.6		10	5.5		0.49	0.45
Grampian	8	14.8		6	10.9		6	7.1		11	5.5		0.48	0.50
Greater Glasgow	6	14.0		9	11.9		5	6.8		7	5.4		0.49	0.45
Highland	14	18.9	*	14	13.3	*	15	9.6	*	15	6.6	*	0.51	0.50
Lanarkshire	4	13.7		2	10.0		9	7.4		6	5.3		0.54	0.52
Lothian	13	17.0	*	11	12.0		11	7.9		13	5.8	*	0.47	0.48
Orkney	5	13.9		1	9.6		14	9.6		3	5.0		0.69	0.52
Shetland	1	5.6	~	3	10.2		1	5.8		1	2.3		1.04	0.23
Tayside	7	14.4		4	10.7		8	7.2		2	4.7		0.50	0.44
Western Isles	15	21.9		5	10.9		13	9.6		8	5.4		0.44	0.50
Northern Ireland		**16.0**	*		**12.6**	*		**7.9**			**5.3**		**0.49**	**0.42**
Eastern	2	15.4		2	12.0	*	2	7.5		3	5.5		0.49	0.46
Northern	1	13.8		3	12.8	*	1	7.4		1	4.5		0.53	0.35
Southern	4	20.2	*	4	14.4	*	4	9.1	*	4	6.0		0.45	0.41
Western	3	16.4		1	11.8		3	8.2		2	4.9		0.50	0.42
Ireland		**13.5**			**10.2**			**7.5**			**5.2**		**0.55**	**0.51**
Eastern	7	14.1		7	10.7		4	7.5		7	5.4		0.53	0.50
Mid-Western	3	12.4		6	10.6		3	7.3		6	5.3		0.58	0.50
Midland	6	13.8		8	10.9		8	8.3		1	3.7		0.60	0.34
North Eastern	4	12.7		1	8.8		1	5.5	~	2	4.7		0.43	0.54
North Western	1	11.9		4	10.4		6	8.1		3	5.0		0.68	0.48
South Eastern	5	13.5		5	10.4		5	7.7		4	5.0		0.57	0.48
Southern	8	14.8		3	9.8		7	8.3		5	5.2		0.56	0.53
Western	2	12.2		2	9.4		2	7.1		8	5.8		0.58	0.61

Table **B16.2**

Non-Hodgkins lymphoma: annual average numbers of registrations and deaths by health authority within country and region of England, UK and Ireland, 1991–2000

Country/ region/ health authority	Number of registrations[1]		Number of deaths[2]	
	Males	Females	Males	Females
UK and Ireland	**4,520**	**4,020**	**2,380**	**2,190**
England	**3,570**	**3,120**	**1,900**	**1,720**
Northern and Yorkshire	**402**	**372**	**227**	**213**
Bradford	29	27	14	15
Calderdale and Kirklees	33	27	18	14
County Durham	43	36	21	20
East Riding	37	36	22	20
Gateshead and South Tyneside	21	19	13	12
Leeds	47	47	24	25
Newcastle and North Tyneside	29	31	20	19
North Cumbria	21	19	14	12
North Yorkshire	55	47	29	28
Northumberland	22	19	13	12
Sunderland	17	16	11	11
Tees	29	31	17	16
Wakefield	19	18	11	9
Trent	**353**	**305**	**190**	**169**
Barnsley	14	14	8	6
Doncaster	21	18	10	9
Leicestershire	53	51	31	30
Lincolnshire	47	39	25	22
North Derbyshire	28	25	13	12
North Nottinghamshire	29	23	19	13
Nottingham	51	41	26	21
Rotherham	19	14	9	8
Sheffield	31	28	18	18
South Humber	22	19	12	11
Southern Derbyshire	37	34	20	19
West Midlands	**344**	**289**	**198**	**172**
Birmingham	65	55	34	30
Coventry	21	15	9	10
Dudley	19	17	13	9
Herefordshire	13	11	9	7
North Staffordshire	28	23	19	16
Sandwell	21	15	13	10
Shropshire	24	21	15	14
Solihull	14	12	8	6
South Staffordshire	38	31	20	18
Walsall	17	14	9	8
Warwickshire	34	31	21	18
Wolverhampton	14	13	8	8
Worcestershire	37	32	22	19
North West	**441**	**394**	**224**	**216**
Bury and Rochdale	26	22	13	12
East Lancashire	36	30	16	16
Liverpool	32	32	17	16
Manchester	27	25	13	15
Morecambe Bay	24	22	11	11
North Cheshire	18	14	10	8
North West Lancashire	40	35	20	18
Salford and Trafford	34	26	16	15
Sefton	19	17	10	11
South Cheshire	45	37	26	24
South Lancashire	22	18	11	9
St Helen's and Knowsley	18	19	8	11
Stockport	21	19	11	10
West Pennine	23	23	14	11
Wigan and Bolton	32	33	15	17
Wirral	23	21	12	14
Eastern	**393**	**338**	**220**	**193**
Bedfordshire	37	32	21	18
Cambridgeshire	51	42	25	23
Hertfordshire	71	62	38	36
Norfolk	69	57	41	32
North Essex	61	58	37	33
South Essex	50	40	29	26
Suffolk	54	47	28	25

1 *England, Wales and Scotland 1991-99; Northern Ireland 1993-99; Ireland 1994-99.*
2 *England, Wales and Northern Ireland 1991-2000; Scotland 1991-99; Ireland 1994-2000.*

Country/ region/ health authority	Number of registrations[1]		Number of deaths[2]	
	Males	Females	Males	Females
London	**517**	**432**	**268**	**232**
Barking and Havering	27	27	14	14
Barnet, Enfield and Haringey	60	54	29	29
Bexley, Bromley and Greenwich	54	46	30	27
Brent and Harrow	32	30	16	16
Camden and Islington	30	26	15	12
Croydon	22	21	13	12
Ealing, Hammersmith and Hounslow	44	32	24	20
East London and The City	43	32	22	17
Hillingdon	16	12	8	7
Kensington and Chelsea, and Westminster	30	23	16	13
Kingston and Richmond	22	18	12	10
Lambeth, Southwark and Lewisham	54	42	31	22
Merton, Sutton and Wandsworth	51	40	23	20
Redbridge and Waltham Forest	32	30	16	14
South East	**665**	**598**	**348**	**326**
Berkshire	53	47	28	24
Buckinghamshire	48	43	21	21
East Kent	49	42	31	28
East Surrey	41	31	19	18
East Sussex, Brighton and Hove	66	61	41	37
Isle of Wight, Portsmouth and South East Hampshire	57	53	28	26
North and Mid-Hampshire	39	33	20	16
Northamptonshire	44	39	23	22
Oxfordshire	44	41	21	19
Southampton and South West Hampshire	45	43	20	22
West Kent	64	56	33	30
West Surrey	51	45	28	27
West Sussex	63	64	36	37
South West	**457**	**394**	**224**	**201**
Avon	80	73	40	40
Cornwall and Isles of Scilly	50	42	24	21
Dorset	82	68	37	34
Gloucestershire	47	41	23	24
North and East Devon	51	42	24	21
Somerset	46	37	25	18
South and West Devon	60	47	30	25
Wiltshire	42	44	21	20
Wales	**231**	**195**	**116**	**102**
Bro Taf	54	41	29	25
Dyfed Powys	41	35	19	19
Gwent	47	38	24	18
Morgannwg	38	33	19	17
North Wales	52	49	24	24
Scotland	**383**	**401**	**189**	**208**
Argyll and Clyde	32	34	15	17
Ayrshire and Arran	27	31	13	15
Borders	10	11	6	5
Dumfries and Galloway	13	13	6	6
Fife	27	29	14	16
Forth Valley	18	21	9	11
Grampian	39	36	19	20
Greater Glasgow	61	76	30	37
Highland	21	18	11	10
Lanarkshire	36	36	19	20
Lothian	63	61	29	32
Orkney	2	1	1	1
Shetland	1	1	1	0
Tayside	31	30	16	16
Western Isles	4	2	2	1
Northern Ireland	**120**	**120**	**59**	**54**
Eastern	48	52	23	25
Northern	27	29	14	11
Southern	26	23	12	10
Western	19	16	9	7
Ireland	**214**	**185**	**118**	**101**
Eastern	70	68	35	36
Mid-Western	18	17	11	9
Midland	13	11	8	4
North Eastern	18	13	7	8
North Western	13	12	9	6
South Eastern	24	20	14	11
Southern	37	27	21	15
Western	22	18	13	11

Table B17.1

Oesophagus: annual average age-standardised incidence and mortality by health authority within country and region of England, UK and Ireland, 1991–2000

Country/ region/ health authority	Incidence[1] Males Rank[4]	Males Rate[5]	Males Signif[6]	Females Rank	Females Rate	Females Signif	Mortality[2] Males Rank	Males Rate	Males Signif	Females Rank	Females Rate	Females Signif	M:I Ratio[3] Males	Females
UK and Ireland		**13.0**			**5.9**			**12.8**			**5.3**		**0.98**	**0.89**
England		**12.7**	~		**5.6**	~		**12.5**			**5.0**	~	**0.99**	**0.90**
Northern and Yorkshire		**12.3**	~		**5.3**	~		**12.5**			**4.9**	~	**1.02**	**0.92**
Bradford	5	12.0		7	5.3		9	12.4		5	4.6		1.03	0.87
Calderdale and Kirklees	2	11.0	~	3	4.8	~	1	11.2	~	3	4.4	~	1.02	0.90
County Durham	8	12.2		8	5.4		5	11.9		8	5.0		0.97	0.92
East Riding	12	13.6		11	5.7		13	15.0	*	11	5.5		1.10	0.98
Gateshead and South Tyneside	9	13.0		5	5.0		11	13.4		4	4.4	~	1.03	0.87
Leeds	4	11.8		1	4.4	~	2	11.5	~	1	3.7	~	0.97	0.86
Newcastle and North Tyneside	6	12.1		10	5.6		3	11.6		6	4.8		0.96	0.85
North Cumbria	11	13.5		2	4.6	~	10	12.8		2	4.3	~	0.95	0.94
North Yorkshire	1	10.3	~	6	5.2	~	6	12.0		9	5.0		1.17	0.97
Northumberland	7	12.1		12	6.6		8	12.4		13	6.5	*	1.02	0.98
Sunderland	10	13.3		4	5.0		7	12.3		7	4.9		0.92	0.98
Tees	13	15.3	*	13	6.7		12	14.5	*	12	5.9		0.95	0.89
Wakefield	3	11.1		9	5.5		4	11.8		10	5.0		1.06	0.92
Trent		**13.5**			**5.9**			**12.3**			**5.1**		**0.91**	**0.87**
Barnsley	3	12.1		1	4.0		3	11.2		1	3.8		0.92	0.96
Doncaster	1	11.2		6	6.0		1	10.5	~	5	5.1		0.94	0.86
Leicestershire	2	11.6	~	9	6.4		2	11.1	~	9	5.5		0.96	0.86
Lincolnshire	9	14.6	*	8	6.1		9	13.2		7	5.3		0.90	0.86
North Derbyshire	8	14.4		5	5.7		6	12.4		8	5.4		0.86	0.94
North Nottinghamshire	11	16.0	*	10	6.5		11	13.6		10	5.6		0.85	0.85
Nottingham	10	14.6	*	11	6.9	*	5	12.0		6	5.2		0.82	0.76
Rotherham	5	12.6		3	5.3		7	12.8		3	4.8		1.02	0.89
Sheffield	4	12.2		2	5.1	~	4	11.8		2	4.5	~	0.97	0.89
South Humber	6	13.7		7	6.0		8	13.1		11	5.8		0.96	0.97
Southern Derbyshire	7	14.1		4	5.5		10	13.4		4	4.9		0.95	0.90
West Midlands		**12.8**			**6.1**			**13.4**	*		**5.5**		**1.04**	**0.91**
Birmingham	13	13.7		6	5.9		5	13.4		6	5.2		0.98	0.88
Coventry	8	13.3		11	6.7		10	14.1		12	6.3		1.06	0.94
Dudley	6	12.9		1	5.3		4	13.0		2	4.8		1.01	0.92
Herefordshire	3	12.3		4	5.6		12	14.5		4	5.0		1.18	0.90
North Staffordshire	10	13.4		9	6.4		11	14.1		11	6.3	*	1.05	0.98
Sandwell	9	13.4		13	7.4	*	6	13.5		13	6.9	*	1.01	0.94
Shropshire	11	13.5		5	5.8		13	14.7	*	8	5.5		1.09	0.96
Solihull	1	10.0	~	8	6.2		1	10.8		3	5.0		1.08	0.81
South Staffordshire	7	13.0		10	6.6		7	13.6		10	6.0		1.04	0.90
Walsall	4	12.5		2	5.4		8	13.8		1	4.6		1.10	0.86
Warwickshire	5	12.7		3	5.6		9	13.8		5	5.2		1.09	0.93
Wolverhampton	12	13.6		12	7.0		3	12.9		9	5.8		0.95	0.82
Worcestershire	2	10.9	~	7	5.9		2	11.7		7	5.3		1.07	0.89
North West		**14.4**	*		**6.6**	*		**15.1**	*		**6.2**	*	**1.05**	**0.93**
Bury and Rochdale	8	14.0		4	6.2		9	15.1	*	3	5.6		1.08	0.90
East Lancashire	2	12.6		1	5.0	~	2	14.2		1	4.7		1.13	0.94
Liverpool	16	17.9	*	16	8.1	*	16	16.7	*	15	7.2	*	0.93	0.88
Manchester	14	16.2	*	15	7.3	*	13	16.0	*	6	6.0		0.98	0.82
Morecambe Bay	11	14.8		2	5.7		4	14.4		2	5.5		0.97	0.96
North Cheshire	7	13.9		12	7.2		10	15.2	*	16	7.4	*	1.09	1.04
North West Lancashire	13	15.6	*	9	6.7		12	15.4	*	12	6.3	*	0.99	0.94
Salford and Trafford	12	14.8	*	10	6.9		14	16.3	*	11	6.2	*	1.10	0.90
Sefton	3	13.5		14	7.3	*	3	14.3		14	7.0	*	1.07	0.95
South Cheshire	10	14.3		3	6.0		8	15.0	*	5	5.9		1.05	0.99
South Lancashire	1	10.5	~	5	6.2		1	13.6		8	6.1		1.29	0.97
St Helen's and Knowsley	15	16.5	*	13	7.2	*	15	16.5	*	7	6.0		1.00	0.83
Stockport	4	13.7		6	6.3		7	14.7		9	6.1		1.07	0.97
West Pennine	9	14.1		7	6.3		5	14.6	*	10	6.2		1.03	0.97
Wigan and Bolton	6	13.7		11	7.2	*	6	14.7	*	13	6.7	*	1.07	0.94
Wirral	5	13.7		8	6.4		11	15.4	*	4	5.8		1.12	0.91
Eastern		**11.3**	~		**5.1**	~		**11.5**	~		**4.6**	~	**1.01**	**0.90**
Bedfordshire	6	12.3		7	5.7		3	10.9	~	6	4.9		0.89	0.87
Cambridgeshire	2	10.9	~	3	5.0	~	4	11.3	~	5	4.8		1.04	0.96
Hertfordshire	3	11.0	~	4	5.1	~	1	10.2	~	2	4.2	~	0.93	0.83
Norfolk	5	11.7		6	5.4		7	13.1		7	5.0		1.11	0.92
North Essex	1	9.9	~	2	4.8	~	2	10.7	~	3	4.5	~	1.08	0.94
South Essex	7	12.5		5	5.3		5	11.5	~	4	4.8		0.92	0.91
Suffolk	4	11.5	~	1	4.7	~	6	12.5		1	4.2	~	1.08	0.89

1 England, Wales and Scotland 1991-99; Northern Ireland 1993-99; Ireland 1994-99.
2 England, Wales and Northern Ireland 1991-2000; Scotland 1991-99; Ireland 1994-2000.
3 Mortality to incidence ratio; age-standardised mortality rate divided by age-standardised incidence rate.
4 Rank within region or country; 1 is lowest.
5 Directly age standardised using the European standard population; per 100,000 population.
6 Rate significantly different from that for UK and Ireland; * significantly higher; ~ significantly lower. See Appendix H.

Country/ region/ health authority	Incidence[1]						Mortality[2]						M:I Ratio[3]	
		Males			Females			Males			Females		Males	Females
	Rank[4]	Rate[5]	Signif[6]	Rank	Rate	Signif	Rank	Rate	Signif	Rank	Rate	Signif		
London		**11.9**	~		**5.1**	~		**11.0**	~		**4.4**	~	**0.92**	**0.85**
Barking and Havering	10	12.9		8	5.3		11	11.9		10	4.7		0.92	0.90
Barnet, Enfield and Haringey	2	10.4	~	2	4.7	~	2	9.5	~	7	4.4	~	0.91	0.94
Bexley, Bromley and Greenwich	5	11.4	~	12	5.4		12	12.0		13	4.9		1.06	0.90
Brent and Harrow	3	10.5	~	4	5.0	~	7	10.7		11	4.8		1.02	0.97
Camden and Islington	11	13.2		14	5.6		8	10.9	~	14	5.3		0.83	0.93
Croydon	8	12.4		11	5.4		4	10.4	~	4	3.8	~	0.84	0.70
Ealing, Hammersmith and Hounslow	9	12.9		3	4.9		9	11.6		5	3.9	~	0.90	0.81
East London and The City	12	13.4		13	5.6		10	11.6		12	4.9		0.87	0.87
Hillingdon	14	14.3		7	5.2		13	12.6		2	3.7	~	0.88	0.71
Kensington and Chelsea, and Westminster	7	12.1		10	5.4		6	10.5		3	3.8	~	0.86	0.70
Kingston and Richmond	4	11.2		5	5.1		1	9.1	~	8	4.4		0.81	0.87
Lambeth, Southwark and Lewisham	13	13.7		9	5.3		14	12.9		6	4.2	~	0.94	0.79
Merton, Sutton and Wandsworth	6	11.5	~	6	5.1		5	10.4	~	9	4.4	~	0.91	0.87
Redbridge and Waltham Forest	1	9.1		1	4.4	~	3	9.8		1	3.2		1.08	0.74
South East		**12.5**	~		**5.2**	~		**12.0**	~		**4.7**	~	**0.96**	**0.90**
Berkshire	9	12.4		5	4.9		5	11.5	~	3	4.2	~	0.92	0.85
Buckinghamshire	2	11.3	~	4	4.9	~	6	11.6		8	4.7		1.03	0.96
East Kent	13	14.9	*	10	5.5		13	13.9		7	4.6		0.94	0.84
East Surrey	4	11.5		7	5.3		1	10.3	~	1	3.9	~	0.90	0.74
East Sussex, Brighton and Hove	6	12.2		12	5.6		10	12.6		10	5.2		1.03	0.93
Isle of Wight, Portsmouth and South East Hampshire	11	13.8		13	6.4		12	13.0		11	5.2		0.95	0.81
North and Mid-Hampshire	3	11.3	~	2	4.8	~	3	11.3	~	5	4.4	~	1.00	0.92
Northamptonshire	10	12.8		11	5.5		11	13.0		13	5.4		1.01	0.99
Oxfordshire	8	12.4		9	5.4		4	11.4	~	12	5.2		0.92	0.97
Southampton and South West Hampshire	1	11.1	~	3	4.8		7	12.0		4	4.3		1.08	0.89
West Kent	7	12.3		6	5.2		8	12.2		9	5.1		0.99	0.98
West Surrey	5	11.7		8	5.3		2	10.4	~	6	4.4	~	0.89	0.84
West Sussex	12	14.0		1	4.5	~	9	12.3		2	4.0	~	0.88	0.88
South West		**12.3**	~		**5.7**			**11.9**	~		**5.0**		**0.97**	**0.88**
Avon	2	11.9		5	6.0		4	12.0		7	5.2		1.01	0.87
Cornwall and Isles of Scilly	7	13.3		8	6.2		8	13.8		8	5.9		1.03	0.95
Dorset	4	12.1		2	5.3		1	10.8	~	1	4.6		0.89	0.87
Gloucestershire	3	12.0		1	5.2		3	11.3	~	2	4.6		0.94	0.89
North and East Devon	1	11.1	~	4	5.6		2	10.9	~	4	4.9		0.98	0.88
Somerset	6	12.8		3	5.3		6	12.3		3	4.8		0.97	0.91
South and West Devon	5	12.6		6	6.0		7	12.6		6	5.2		1.00	0.86
Wiltshire	8	13.4		7	6.2		5	12.1		5	4.9		0.91	0.79
Wales		**13.0**			**6.6**	*		**12.9**			**5.6**		**0.99**	**0.85**
Bro Taf	4	13.7		3	6.2		4	12.8		2	5.1		0.93	0.82
Dyfed Powys	1	11.3	~	2	5.7		2	11.8		3	5.3		1.04	0.92
Gwent	3	12.3		4	7.5	*	3	11.9		4	6.3	*	0.97	0.84
Morgannwg	2	11.5	~	1	5.4		1	11.3	~	1	4.4	~	0.99	0.83
North Wales	5	15.4	*	5	7.9	*	5	15.9	*	5	6.6	*	1.03	0.84
Scotland		**16.7**	*		**8.2**	*		**16.1**	*		**7.4**	*	**0.97**	**0.90**
Argyll and Clyde	11	18.2	*	12	9.0	*	9	16.1	*	13	8.0	*	0.88	0.88
Ayrshire and Arran	9	16.8	*	14	9.4	*	4	14.4		12	7.8	*	0.86	0.83
Borders	8	16.4		7	7.8		8	15.9		10	7.4		0.96	0.94
Dumfries and Galloway	3	13.3		11	8.7	*	3	14.2		5	7.0	*	1.07	0.80
Fife	1	11.8		5	7.6	*	5	15.2	*	9	7.3	*	1.28	0.96
Forth Valley	2	12.8		3	6.6		2	13.5		3	6.4		1.05	0.97
Grampian	5	15.9	*	4	7.3	*	7	15.9	*	4	6.9	*	1.00	0.94
Greater Glasgow	14	19.5	*	6	7.7	*	14	18.6	*	6	7.1	*	0.95	0.92
Highland	7	16.3	*	8	8.1	*	12	16.7	*	8	7.3	*	1.02	0.89
Lanarkshire	10	18.1	*	13	9.2	*	10	16.1	*	14	8.4	*	0.89	0.91
Lothian	4	15.2	*	10	8.6	*	6	15.8	*	11	7.6	*	1.04	0.88
Orkney	6	15.9		2	3.9		1	12.7		1	1.7	~	0.80	0.44
Shetland	13	19.0		1	2.7	~	15	20.1		2	3.9		1.06	1.45
Tayside	12	18.8	*	9	8.4	*	11	16.4	*	7	7.2	*	0.87	0.86
Western Isles	15	20.9	*	15	10.5		13	18.2		15	8.7		0.87	0.83
Northern Ireland		**13.1**			**5.3**			**11.4**	~		**4.7**	~	**0.87**	**0.89**
Eastern	4	15.1	*	3	5.3		4	12.7		3	4.8		0.84	0.90
Northern	1	11.2		2	5.3		1	9.9	~	2	4.8		0.88	0.90
Southern	2	11.2		4	5.5		3	11.2		4	4.8		1.00	0.88
Western	3	13.0		1	5.2		2	10.6	~	1	4.1		0.81	0.80
Ireland		**11.5**	~		**5.6**			**12.1**			**5.3**		**1.05**	**0.95**
Eastern	7	12.7		8	6.7		8	14.1		8	6.1	*	1.11	0.91
Mid-Western	3	9.8		3	5.1		4	11.3		4	5.4		1.15	1.07
Midland	4	11.4		7	6.4		7	13.1		7	6.0		1.15	0.93
North Eastern	8	13.6		4	5.2		5	11.8		3	5.4		0.87	1.04
North Western	1	8.8		2	3.5	~	1	9.8	~	2	3.5	~	1.11	0.99
South Eastern	5	11.5		5	5.5		6	12.4		5	5.5		1.08	0.99
Southern	6	12.1		6	6.4		3	11.1		6	5.5		0.92	0.86
Western	2	9.5	~	1	3.2	~	2	10.3	~	1	3.4	~	1.09	1.07

Table **B17.2**

Oesophagus: annual average numbers of registrations and deaths by health authority within country and region of England, UK and Ireland, 1991–2000

Country/ region/ health authority	Number of registrations[1]		Number of deaths[2]	
	Males	Females	Males	Females
UK and Ireland	**4,240**	**2,890**	**4,220**	**2,690**
England	**3,310**	**2,220**	**3,310**	**2,060**
Northern and Yorkshire	**415**	**273**	**427**	**260**
Bradford	27	19	29	17
Calderdale and Kirklees	32	22	33	21
County Durham	40	26	39	25
East Riding	43	26	48	26
Gateshead and South Tyneside	25	16	27	14
Leeds	44	24	43	22
Newcastle and North Tyneside	31	23	30	20
North Cumbria	25	14	24	13
North Yorkshire	46	35	54	35
Northumberland	22	18	23	18
Sunderland	20	11	19	11
Tees	42	27	40	25
Wakefield	18	13	19	13
Trent	**377**	**239**	**348**	**215**
Barnsley	15	7	14	7
Doncaster	17	13	16	12
Leicestershire	55	43	53	38
Lincolnshire	57	34	53	30
North Derbyshire	31	18	27	17
North Nottinghamshire	35	20	30	18
Nottingham	50	33	40	27
Rotherham	16	10	17	9
Sheffield	35	23	35	21
South Humber	23	14	23	14
Southern Derbyshire	43	24	41	22
West Midlands	**365**	**252**	**385**	**236**
Birmingham	68	45	67	41
Coventry	22	15	23	14
Dudley	22	13	22	12
Herefordshire	14	9	16	9
North Staffordshire	34	25	37	25
Sandwell	21	17	21	16
Shropshire	32	19	36	19
Solihull	11	9	12	8
South Staffordshire	39	28	41	26
Walsall	17	11	19	9
Warwickshire	35	23	39	22
Wolverhampton	18	13	18	11
Worcestershire	33	25	35	24
North West	**496**	**344**	**526**	**332**
Bury and Rochdale	27	17	29	16
East Lancashire	33	21	38	20
Liverpool	40	29	38	27
Manchester	31	20	30	18
Morecambe Bay	28	18	27	18
North Cheshire	21	15	23	17
North West Lancashire	43	30	43	30
Salford and Trafford	35	25	39	23
Sefton	23	19	25	20
South Cheshire	53	33	56	34
South Lancashire	17	14	22	14
St Helen's and Knowsley	27	17	27	15
Stockport	21	15	23	15
West Pennine	33	22	35	22
Wigan and Bolton	38	29	42	28
Wirral	25	18	29	17
Eastern	**334**	**220**	**343**	**205**
Bedfordshire	31	19	28	18
Cambridgeshire	39	26	41	26
Hertfordshire	58	40	55	34
Norfolk	60	39	68	37
North Essex	51	37	55	36
South Essex	48	31	45	28
Suffolk	47	28	51	26

1 *England, Wales and Scotland 1991-99; Northern Ireland 1993-99; Ireland 1994-99.*
2 *England, Wales and Northern Ireland 1991-2000; Scotland 1991-99; Ireland 1994-2000.*

Country/ region/ health authority	Number of registrations[1]		Number of deaths[2]	
	Males	Females	Males	Females
London	**382**	**250**	**356**	**223**
Barking and Havering	27	17	25	16
Barnet, Enfield and Haringey	38	27	35	27
Bexley, Bromley and Greenwich	43	31	46	30
Brent and Harrow	22	17	23	16
Camden and Islington	20	12	17	12
Croydon	19	12	16	9
Ealing, Hammersmith and Hounslow	37	20	33	17
East London and The City	31	18	27	16
Hillingdon	17	10	15	8
Kensington and Chelsea, and Westminster	18	12	16	9
Kingston and Richmond	17	13	15	12
Lambeth, Southwark and Lewisham	41	24	39	20
Merton, Sutton and Wandsworth	32	23	29	21
Redbridge and Waltham Forest	19	15	21	12
South East	**580**	**375**	**566**	**350**
Berkshire	44	26	41	22
Buckinghamshire	35	22	37	21
East Kent	54	33	52	29
East Surrey	28	20	26	17
East Sussex, Brighton and Hove	58	47	62	46
Isle of Wight, Portsmouth and South East Hampshire	52	37	50	32
North and Mid-Hampshire	30	19	30	19
Northamptonshire	39	24	40	24
Oxfordshire	37	24	35	23
Southampton and South West Hampshire	35	21	38	21
West Kent	61	39	61	38
West Surrey	40	26	36	23
West Sussex	66	37	59	35
South West	**365**	**265**	**361**	**244**
Avon	63	50	64	45
Cornwall and Isles of Scilly	43	30	46	30
Dorset	57	43	53	38
Gloucestershire	38	25	37	23
North and East Devon	36	30	37	29
Somerset	39	24	39	24
South and West Devon	46	36	47	33
Wiltshire	42	28	39	23
Wales	**217**	**168**	**217**	**148**
Bro Taf	51	36	47	31
Dyfed Powys	35	25	37	24
Gwent	37	33	36	29
Morgannwg	33	25	33	21
North Wales	60	49	63	44
Scotland	**434**	**330**	**421**	**307**
Argyll and Clyde	40	31	35	28
Ayrshire and Arran	34	28	29	25
Borders	11	9	10	8
Dumfries and Galloway	12	11	13	10
Fife	22	21	28	22
Forth Valley	18	14	19	14
Grampian	41	30	41	29
Greater Glasgow	86	56	81	52
Highland	18	14	19	13
Lanarkshire	47	34	42	31
Lothian	56	49	58	46
Orkney	2	1	1	0
Shetland	2	1	2	1
Tayside	42	29	37	26
Western Isles	3	3	3	2
Northern Ireland	**96**	**59**	**84**	**54**
Eastern	47	28	40	26
Northern	21	15	19	14
Southern	14	10	14	9
Western	14	7	11	6
Ireland	**176**	**116**	**190**	**115**
Eastern	57	45	64	43
Mid-Western	14	9	16	10
Midland	11	7	13	7
North Eastern	18	9	16	9
North Western	10	4	11	5
South Eastern	21	13	22	13
Southern	29	21	28	19
Western	17	7	19	8

Table B18.1

Ovary: annual average age-standardised incidence and mortality by health authority within country and region of England, UK and Ireland, 1991–2000

Country/ region/ health authority	Incidence[1]						Mortality[2]						M:I Ratio[3]	
	Males			Females			Males			Females			Males	Females
	Rank[4]	Rate[5]	Signif[6]	Rank	Rate	Signif	Rank	Rate	Signif	Rank	Rate	Signif		
UK and Ireland					18.0						11.5			**0.64**
England					17.7						11.5			**0.65**
Northern and Yorkshire					16.8	~					10.7	~		**0.64**
Bradford				11	17.6					10	11.6			0.66
Calderdale and Kirklees				6	16.7					4	10.4			0.62
County Durham				3	16.1	~				3	10.3	~		0.64
East Riding				4	16.3					8	10.7			0.66
Gateshead and South Tyneside				7	16.7					7	10.5			0.63
Leeds				10	17.4					6	10.5			0.60
Newcastle and North Tyneside				2	16.0	~				5	10.4			0.65
North Cumbria				12	18.0					11	11.8			0.65
North Yorkshire				9	17.3					9	10.8			0.62
Northumberland				1	13.5	~				1	9.7	~		0.71
Sunderland				8	16.8					12	11.8			0.70
Tees				5	16.6					2	9.9	~		0.60
Wakefield				13	18.8					13	12.7			0.67
Trent					17.9						11.7			**0.66**
Barnsley				5	17.7					3	10.9			0.61
Doncaster				9	18.6					4	11.2			0.60
Leicestershire				1	16.7					8	12.0			0.72
Lincolnshire				11	18.9					10	12.2			0.64
North Derbyshire				7	18.4					11	13.4	*		0.73
North Nottinghamshire				4	17.0					2	10.8			0.64
Nottingham				8	18.4					6	11.8			0.64
Rotherham				6	18.3					5	11.5			0.63
Sheffield				3	17.0					1	10.3			0.61
South Humber				2	16.9					9	12.0			0.71
Southern Derbyshire				10	18.9					7	11.9			0.63
West Midlands					19.2	*					12.0			**0.62**
Birmingham				3	17.5					1	10.2	~		0.58
Coventry				9	20.4					9	12.3			0.60
Dudley				13	21.2	*				11	12.6			0.59
Herefordshire				1	15.7					3	11.3			0.72
North Staffordshire				7	19.7					13	14.3	*		0.72
Sandwell				8	20.0					2	11.2			0.56
Shropshire				4	18.0					8	12.2			0.68
Solihull				11	20.6					10	12.6			0.61
South Staffordshire				12	20.9	*				12	13.1	*		0.63
Walsall				10	20.5					6	11.9			0.58
Warwickshire				5	19.2					7	12.1			0.63
Wolverhampton				2	17.3					5	11.7			0.68
Worcestershire				6	19.5					4	11.5			0.59
North West					17.4						11.3			**0.65**
Bury and Rochdale				5	16.7					5	10.9			0.65
East Lancashire				14	18.5					10	11.3			0.61
Liverpool				3	16.2					12	11.7			0.72
Manchester				9	17.1					7	11.1			0.65
Morecambe Bay				4	16.5					14	12.0			0.72
North Cheshire				11	17.9					15	12.0			0.67
North West Lancashire				12	18.5					4	10.9			0.59
Salford and Trafford				16	19.5					9	11.3			0.58
Sefton				2	16.2					8	11.3			0.69
South Cheshire				8	17.0					13	11.9			0.70
South Lancashire				6	16.7					2	10.4			0.62
St Helen's and Knowsley				1	15.4	~				1	10.3			0.67
Stockport				13	18.5					3	10.8			0.58
West Pennine				10	17.9					11	11.3			0.63
Wigan and Bolton				7	16.9					6	11.1			0.65
Wirral				15	18.6					16	12.2			0.65
Eastern					17.8						11.8			**0.66**
Bedfordshire				2	17.4					2	11.5			0.66
Cambridgeshire				7	19.0					3	11.5			0.61
Hertfordshire				4	17.4					5	12.0			0.69
Norfolk				6	18.7					4	11.7			0.63
North Essex				1	16.7					6	12.2			0.73
South Essex				5	18.4					7	12.5			0.68
Suffolk				3	17.4					1	11.1			0.64

1 England, Wales and Scotland 1991-99; Northern Ireland 1993-99; Ireland 1994-99.
2 England, Wales and Northern Ireland 1991-2000; Scotland 1991-99; Ireland 1994-2000.
3 Mortality to incidence ratio; age-standardised mortality rate divided by age-standardised incidence rate.
4 Rank within region or country; 1 is lowest.
5 Directly age standardised using the European standard population; per 100,000 population.
6 Rate significantly different from that for UK and Ireland; * significantly higher; ~ significantly lower. See Appendix H.

Country/ region/ health authority	Incidence[1]						Mortality[2]						M:I Ratio[3]	
	Males			Females			Males			Females			Males	Females
	Rank[4]	Rate[5]	Signif[6]	Rank	Rate	Signif	Rank	Rate	Signif	Rank	Rate	Signif		
London					16.8	~					10.9	~		0.65
Barking and Havering				10	17.9					6	10.6			0.59
Barnet, Enfield and Haringey				9	17.5					11	11.9			0.68
Bexley, Bromley and Greenwich				6	16.2	~				8	11.0			0.68
Brent and Harrow				3	15.5	~				4	10.1			0.65
Camden and Islington				8	17.3					7	10.7			0.62
Croydon				1	13.4	~				2	9.6	~		0.72
Ealing, Hammersmith and Hounslow				5	16.0	~				10	11.4			0.71
East London and The City				2	15.2	~				1	9.5	~		0.63
Hillingdon				11	18.2					14	13.0			0.72
Kensington and Chelsea, and Westminster				7	16.9					3	9.8	~		0.58
Kingston and Richmond				14	19.2					13	12.2			0.64
Lambeth, Southwark and Lewisham				4	15.8	~				5	10.1	~		0.64
Merton, Sutton and Wandsworth				13	18.6					12	12.0			0.65
Redbridge and Waltham Forest				12	18.4					9	11.1			0.60
South East					18.4						11.8			0.64
Berkshire				2	17.2					2	10.8			0.63
Buckinghamshire				13	21.0	*				9	12.0			0.57
East Kent				7	18.5					13	12.9	*		0.70
East Surrey				8	18.6					8	11.9			0.64
East Sussex, Brighton and Hove				1	16.7					5	11.6			0.70
Isle of Wight, Portsmouth and South East Hampshire				11	19.6					7	11.8			0.60
North and Mid-Hampshire				4	17.7					11	12.4			0.70
Northamptonshire				10	18.9					3	11.3			0.60
Oxfordshire				12	20.6	*				10	12.2			0.59
Southampton and South West Hampshire				9	18.7					1	10.7			0.57
West Kent				3	17.4					4	11.6			0.67
West Surrey				6	18.4					12	12.5			0.68
West Sussex				5	17.7					6	11.8			0.66
South West					17.6						11.7			0.66
Avon				3	16.7	~				2	11.5			0.69
Cornwall and Isles of Scilly				4	17.4					3	11.7			0.67
Dorset				7	18.9					7	11.8			0.63
Gloucestershire				1	16.4					5	11.7			0.71
North and East Devon				6	17.8					4	11.7			0.66
Somerset				2	16.5					1	11.0			0.66
South and West Devon				5	17.6					8	12.0			0.69
Wiltshire				8	19.8	*				6	11.8			0.59
Wales					19.9	*					11.6			0.58
Bro Taf				4	20.4	*				2	11.5			0.56
Dyfed Powys				3	20.4	*				3	11.7			0.57
Gwent				1	18.0					1	10.7			0.60
Morgannwg				2	19.9					5	12.3			0.62
North Wales				5	20.6	*				4	11.8			0.57
Scotland					19.1	*					11.5			0.61
Argyll and Clyde				4	17.5					8	11.7			0.67
Ayrshire and Arran				7	17.8					4	10.8			0.60
Borders				12	20.4					15	13.8			0.68
Dumfries and Galloway				3	17.2					2	9.9			0.58
Fife				14	21.5	*				10	12.1			0.56
Forth Valley				5	17.7					12	12.2			0.69
Grampian				10	19.5					5	10.9			0.56
Greater Glasgow				8	19.1					7	11.3			0.59
Highland				11	19.9					13	12.4			0.62
Lanarkshire				2	17.2					9	11.8			0.68
Lothian				13	20.9	*				11	12.1			0.58
Orkney				15	22.4					14	13.2			0.59
Shetland				1	16.2					1	8.0			0.49
Tayside				9	19.3					6	11.2			0.58
Western Isles				6	17.8					3	10.3			0.58
Northern Ireland					19.5	*					10.2	~		0.52
Eastern				2	19.4					4	10.7			0.55
Northern				4	21.0	*				3	10.4			0.49
Southern				3	19.5					2	9.9			0.51
Western				1	16.9					1	8.5	~		0.51
Ireland					18.2						12.1			0.67
Eastern				2	17.5					5	12.1			0.69
Mid-Western				3	17.9					8	14.2	*		0.79
Midland				4	18.1					3	11.2			0.62
North Eastern				5	18.1					1	9.6			0.53
North Western				6	18.5					6	13.1			0.71
South Eastern				8	20.1					4	11.8			0.59
Southern				7	19.9					7	13.9	*		0.70
Western				1	16.6					2	10.4			0.63

Table B18.2

Ovary: annual average numbers of registrations and deaths by health authority within country and region of England, UK and Ireland, 1991–2000

Country/ region/ health authority	Number of registrations[1]		Number of deaths[2]	
	Males	Females	Males	Females
UK and Ireland		**6,700**		**4,610**
England		**5,240**		**3,660**
Northern and Yorkshire		**648**		**452**
Bradford		48		35
Calderdale and Kirklees		56		37
County Durham		61		41
East Riding		57		40
Gateshead and South Tyneside		38		27
Leeds		75		50
Newcastle and North Tyneside		48		33
North Cumbria		37		27
North Yorkshire		84		58
Northumberland		28		22
Sunderland		29		22
Tees		53		34
Wakefield		36		26
Trent		**548**		**386**
Barnsley		24		16
Doncaster		32		21
Leicestershire		89		67
Lincolnshire		75		52
North Derbyshire		43		34
North Nottinghamshire		41		28
Nottingham		69		47
Rotherham		28		18
Sheffield		54		37
South Humber		31		24
Southern Derbyshire		63		43
West Midlands		**605**		**408**
Birmingham		99		63
Coventry		35		23
Dudley		41		26
Herefordshire		18		14
North Staffordshire		58		44
Sandwell		36		21
Shropshire		46		34
Solihull		26		17
South Staffordshire		70		47
Walsall		32		20
Warwickshire		58		40
Wolverhampton		25		18
Worcestershire		62		41
North West		**695**		**485**
Bury and Rochdale		38		27
East Lancashire		56		37
Liverpool		44		33
Manchester		37		26
Morecambe Bay		34		27
North Cheshire		31		22
North West Lancashire		57		37
Salford and Trafford		53		33
Sefton		31		23
South Cheshire		72		54
South Lancashire		32		20
St Helen's and Knowsley		30		21
Stockport		33		22
West Pennine		50		34
Wigan and Bolton		57		40
Wirral		39		28
Eastern		**573**		**412**
Bedfordshire		51		36
Cambridgeshire		73		48
Hertfordshire		104		75
Norfolk		100		70
North Essex		95		74
South Essex		78		57
Suffolk		71		51

1 England, Wales and Scotland 1991-99; Northern Ireland 1993-99; Ireland 1994-99.
2 England, Wales and Northern Ireland 1991-2000; Scotland 1991-99; Ireland 1994-2000.

Country/ region/ health authority	Number of registrations[1]		Number of deaths[2]	
	Males	Females	Males	Females
London		640		436
Barking and Havering		43		27
Barnet, Enfield and Haringey		76		53
Bexley, Bromley and Greenwich		72		52
Brent and Harrow		39		26
Camden and Islington		31		20
Croydon		26		19
Ealing, Hammersmith and Hounslow		54		40
East London and The City		43		27
Hillingdon		25		19
Kensington and Chelsea, and Westminster		31		19
Kingston and Richmond		36		25
Lambeth, Southwark and Lewisham		58		38
Merton, Sutton and Wandsworth		62		43
Redbridge and Waltham Forest		45		29
South East		956		663
Berkshire		71		46
Buckinghamshire		74		44
East Kent		73		57
East Surrey		50		36
East Sussex, Brighton and Hove		93		71
Isle of Wight, Portsmouth and South East Hampshire		86		57
North and Mid-Hampshire		53		39
Northamptonshire		64		40
Oxfordshire		67		43
Southampton and South West Hampshire		63		41
West Kent		99		70
West Surrey		71		51
West Sussex		94		70
South West		572		420
Avon		102		75
Cornwall and Isles of Scilly		61		45
Dorset		97		71
Gloucestershire		58		45
North and East Devon		61		47
Somerset		53		40
South and West Devon		72		55
Wiltshire		68		43
Wales		374		234
Bro Taf		87		52
Dyfed Powys		66		42
Gwent		61		39
Morgannwg		64		42
North Wales		94		58
Scotland		607		401
Argyll and Clyde		48		35
Ayrshire and Arran		45		29
Borders		15		11
Dumfries and Galloway		16		11
Fife		46		28
Forth Valley		32		23
Grampian		61		37
Greater Glasgow		108		70
Highland		26		18
Lanarkshire		55		40
Lothian		97		62
Orkney		3		2
Shetland		2		1
Tayside		51		33
Western Isles		3		2
Northern Ireland		167		94
Eastern		73		45
Northern		45		24
Southern		28		15
Western		20		11
Ireland		316		221
Eastern		107		76
Mid-Western		26		22
Midland		17		11
North Eastern		26		14
North Western		21		15
South Eastern		38		23
Southern		53		39
Western		29		20

Table B19.1

Pancreas: annual average age-standardised incidence and mortality by health authority within country and region of England, UK and Ireland, 1991–2000

Country/ region/ health authority	Incidence[1]						Mortality[2]						M:I Ratio[3]	
	Rank[4]	Males Rate[5]	Signif[6]	Rank	Females Rate	Signif	Rank	Males Rate	Signif	Rank	Females Rate	Signif	Males	Females
UK and Ireland		10.5			7.8			10.1			7.4		0.96	0.95
England		10.5			7.7			10.0			7.4		0.96	0.95
Northern and Yorkshire		10.4			7.5			10.0			7.2		0.96	0.96
Bradford	8	10.2		9	7.8		8	10.0		12	7.7		0.98	0.99
Calderdale and Kirklees	4	9.5		11	8.1		5	9.3		11	7.7		0.98	0.95
County Durham	11	12.0	*	8	7.7		9	10.9		9	7.5		0.91	0.98
East Riding	6	9.9		6	7.5		7	9.5		7	7.2		0.96	0.96
Gateshead and South Tyneside	9	10.9		7	7.5		11	10.9		5	7.0		1.00	0.93
Leeds	1	8.8	~	5	7.3		1	8.9	~	3	6.7		1.02	0.92
Newcastle and North Tyneside	7	10.1		10	8.0		4	9.2		8	7.3		0.91	0.92
North Cumbria	3	9.1		4	7.2		3	9.1		6	7.1		1.00	0.99
North Yorkshire	5	9.7		2	6.7	~	2	8.9	~	4	6.9		0.92	1.03
Northumberland	2	8.9	~	1	6.6		6	9.4		1	6.1	~	1.06	0.93
Sunderland	10	11.4		13	9.1		10	10.9		13	8.4		0.96	0.92
Tees	13	13.5	*	12	8.3		13	12.6	*	10	7.6		0.94	0.92
Wakefield	12	12.6	*	3	6.8		12	12.2	*	2	6.4		0.97	0.94
Trent		10.5			8.1			9.8			7.5		0.94	0.92
Barnsley	9	12.6		10	9.0		10	12.3	*	11	8.9		0.98	0.99
Doncaster	5	10.1		1	6.5	~	4	9.2		2	6.7		0.91	1.02
Leicestershire	3	9.4	~	2	7.3		2	8.9	~	1	6.4	~	0.95	0.87
Lincolnshire	6	10.3		7	8.5		7	10.0		6	7.7		0.96	0.90
North Derbyshire	1	9.2		3	7.7		1	8.5	~	5	7.4		0.93	0.96
North Nottinghamshire	10	12.6	*	9	8.8		9	10.6		8	8.1		0.84	0.92
Nottingham	8	11.4		5	8.3		6	9.8		7	7.8		0.86	0.93
Rotherham	11	13.6	*	6	8.4		11	13.6	*	3	7.1		1.00	0.84
Sheffield	7	10.5		8	8.6		8	10.0		9	8.2		0.95	0.95
South Humber	4	10.0		11	9.1		5	9.7		10	8.4		0.97	0.92
Southern Derbyshire	2	9.3		4	7.8		3	9.1		4	7.4		0.97	0.95
West Midlands		10.3			7.5			10.0			7.2		0.97	0.96
Birmingham	12	11.7	*	9	7.8		10	10.7		9	7.3		0.91	0.94
Coventry	3	9.5		5	7.3		2	8.8		3	6.6		0.92	0.91
Dudley	5	9.8		7	7.5		6	9.7		8	7.3		0.99	0.97
Herefordshire	11	11.3		1	5.5	~	8	10.2		1	5.2	~	0.90	0.94
North Staffordshire	10	10.9		11	8.0		9	10.4		10	7.5		0.95	0.94
Sandwell	9	10.8		2	6.6		12	11.4		5	7.0		1.06	1.06
Shropshire	2	9.5		4	7.2		4	9.7		4	6.7		1.02	0.94
Solihull	4	9.7		3	6.8		3	9.1		2	6.0	~	0.94	0.87
South Staffordshire	7	10.0		10	7.8		7	9.8		12	7.8		0.97	1.00
Walsall	8	10.4		6	7.4		11	10.9		7	7.1		1.05	0.97
Warwickshire	6	10.0		12	8.1		5	9.7		11	7.8		0.97	0.96
Wolverhampton	13	11.8		13	8.2		13	11.7		13	8.6		0.99	1.06
Worcestershire	1	8.2	~	8	7.6		1	8.1	~	6	7.1		0.99	0.93
North West		10.3			7.7			10.1			7.4		0.98	0.96
Bury and Rochdale	7	10.2		12	8.1		8	9.9		12	8.0		0.97	0.99
East Lancashire	14	11.8		5	7.4		16	12.4	*	5	7.2		1.05	0.98
Liverpool	11	11.0		14	8.6		13	10.9		15	8.2		1.00	0.96
Manchester	16	12.7	*	16	10.0	*	15	11.7		16	8.6		0.92	0.86
Morecambe Bay	1	8.3	~	4	7.4		2	8.8		4	7.1		1.06	0.96
North Cheshire	15	12.5		7	7.5		14	11.7		7	7.5		0.94	1.00
North West Lancashire	10	10.8		10	7.9		7	9.6		6	7.5		0.89	0.95
Salford and Trafford	6	9.4		8	7.6		1	8.8	~	10	7.6		0.93	0.99
Sefton	9	10.7		15	8.6		12	10.9		14	8.0		1.01	0.93
South Cheshire	4	9.2	~	2	6.6	~	6	9.5		1	6.3	~	1.03	0.94
South Lancashire	3	9.2		6	7.5		5	9.2		11	7.7		1.01	1.03
St Helen's and Knowsley	5	9.3		11	8.0		3	9.0		9	7.6		0.97	0.95
Stockport	13	11.1		1	6.6		9	9.9		2	6.6		0.89	1.00
West Pennine	12	11.1		9	7.7		11	10.5		8	7.5		0.95	0.98
Wigan and Bolton	2	8.5	~	3	7.0		4	9.2		3	7.0		1.08	1.01
Wirral	8	10.3		13	8.6		10	10.4		13	8.0		1.00	0.93
Eastern		10.0	~		7.5			9.6	~		7.0	~	0.96	0.94
Bedfordshire	3	9.6		2	7.2		5	9.9		6	7.4		1.03	1.03
Cambridgeshire	2	9.3	~	6	8.1		2	9.0		5	7.3		0.97	0.90
Hertfordshire	6	10.7		7	8.3		3	9.7		7	7.8		0.90	0.94
Norfolk	1	8.8	~	1	6.5	~	1	8.8	~	1	6.2	~	1.00	0.96
North Essex	4	10.3		3	7.3		4	9.9		3	7.0		0.96	0.96
South Essex	7	10.8		5	7.9		6	10.1		4	7.1		0.94	0.89
Suffolk	5	10.4		4	7.3		7	10.2		2	6.6	~	0.98	0.90

1 England, Wales and Scotland 1991-99; Northern Ireland 1993-99; Ireland 1994-99.
2 England, Wales and Northern Ireland 1991-2000; Scotland 1991-99; Ireland 1994-2000.
3 Mortality to incidence ratio; age-standardised mortality rate divided by age-standardised incidence rate.
4 Rank within region or country; 1 is lowest.
5 Directly age standardised using the European standard population; per 100,000 population.
6 Rate significantly different from that for UK and Ireland; * significantly higher; ~ significantly lower. See Appendix H.

Country/ region/ health authority	Incidence[1]						Mortality[2]						M:I Ratio[3]	
	Rank[4]	Males Rate[5]	Signif[6]	Rank	Females Rate	Signif	Rank	Males Rate	Signif	Rank	Females Rate	Signif	Males	Females
London		**11.7**	*		**8.3**	*		**10.6**			**7.6**		**0.90**	**0.92**
Barking and Havering	13	13.3	*	11	8.8		14	11.7		14	8.5		0.88	0.97
Barnet, Enfield and Haringey	7	11.3		4	7.9		7	10.3		3	6.9		0.92	0.88
Bexley, Bromley and Greenwich	9	11.9	*	12	8.8	*	8	10.7		12	8.0		0.90	0.92
Brent and Harrow	6	11.1		3	7.7		5	9.8		7	7.5		0.89	0.98
Camden and Islington	3	10.0		1	6.9		4	9.6		1	6.8		0.96	0.98
Croydon	8	11.6		6	8.2		9	11.0		6	7.5		0.94	0.91
Ealing, Hammersmith and Hounslow	5	11.0		7	8.3		6	10.2		9	7.5		0.93	0.91
East London and The City	10	12.1		9	8.4		10	11.0		11	7.9		0.91	0.94
Hillingdon	2	9.9		8	8.3		2	8.9		8	7.5		0.90	0.90
Kensington and Chelsea, and Westminster	11	12.8	*	5	8.1		13	11.7		4	7.2		0.91	0.88
Kingston and Richmond	1	9.9		2	7.5		1	8.6		2	6.9		0.86	0.91
Lambeth, Southwark and Lewisham	12	13.2	*	14	8.9	*	11	11.6	*	13	8.5	*	0.88	0.95
Merton, Sutton and Wandsworth	14	13.3	*	10	8.5		12	11.6	*	5	7.4		0.87	0.87
Redbridge and Waltham Forest	4	10.5		13	8.8		3	9.5		10	7.6		0.91	0.86
South East		**10.8**			**7.9**			**10.2**			**7.6**		**0.95**	**0.97**
Berkshire	3	10.3		12	8.6		6	10.1		12	8.2		0.97	0.96
Buckinghamshire	4	10.4		1	6.9		2	9.6		1	6.7		0.92	0.98
East Kent	12	11.8		8	7.7		13	11.7	*	10	7.9		0.99	1.02
East Surrey	10	10.8		6	7.6		4	9.8		3	7.2		0.91	0.95
East Sussex, Brighton and Hove	8	10.7		5	7.6		3	9.7		6	7.5		0.91	0.99
Isle of Wight, Portsmouth and South East Hampshire	1	10.0		4	7.6		5	10.0		7	7.6		1.00	1.01
North and Mid-Hampshire	11	10.9		2	7.5		11	10.9		9	7.8		1.00	1.05
Northamptonshire	5	10.6		13	9.3	*	8	10.4		13	9.2	*	0.98	0.99
Oxfordshire	9	10.7		7	7.6		7	10.1		4	7.4		0.94	0.96
Southampton and South West Hampshire	6	10.6		10	8.2		10	10.5		11	8.0		0.99	0.98
West Kent	2	10.2		11	8.3		1	9.4		8	7.7		0.93	0.92
West Surrey	13	12.7	*	3	7.5		12	11.4	*	2	7.0		0.90	0.93
West Sussex	7	10.7		9	7.8		9	10.4		5	7.4		0.98	0.94
South West		**9.6**	~		**7.1**	~		**9.8**			**7.1**		**1.02**	**1.00**
Avon	2	9.4	~	3	6.8	~	2	9.3		4	7.0		0.99	1.03
Cornwall and Isles of Scilly	8	10.1		7	7.5		8	10.6		7	7.6		1.05	1.01
Dorset	6	9.9		5	7.0		3	9.4		2	6.6	~	0.95	0.95
Gloucestershire	4	9.7		6	7.2		6	10.3		6	7.4		1.06	1.04
North and East Devon	1	8.7	~	2	6.8	~	1	8.9	~	3	6.9		1.02	1.03
Somerset	5	9.8		4	6.9		5	10.2		5	7.1		1.04	1.02
South and West Devon	7	9.9		1	6.7	~	7	10.5		1	6.5	~	1.05	0.97
Wiltshire	3	9.5		8	8.2		4	9.5		8	7.7		1.00	0.94
Wales		**11.1**			**8.1**			**9.7**			**7.2**		**0.88**	**0.89**
Bro Taf	3	10.9		2	7.6		3	9.6		2	7.1		0.88	0.93
Dyfed Powys	1	9.9		1	7.4		1	8.8	~	1	6.6		0.89	0.90
Gwent	2	10.7		3	7.9		2	9.5		3	7.2		0.89	0.92
Morgannwg	4	11.9		4	8.3		4	10.3		4	7.2		0.87	0.87
North Wales	5	12.0	*	5	9.0	*	5	10.4		5	7.6		0.87	0.84
Scotland		**10.9**			**8.2**	*		**10.4**			**7.7**		**0.95**	**0.93**
Argyll and Clyde	11	11.5		1	7.1		12	11.2		2	6.9		0.97	0.98
Ayrshire and Arran	13	12.3		12	8.9		13	11.5		10	7.9		0.94	0.89
Borders	12	12.0		7	8.1		11	10.9		12	8.3		0.91	1.02
Dumfries and Galloway	7	10.7		15	10.3	*	7	10.4		14	10.0	*	0.97	0.97
Fife	8	11.1		8	8.3		6	10.3		3	7.0		0.92	0.84
Forth Valley	3	9.8		2	7.4		4	9.5		4	7.3		0.97	0.98
Grampian	4	9.9		4	7.7		2	8.8		7	7.5		0.89	0.97
Greater Glasgow	9	11.1		11	8.5		10	10.5		8	7.6		0.95	0.89
Highland	2	9.2		5	7.8		3	9.0		6	7.5		0.98	0.96
Lanarkshire	5	10.5		10	8.4		9	10.5		11	8.1		1.00	0.97
Lothian	10	11.1		9	8.3		5	10.1		9	7.7		0.91	0.93
Orkney	1	8.8		14	9.5		1	7.6		15	10.2		0.86	1.07
Shetland	14	16.7		3	7.4		14	16.6		1	5.7		1.00	0.76
Tayside	6	10.7		6	7.9		8	10.4		5	7.4		0.97	0.94
Western Isles	15	17.7	*	13	9.0		15	22.0	*	13	8.9		1.24	0.98
Northern Ireland		**9.6**			**6.6**	~		**10.2**			**7.0**		**1.06**	**1.05**
Eastern	4	10.0		2	6.3		3	10.1		1	6.4	~	1.01	1.02
Northern	2	9.5		1	6.3		2	10.0		2	7.0		1.05	1.11
Southern	1	9.2		3	7.1		1	9.7		3	7.0		1.06	0.98
Western	3	9.5		4	7.8		4	11.2		4	8.6		1.18	1.10
Ireland		**10.3**			**8.0**			**11.6**	*		**8.2**	*	**1.13**	**1.03**
Eastern	1	9.3		3	7.4		2	11.2		2	7.8		1.20	1.05
Mid-Western	7	11.1		6	8.2		8	12.8	*	7	8.7		1.15	1.06
Midland	8	11.7		2	7.2		6	12.1		3	8.1		1.03	1.12
North Eastern	6	10.9		5	8.1		7	12.7	*	1	7.2		1.16	0.90
North Western	5	10.9		1	7.1		1	10.7		5	8.6		0.98	1.21
South Eastern	3	10.7		4	7.4		3	11.5		4	8.4		1.07	1.12
Southern	4	10.7		7	9.1		4	11.5		6	8.6		1.08	0.95
Western	2	9.6		8	9.3		5	11.8		8	9.3	*	1.23	1.01

Table B19.2

Pancreas: annual average numbers of registrations and deaths by health authority within country and region of England, UK and Ireland, 1991–2000

Country/ region/ health authority	Number of registrations[1]		Number of deaths[2]	
	Males	Females	Males	Females
UK and Ireland	**3,470**	**3,730**	**3,370**	**3,600**
England	**2,770**	**2,970**	**2,670**	**2,860**
Northern and Yorkshire	**353**	**370**	**343**	**359**
Bradford	23	27	23	26
Calderdale and Kirklees	28	36	28	34
County Durham	40	36	36	36
East Riding	31	35	30	34
Gateshead and South Tyneside	21	21	22	21
Leeds	33	41	33	38
Newcastle and North Tyneside	26	29	24	28
North Cumbria	17	19	17	19
North Yorkshire	42	43	40	45
Northumberland	16	16	17	16
Sunderland	17	18	16	17
Tees	38	33	36	31
Wakefield	21	16	20	15
Trent	**296**	**318**	**280**	**297**
Barnsley	15	15	15	15
Doncaster	16	14	15	15
Leicestershire	45	49	43	43
Lincolnshire	41	44	41	40
North Derbyshire	20	23	19	22
North Nottinghamshire	27	26	24	24
Nottingham	38	39	33	36
Rotherham	18	16	18	13
Sheffield	30	39	29	36
South Humber	18	21	17	20
Southern Derbyshire	29	33	28	32
West Midlands	**294**	**302**	**287**	**293**
Birmingham	58	56	54	53
Coventry	15	17	14	16
Dudley	17	18	17	18
Herefordshire	12	9	11	9
North Staffordshire	28	29	27	28
Sandwell	17	16	18	17
Shropshire	22	24	23	23
Solihull	11	10	10	9
South Staffordshire	30	32	30	33
Walsall	14	14	15	14
Warwickshire	28	31	28	30
Wolverhampton	16	15	16	16
Worcestershire	24	30	25	29
North West	**358**	**398**	**353**	**386**
Bury and Rochdale	20	23	19	22
East Lancashire	31	30	33	30
Liverpool	25	30	25	29
Manchester	24	28	22	24
Morecambe Bay	16	21	17	21
North Cheshire	19	16	18	16
North West Lancashire	31	35	28	33
Salford and Trafford	23	27	22	26
Sefton	18	23	19	22
South Cheshire	34	36	36	34
South Lancashire	14	17	15	18
St Helen's and Knowsley	15	19	15	18
Stockport	17	16	16	16
West Pennine	26	27	25	27
Wigan and Bolton	24	29	26	29
Wirral	19	23	19	23
Eastern	**294**	**313**	**287**	**298**
Bedfordshire	25	24	26	25
Cambridgeshire	33	40	33	37
Hertfordshire	56	63	52	60
Norfolk	45	47	46	47
North Essex	52	53	51	52
South Essex	40	44	38	40
Suffolk	42	42	42	38

1 England, Wales and Scotland 1991-99; Northern Ireland 1993-99; Ireland 1994-99.
2 England, Wales and Northern Ireland 1991-2000; Scotland 1991-99; Ireland 1994-2000.

Country/ region/ health authority	Number of registrations[1]		Number of deaths[2]	
	Males	Females	Males	Females
London	**377**	**400**	**342**	**371**
Barking and Havering	27	26	25	25
Rarnet, Enfield and Haringey	42	45	39	41
Bexley, Bromley and Greenwich	45	51	41	48
Brent and Harrow	23	24	21	24
Camden and Islington	16	15	15	15
Croydon	17	18	17	17
Ealing, Hammersmith and Hounslow	31	34	29	31
East London and The City	28	28	25	27
Hillingdon	12	15	11	13
Kensington and Chelsea, and Westminster	20	19	19	17
Kingston and Richmond	16	20	14	19
Lambeth, Southwark and Lewisham	39	40	35	37
Merton, Sutton and Wandsworth	37	37	33	33
Redbridge and Waltham Forest	23	29	21	25
South East	**503**	**548**	**484**	**534**
Berkshire	37	44	36	42
Buckinghamshire	32	30	30	30
East Kent	45	45	44	45
East Surrey	27	28	24	26
East Sussex, Brighton and Hove	53	63	49	60
Isle of Wight, Portsmouth and South East Hampshire	39	43	39	44
North and Mid-Hampshire	28	28	29	30
Northamptonshire	33	40	33	40
Oxfordshire	32	32	31	32
Southampton and South West Hampshire	33	38	33	37
West Kent	50	59	47	55
West Surrey	44	37	40	36
West Sussex	51	61	50	58
South West	**292**	**316**	**299**	**321**
Avon	31	32	32	33
Cornwall and Isles of Scilly	51	55	51	57
Dorset	33	35	35	37
Gloucestershire	48	53	46	51
North and East Devon	31	33	34	35
Somerset	28	33	30	35
South and West Devon	37	39	39	39
Wiltshire	31	38	31	36
Wales	**187**	**204**	**165**	**185**
Bro Taf	33	34	30	32
Dyfed Powys	40	43	35	41
Gwent	31	33	28	31
Morgannwg	34	37	30	33
North Wales	48	57	42	49
Scotland	**286**	**324**	**272**	**305**
Argyll and Clyde	25	24	25	24
Ayrshire and Arran	25	26	23	24
Borders	8	8	7	8
Dumfries and Galloway	10	13	10	13
Fife	21	23	19	20
Forth Valley	14	15	14	15
Grampian	26	30	23	29
Greater Glasgow	50	60	46	53
Highland	10	12	10	12
Lanarkshire	27	32	27	31
Lothian	42	48	38	45
Orkney	1	1	1	1
Shetland	2	1	2	1
Tayside	24	27	24	26
Western Isles	3	3	4	3
Northern Ireland	**73**	**73**	**76**	**76**
Eastern	31	32	32	32
Northern	19	17	19	18
Southern	12	13	13	13
Western	11	11	13	13
Ireland	**160**	**162**	**182**	**175**
Eastern	42	51	51	55
Mid-Western	16	15	18	17
Midland	11	8	12	10
North Eastern	15	13	17	12
North Western	13	10	13	12
South Eastern	19	16	20	19
Southern	26	29	29	28
Western	18	21	22	22

Table **B20.1**

Prostate: annual average age-standardised incidence and mortality by health authority within country and region of England, UK and Ireland, 1991–2000

Country/ region/ health authority	Incidence[1]						Mortality[2]						M:I Ratio[3]	
		Males			Females			Males			Females		Males	Females
	Rank[4]	Rate[5]	Signif[6]	Rank	Rate	Signif	Rank	Rate	Signif	Rank	Rate	Signif		
UK and Ireland		**64.9**						**28.7**					**0.44**	
England		**64.3**						**28.7**					**0.45**	
Northern and Yorkshire		**58.5**	~					**26.8**	~				**0.46**	
Bradford	3	52.2	~				5	26.2	~				0.50	
Calderdale and Kirklees	12	67.5					9	27.6					0.41	
County Durham	6	56.7	~				7	26.9					0.48	
East Riding	10	61.2	~				13	28.6					0.47	
Gateshead and South Tyneside	1	47.7	~				3	25.1	~				0.53	
Leeds	8	59.6	~				2	24.8	~				0.42	
Newcastle and North Tyneside	5	54.0	~				4	25.9	~				0.48	
North Cumbria	4	53.5	~				10	28.0					0.52	
North Yorkshire	11	63.2					12	28.4					0.45	
Northumberland	2	48.4	~				1	24.0	~				0.50	
Sunderland	7	59.3	~				8	27.1					0.46	
Tees	9	60.2	~				6	26.8					0.44	
Wakefield	13	70.3	*				11	28.1					0.40	
Trent		**56.1**	~					**28.7**					**0.51**	
Barnsley	3	50.0	~				1	24.8	~				0.50	
Doncaster	8	58.4	~				10	30.8					0.53	
Leicestershire	4	53.5	~				2	26.5	~				0.50	
Lincolnshire	11	66.7					7	29.2					0.44	
North Derbyshire	7	56.5	~				11	31.6	*				0.56	
North Nottinghamshire	10	62.0					4	28.7					0.46	
Nottingham	9	58.9	~				6	28.9					0.49	
Rotherham	2	47.2	~				3	27.9					0.59	
Sheffield	1	45.2	~				8	29.2					0.65	
South Humber	6	56.2	~				9	29.9					0.53	
Southern Derbyshire	5	54.6	~				5	28.7					0.53	
West Midlands		**64.8**						**28.9**					**0.45**	
Birmingham	9	67.1					3	26.7	~				0.40	
Coventry	11	70.2	*				1	26.3					0.37	
Dudley	1	53.9	~				8	29.1					0.54	
Herefordshire	5	60.1					13	32.1	*				0.53	
North Staffordshire	2	56.6	~				7	29.0					0.51	
Sandwell	4	59.8	~				2	26.7					0.45	
Shropshire	12	75.7	*				11	31.0					0.41	
Solihull	13	80.7	*				10	30.9					0.38	
South Staffordshire	7	62.7					12	31.4	*				0.50	
Walsall	3	57.6	~				4	27.4					0.48	
Warwickshire	10	67.5					6	28.4					0.42	
Wolverhampton	6	61.6					5	28.4					0.46	
Worcestershire	8	65.2					9	30.0					0.46	
North West		**62.5**	~					**27.5**	~				**0.44**	
Bury and Rochdale	8	62.9					15	29.0					0.46	
East Lancashire	9	63.8					14	28.9					0.45	
Liverpool	2	55.8	~				2	26.2	~				0.47	
Manchester	4	57.9	~				4	26.4					0.46	
Morecambe Bay	6	60.7	~				13	28.7					0.47	
North Cheshire	10	65.4					16	29.5					0.45	
North West Lancashire	14	70.0	*				6	27.5					0.39	
Salford and Trafford	13	68.7					12	28.4					0.41	
Sefton	12	65.9					10	27.8					0.42	
South Cheshire	3	56.2	~				5	27.2					0.48	
South Lancashire	16	73.7	*				9	27.8					0.38	
St Helen's and Knowsley	11	65.5					11	28.2					0.43	
Stockport	15	70.9	*				7	27.7					0.39	
West Pennine	7	61.6					8	27.8					0.45	
Wigan and Bolton	1	52.0	~				1	24.6	~				0.47	
Wirral	5	60.1					3	26.2					0.44	
Eastern		**65.7**						**28.8**					**0.44**	
Bedfordshire	1	56.0	~				3	28.6					0.51	
Cambridgeshire	6	67.0					2	28.3					0.42	
Hertfordshire	4	66.0					4	28.9					0.44	
Norfolk	5	66.7					5	28.9					0.43	
North Essex	2	60.0	~				6	29.2					0.49	
South Essex	3	65.7					7	30.7	*				0.47	
Suffolk	7	76.4	*				1	26.8	~				0.35	

1 England, Wales and Scotland 1991-99; Northern Ireland 1993-99; Ireland 1994-99.
2 England, Wales and Northern Ireland 1991-2000; Scotland 1991-99; Ireland 1994-2000.
3 Mortality to incidence ratio; age-standardised mortality rate divided by age-standardised incidence rate.
4 Rank within region or country; 1 is lowest.
5 Directly age standardised using the European standard population; per 100,000 population.
6 Rate significantly different from that for UK and Ireland; * significantly higher; ~ significantly lower. See Appendix H.

Country/ region/ health authority	Incidence[1]						Mortality[2]						M:I Ratio[3]	
		Males			Females			Males			Females		Males	Females
	Rank[4]	Rate[5]	Signif[6]	Rank	Rate	Signif	Rank	Rate	Signif	Rank	Rate	Signif		
London		**66.2**	*					**27.6**	~				**0.42**	
Barking and Havering	10	71.5	*				14	31.0					0.43	
Barnet, Enfield and Haringey	7	69.1	*				7	27.9					0.40	
Bexley, Bromley and Greenwich	2	55.8	~				5	27.1					0.49	
Brent and Harrow	13	75.2	*				2	24.8	~				0.33	
Camden and Islington	14	76.6	*				3	25.1	~				0.33	
Croydon	4	62.4					4	26.8					0.43	
Ealing, Hammersmith and Hounslow	1	53.8	~				1	24.6	~				0.46	
East London and The City	5	63.2					8	28.0					0.44	
Hillingdon	6	64.3					11	28.4					0.44	
Kensington and Chelsea, and Westminster	9	71.3	*				10	28.2					0.40	
Kingston and Richmond	8	70.8	*				12	29.2					0.41	
Lambeth, Southwark and Lewisham	3	61.0	~				9	28.2					0.46	
Merton, Sutton and Wandsworth	12	73.2	*				13	29.5					0.40	
Redbridge and Waltham Forest	11	72.3	*				6	27.1					0.37	
South East		**69.7**	*					**30.6**	*				**0.44**	
Berkshire	12	85.2	*				4	30.1					0.35	
Buckinghamshire	11	75.1	*				5	30.2					0.40	
East Kent	1	52.8	~				12	32.1	*				0.61	
East Surrey	6	67.0					2	29.8					0.45	
East Sussex, Brighton and Hove	7	67.5					6	30.3					0.45	
Isle of Wight, Portsmouth and South East Hampshire	5	66.8					10	30.7	*				0.46	
North and Mid-Hampshire	3	63.0					11	30.9					0.49	
Northamptonshire	8	70.5	*				8	30.6					0.43	
Oxfordshire	2	60.7	~				9	30.6					0.50	
Southampton and South West Hampshire	13	86.1	*				3	29.9					0.35	
West Kent	10	73.6	*				13	32.4	*				0.44	
West Surrey	9	71.9	*				7	30.5					0.42	
West Sussex	4	66.7					1	29.8					0.45	
South West		**68.4**	*					**30.1**	*				**0.44**	
Avon	3	63.9					4	29.4					0.46	
Cornwall and Isles of Scilly	7	69.3	*				7	30.7					0.44	
Dorset	8	86.1	*				1	28.7					0.33	
Gloucestershire	4	65.5					2	29.0					0.44	
North and East Devon	1	60.3	~				8	33.5	*				0.55	
Somerset	2	63.0					5	30.4					0.48	
South and West Devon	5	65.6					3	29.3					0.45	
Wiltshire	6	67.2					6	30.6					0.45	
Wales		**67.4**	*					**28.7**					**0.43**	
Bro Taf	2	63.5					5	29.5					0.46	
Dyfed Powys	3	65.7					3	29.1					0.44	
Gwent	1	59.5	~				1	27.2					0.46	
Morgannwg	4	71.8	*				2	27.7					0.39	
North Wales	5	75.0	*				4	29.4					0.39	
Scotland		**66.7**	*					**27.2**	~				**0.41**	
Argyll and Clyde	8	62.4					10	28.9					0.46	
Ayrshire and Arran	1	56.0	~				6	27.2					0.48	
Borders	9	66.9					12	31.1					0.47	
Dumfries and Galloway	6	61.6					7	27.3					0.44	
Fife	11	70.6	*				8	28.3					0.40	
Forth Valley	15	85.9	*				4	26.3					0.31	
Grampian	13	74.3	*				11	29.0					0.39	
Greater Glasgow	3	59.1	~				3	25.7	~				0.43	
Highland	12	73.9	*				13	31.5					0.43	
Lanarkshire	2	57.8	~				2	25.7	~				0.44	
Lothian	14	77.9	*				1	25.1	~				0.32	
Orkney	5	60.0					9	28.7					0.48	
Shetland	10	69.3					15	37.4					0.54	
Tayside	7	62.1					5	26.9					0.43	
Western Isles	4	59.8					14	31.8					0.53	
Northern Ireland		**58.9**	~					**26.6**	~				**0.45**	
Eastern	1	52.3	~				1	24.5	~				0.47	
Northern	2	58.7	~				3	28.1					0.48	
Southern	3	65.1					2	25.8					0.40	
Western	4	71.5	*				4	31.1					0.44	
Ireland		**74.9**	*					**32.4**	*				**0.43**	
Eastern	8	83.2	*				4	31.8	*				0.38	
Mid-Western	1	60.6					3	31.3					0.52	
Midland	5	74.9					7	34.2	*				0.46	
North Eastern	3	69.2					2	30.8					0.45	
North Western	4	70.5					5	33.1	*				0.47	
South Eastern	7	83.1	*				8	35.4	*				0.43	
Southern	6	75.4	*				1	30.5					0.40	
Western	2	65.3					6	34.1	*				0.52	

Table **B20.2**

Prostate: annual average numbers of registrations and deaths by health authority within country and region of England, UK and Ireland, 1991–2000

Country/ region/ health authority	Number of registrations[1]		Number of deaths[2]	
	Males	Females	Males	Females
UK and Ireland	**22,500**		**10,000**	
England	**17,800**		**8,090**	
Northern and Yorkshire	**2,070**		**942**	
Bradford	125		62	
Calderdale and Kirklees	206		85	
County Durham	192		89	
East Riding	205		95	
Gateshead and South Tyneside	97		50	
Leeds	235		100	
Newcastle and North Tyneside	143		68	
North Cumbria	103		53	
North Yorkshire	296		138	
Northumberland	92		45	
Sunderland	88		39	
Tees	168		72	
Wakefield	116		46	
Trent	**1,660**		**855**	
Barnsley	63		31	
Doncaster	94		49	
Leicestershire	266		134	
Lincolnshire	281		125	
North Derbyshire	125		70	
North Nottinghamshire	140		65	
Nottingham	209		104	
Rotherham	63		37	
Sheffield	142		93	
South Humber	101		55	
Southern Derbyshire	177		94	
West Midlands	**1,920**		**854**	
Birmingham	355		141	
Coventry	122		46	
Dudley	92		48	
Herefordshire	71		39	
North Staffordshire	151		76	
Sandwell	100		44	
Shropshire	185		76	
Solihull	91		35	
South Staffordshire	187		91	
Walsall	82		39	
Warwickshire	197		84	
Wolverhampton	88		40	
Worcestershire	201		93	
North West	**2,250**		**986**	
Bury and Rochdale	124		56	
East Lancashire	176		80	
Liverpool	136		63	
Manchester	118		53	
Morecambe Bay	124		60	
North Cheshire	97		43	
North West Lancashire	216		86	
Salford and Trafford	171		70	
Sefton	118		50	
South Cheshire	216		104	
South Lancashire	118		44	
St Helen's and Knowsley	105		43	
Stockport	114		45	
West Pennine	150		68	
Wigan and Bolton	150		70	
Wirral	118		52	
Eastern	**2,050**		**914**	
Bedfordshire	149		75	
Cambridgeshire	249		108	
Hertfordshire	361		158	
Norfolk	378		171	
North Essex	321		161	
South Essex	260		120	
Suffolk	332		122	

1 *England, Wales and Scotland 1991-99; Northern Ireland 1993-99; Ireland 1994-99.*
2 *England, Wales and Northern Ireland 1991-2000; Scotland 1991-99; Ireland 1994-2000.*

Country/ region/ health authority	Number of registrations[1]		Number of deaths[2]	
	Males	Females	Males	Females
London	**2,210**		**944**	
Barking and Havering	160		68	
Barnet, Enfield and Haringey	263		112	
Bexley, Bromley and Greenwich	221		110	
Brent and Harrow	163		55	
Camden and Islington	121		41	
Croydon	96		43	
Ealing, Hammersmith and Hounslow	157		73	
East London and The City	154		68	
Hillingdon	79		36	
Kensington and Chelsea, and Westminster	114		47	
Kingston and Richmond	119		51	
Lambeth, Southwark and Lewisham	188		88	
Merton, Sutton and Wandsworth	213		89	
Redbridge and Waltham Forest	164		63	
South East	**3,430**		**1,570**	
Berkshire	305		109	
Buckinghamshire	236		96	
East Kent	216		141	
East Surrey	175		82	
East Sussex, Brighton and Hove	372		182	
Isle of Wight, Portsmouth and South East Hampshire	278		134	
North and Mid-Hampshire	166		82	
Northamptonshire	228		100	
Oxfordshire	189		98	
Southampton and South West Hampshire	284		102	
West Kent	369		163	
West Surrey	257		112	
West Sussex	356		169	
South West	**2,240**		**1,030**	
Avon	366		173	
Cornwall and Isles of Scilly	245		113	
Dorset	475		172	
Gloucestershire	228		102	
North and East Devon	220		129	
Somerset	214		109	
South and West Devon	269		125	
Wiltshire	226		105	
Wales	**1,190**		**509**	
Bro Taf	249		115	
Dyfed Powys	219		98	
Gwent	185		84	
Morgannwg	218		85	
North Wales	323		128	
Scotland	**1,800**		**723**	
Argyll and Clyde	138		63	
Ayrshire and Arran	117		55	
Borders	48		23	
Dumfries and Galloway	60		27	
Fife	135		53	
Forth Valley	123		37	
Grampian	199		76	
Greater Glasgow	271		116	
Highland	85		35	
Lanarkshire	149		62	
Lothian	300		97	
Orkney	7		4	
Shetland	7		4	
Tayside	148		64	
Western Isles	11		6	
Northern Ireland	**454**		**203**	
Eastern	172		81	
Northern	114		53	
Southern	86		34	
Western	82		35	
Ireland	**1,190**		**515**	
Eastern	365		132	
Mid-Western	89		47	
Midland	76		35	
North Eastern	95		42	
North Western	86		43	
South Eastern	150		63	
Southern	194		80	
Western	134		74	

Table B21.1

Stomach: annual average age-standardised incidence and mortality by health authority within country and region of England, UK and Ireland, 1991–2000

Country/ region/ health authority	Incidence[1] Males Rank[4]	Males Rate[5]	Signif[6]	Females Rank	Females Rate	Signif	Mortality[2] Males Rank	Males Rate	Signif	Females Rank	Females Rate	Signif	M:I Ratio[3] Males	Females
UK and Ireland		20.9			8.3			15.0			6.2		0.72	0.75
England		20.4	~		8.0	~		14.7	~		6.0	~	0.72	0.75
Northern and Yorkshire		23.5	*		10.0	*		17.0	*		7.4	*	0.72	0.75
Bradford	3	22.2		5	9.7	*	5	16.2		6	7.3	*	0.73	0.76
Calderdale and Kirklees	2	22.1		3	8.7		4	16.1		2	6.2		0.73	0.71
County Durham	10	25.6	*	12	11.3	*	11	20.2	*	12	9.0	*	0.79	0.80
East Riding	7	23.6	*	4	9.0		3	16.0		4	6.4		0.68	0.72
Gateshead and South Tyneside	12	26.4	*	6	10.8	*	12	20.2	*	10	8.5	*	0.77	0.78
Leeds	9	25.2	*	8	10.8	*	2	15.9		5	7.3	*	0.63	0.67
Newcastle and North Tyneside	8	24.4	*	7	10.8	*	10	19.2	*	8	8.1	*	0.79	0.75
North Cumbria	6	23.3		10	11.0	*	7	17.1	*	7	7.8	*	0.73	0.72
North Yorkshire	1	18.0	~	1	7.5		1	11.6	~	1	5.4	~	0.65	0.72
Northumberland	5	23.3		2	8.5		6	17.0		3	6.4		0.73	0.75
Sunderland	13	28.6	*	13	12.1	*	13	21.6	*	13	9.4	*	0.76	0.78
Tees	4	23.3	*	11	11.0	*	9	18.8	*	11	8.9	*	0.81	0.81
Wakefield	11	26.2	*	9	10.9	*	8	18.2	*	9	8.3	*	0.69	0.76
Trent		22.0	*		8.6			16.4	*		6.6	*	0.75	0.77
Barnsley	11	29.0	*	11	13.0	*	10	20.4	*	11	10.3	*	0.71	0.79
Doncaster	9	26.4	*	10	10.9	*	8	18.6	*	10	9.1	*	0.70	0.83
Leicestershire	3	20.1		1	7.2	~	3	14.5		1	5.1	~	0.72	0.70
Lincolnshire	1	18.0	~	2	7.2	~	1	12.8	~	2	5.4	~	0.71	0.74
North Derbyshire	5	21.9		7	8.4		7	17.5	*	7	7.1	*	0.80	0.84
North Nottinghamshire	6	23.1		4	7.7		6	17.4	*	3	5.8		0.75	0.75
Nottingham	4	21.1		6	8.2		4	16.1		4	6.1		0.76	0.74
Rotherham	10	26.7	*	8	10.5	*	11	20.8	*	9	8.4	*	0.78	0.79
Sheffield	8	24.8	*	9	10.6	*	9	20.0	*	8	8.0	*	0.80	0.75
South Humber	7	23.5	*	3	7.5		5	17.3	*	6	6.3		0.74	0.84
Southern Derbyshire	2	19.7		5	8.0		2	14.3		5	6.2		0.72	0.77
West Midlands		23.1	*		8.7	*		16.2	*		6.7	*	0.70	0.76
Birmingham	10	24.7	*	10	10.1	*	8	16.4		8	7.5	*	0.66	0.74
Coventry	5	21.2		7	8.3		3	13.2		3	5.2		0.62	0.63
Dudley	9	24.4	*	8	10.0	*	9	16.9		10	8.0	*	0.69	0.80
Herefordshire	1	17.4	~	1	5.2	~	1	12.6	~	1	4.2	~	0.72	0.81
North Staffordshire	13	30.4	*	13	11.8	*	13	24.0	*	13	9.2	*	0.79	0.77
Sandwell	12	27.6	*	11	10.4	*	12	21.0	*	12	9.0	*	0.76	0.86
Shropshire	3	19.8		3	6.9	~	5	13.4		5	5.6		0.68	0.80
Solihull	6	21.6		5	7.7		6	14.5		7	6.3		0.67	0.81
South Staffordshire	7	22.8		4	7.0	~	7	16.4		4	5.4	~	0.72	0.77
Walsall	11	26.9	*	12	11.6	*	11	19.1	*	11	8.4	*	0.71	0.72
Warwickshire	4	20.0		6	7.9		2	12.8	~	6	5.8		0.64	0.73
Wolverhampton	8	23.6		9	10.1		10	18.1	*	9	7.7	*	0.77	0.76
Worcestershire	2	18.2	~	2	5.9	~	4	13.2	~	2	4.7	~	0.73	0.79
North West		24.2	*		9.6	*		16.9	*		7.2	*	0.70	0.75
Bury and Rochdale	7	23.1		7	8.9		8	16.2		8	6.8		0.70	0.76
East Lancashire	8	23.9	*	8	9.1		10	17.2	*	9	7.1		0.72	0.78
Liverpool	16	31.0	*	15	11.7	*	16	22.1	*	13	8.8	*	0.71	0.76
Manchester	11	25.6	*	14	11.6	*	11	17.9	*	16	9.2	*	0.70	0.79
Morecambe Bay	1	19.1		6	8.6		1	13.9		6	6.6		0.73	0.76
North Cheshire	13	26.5	*	13	11.5	*	13	18.9	*	12	8.2	*	0.71	0.71
North West Lancashire	4	22.0		4	8.4		4	14.5		4	6.0		0.66	0.71
Salford and Trafford	9	24.3	*	10	10.2	*	7	16.2		10	7.4	*	0.67	0.73
Sefton	3	21.2		1	7.3		3	14.5		2	5.6		0.68	0.77
South Cheshire	2	20.2		2	7.8		2	14.3		7	6.7		0.71	0.85
South Lancashire	10	25.3	*	9	9.2		9	16.5		3	5.9		0.65	0.65
St Helen's and Knowsley	15	29.6	*	16	13.0	*	15	21.6	*	14	8.9	*	0.73	0.68
Stockport	5	22.3		3	7.9		5	14.9		1	5.5		0.67	0.69
West Pennine	14	26.6	*	12	11.2	*	14	19.6	*	15	9.0	*	0.74	0.81
Wigan and Bolton	12	25.7	*	11	10.6	*	12	18.1	*	11	7.8	*	0.71	0.74
Wirral	6	22.8		5	8.4		6	15.3		5	6.4		0.67	0.76
Eastern		18.2	~		6.4	~		13.2	~		4.9	~	0.72	0.76
Bedfordshire	7	20.9		6	7.8		6	14.5		6	5.2	~	0.70	0.67
Cambridgeshire	3	18.3	~	4	6.2	~	4	13.1	~	4	4.8	~	0.72	0.77
Hertfordshire	1	16.0	~	5	6.7	~	1	12.1	~	5	5.2	~	0.76	0.77
Norfolk	5	18.4	~	2	5.7	~	5	13.1	~	2	4.6	~	0.71	0.79
North Essex	4	18.3	~	3	5.9	~	3	12.5	~	3	4.6	~	0.68	0.78
South Essex	6	19.8		7	7.8		7	16.1		7	5.8		0.81	0.75
Suffolk	2	17.4	~	1	5.6	~	2	12.1	~	1	4.4	~	0.69	0.79

1 England, Wales and Scotland 1991-99; Northern Ireland 1993-99; Ireland 1994-99.
2 England, Wales and Northern Ireland 1991-2000; Scotland 1991-99; Ireland 1994-2000.
3 Mortality to incidence ratio; age-standardised mortality rate divided by age-standardised incidence rate.
4 Rank within region or country; 1 is lowest.
5 Directly age standardised using the European standard population; per 100,000 population.
6 Rate significantly different from that for UK and Ireland; * significantly higher; ~ significantly lower. See Appendix H.

Country/ region/ health authority	Incidence[1]						Mortality[2]						M:I Ratio[3]	
	Males			Females			Males			Females			Males	Females
	Rank[4]	Rate[5]	Signif[6]	Rank	Rate	Signif	Rank	Rate	Signif	Rank	Rate	Signif		
London		**19.6**	~		**7.9**	~		**14.6**			**6.0**		**0.74**	**0.76**
Barking and Havering	13	24.4	*	12	8.7		13	18.6	*	12	6.5		0.77	0.75
Barnet, Enfield and Haringey	6	18.3	~	8	7.8		5	13.2	~	9	5.9		0.72	0.75
Bexley, Bromley and Greenwich	8	19.1	~	5	7.4	~	8	13.6	..	8	5.7		0.71	0.78
Brent and Harrow	4	16.5	~	6	7.5		2	11.8	~	4	5.1	~	0.72	0.68
Camden and Islington	11	21.1		11	8.3		11	15.6		7	5.7		0.74	0.68
Croydon	5	17.6	~	3	6.5	~	6	13.5		5	5.4		0.77	0.84
Ealing, Hammersmith and Hounslow	7	18.7	~	9	7.9		9	14.0		10	6.0		0.75	0.76
East London and The City	14	25.9	*	14	10.7	*	14	19.2	*	14	7.9	*	0.74	0.74
Hillingdon	2	15.5	~	2	6.4	~	3	12.1	~	3	5.1	~	0.78	0.79
Kensington and Chelsea, and Westminster	3	15.9	~	1	6.4	~	4	12.4	~	1	5.0	~	0.78	0.78
Kingston and Richmond	1	15.4	~	4	6.8	~	1	11.8	~	2	5.0	~	0.77	0.74
Lambeth, Southwark and Lewisham	12	22.0		13	9.2		12	16.8	*	13	7.7	*	0.77	0.83
Merton, Sutton and Wandsworth	9	20.2		7	7.7		10	15.3		11	6.0		0.76	0.78
Redbridge and Waltham Forest	10	20.4		10	7.9		7	13.6		6	5.6		0.66	0.71
South East		**17.1**	~		**6.1**	~		**12.3**	~		**4.6**	~	**0.72**	**0.74**
Berkshire	8	17.3	~	7	6.3	~	6	11.7	~	5	4.5	~	0.68	0.71
Buckinghamshire	7	16.6	~	8	6.3	~	8	12.5	~	6	4.5	~	0.75	0.71
East Kent	5	16.0	~	3	5.4	~	12	14.2	~	9	4.9	~	0.88	0.90
East Surrey	4	15.4	~	1	4.9	~	1	10.6		2	3.9	~	0.69	0.79
East Sussex, Brighton and Hove	6	16.5	~	4	5.5	~	5	11.3	~	3	4.1	~	0.68	0.73
Isle of Wight, Portsmouth and South East Hampshire	12	21.0		13	7.5		9	12.9	~	7	4.5	~	0.62	0.60
North and Mid-Hampshire	9	17.3	~	6	6.1	~	7	12.1	~	8	4.5	~	0.69	0.74
Northamptonshire	11	19.5		11	7.0	~	13	14.7		13	5.4	~	0.75	0.77
Oxfordshire	2	14.7	~	9	6.5	~	3	10.8	~	10	4.9	~	0.73	0.76
Southampton and South West Hampshire	13	21.1		12	7.1	~	10	13.3	~	11	4.9	~	0.63	0.69
West Kent	10	18.2	~	10	6.6	~	11	13.5	~	12	5.2	~	0.74	0.78
West Surrey	3	15.2	~	2	5.2	~	4	11.1	~	1	3.7	~	0.73	0.72
West Sussex	1	14.5	~	5	5.7	~	2	10.8	~	4	4.4	~	0.74	0.77
South West		**16.6**	~		**6.5**	~		**11.9**	~		**4.8**	~	**0.72**	**0.73**
Avon	8	18.6	~	8	7.4	~	8	13.7	~	8	5.5	~	0.74	0.74
Cornwall and Isles of Scilly	4	16.7	~	2	6.0	~	6	11.7	~	3	4.5	~	0.70	0.76
Dorset	5	16.9	~	6	6.7	~	2	11.1	~	1	4.2	~	0.66	0.63
Gloucestershire	3	16.0	~	4	6.2	~	3	11.3	~	6	4.9	~	0.71	0.78
North and East Devon	1	14.3	~	1	6.0	~	5	11.4	~	2	4.3	~	0.80	0.72
Somerset	2	14.4	~	3	6.2	~	1	10.7	~	5	4.8	~	0.74	0.77
South and West Devon	7	17.9	~	7	6.8	~	7	13.5	~	7	5.1	~	0.75	0.75
Wiltshire	6	16.9	~	5	6.4	~	4	11.3	~	4	4.6	~	0.67	0.72
Wales		**26.2**	*		**10.3**	*		**18.2**	*		**7.3**	*	**0.69**	**0.71**
Bro Taf	2	24.8	*	4	10.8	*	3	18.0	*	4	7.9	*	0.73	0.73
Dyfed Powys	1	24.4	*	1	9.1		1	16.0		1	6.2		0.66	0.69
Gwent	3	25.4	*	3	10.6	*	4	18.6	*	5	8.0	*	0.73	0.75
Morgannwg	4	26.1	*	5	11.0	*	2	18.0	*	3	7.5	*	0.69	0.69
North Wales	5	29.7	*	2	9.8	*	5	19.9	*	2	6.8		0.67	0.70
Scotland		**22.6**	*		**10.3**	*		**15.9**	*		**7.7**	*	**0.71**	**0.74**
Argyll and Clyde	12	24.4	*	12	10.9	*	14	18.8	*	12	8.4	*	0.77	0.77
Ayrshire and Arran	6	19.5		6	9.6		10	15.7		9	7.0		0.81	0.73
Borders	4	18.2		2	7.5		4	11.3	~	4	6.2		0.62	0.83
Dumfries and Galloway	3	15.3	~	11	10.8	*	2	10.6	~	10	8.3	*	0.70	0.77
Fife	14	25.0	*	9	10.4	*	7	14.2		5	6.7		0.57	0.64
Forth Valley	13	24.6	*	8	9.8		9	15.6		7	6.9		0.64	0.70
Grampian	9	20.9		7	9.8	*	5	13.8		8	7.0		0.66	0.71
Greater Glasgow	15	26.9	*	15	11.8	*	15	19.4	*	14	8.7	*	0.72	0.74
Highland	8	20.2		3	8.1		6	13.9		3	5.8		0.69	0.72
Lanarkshire	10	22.1		10	10.8	*	12	16.9	*	13	8.6	*	0.77	0.80
Lothian	11	23.4	*	13	11.0	*	11	16.0		11	8.3	*	0.68	0.75
Orkney	1	10.8	~	1	1.8	~	3	11.0		2	3.4		1.02	1.89
Shetland	2	11.7	~	4	8.2		1	6.9	~	1	2.7	~	0.59	0.32
Tayside	7	20.0		5	8.5		8	14.5		6	6.8		0.73	0.79
Western Isles	5	19.0		14	11.2		13	17.6		15	9.5		0.93	0.84
Northern Ireland		**21.8**			**9.2**	*		**15.7**			**6.8**	*	**0.72**	**0.74**
Eastern	3	21.9		4	9.7	*	2	15.5		4	7.2	*	0.71	0.74
Northern	1	19.0		1	8.4		1	14.3		1	6.0		0.75	0.71
Southern	4	26.6	*	2	8.7		4	17.5		2	6.9		0.66	0.79
Western	2	21.0		3	9.9		3	17.0		3	7.1		0.81	0.72
Ireland		**18.7**	~		**8.6**			**14.4**			**7.1**	*	**0.77**	**0.83**
Eastern	8	23.3	*	7	10.2	*	8	17.7	*	8	8.1	*	0.76	0.79
Mid-Western	1	13.3	~	2	7.2		1	10.4	~	5	7.4		0.78	1.03
Midland	6	19.5		3	7.6		6	15.3		7	7.9		0.79	1.05
North Eastern	7	20.7		8	11.5	*	7	16.0		6	7.7		0.78	0.67
North Western	5	18.3		4	7.6		5	14.4		3	6.6		0.79	0.87
South Eastern	4	16.8	~	6	8.0		4	12.9		2	6.2		0.77	0.78
Southern	2	15.7		1	6.5	~	3	12.6	~	1	5.5		0.81	0.85
Western	3	15.9	~	5	7.6		2	11.9	~	4	6.9		0.75	0.90

Table **B21.2**

Stomach: annual average numbers of registrations and deaths by health authority within country and region of England, UK and Ireland, 1991–2000

Country/ region/ health authority	Number of registrations[1]		Number of deaths[2]	
	Males	Females	Males	Females
UK and Ireland	**6,980**	**4,200**	**5,090**	**3,280**
England	**5,480**	**3,240**	**4,000**	**2,540**
Northern and Yorkshire	**808**	**517**	**591**	**400**
Bradford	52	36	37	28
Calderdale and Kirklees	66	41	49	31
County Durham	85	56	68	46
East Riding	75	42	52	32
Gateshead and South Tyneside	53	33	40	26
Leeds	96	61	62	43
Newcastle and North Tyneside	63	44	50	34
North Cumbria	44	30	33	23
North Yorkshire	80	53	53	41
Northumberland	42	23	31	18
Sunderland	43	27	32	21
Tees	65	45	54	37
Wakefield	42	26	30	21
Trent	**632**	**361**	**477**	**286**
Barnsley	36	22	26	18
Doncaster	42	25	30	21
Leicestershire	99	52	72	39
Lincolnshire	72	41	53	31
North Derbyshire	48	27	39	23
North Nottinghamshire	51	24	39	19
Nottingham	72	41	55	32
Rotherham	35	21	27	17
Sheffield	74	53	61	41
South Humber	42	19	31	16
Southern Derbyshire	61	36	45	29
West Midlands	**666**	**372**	**475**	**297**
Birmingham	127	76	85	59
Coventry	35	19	22	13
Dudley	42	25	30	21
Herefordshire	19	9	14	7
North Staffordshire	80	47	64	39
Sandwell	44	26	34	22
Shropshire	47	25	32	21
Solihull	24	13	16	10
South Staffordshire	68	30	50	24
Walsall	38	24	27	17
Warwickshire	56	32	37	25
Wolverhampton	32	19	25	15
Worcestershire	55	27	40	23
North West	**843**	**532**	**594**	**414**
Bury and Rochdale	40	22	27	15
East Lancashire	45	27	31	21
Liverpool	63	39	46	31
Manchester	72	44	51	34
Morecambe Bay	50	37	35	29
North Cheshire	37	26	27	21
North West Lancashire	40	25	28	18
Salford and Trafford	64	40	43	31
Sefton	58	38	39	28
South Cheshire	36	21	25	17
South Lancashire	76	46	54	40
St Helen's and Knowsley	48	31	35	23
Stockport	35	20	24	15
West Pennine	63	45	47	37
Wigan and Bolton	74	45	52	35
Wirral	43	26	29	20
Eastern	**547**	**283**	**404**	**226**
Bedfordshire	54	28	38	20
Cambridgeshire	66	32	48	26
Hertfordshire	86	54	65	43
Norfolk	98	45	72	38
North Essex	95	45	67	37
South Essex	76	45	62	36
Suffolk	72	34	51	28

1 England, Wales and Scotland 1991-99; Northern Ireland 1993-99; Ireland 1994-99.
2 England, Wales and Northern Ireland 1991-2000; Scotland 1991-99; Ireland 1994-2000.

Country/ region/ health authority	Number of registrations[1]		Number of deaths[2]	
	Males	Females	Males	Females
London	**646**	**405**	**483**	**316**
Barking and Havering	53	29	41	22
Barnet, Enfield and Haringey	69	44	51	35
Bexley, Bromley and Greenwich	73	47	53	37
Brent and Harrow	36	24	26	17
Camden and Islington	34	20	25	14
Croydon	27	16	21	14
Ealing, Hammersmith and Hounslow	54	33	41	27
East London and The City	61	36	46	27
Hillingdon	19	13	15	10
Kensington and Chelsea, and Westminster	25	14	20	12
Kingston and Richmond	25	18	19	14
Lambeth, Southwark and Lewisham	68	44	52	37
Merton, Sutton and Wandsworth	58	37	45	29
Redbridge and Waltham Forest	45	29	30	22
South East	**817**	**457**	**600**	**359**
Berkshire	62	33	43	25
Buckinghamshire	52	29	40	21
East Kent	62	36	56	33
East Surrey	39	20	27	16
East Sussex, Brighton and Hove	83	52	60	40
Isle of Wight, Portsmouth and South East Hampshire	84	44	53	31
North and Mid-Hampshire	45	24	32	18
Northamptonshire	62	32	48	26
Oxfordshire	45	29	34	23
Southampton and South West Hampshire	67	34	43	25
West Kent	91	51	68	43
West Surrey	54	28	40	21
West Sussex	72	46	56	38
South West	**518**	**310**	**381**	**238**
Avon	104	66	79	51
Cornwall and Isles of Scilly	56	31	40	25
Dorset	85	50	58	36
Gloucestershire	53	31	38	24
North and East Devon	49	30	39	24
Somerset	46	30	36	24
South and West Devon	70	42	54	31
Wiltshire	54	31	37	23
Wales	**449**	**265**	**317**	**195**
Bro Taf	95	63	70	48
Dyfed Powys	78	42	52	31
Gwent	78	47	58	36
Morgannwg	77	49	54	35
North Wales	120	65	83	46
Scotland	**597**	**420**	**421**	**321**
Argyll and Clyde	54	39	42	31
Ayrshire and Arran	40	29	32	22
Borders	12	8	8	7
Dumfries and Galloway	14	15	10	11
Fife	46	30	26	20
Forth Valley	35	21	22	15
Grampian	55	37	36	28
Greater Glasgow	121	87	88	66
Highland	23	14	16	10
Lanarkshire	57	41	43	34
Lothian	88	65	61	50
Orkney	1	0	1	1
Shetland	1	1	1	0
Tayside	45	30	33	24
Western Isles	3	3	3	3
Northern Ireland	**163**	**101**	**117**	**79**
Eastern	69	48	50	37
Northern	36	22	27	17
Southern	34	16	22	14
Western	24	15	19	11
Ireland	**293**	**174**	**226**	**153**
Eastern	106	68	79	56
Mid-Western	19	14	15	14
Midland	19	9	15	9
North Eastern	29	19	22	13
North Western	21	10	17	10
South Eastern	30	17	23	14
Southern	40	21	32	19
Western	30	17	24	16

Table **B22.1**

Testis: annual average age-standardised incidence and mortality by health authority within country and region of England, UK and Ireland, 1991–2000

Country/ region/ health authority	Incidence[1]						Mortality[2]						M:I Ratio[3]	
	Males			Females			Males			Females			Males	Females
	Rank[4]	Rate[5]	Signif[6]	Rank	Rate	Signif	Rank	Rate	Signif	Rank	Rate	Signif		
UK and Ireland		5.9						0.3					0.06	
England		5.8						0.3					0.06	
Northern and Yorkshire		5.5	~					0.4					0.07	
Bradford	6	5.0					6	0.3					0.06	
Calderdale and Kirklees	10	6.0					2	0.2	~				0.03	
County Durham	5	5.0					11	0.6					0.11	
East Riding	13	6.7					13	0.6					0.09	
Gateshead and South Tyneside	2	4.3	~				9	0.4					0.08	
Leeds	11	6.0					3	0.2					0.04	
Newcastle and North Tyneside	3	4.8	~				1	0.1	~				0.03	
North Cumbria	9	5.9					4	0.2					0.04	
North Yorkshire	12	6.2					7	0.3					0.05	
Northumberland	4	4.9					5	0.3					0.06	
Sunderland	1	4.0	~				8	0.4					0.09	
Tees	8	5.7					12	0.6					0.10	
Wakefield	7	5.1					10	0.4					0.07	
Trent		5.2	~					0.4					0.07	
Barnsley	6	5.4					5	0.4					0.07	
Doncaster	9	6.0					10	0.5					0.09	
Leicestershire	1	4.4	~				1	0.2					0.05	
Lincolnshire	8	5.5					7	0.4					0.07	
North Derbyshire	11	6.2					8	0.4					0.07	
North Nottinghamshire	7	5.4					3	0.3					0.05	
Nottingham	5	5.2					2	0.3					0.05	
Rotherham	3	4.7					4	0.3					0.07	
Sheffield	2	4.7	~				9	0.5					0.10	
South Humber	4	4.9					11	0.7					0.14	
Southern Derbyshire	10	6.1					6	0.4					0.06	
West Midlands		6.0						0.4					0.06	
Birmingham	6	5.3					10	0.4					0.07	
Coventry	8	6.0					5	0.3					0.05	
Dudley	5	5.3					11	0.5					0.09	
Herefordshire	4	5.2					1	0.1	~				0.02	
North Staffordshire	1	4.5	~				12	0.6					0.13	
Sandwell	10	6.8					7	0.4					0.05	
Shropshire	12	7.5	*				6	0.3					0.05	
Solihull	13	7.8					2	0.2					0.03	
South Staffordshire	9	6.3					4	0.3					0.05	
Walsall	3	5.1					8	0.4					0.07	
Warwickshire	7	5.8					9	0.4					0.07	
Wolverhampton	2	5.1					13	0.6					0.12	
Worcestershire	11	7.1	*				3	0.3					0.04	
North West		5.9						0.5	*				0.08	
Bury and Rochdale	4	5.6					2	0.3					0.05	
East Lancashire	5	5.7					12	0.5					0.09	
Liverpool	1	4.8	~				15	0.7					0.14	
Manchester	9	6.3					14	0.6					0.10	
Morecambe Bay	7	5.9					8	0.4					0.08	
North Cheshire	13	6.4					6	0.4					0.06	
North West Lancashire	10	6.3					11	0.5					0.07	
Salford and Trafford	16	6.7					1	0.3					0.04	
Sefton	11	6.3					16	0.7					0.12	
South Cheshire	8	6.0					9	0.5					0.08	
South Lancashire	6	5.8					5	0.4					0.07	
St Helen's and Knowsley	15	6.7					3	0.3					0.05	
Stockport	12	6.3					7	0.4					0.07	
West Pennine	2	4.8	~				10	0.5					0.10	
Wigan and Bolton	3	5.6					4	0.3					0.06	
Wirral	14	6.5					13	0.6					0.09	
Eastern		6.1						0.3					0.05	
Bedfordshire	3	5.7					3	0.3					0.05	
Cambridgeshire	2	5.4					2	0.3					0.05	
Hertfordshire	4	6.0					1	0.2	~				0.03	
Norfolk	6	6.7					5	0.3					0.05	
North Essex	1	5.4					6	0.4					0.07	
South Essex	5	6.3					4	0.3					0.05	
Suffolk	7	7.1	*				7	0.4					0.06	

1 *England, Wales and Scotland 1991-99; Northern Ireland 1993-99; Ireland 1994-99.*
2 *England, Wales and Northern Ireland 1991-2000; Scotland 1991-99; Ireland 1994-2000.*
3 *Mortality to incidence ratio; age-standardised mortality rate divided by age-standardised incidence rate.*
4 *Rank within region or country; 1 is lowest.*
5 *Directly age standardised using the European standard population; per 100,000 population.*
6 *Rate significantly different from that for UK and Ireland; * significantly higher; ~ significantly lower. See Appendix H.*

Country/ region/ health authority	Incidence[1]						Mortality[2]						M:I Ratio[3]	
	Males			Females			Males			Females			Males	Females
	Rank[4]	Rate[5]	Signif[6]	Rank	Rate	Signif	Rank	Rate	Signif	Rank	Rate	Signif		
London		**4.7**	~					**0.2**	~				**0.04**	
Barking and Havering	10	5.1					14	0.4					0.08	
Barnet, Enfield and Haringey	7	4.8	~				12	0.3					0.06	
Bexley, Bromley and Greenwich	14	6.2					9	0.2					0.03	
Brent and Harrow	2	3.4	~				1	0.0	~				0.01	
Camden and Islington	4	4.2	~				2	0.0	~				0.01	
Croydon	6	4.7	~				8	0.2					0.04	
Ealing, Hammersmith and Hounslow	3	3.9	~				6	0.2	~				0.04	
East London and The City	1	2.9	~				4	0.1	~				0.04	
Hillingdon	8	4.9					3	0.1	~				0.02	
Kensington and Chelsea, and Westminster	11	5.3					7	0.2					0.03	
Kingston and Richmond	13	5.9					5	0.1	~				0.02	
Lambeth, Southwark and Lewisham	5	4.4	~				10	0.2					0.05	
Merton, Sutton and Wandsworth	12	5.8					11	0.2					0.04	
Redbridge and Waltham Forest	9	4.9					13	0.3					0.06	
South East		**6.6**	*					**0.3**					**0.04**	
Berkshire	7	6.6					4	0.2	~				0.03	
Buckinghamshire	11	7.5	*				12	0.4					0.05	
East Kent	5	6.3					10	0.4					0.06	
East Surrey	13	7.9	*				2	0.1	~				0.02	
East Sussex, Brighton and Hove	2	5.5					9	0.3					0.06	
Isle of Wight, Portsmouth and South East Hampshire	8	7.0	*				1	0.1	~				0.02	
North and Mid-Hampshire	4	6.0					6	0.2					0.04	
Northamptonshire	12	7.5	*				11	0.4					0.05	
Oxfordshire	6	6.4					7	0.3					0.04	
Southampton and South West Hampshire	10	7.3	*				8	0.3					0.04	
West Kent	3	5.9					13	0.4					0.07	
West Surrey	9	7.1	*				5	0.2					0.03	
West Sussex	1	5.4					3	0.2	~				0.03	
South West		**6.9**	*					**0.4**					**0.06**	
Avon	5	7.1	*				2	0.3					0.05	
Cornwall and Isles of Scilly	1	6.0					6	0.5					0.08	
Dorset	6	7.2	*				1	0.3					0.04	
Gloucestershire	2	6.2					3	0.4					0.06	
North and East Devon	8	7.6	*				7	0.7					0.09	
Somerset	4	7.1					8	0.7	*				0.10	
South and West Devon	3	6.3					4	0.4					0.06	
Wiltshire	7	7.2	*				5	0.4					0.06	
Wales		**5.9**						**0.4**					**0.07**	
Bro Taf	2	5.3					1	0.3					0.05	
Dyfed Powys	3	6.3					3	0.4					0.06	
Gwent	1	5.0					4	0.5					0.09	
Morgannwg	4	6.4					5	0.7	*				0.11	
North Wales	5	6.8					2	0.4					0.05	
Scotland		**7.0**	*					**0.4**					**0.05**	
Argyll and Clyde	6	6.5					4	0.1	~				0.02	
Ayrshire and Arran	9	6.7					8	0.4					0.05	
Borders	1	5.0					2	0.0	~				0.00	
Dumfries and Galloway	14	9.5	*				14	0.6					0.06	
Fife	5	6.5					9	0.4					0.06	
Forth Valley	10	6.7					5	0.2					0.03	
Grampian	12	8.2	*				13	0.6					0.07	
Greater Glasgow	8	6.6					12	0.4					0.07	
Highland	13	9.3	*				10	0.4					0.04	
Lanarkshire	7	6.6					7	0.4					0.05	
Lothian	11	7.3	*				6	0.3					0.05	
Orkney	2	5.4					1	0.0	~				0.00	
Shetland	15	11.4					15	1.8					0.16	
Tayside	3	6.1					11	0.4					0.07	
Western Isles	4	6.2					3	0.4	~				0.00	
Northern Ireland		**6.0**						**0.3**					**0.05**	
Eastern	2	5.9					1	0.2					0.04	
Northern	1	5.6					3	0.4					0.07	
Southern	4	6.4					2	0.3					0.04	
Western	3	6.2					4	0.6					0.09	
Ireland		**4.9**	~					**0.4**					**0.09**	
Eastern	7	5.0	~				3	0.4					0.07	
Mid-Western	3	4.0	~				7	0.5					0.14	
Midland	6	4.8					2	0.3					0.07	
North Eastern	4	4.5					4	0.4					0.09	
North Western	1	3.1	~				1	0.0	~				0.00	
South Eastern	2	3.8	~				6	0.4					0.11	
Southern	8	7.0					8	0.7					0.10	
Western	5	4.7					5	0.4					0.09	

Table B22.2

Testis: annual average numbers of registrations and deaths by health authority within country and region of England, UK and Ireland, 1991–2000

Country/ region/ health authority	Number of registrations[1]		Number of deaths[2]	
	Males	Females	Males	Females
UK and Ireland	**1,830**		**109**	
England	**1,430**		**83**	
Northern and Yorkshire	**171**		**11**	
Bradford	12		1	
Calderdale and Kirklees	17		1	
County Durham	15		2	
East Riding	18		2	
Gateshead and South Tyneside	8		1	
Leeds	22		1	
Newcastle and North Tyneside	11		0	
North Cumbria	9		0	
North Yorkshire	22		1	
Northumberland	7		1	
Sunderland	6		1	
Tees	15		2	
Wakefield	8		1	
Trent	**134**		**10**	
Barnsley	6		0	
Doncaster	9		1	
Leicestershire	20		1	
Lincolnshire	16		1	
North Derbyshire	11		1	
North Nottinghamshire	11		1	
Nottingham	17		1	
Rotherham	6		0	
Sheffield	13		1	
South Humber	8		1	
Southern Derbyshire	17		1	
West Midlands	**156**		**10**	
Birmingham	26		2	
Coventry	10		1	
Dudley	8		1	
Herefordshire	4		0	
North Staffordshire	10		1	
Sandwell	10		1	
Shropshire	16		1	
Solihull	7		0	
South Staffordshire	19		1	
Walsall	7		1	
Warwickshire	14		1	
Wolverhampton	6		1	
Worcestershire	19		1	
North West	**190**		**15**	
Bury and Rochdale	9		1	
East Lancashire	11		1	
Liverpool	14		1	
Manchester	11		1	
Morecambe Bay	13		1	
North Cheshire	9		1	
North West Lancashire	10		1	
Salford and Trafford	13		1	
Sefton	15		1	
South Cheshire	8		1	
South Lancashire	20		2	
St Helen's and Knowsley	11		1	
Stockport	9		1	
West Pennine	11		1	
Wigan and Bolton	16		1	
Wirral	10		1	
Eastern	**160**		**9**	
Bedfordshire	17		1	
Cambridgeshire	19		1	
Hertfordshire	31		1	
Norfolk	25		1	
North Essex	24		2	
South Essex	21		1	
Suffolk	23		1	

1 England, Wales and Scotland 1991-99; Northern Ireland 1993-99; Ireland 1994-99.
2 England, Wales and Northern Ireland 1991-2000; Scotland 1991-99; Ireland 1994-2000.

Country/ region/ health authority	Number of registrations[1]		Number of deaths[2]	
	Males	Females	Males	Females
London	**176**		**7**	
Barking and Havering	10		1	
Barnet, Enfield and Haringey	20		1	
Bexley, Bromley and Greenwich	22		1	
Brent and Harrow	8		0	
Camden and Islington	9		0	
Croydon	8		0	
Ealing, Hammersmith and Hounslow	14		1	
East London and The City	9		0	
Hillingdon	6		0	
Kensington and Chelsea, and Westminster	10		0	
Kingston and Richmond	10		0	
Lambeth, Southwark and Lewisham	19		1	
Merton, Sutton and Wandsworth	20		1	
Redbridge and Waltham Forest	11		1	
South East	**279**		**12**	
Berkshire	28		1	
Buckinghamshire	25		1	
East Kent	17		1	
East Surrey	16		0	
East Sussex, Brighton and Hove	18		1	
Isle of Wight, Portsmouth and South East Hampshire	23		1	
North and Mid-Hampshire	17		1	
Northamptonshire	23		1	
Oxfordshire	20		1	
Southampton and South West Hampshire	19		1	
West Kent	29		2	
West Surrey	23		1	
West Sussex	19		1	
South West	**160**		**10**	
Avon	35		2	
Cornwall and Isles of Scilly	13		1	
Dorset	22		1	
Gloucestershire	17		1	
North and East Devon	16		2	
Somerset	16		2	
South and West Devon	18		1	
Wiltshire	22		1	
Wales	**82**		**6**	
Bro Taf	19		1	
Dyfed Powys	14		1	
Gwent	13		1	
Morgannwg	15		2	
North Wales	21		1	
Scotland	**180**		**10**	
Argyll and Clyde	14		0	
Ayrshire and Arran	12		1	
Borders	2		0	
Dumfries and Galloway	7		0	
Fife	11		1	
Forth Valley	9		0	
Grampian	23		2	
Greater Glasgow	30		2	
Highland	10		0	
Lanarkshire	19		1	
Lothian	29		1	
Orkney	1		0	
Shetland	1		0	
Tayside	12		1	
Western Isles	1		0	
Northern Ireland	**49**		**3**	
Eastern	19		1	
Northern	12		1	
Southern	10		0	
Western	8		1	
Ireland	**92**		**8**	
Eastern	34		2	
Mid-Western	7		1	
Midland	5		0	
North Eastern	7		1	
North Western	3		0	
South Eastern	8		1	
Southern	20		2	
Western	8		1	

Table **B23.1**

Uterus: annual average age-standardised incidence and mortality by health authority within country and region of England, UK and Ireland, 1991–2000

Country/ region/ health authority	Incidence[1]						Mortality[2]						M:I Ratio[3]	
	Males			Females			Males			Females			Males	Females
	Rank[4]	Rate[5]	Signif[6]	Rank	Rate	Signif	Rank	Rate	Signif	Rank	Rate	Signif		
UK and Ireland					13.0						3.3			0.25
England					13.1						3.3			0.25
Northern and Yorkshire					10.9	~					2.9	~		0.27
Bradford				9	11.5					4	2.6	~		0.23
Calderdale and Kirklees				11	11.9					10	3.1			0.26
County Durham				7	11.0	~				5	2.9			0.26
East Riding				12	13.3					13	3.6			0.27
Gateshead and South Tyneside				5	10.2	~				8	2.9			0.29
Leeds				6	11.0	~				2	2.4	~		0.22
Newcastle and North Tyneside				4	9.6	~				3	2.5	~		0.26
North Cumbria				10	11.6					11	3.2			0.27
North Yorkshire				8	11.3	~				9	2.9			0.26
Northumberland				2	7.9	~				1	2.4	~		0.31
Sunderland				1	7.0	~				7	2.9			0.41
Tees				3	9.4	~				12	3.2			0.34
Wakefield				13	13.7					6	2.9			0.21
Trent					13.0						3.5			0.27
Barnsley				11	15.6	*				6	3.7			0.24
Doncaster				2	11.1	~				8	3.7			0.33
Leicestershire				5	12.7					5	3.6			0.28
Lincolnshire				7	13.8					7	3.7			0.27
North Derbyshire				8	14.3					11	3.9			0.27
North Nottinghamshire				10	15.0	*				9	3.8			0.25
Nottingham				4	12.5					2	3.1			0.25
Rotherham				1	10.3	~				3	3.2			0.31
Sheffield				3	11.2	~				1	3.0			0.26
South Humber				9	14.3					10	3.8			0.27
Southern Derbyshire				6	13.1					4	3.4			0.26
West Midlands					14.3	*					3.5			0.25
Birmingham				7	14.3	*				7	3.5			0.25
Coventry				10	14.9					6	3.5			0.24
Dudley				13	15.6	*				12	4.0			0.25
Herefordshire				9	14.8					11	3.9			0.26
North Staffordshire				11	14.9	*				13	4.0			0.27
Sandwell				8	14.5					3	3.0			0.21
Shropshire				5	14.1					9	3.7			0.26
Solihull				2	13.2					1	2.7			0.20
South Staffordshire				12	15.0	*				4	3.1			0.21
Walsall				1	12.2					2	2.9			0.24
Warwickshire				3	13.5					5	3.5			0.26
Wolverhampton				6	14.3					10	3.7			0.26
Worcestershire				4	14.0					8	3.6			0.26
North West					12.2	~					3.0	~		0.25
Bury and Rochdale				2	11.1	~				7	2.9			0.26
East Lancashire				3	11.5	~				6	2.9			0.25
Liverpool				7	11.9					13	3.4			0.29
Manchester				8	12.1					16	3.7			0.30
Morecambe Bay				10	12.3					8	2.9			0.24
North Cheshire				16	14.0					10	3.1			0.22
North West Lancashire				9	12.2					1	2.4	~		0.20
Salford and Trafford				1	10.9	~				5	2.8			0.26
Sefton				4	11.6					14	3.5			0.30
South Cheshire				12	12.4					15	3.5			0.28
South Lancashire				14	13.1					4	2.7			0.21
St Helen's and Knowsley				13	12.9					2	2.5			0.20
Stockport				5	11.7					11	3.3			0.28
West Pennine				15	14.0					12	3.4			0.24
Wigan and Bolton				11	12.4					3	2.6	~		0.21
Wirral				6	11.9					9	3.0			0.25
Eastern					14.2	*					3.2			0.23
Bedfordshire				3	14.3					6	3.5			0.24
Cambridgeshire				6	15.7	*				3	3.2			0.20
Hertfordshire				1	11.7	~				1	3.0			0.25
Norfolk				5	15.7	*				7	3.5			0.22
North Essex				2	12.3					2	3.1			0.25
South Essex				4	14.7	*				5	3.4			0.23
Suffolk				7	16.2	*				4	3.3			0.20

1 England, Wales and Scotland 1991-99; Northern Ireland 1993-99; Ireland 1994-99.
2 England, Wales and Northern Ireland 1991-2000; Scotland 1991-99; Ireland 1994-2000.
3 Mortality to incidence ratio; age-standardised mortality rate divided by age-standardised incidence rate.
4 Rank within region or country; 1 is lowest.
5 Directly age standardised using the European standard population; per 100,000 population.
6 Rate significantly different from that for UK and Ireland; * significantly higher; ~ significantly lower. See Appendix H.

Country/ region/ health authority	Incidence[1]						Mortality[2]						M:I Ratio[3]	
	Males			Females			Males			Females			Males	Females
	Rank[4]	Rate[5]	Signif[6]	Rank	Rate	Signif	Rank	Rate	Signif	Rank	Rate	Signif		
London					12.8						3.4			0.26
Barking and Havering				13	14.2					11	3.7			0.26
Barnet, Enfield and Haringey				10	13.3					6	3.3			0.25
Bexley, Bromley and Greenwich				11	13.4					7	3.4			0.25
Brent and Harrow				8	13.2					2	2.8			0.21
Camden and Islington				6	12.2					9	3.5			0.28
Croydon				9	13.2					14	4.2			0.32
Ealing, Hammersmith and Hounslow				4	11.7					13	4.2	*		0.36
East London and The City				7	12.2					5	3.2			0.26
Hillingdon				1	10.7	~				10	3.6			0.33
Kensington and Chelsea, and Westminster				3	11.4					1	2.7			0.23
Kingston and Richmond				2	11.3					4	2.9			0.25
Lambeth, Southwark and Lewisham				5	12.0					12	3.8			0.32
Merton, Sutton and Wandsworth				14	14.5					8	3.4			0.23
Redbridge and Waltham Forest				12	14.1					3	2.8			0.20
South East					13.5	*					3.3			0.24
Berkshire				10	14.6	*				9	3.5			0.24
Buckinghamshire				5	12.8					3	3.1			0.24
East Kent				4	12.7					12	3.6			0.28
East Surrey				7	13.0					2	2.9			0.22
East Sussex, Brighton and Hove				2	12.2					5	3.2			0.26
Isle of Wight, Portsmouth and South East Hampshire				9	14.3					7	3.2			0.23
North and Mid-Hampshire				3	12.3					8	3.3			0.27
Northamptonshire				13	16.4	*				13	3.8			0.23
Oxfordshire				11	14.6					1	2.7			0.19
Southampton and South West Hampshire				12	15.1	*				11	3.6			0.24
West Kent				8	13.6					10	3.6			0.26
West Surrey				1	11.8					6	3.2			0.27
West Sussex				6	13.0					4	3.2			0.25
South West					14.4	*					3.6	*		0.25
Avon				5	14.5	*				2	3.3			0.23
Cornwall and Isles of Scilly				7	15.5	*				8	4.2	*		0.27
Dorset				6	14.7	*				5	3.6			0.24
Gloucestershire				1	12.7					3	3.4			0.27
North and East Devon				4	14.4					4	3.5			0.25
Somerset				3	14.0					7	3.8			0.27
South and West Devon				2	13.4					6	3.6			0.27
Wiltshire				8	15.7	*				1	3.2			0.21
Wales					13.5						3.8	*		0.28
Bro Taf				1	12.1					5	4.0	*		0.33
Dyfed Powys				4	14.1					4	3.9			0.27
Gwent				2	12.2					1	3.5			0.28
Morgannwg				5	15.7	*				3	3.8			0.25
North Wales				3	13.6					2	3.6			0.26
Scotland					12.3	~					3.0			0.25
Argyll and Clyde				1	10.6	~				8	3.1			0.29
Ayrshire and Arran				4	11.0	~				11	3.6			0.33
Borders				13	15.9					1	2.4			0.15
Dumfries and Galloway				11	13.9					12	3.7			0.27
Fife				10	13.7					5	2.8			0.20
Forth Valley				7	12.4					4	2.8			0.22
Grampian				9	13.5					6	2.9			0.21
Greater Glasgow				5	11.4	~				7	3.0			0.26
Highland				12	14.1					2	2.7			0.19
Lanarkshire				2	10.8	~				9	3.1			0.29
Lothian				8	13.1					3	2.7	~		0.21
Orkney				3	11.0					14	5.0			0.46
Shetland				15	17.8					13	4.1			0.23
Tayside				6	12.3					10	3.3			0.27
Western Isles				14	17.7					15	8.9	*		0.50
Northern Ireland					11.9	~					2.8	~		0.23
Eastern				2	11.9					3	2.9			0.24
Northern				3	12.1					2	2.5	~		0.20
Southern				4	12.5					4	3.4			0.27
Western				1	10.8					1	2.4	~		0.22
Ireland					12.7						3.0			0.23
Eastern				3	12.1					1	2.5	~		0.20
Mid-Western				5	13.2					7	3.5			0.26
Midland				2	11.3					3	2.8			0.25
North Eastern				1	11.1					4	2.8			0.26
North Western				7	13.7					5	2.9			0.21
South Eastern				4	12.7					8	3.9			0.31
Southern				8	14.5					6	3.4			0.24
Western				6	13.7					2	2.7			0.20

Table **B23.2**

Uterus: annual average numbers of registrations and deaths by health authority within country and region of England, UK and Ireland, 1991–2000

Country/ region/ health authority	Number of registrations[1]		Number of deaths[2]	
	Males	Females	Males	Females
UK and Ireland		4,880		1,530
England		3,930		1,230
Northern and Yorkshire		428		143
Bradford		30		10
Calderdale and Kirklees		42		13
County Durham		42		13
East Riding		47		15
Gateshead and South Tyneside		24		8
Leeds		48		14
Newcastle and North Tyneside		29		10
North Cumbria		24		9
North Yorkshire		57		20
Northumberland		17		6
Sunderland		12		6
Tees		31		13
Wakefield		26		7
Trent		407		137
Barnsley		21		6
Doncaster		20		8
Leicestershire		67		24
Lincolnshire		57		19
North Derbyshire		34		12
North Nottinghamshire		37		11
Nottingham		47		15
Rotherham		16		6
Sheffield		37		13
South Humber		27		9
Southern Derbyshire		45		15
West Midlands		455		138
Birmingham		79		24
Coventry		27		8
Dudley		30		10
Herefordshire		17		6
North Staffordshire		43		14
Sandwell		26		7
Shropshire		37		12
Solihull		17		4
South Staffordshire		51		13
Walsall		19		6
Warwickshire		42		13
Wolverhampton		20		7
Worcestershire		46		14
North West		495		155
Bury and Rochdale		25		6
East Lancashire		25		9
Liverpool		36		12
Manchester		32		12
Morecambe Bay		26		10
North Cheshire		26		7
North West Lancashire		24		7
Salford and Trafford		39		11
Sefton		31		10
South Cheshire		25		9
South Lancashire		52		18
St Helen's and Knowsley		26		6
Stockport		22		8
West Pennine		39		12
Wigan and Bolton		41		11
Wirral		27		9
Eastern		465		131
Bedfordshire		41		12
Cambridgeshire		60		15
Hertfordshire		71		22
Norfolk		87		24
North Essex		71		22
South Essex		66		18
Suffolk		69		18

1 England, Wales and Scotland 1991-99; Northern Ireland 1993-99; Ireland 1994-99.
2 England, Wales and Northern Ireland 1991-2000; Scotland 1991-99; Ireland 1994-2000.

Country/ region/ health authority	Number of registrations[1]		Number of deaths[2]	
	Males	Females	Males	Females
London		481		150
Barking and Havering		35		10
Barnet, Enfield and Haringey		56		17
Bexley, Bromley and Greenwich		60		19
Brent and Harrow		33		8
Camden and Islington		22		8
Croydon		24		9
Ealing, Hammersmith and Hounslow		39		15
East London and The City		32		10
Hillingdon		15		6
Kensington and Chelsea, and Westminster		21		6
Kingston and Richmond		21		7
Lambeth, Southwark and Lewisham		41		15
Merton, Sutton and Wandsworth		48		13
Redbridge and Waltham Forest		35		9
South East		716		222
Berkshire		59		17
Buckinghamshire		46		13
East Kent		55		20
East Surrey		35		11
East Sussex, Brighton and Hove		73		26
Isle of Wight, Portsmouth and South East Hampshire		63		18
North and Mid-Hampshire		38		12
Northamptonshire		56		16
Oxfordshire		45		11
Southampton and South West Hampshire		50		15
West Kent		80		26
West Surrey		47		15
West Sussex		72		23
South West		478		154
Avon		87		27
Cornwall and Isles of Scilly		54		18
Dorset		79		26
Gloucestershire		45		15
North and East Devon		49		16
Somerset		49		17
South and West Devon		58		20
Wiltshire		56		15
Wales		253		90
Bro Taf		51		21
Dyfed Powys		47		17
Gwent		42		15
Morgannwg		50		16
North Wales		63		21
Scotland		387		117
Argyll and Clyde		28		10
Ayrshire and Arran		27		11
Borders		12		3
Dumfries and Galloway		15		4
Fife		29		7
Forth Valley		21		6
Grampian		40		11
Greater Glasgow		63		20
Highland		18		4
Lanarkshire		36		12
Lothian		57		14
Orkney		2		1
Shetland		2		1
Tayside		32		12
Western Isles		3		2
Northern Ireland		104		31
Eastern		45		15
Northern		26		7
Southern		19		6
Western		13		4
Ireland		212		59
Eastern		70		16
Mid-Western		20		6
Midland		11		3
North Eastern		15		4
North Western		14		4
South Eastern		24		8
Southern		38		11
Western		22		6

Appendix C

Population estimates

This appendix gives mid-year population estimates for 1996, for the UK and Ireland, by country, by health regional office (within England), and by health authority. For the countries of the UK, mid-year population estimates (revised in light of the 2001 UK Census) were used. For Ireland, the 1996 populations are official national census figures published by the Central Statistics Office.

The population figures used in the calculation of the incidence and mortality rates for this volume were the summed populations of each country for the various time periods of cancer incidence or mortality data provided by that country. The mid-year population estimates for 1996 are given here as an estimate of the annual population during the period covered by this atlas, 1996 being approximately the mid-point of the period covered by most of the data.

Tables C1-C7 give rounded population estimates (to the nearest 100), by sex and five-year age group for: UK and Ireland combined; UK; England; Wales; Scotland; Northern Ireland; and Ireland. The European standard population is given in Table C8. Figures C1-C8 are the corresponding population pyramids.

Table C9 gives rounded population estimates (to the nearest 100) for all ages combined, by sex, country, health regional office (within England), and health authority, for the UK and Ireland.

Table C1
UK and Ireland

	Males	Females
0-4	2,045,500	1,943,800
5-9	2,125,400	2,032,400
10-14	2,030,300	1,945,400
15-19	1,924,400	1,883,900
20-24	1,999,500	2,007,200
25-29	2,327,000	2,367,000
30-34	2,434,700	2,485,000
35-39	2,206,100	2,238,200
40-44	1,993,100	2,014,700
45-49	2,148,700	2,165,100
50-54	1,820,500	1,832,200
55-59	1,560,600	1,589,300
60-64	1,427,000	1,493,400
65-69	1,302,400	1,471,600
70-74	1,117,400	1,408,800
75-79	770,100	1,142,300
80-84	473,000	881,900
85+	276,300	785,100
All ages	29,981,900	31,687,200

Table C2
UK

	Males	Females
0-4	1,916,800	1,822,100
5-9	1,980,000	1,894,800
10-14	1,862,900	1,786,700
15-19	1,750,400	1,718,300
20-24	1,850,400	1,863,000
25-29	2,197,600	2,237,300
30-34	2,307,000	2,351,800
35-39	2,080,000	2,108,700
40-44	1,873,100	1,894,400
45-49	2,034,900	2,053,600
50-54	1,725,600	1,740,300
55-59	1,482,800	1,513,300
60-64	1,358,300	1,424,200
65-69	1,242,100	1,405,000
70-74	1,067,300	1,346,400
75-79	734,900	1,093,100
80-84	451,900	847,200
85+	265,700	761,000
All ages	28,181,700	29,861,300

Table C3
England

	Males	Females
0-4	1,603,700	1,522,800
5-9	1,646,800	1,577,800
10-14	1,538,100	1,473,700
15-19	1,442,600	1,416,800
20-24	1,530,900	1,546,200
25-29	1,841,400	1,875,800
30-34	1,941,300	1,971,900
35-39	1,739,500	1,754,800
40-44	1,562,300	1,579,100
45-49	1,707,000	1,721,300
50-54	1,446,100	1,454,300
55-59	1,236,600	1,254,300
60-64	1,131,600	1,175,400
65-69	1,037,100	1,162,900
70-74	892,900	1,118,800
75-79	619,000	913,100
80-84	383,500	713,200
85+	226,700	643,000
All ages	23,526,800	24,875,300

Table C4
Wales

	Males	Females
0-4	91,700	87,800
5-9	99,400	94,400
10-14	94,600	91,200
15-19	88,100	86,700
20-24	88,800	87,100
25-29	98,400	99,600
30-34	103,500	106,400
35-39	95,700	99,200
40-44	91,200	92,800
45-49	102,700	103,600
50-54	88,600	89,600
55-59	77,400	78,500
60-64	71,700	75,400
65-69	67,400	75,600
70-74	60,400	75,500
75-79	40,800	61,300
80-84	23,600	44,800
85+	13,600	40,000
All ages	1,397,500	1,489,500

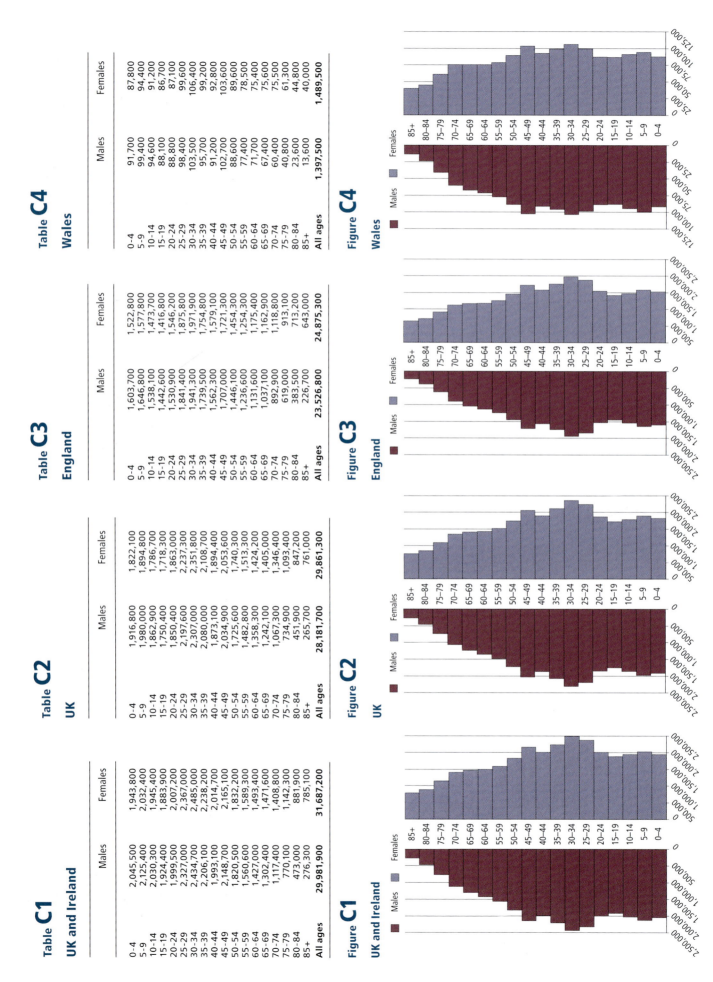

Figure C1
UK and Ireland

Figure C2
UK

Figure C3
England

Figure C4
Wales

Table C5
Scotland

	Males	Females
0-4	158,500	151,600
5-9	165,700	157,900
10-14	162,500	156,700
15-19	157,300	153,100
20-24	169,500	171,000
25-29	193,300	198,100
30-34	198,500	208,100
35-39	187,100	195,000
40-44	168,200	171,700
45-49	176,500	178,700
50-54	146,200	150,400
55-59	131,700	141,300
60-64	122,000	136,700
65-69	108,500	130,800
70-74	89,300	119,200
75-79	58,000	93,300
80-84	34,700	70,200
85+	19,700	61,400
All ages	2,447,000	2,645,200

Table C6
Northern Ireland

	Males	Females
0-4	62,900	59,900
5-9	68,100	64,700
10-14	67,700	65,100
15-19	62,500	61,700
20-24	61,200	58,700
25-29	64,500	63,700
30-34	63,700	65,300
35-39	57,700	59,600
40-44	51,400	50,800
45-49	48,800	50,000
50-54	44,800	46,100
55-59	37,100	39,200
60-64	33,000	36,600
65-69	29,300	35,800
70-74	24,700	32,900
75-79	17,200	25,700
80-84	10,000	19,000
85+	5,700	16,600
All ages	810,300	851,400

Table C7
Ireland

	Males	Females
0-4	128,700	121,700
5-9	145,300	137,600
10-14	167,400	158,700
15-19	174,000	165,600
20-24	149,100	144,200
25-29	129,400	129,700
30-34	127,700	133,200
35-39	126,100	129,500
40-44	120,100	120,400
45-49	113,800	111,600
50-54	94,800	91,800
55-59	77,800	76,000
60-64	68,700	69,300
65-69	60,300	66,600
70-74	50,100	62,400
75-79	35,200	48,900
80-84	21,100	34,700
85+	10,600	24,100
All ages	1,800,200	1,825,900

Table C8
European standard

	Males	Females
0-4	8,000	8,000
5-9	7,000	7,000
10-14	7,000	7,000
15-19	7,000	7,000
20-24	7,000	7,000
25-29	7,000	7,000
30-34	7,000	7,000
35-39	7,000	7,000
40-44	7,000	7,000
45-49	7,000	7,000
50-54	7,000	7,000
55-59	6,000	6,000
60-64	5,000	5,000
65-69	4,000	4,000
70-74	3,000	3,000
75-79	2,000	2,000
80-84	1,000	1,000
85+	1,000	1,000
All ages	100,000	100,000

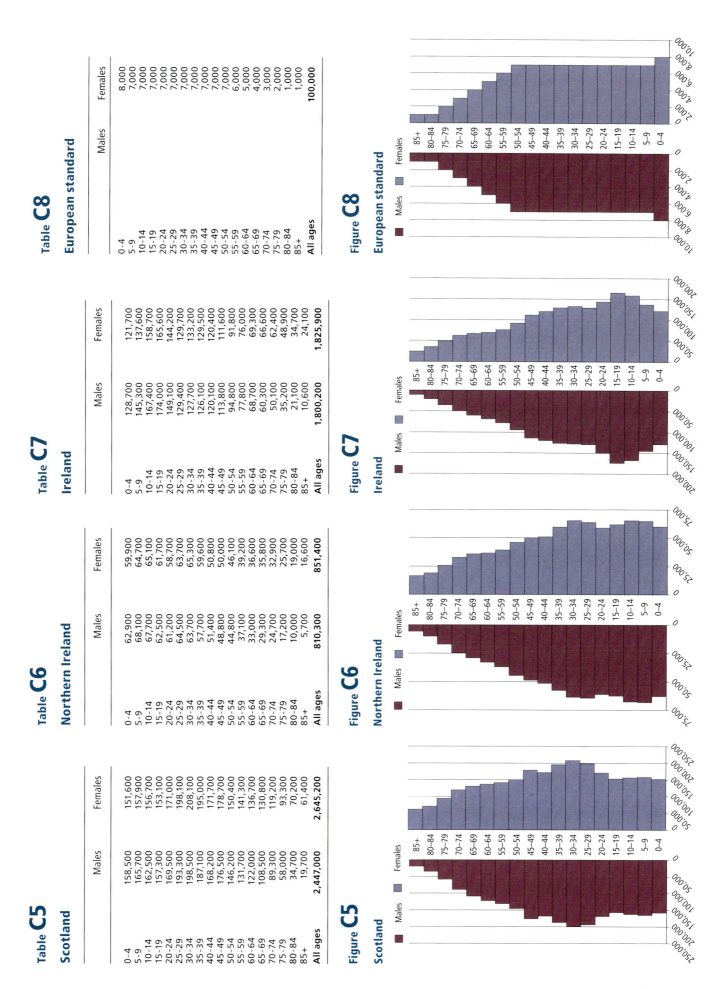

Figure C5
Scotland

Figure C6
Northern Ireland

Figure C7
Ireland

Figure C8
European standard

Table C9

Population estimates: UK and Ireland, by country, region of England, and health authority, 1996

Country/ region/ health authority	Males	Females
UK and Ireland	29,981,900	31,687,200
England	23,526,800	24,875,300
Northern and Yorkshire	3,028,700	3,214,700
Bradford	226,800	242,000
Calderdale and Kirklees	278,400	295,700
County Durham	291,300	308,100
East Riding	275,300	289,500
Gateshead and South Tyneside	171,200	182,600
Leeds	345,900	368,400
Newcastle and North Tyneside	224,800	242,100
North Cumbria	154,400	161,200
North Yorkshire	351,900	376,400
Northumberland	148,700	157,000
Sunderland	140,800	148,900
Tees	266,400	282,600
Wakefield	152,900	160,100
Trent	2,478,800	2,581,600
Barnsley	107,500	113,600
Doncaster	141,500	147,900
Leicestershire	450,100	465,000
Lincolnshire	302,100	316,300
North Derbyshire	180,100	187,200
North Nottinghamshire	191,800	199,400
Nottingham	309,500	320,200
Rotherham	122,700	128,500
Sheffield	252,200	265,100
South Humber	152,200	159,300
Southern Derbyshire	269,200	279,100
West Midlands	2,575,500	2,684,700
Birmingham	485,600	512,600
Coventry	150,500	154,400
Dudley	150,900	156,400
Herefordshire	81,300	85,300
North Staffordshire	226,400	237,200
Sandwell	139,900	148,700
Shropshire	209,000	214,000
Solihull	97,100	103,100
South Staffordshire	287,000	294,400
Walsall	126,500	132,000
Warwickshire	243,300	251,900
Wolverhampton	118,700	123,800
Worcestershire	259,400	270,800
North West	3,149,200	3,353,600
Bury and Rochdale	187,000	197,000
East Lancashire	251,400	263,400
Liverpool	219,400	238,900
Manchester	197,700	208,700
Morecambe Bay	147,800	158,800
North Cheshire	152,000	158,400
North West Lancashire	219,000	236,500
Salford and Trafford	213,400	224,800
Sefton	135,200	151,500
South Cheshire	324,800	341,500
South Lancashire	150,300	157,500
St Helen's and Knowsley	158,700	171,200
Stockport	138,200	147,900
West Pennine	226,800	240,000
Wigan and Bolton	275,900	288,300
Wirral	151,700	169,200
Eastern	2,563,800	2,672,400
Bedfordshire	271,300	275,000
Cambridgeshire	340,000	349,300
Hertfordshire	490,300	511,900
Norfolk	374,400	395,400
North Essex	430,200	450,100
South Essex	335,800	356,200
Suffolk	321,800	334,600

Country/ region/ health authority	Males	Females
London	3,308,300	3,593,000
Barking and Havering	185,300	198,700
Barnet, Enfield and Haringey	365,700	405,900
Bexley, Bromley and Greenwich	343,000	372,400
Brent and Harrow	218,700	236,700
Camden and Islington	168,300	187,800
Croydon	157,000	169,600
Ealing, Hammersmith and Hounslow	310,200	336,800
East London and The City	294,400	310,100
Hillingdon	115,900	123,500
Kensington and Chelsea and Westminster	140,700	161,600
Kingston and Richmond	145,400	158,300
Lambeth, Southwark and Lewisham	353,300	379,000
Merton, Sutton and Wandsworth	292,100	320,000
Redbridge and Waltham Forest	218,300	232,600
South East	4,099,500	4,308,700
Berkshire	392,100	395,300
Buckinghamshire	327,200	337,900
East Kent	276,600	301,000
East Surrey	197,400	210,500
East Sussex, Brighton and Hove	340,700	380,900
Isle of Wight, Portsmouth and SE Hampshire	323,200	340,800
North and Mid Hampshire	268,400	273,300
Northamptonshire	297,200	306,000
Oxfordshire	290,300	297,000
Southampton and SW Hampshire	262,100	272,700
West Kent	471,700	492,200
West Surrey	304,800	319,500
West Sussex	347,900	381,600
South West	2,322,900	2,466,600
Avon	469,200	493,400
Cornwall and Isles of Scilly	232,200	249,000
Dorset	323,600	352,500
Gloucestershire	269,800	282,000
North and East Devon	223,800	241,800
Somerset	233,200	247,400
South and West Devon	280,300	301,900
Wiltshire	290,700	298,700
Wales	1,397,500	1,489,500
Bro Taf	342,900	368,000
Dyfed Powys	232,100	244,900
Gwent	268,300	283,600
Morgannwg	238,500	254,100
North Wales	315,700	339,000
Scotland	2,447,000	2,645,200
Argyll and Clyde	205,900	222,700
Ayrshire and Arran	178,000	195,100
Borders	50,800	55,100
Dumfries and Galloway	71,800	76,800
Fife	166,800	180,600
Forth Valley	132,200	142,500
Grampian	261,800	271,000
Greater Glasgow	419,500	468,700
Highland	102,000	106,700
Lanarkshire	267,700	288,600
Lothian	365,100	395,900
Orkney	9,700	10,000
Shetland	11,600	11,300
Tayside	189,900	206,000
Western Isles	14,100	14,500
Northern Ireland	810,300	851,400
Eastern	320,900	348,600
Northern	204,200	212,600
Southern	148,500	152,100
Western	136,800	138,100
Ireland	1,800,200	1,825,900
Eastern	627,800	668,100
Mid Western	159,600	157,400
Midland	104,200	101,300
North Eastern	154,400	151,700
North Western	106,300	104,600
South Eastern	197,300	194,200
Southern	273,000	273,700
Western	177,600	174,800

Appendix D

Figures and maps for all cancers combined

This appendix gives charts and maps for all cancers combined (see Appendix G for details), equivalent to those for the individual cancers in Chapters 3-23.

Figures D2.1 and D2.2 illustrate age-standardised incidence and mortality rates, respectively, by country, and region of England, by sex. Figures D2.3 and D2.4 illustrate age-standardised incidence and mortality rates, respectively, by health authority within each country, and region of England, with separate charts for each sex.

Maps D2.1 and D2.2 illustrate the ratios of the incidence and mortality rates, respectively, in each health authority to the overall rates for the UK and Ireland, with separate maps for each sex. Details of the scale used in the maps are given in Appendix H.

Figure **D2.1**

All cancers[1]: incidence by sex, country, and region of England
UK and Ireland 1991-99[2]

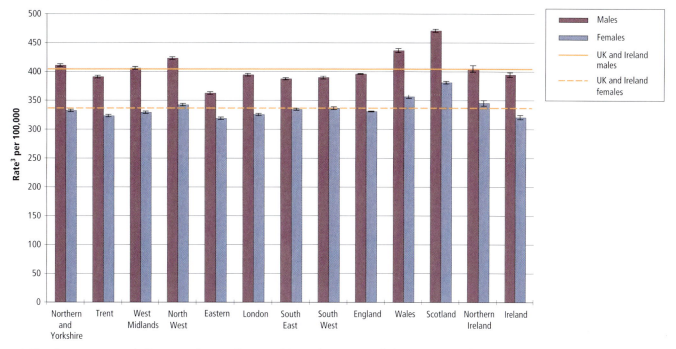

1 All malignant cancers excluding non-melanoma skin cancer (ICD9 codes 140-208 excluding 173; ICD10 codes C00-C97 excluding C44)

2 Northern Ireland 1993-99, Ireland 1994-99

3 Age standardised using the European standard population, with 95% confidence interval

Figure **D2.2**

All cancers[1]: mortality by sex, country, and region of England
UK and Ireland 1991-2000[2]

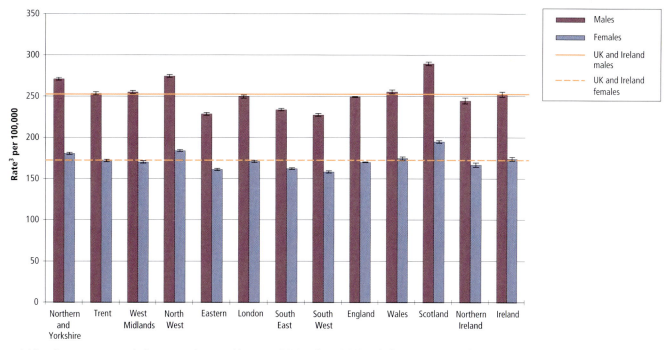

1 All malignant cancers excluding non-melanoma skin cancer (ICD9 codes 140-208 excluding 173; ICD10 codes C00-C97 excluding C44)

2 Scotland 1991-99, Ireland 1994-2000

3 Age standardised using the European standard population, with 95% confidence interval

Figure **D2.3a**

All cancers[1]: incidence by health authority within country, and region of England
Males, UK and Ireland 1991-99[2]

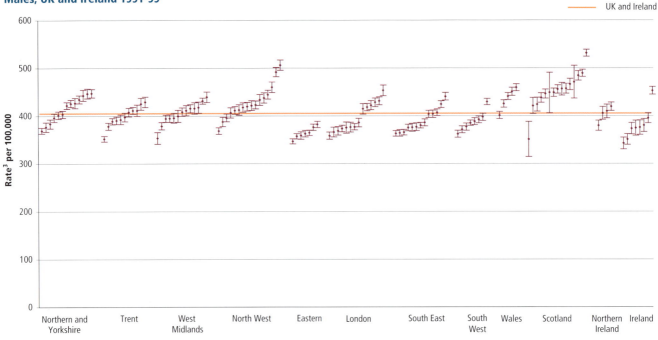

1 All malignant cancers excluding non-melanoma skin cancer (ICD9 codes 140-208 excluding 173; ICD10 codes C00-C97 excluding C44)

2 Northern Ireland 1993-99, Ireland 1994-99

3 Age standardised using the European standard population, with 95% confidence interval

Figure **D2.3b**

All cancers[1]: incidence by health authority within country, and region of England
Females, UK and Ireland 1991-99[2]

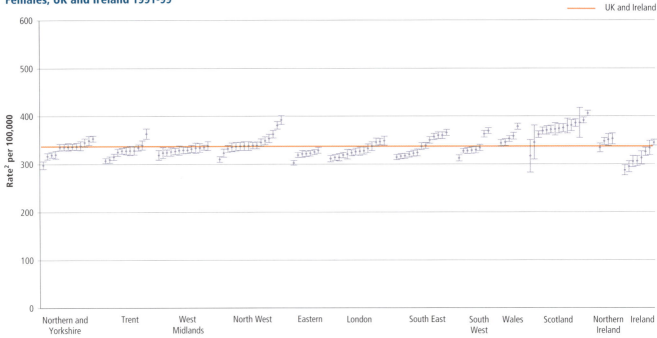

1 All malignant cancers excluding non-melanoma skin cancer (ICD9 codes 140-208 excluding 173; ICD10 codes C00-C97 excluding C44)

2 Northern Ireland 1993-99, Ireland 1994-99

3 Age standardised using the European standard population, with 95% confidence interval

Figure **D2.4a**

All cancers[1]: mortality by health authority within country, and region of England
Males, UK and Ireland 1991-2000[2]

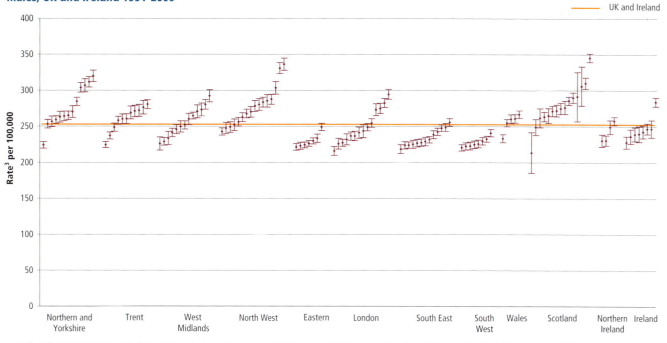

1 All malignant cancers excluding non-melanoma skin cancer (ICD9 codes 140-208 excluding 173; ICD10 codes C00-C97 excluding C44)

2 Scotland 1991-99, Ireland 1994-2000

3 Age standardised using the European standard population, with 95% confidence interval

Figure **D2.4b**

All cancers[1]: mortality by health authority within country, and region of England
Females, UK and Ireland 1991-2000[2]

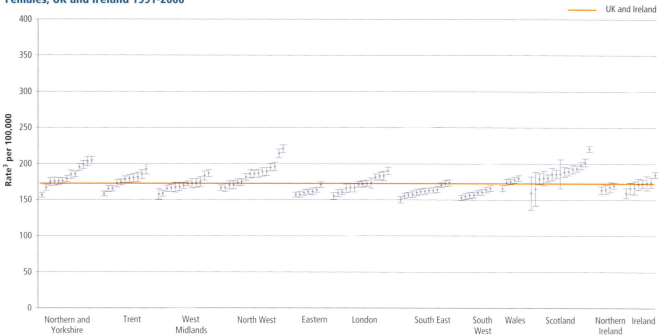

1 All malignant cancers excluding non-melanoma skin cancer (ICD9 codes 140-208 excluding 173; ICD10 codes C00-C97 excluding C44)

2 Scotland 1991-99, Ireland 1994-2000

3 Age standardised using the European standard population, with 95% confidence interval

Map **D2.1a**

All cancers[1]: incidence[2] by health authority
Males, UK and Ireland 1991-99

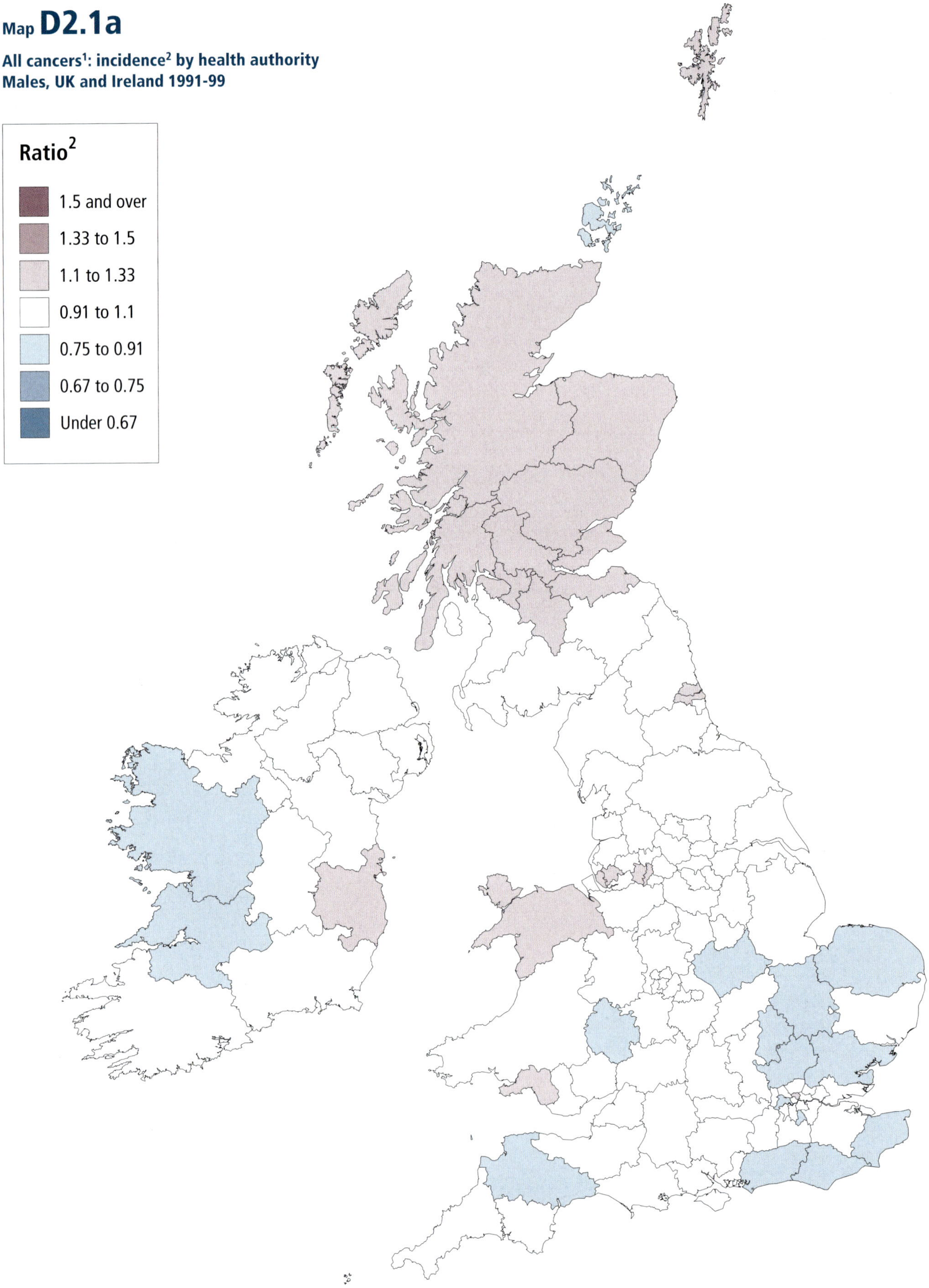

Ratio[2]

	1.5 and over
	1.33 to 1.5
	1.1 to 1.33
	0.91 to 1.1
	0.75 to 0.91
	0.67 to 0.75
	Under 0.67

1 All malignant cancers excluding non-melanoma skin cancer (ICD9 codes 140-208 excluding 173; IDC10 codes C00-C97 excluding C44)
2 Ratio of directly age-standardised rate in health authority to UK and Ireland average

Figure **D2.4a**

All cancers[1]: mortality by health authority within country, and region of England
Males, UK and Ireland 1991-2000[2]

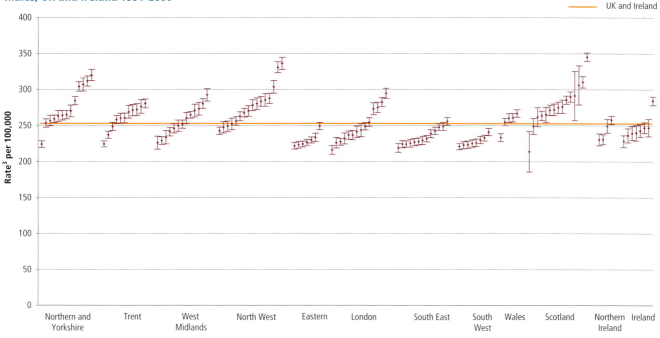

1 All malignant cancers excluding non-melanoma skin cancer (ICD9 codes 140-208 excluding 173; ICD10 codes C00-C97 excluding C44)

2 Scotland 1991-99, Ireland 1994-2000

3 Age standardised using the European standard population, with 95% confidence interval

Figure **D2.4b**

All cancers[1]: mortality by health authority within country, and region of England
Females, UK and Ireland 1991-2000[2]

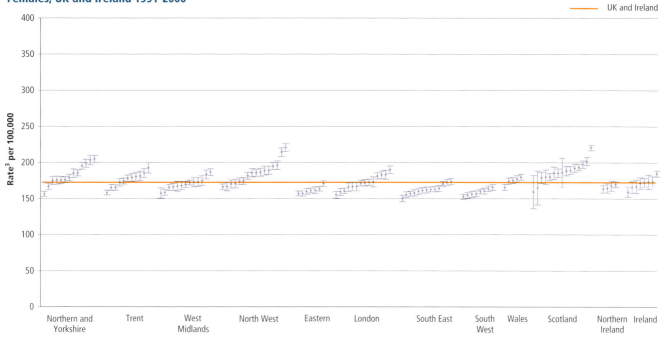

1 All malignant cancers excluding non-melanoma skin cancer (ICD9 codes 140-208 excluding 173; ICD10 codes C00-C97 excluding C44)

2 Scotland 1991-99, Ireland 1994-2000

3 Age standardised using the European standard population, with 95% confidence interval

Map **D2.1a**

All cancers[1]: incidence[2] by health authority
Males, UK and Ireland 1991-99

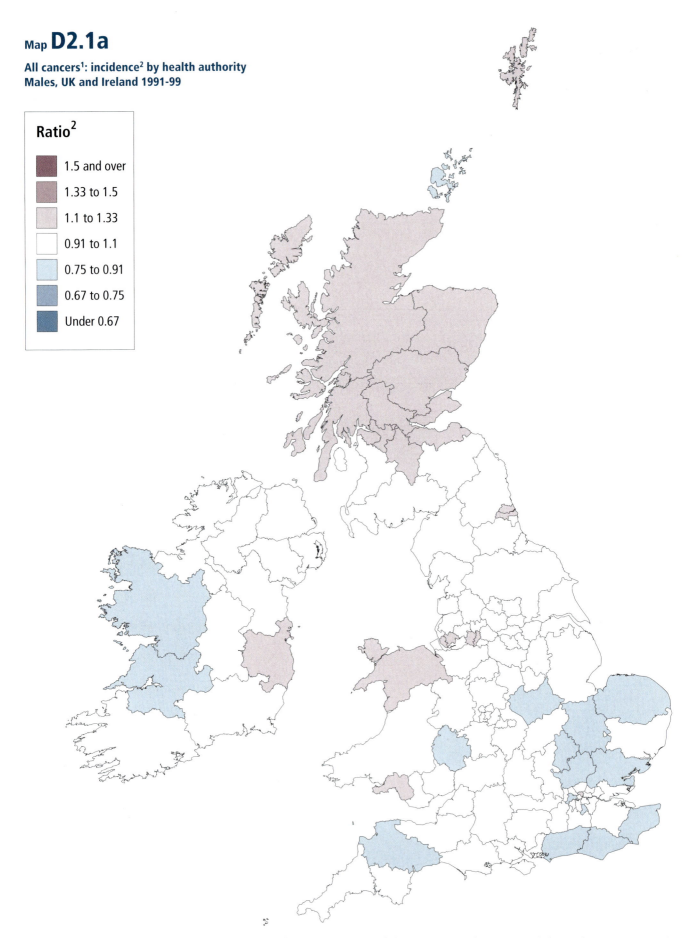

Ratio[2]

- 1.5 and over
- 1.33 to 1.5
- 1.1 to 1.33
- 0.91 to 1.1
- 0.75 to 0.91
- 0.67 to 0.75
- Under 0.67

1 All malignant cancers excluding non-melanoma skin cancer (ICD9 codes 140-208 excluding 173; IDC10 codes C00-C97 excluding C44)
2 Ratio of directly age-standardised rate in health authority to UK and Ireland average

Map **D2.1b**

**All cancers[1]: incidence[2] by health authority
Females, UK and Ireland 1991-99**

Ratio[2]

	1.5 and over
	1.33 to 1.5
	1.1 to 1.33
	0.91 to 1.1
	0.75 to 0.91
	0.67 to 0.75
	Under 0.67

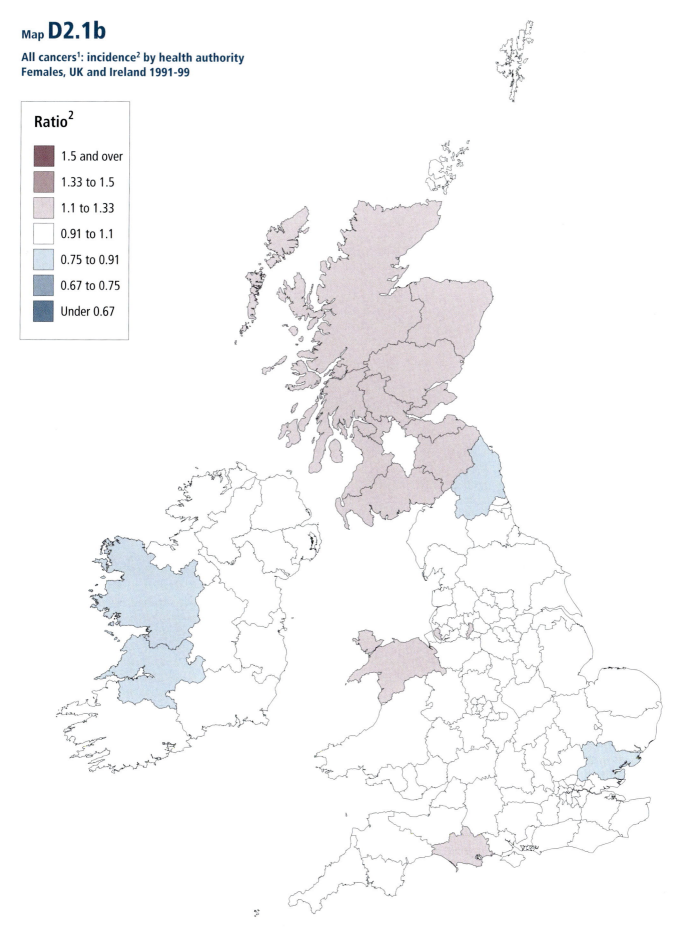

1 All malignant cancers excluding non-melanoma skin cancer (ICD9 codes 140-208 excluding 173; IDC10 codes C00-C97 excluding C44)
2 Ratio of directly age-standardised rate in health authority to UK and Ireland average

Map **D2.2a**

All cancers[1]: mortality[2] by health authority
Males, UK and Ireland 1991-2000

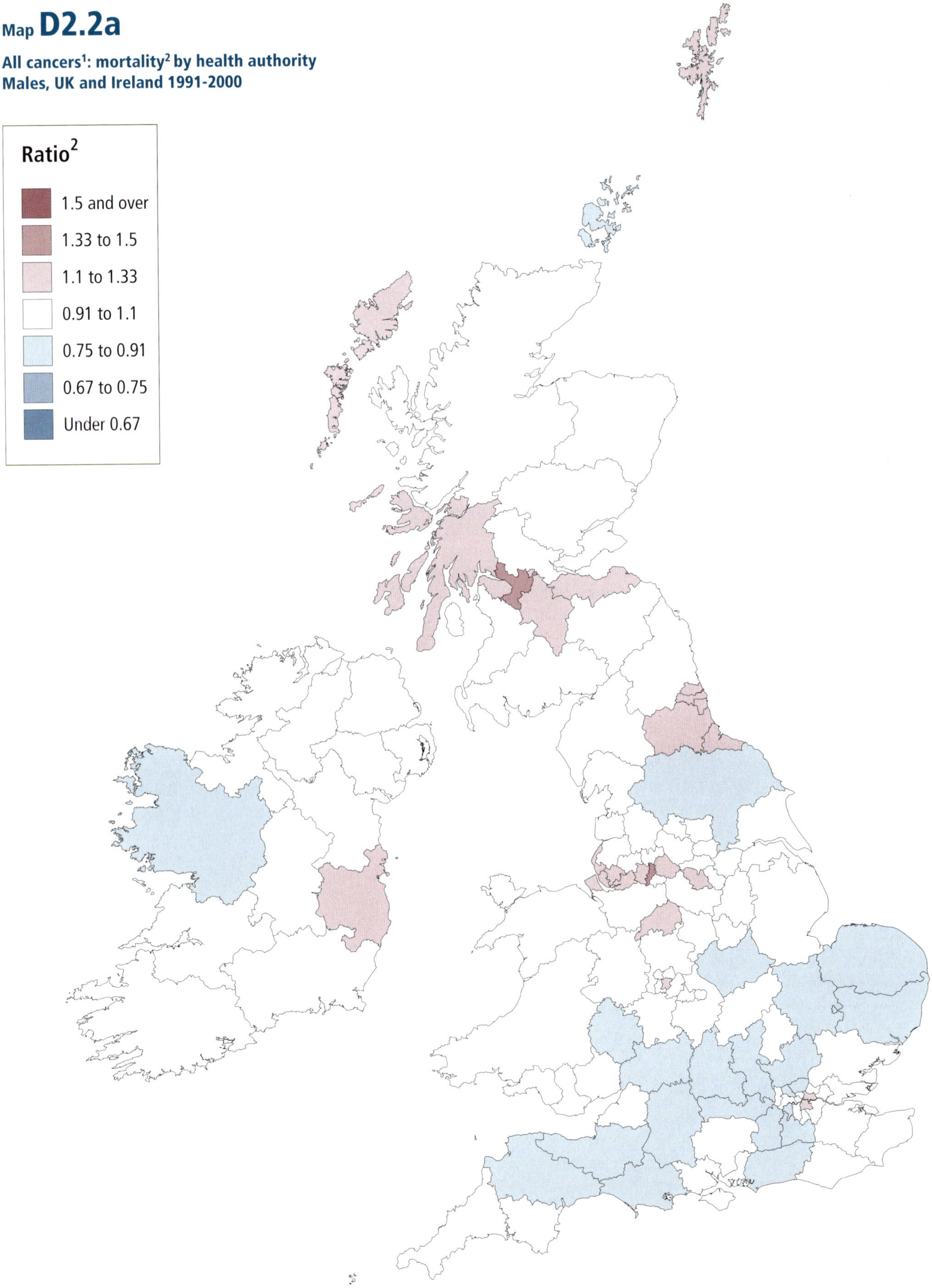

Ratio[2]

	1.5 and over
	1.33 to 1.5
	1.1 to 1.33
	0.91 to 1.1
	0.75 to 0.91
	0.67 to 0.75
	Under 0.67

1 All malignant cancers excluding non-melanoma skin cancer (ICD9 codes 140-208 excluding 173; IDC10 codes C00-C97 excluding C44)
2 Ratio of directly age-standardised rate in health authority to UK and Ireland average

Map **D2.2b**

All cancers[1]: mortality[2] by health authority
Females, UK and Ireland, 1991-2000

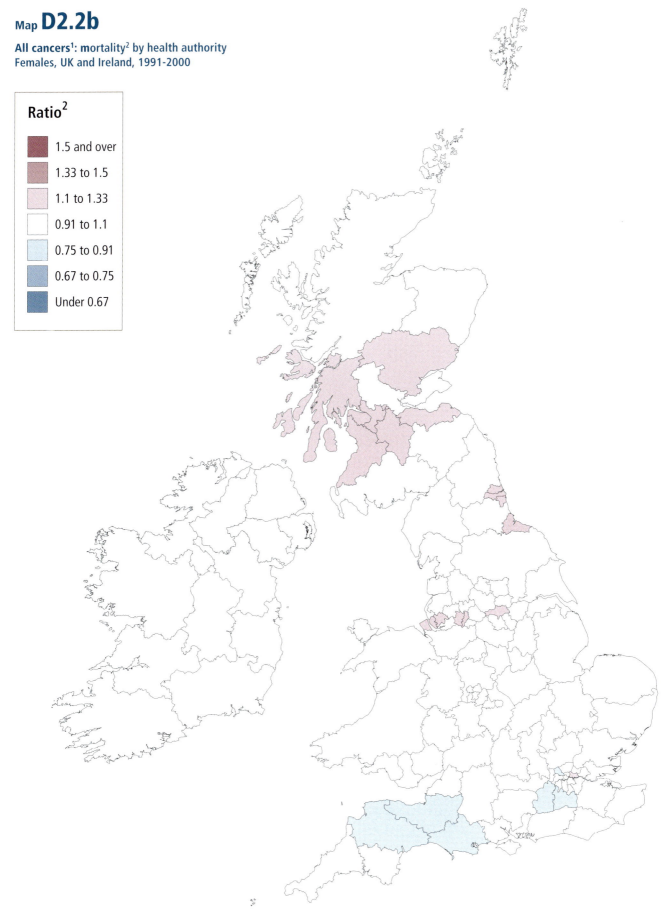

Ratio[2]

- 1.5 and over
- 1.33 to 1.5
- 1.1 to 1.33
- 0.91 to 1.1
- 0.75 to 0.91
- 0.67 to 0.75
- Under 0.67

1 All malignant cancers excluding non-melanoma skin cancer (ICD9 codes 140-208 excluding 173; IDC10 codes C00-C97 excluding C44)
2 Ratio of directly age-standardised rate in health authority to UK and Ireland average

Appendix E

Cancer maps - 'absolute' scale

This appendix gives maps illustrating the age-standardised incidence and mortality rates at the level of health authority, by sex, for each of the 21 cancers discussed in detail in this atlas.

The maps included in Chapters 3-23, which illustrate the ratio of the age-standardised rates in each health authority to the average for the UK and Ireland, can be used to examine regional variations in incidence and mortality for a particular cancer relative to the overall average for that cancer.

In contrast, the maps given in this appendix are all based on the same scale, in 16 intervals from 0 to a rate of 180 per 100,000 population. They can be used to examine not only the regional variations for a particular cancer but to compare both the overall level of incidence with that of mortality, and the levels of both incidence and mortality with those for other cancers. These maps therefore give perspective to the ratio maps in terms of the overall cancer burden. This is particularly useful when assessing the geographical variations in the less common cancers because although rates for some areas may be relatively high (coloured purple in the maps in Chapters 3-23) it can be seen from the maps in this appendix that these 'high' rates are in fact low in absolute terms – for example, when compared with rates for the major cancers such as lung, breast and colorectal.

Details of the scale used in these maps are given in Appendix H.

Bladder

**by health authority,
UK and Ireland, 1991-2000[1]**

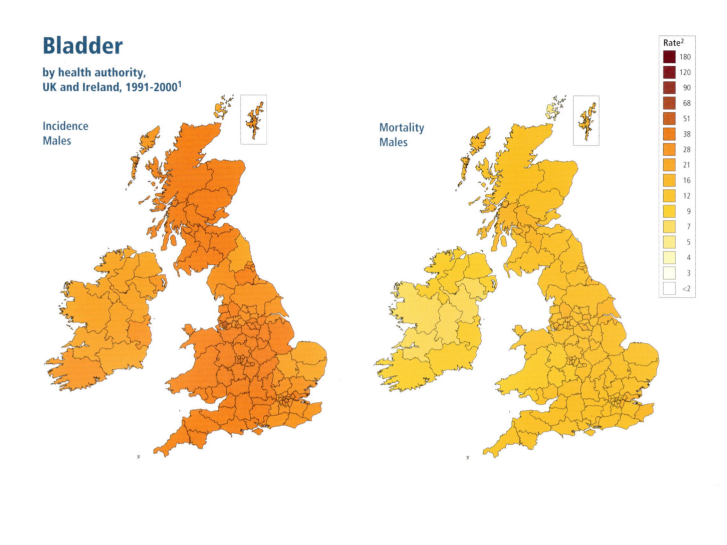

Incidence
Males

Mortality
Males

Rate[2]	
	180
	120
	90
	68
	51
	38
	28
	21
	16
	12
	9
	7
	5
	4
	3
	<2

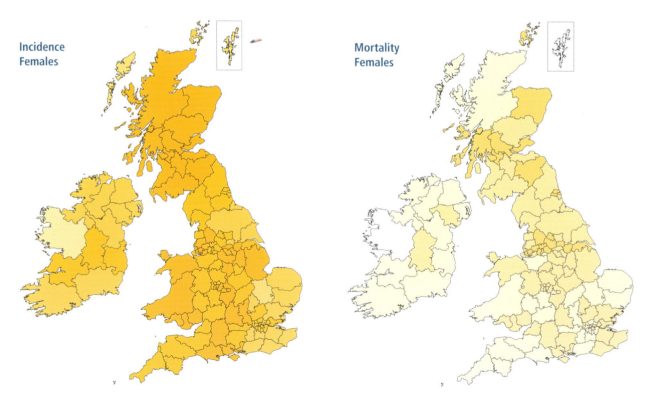

Incidence
Females

Mortality
Females

1 Incidence 1991-99, except Northern Ireland 1993-99, Ireland 1994-99. Mortality 1991-2000, except Scotland 1991-99, Ireland 1994-2000.
2 Rate per 100,000 population directly age standardised using the European standard population.

Brain

**by health authority,
UK and Ireland, 1991-2000[1]**

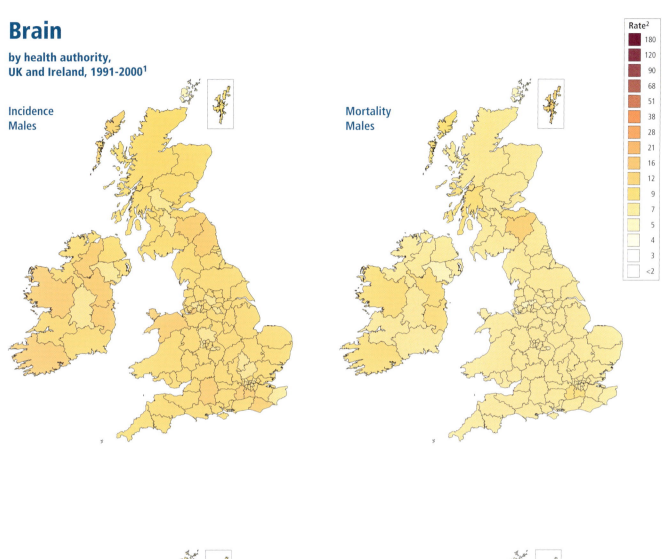

Incidence
Males

Mortality
Males

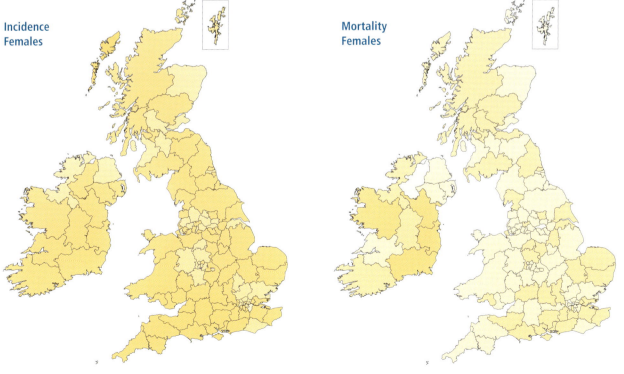

Incidence
Females

Mortality
Females

Rate[2]	
	180
	120
	90
	68
	51
	38
	28
	21
	16
	12
	9
	7
	5
	4
	3
	<2

1 Incidence 1991-99, except Northern Ireland 1993-99, Ireland 1994-99. Mortality 1991-2000, except Scotland 1991-99, Ireland 1994-2000.
2 Rate per 100,000 population directly age standardised using the European standard population.

Breast

**by health authority,
UK and Ireland, 1991-2000[1]**

Incidence

Mortality

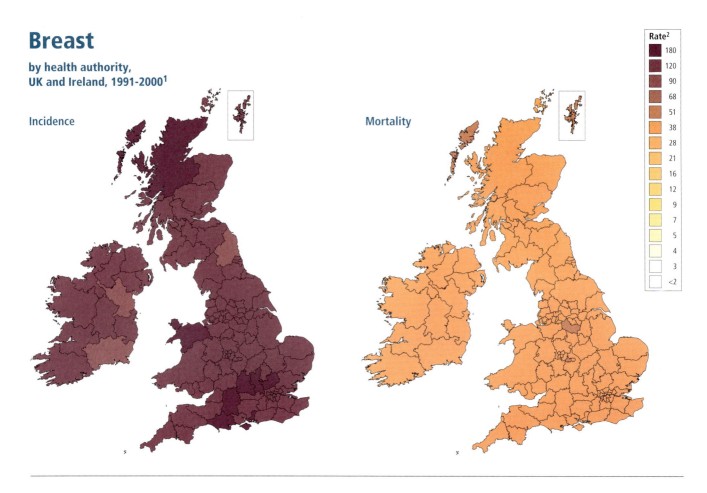

Rate[2]	
	180
	120
	90
	68
	51
	38
	28
	21
	16
	12
	9
	7
	5
	4
	3
	<2

Cervix

**by health authority,
UK and Ireland, 1991-2000[1]**

Incidence

Mortality

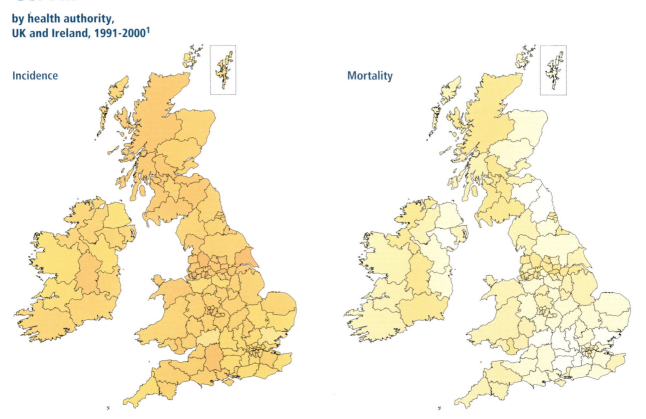

1 Incidence 1991-99, except Northern Ireland 1993-99, Ireland 1994-99. Mortality 1991-2000, except Scotland 1991-99, Ireland 1994-2000.
2 Rate per 100,000 population directly age standardised using the European standard population.

Colorectal

**by health authority,
UK and Ireland, 1991-2000[1]**

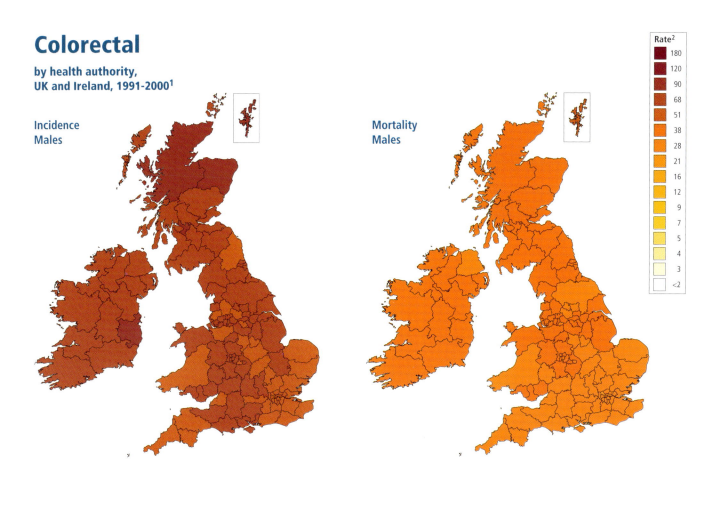

Incidence
Males

Mortality
Males

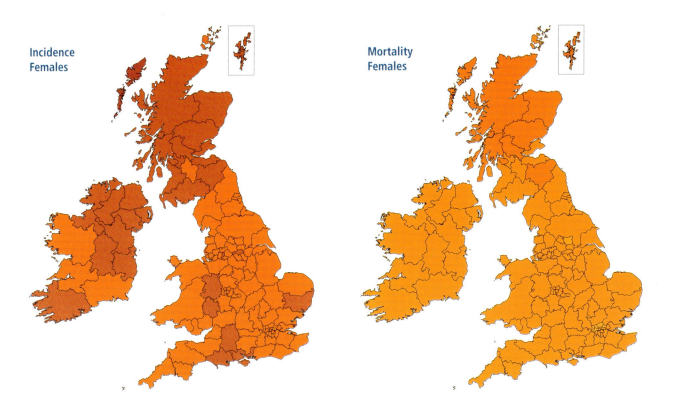

Incidence
Females

Mortality
Females

Rate[2]

	180
	120
	90
	68
	51
	38
	28
	21
	16
	12
	9
	7
	5
	4
	3
	<2

1 Incidence 1991-99, except Northern Ireland 1993-99, Ireland 1994-99. Mortality 1991-2000, except Scotland 1991-99, Ireland 1994-2000.
2 Rate per 100,000 population directly age standardised using the European standard population.

Hodgkin's disease

**by health authority,
UK and Ireland, 1991-2000[1]**

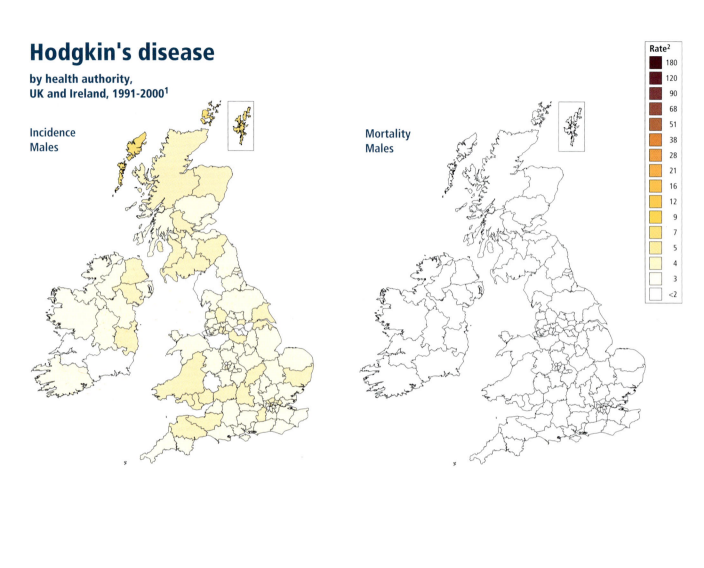

Incidence
Males

Mortality
Males

Rate[2]

	180
	120
	90
	68
	51
	38
	28
	21
	16
	12
	9
	7
	5
	4
	3
	<2

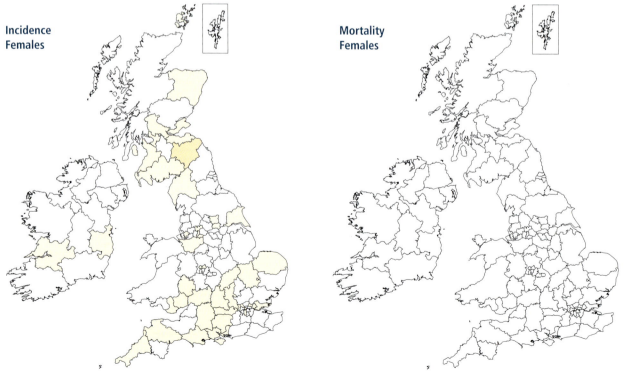

Incidence
Females

Mortality
Females

1 Incidence 1991-99, except Northern Ireland 1993-99, Ireland 1994-99. Mortality 1991-2000, except Scotland 1991-99, Ireland 1994-2000.
2 Rate per 100,000 population directly age standardised using the European standard population.

Kidney

**by health authority,
UK and Ireland, 1991-2000[1]**

**Incidence
Males**

**Mortality
Males**

**Incidence
Females**

**Mortality
Females**

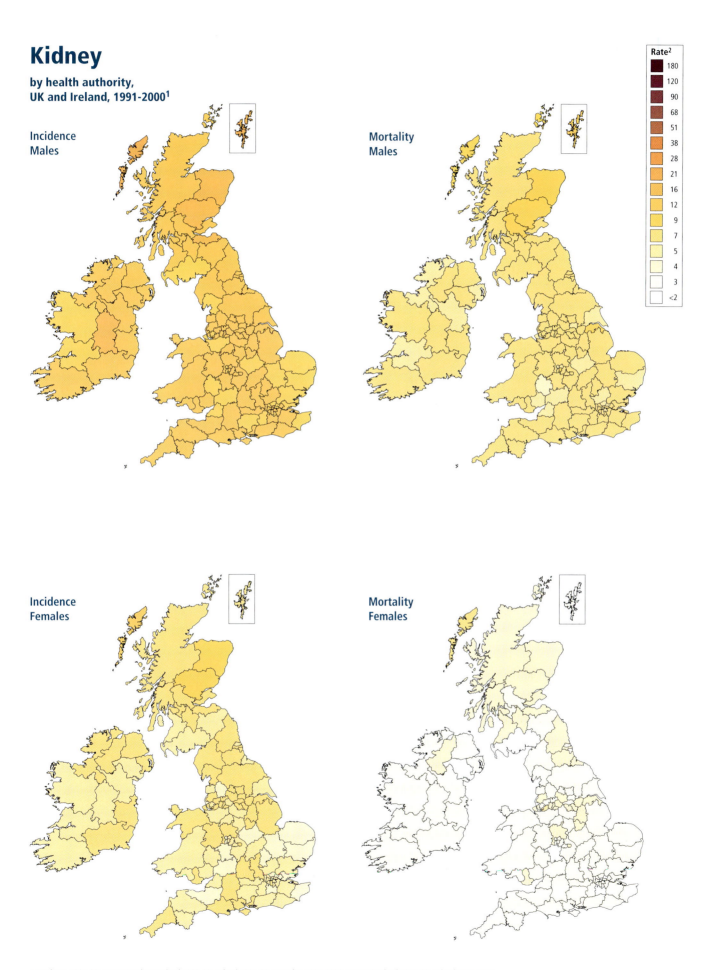

Rate[2]

	180
	120
	90
	68
	51
	38
	28
	21
	16
	12
	9
	7
	5
	4
	3
	<2

1 Incidence 1991-99, except Northern Ireland 1993-99, Ireland 1994-99. Mortality 1991-2000, except Scotland 1991-99, Ireland 1994-2000.
2 Rate per 100,000 population directly age standardised using the European standard population.

Larynx

**by health authority,
UK and Ireland, 1991-2000[1]**

**Incidence
Males**

**Mortality
Males**

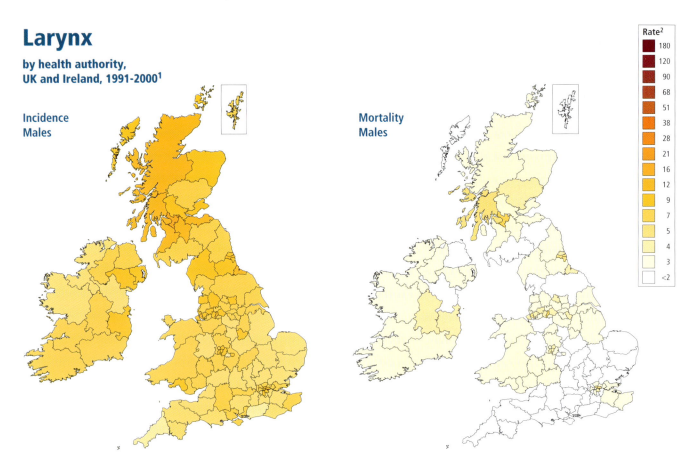

Rate[2]

| 180 |
| 120 |
| 90 |
| 68 |
| 51 |
| 38 |
| 28 |
| 21 |
| 16 |
| 12 |
| 9 |
| 7 |
| 5 |
| 4 |
| 3 |
| <2 |

1 Incidence 1991-99, except Northern Ireland 1993-99, Ireland 1994-99. Mortality 1991-2000, except Scotland 1991-99, Ireland 1994-2000.
2 Rate per 100,000 population directly age standardised using the European standard population.

Leukaemia

**by health authority,
UK and Ireland, 1991-2000[1]**

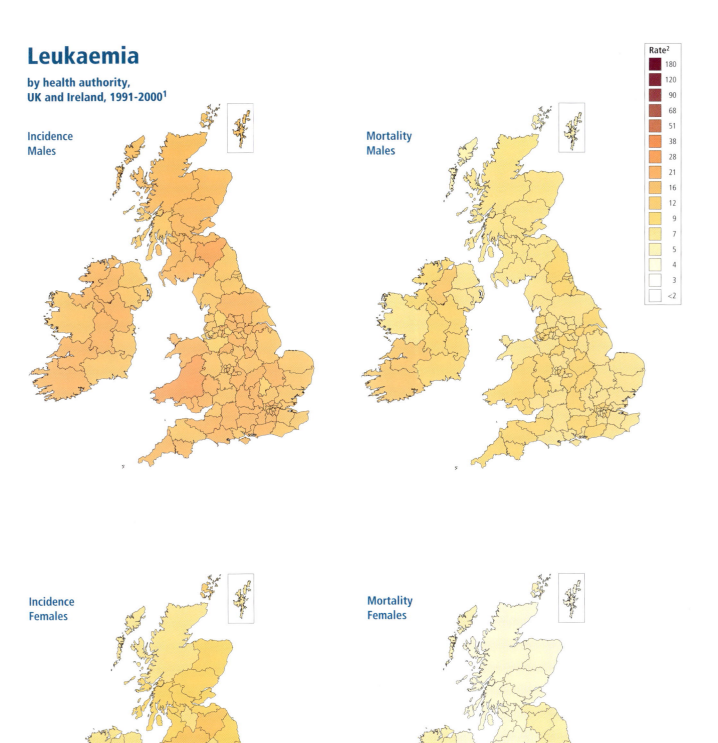

Incidence
Males

Mortality
Males

Incidence
Females

Mortality
Females

Rate[2]
180
120
90
68
51
38
28
21
16
12
9
7
5
4
3
<2

1 Incidence 1991-99, except Northern Ireland 1993-99, Ireland 1994-99. Mortality 1991-2000, except Scotland 1991-99, Ireland 1994-2000.
2 Rate per 100,000 population directly age standardised using the European standard population.

Lip, mouth and pharynx

**by health authority,
UK and Ireland, 1991-2000[1]**

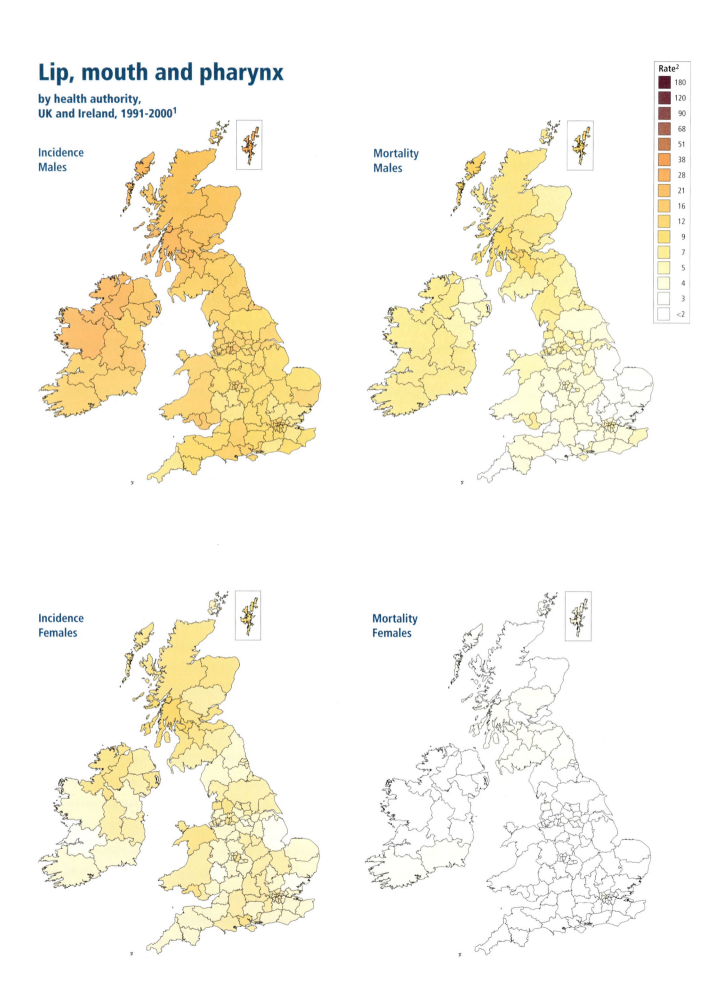

Incidence
Males

Mortality
Males

Incidence
Females

Mortality
Females

Rate[2]

180
120
90
68
51
38
28
21
16
12
9
7
5
4
3
<2

1 Incidence 1991-99, except Northern Ireland 1993-99, Ireland 1994-99. Mortality 1991-2000, except Scotland 1991-99, Ireland 1994-2000.
2 Rate per 100,000 population directly age standardised using the European standard population.

Lung

**by health authority,
UK and Ireland, 1991-2000[1]**

Incidence
Males

Mortality
Males

Rate[2]	
	180
	120
	90
	68
	51
	38
	28
	21
	16
	12
	9
	7
	5
	4
	3
	<2

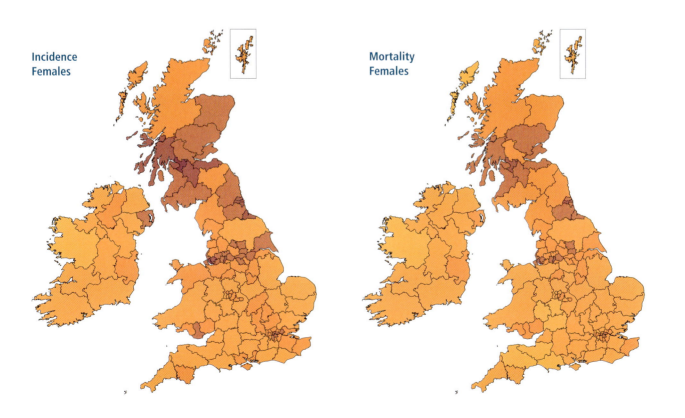

Incidence
Females

Mortality
Females

1 Incidence 1991-99, except Northern Ireland 1993-99, Ireland 1994-99. Mortality 1991-2000, except Scotland 1991-99, Ireland 1994-2000.
2 Rate per 100,000 population directly age standardised using the European standard population.

Melanoma of skin

**by health authority,
UK and Ireland, 1991-2000[1]**

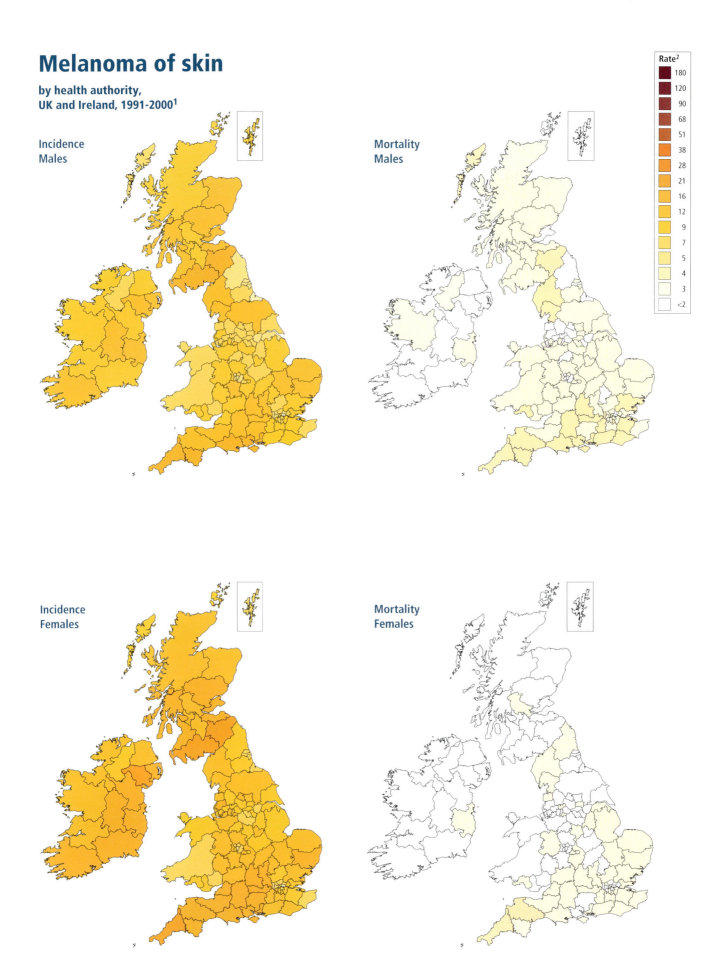

Incidence
Males

Mortality
Males

Incidence
Females

Mortality
Females

Rate[2]

180
120
90
68
51
38
28
21
16
12
9
7
5
4
3
<2

1 Incidence 1991-99, except Northern Ireland 1993-99, Ireland 1994-99. Mortality 1991-2000, except Scotland 1991-99, Ireland 1994-2000.
2 Rate per 100,000 population directly age standardised using the European standard population.

Multiple myeloma

**by health authority,
UK and Ireland, 1991-2000[1]**

Incidence
Males

Mortality
Males

Incidence
Females

Mortality
Females

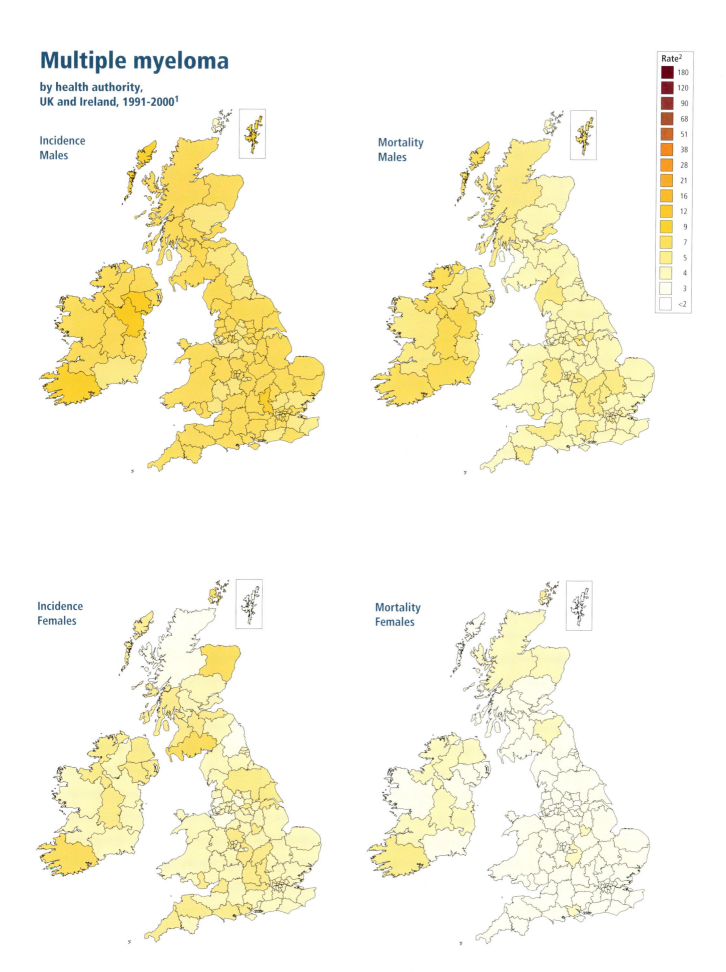

Rate[2]	
	180
	120
	90
	68
	51
	38
	28
	21
	16
	12
	9
	7
	5
	4
	3
	<2

1 Incidence 1991-99, except Northern Ireland 1993-99, Ireland 1994-99. Mortality 1991-2000, except Scotland 1991-99, Ireland 1994-2000.
2 Rate per 100,000 population directly age standardised using the European standard population.

Non-Hodgkin's lymphoma

**by health authority,
UK and Ireland, 1991-2000[1]**

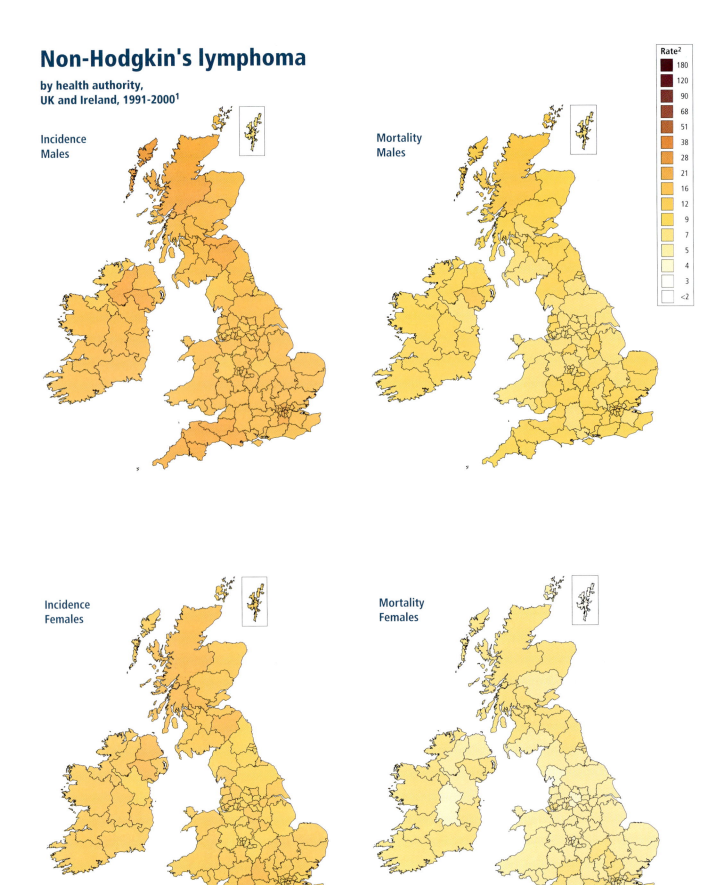

Incidence
Males

Mortality
Males

Incidence
Females

Mortality
Females

Rate[2]
180
120
90
68
51
38
28
21
16
12
9
7
5
4
3
<2

1 Incidence 1991-99, except Northern Ireland 1993-99, Ireland 1994-99. Mortality 1991-2000, except Scotland 1991-99, Ireland 1994-2000.
2 Rate per 100,000 population directly age standardised using the European standard population.

Oesophagus

**by health authority,
UK and Ireland, 1991-2000[1]**

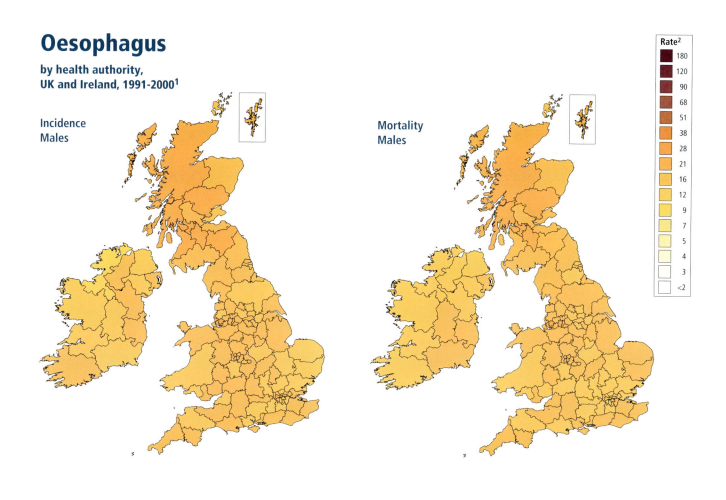

Incidence
Males

Mortality
Males

Rate[2]
180
120
90
68
51
38
28
21
16
12
9
7
5
4
3
<2

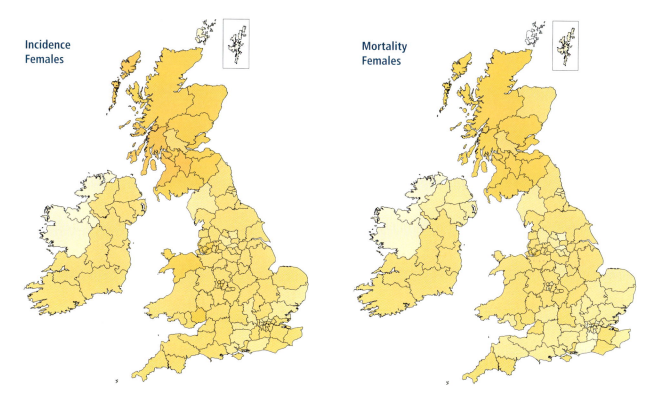

Incidence
Females

Mortality
Females

1 Incidence 1991-99, except Northern Ireland 1993-99, Ireland 1994-99. Mortality 1991-2000, except Scotland 1991-99, Ireland 1994-2000.
2 Rate per 100,000 population directly age standardised using the European standard population.

Ovary

**by health authority,
UK and Ireland, 1991-2000[1]**

Incidence

Mortality

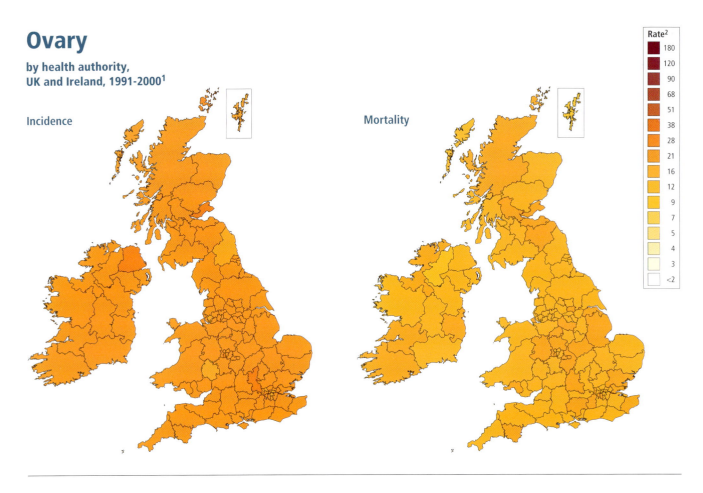

Rate[2]	
	180
	120
	90
	68
	51
	38
	28
	21
	16
	12
	9
	7
	5
	4
	3
	<2

Prostate

**by health authority,
UK and Ireland, 1991-2000[1]**

Incidence

Mortality

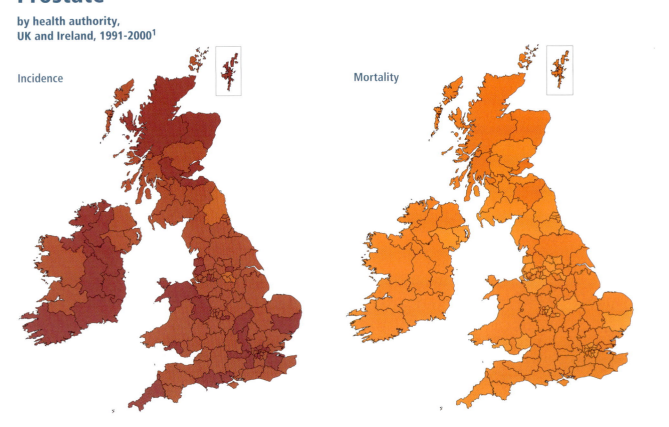

1 Incidence 1991-99, except Northern Ireland 1993-99, Ireland 1994-99. Mortality 1991-2000, except Scotland 1991-99, Ireland 1994-2000.
2 Rate per 100,000 population directly age standardised using the European standard population.

Pancreas

**by health authority,
UK and Ireland, 1991-2000[1]**

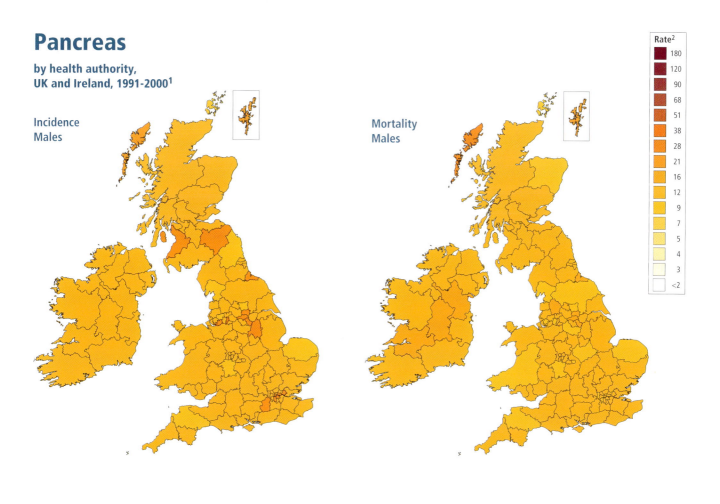

Incidence
Males

Mortality
Males

Rate[2]	
	180
	120
	90
	68
	51
	38
	28
	21
	16
	12
	9
	7
	5
	4
	3
	<2

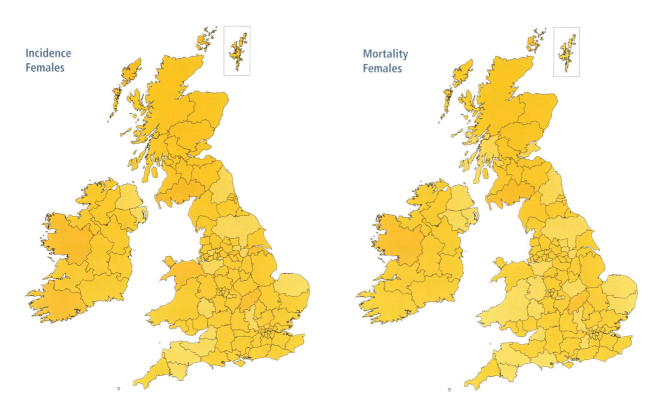

Incidence
Females

Mortality
Females

1 Incidence 1991-99, except Northern Ireland 1993-99, Ireland 1994-99. Mortality 1991-2000, except Scotland 1991-99, Ireland 1994-2000.
2 Rate per 100,000 population directly age standardised using the European standard population.

Stomach

**by health authority,
UK and Ireland, 1991-2000[1]**

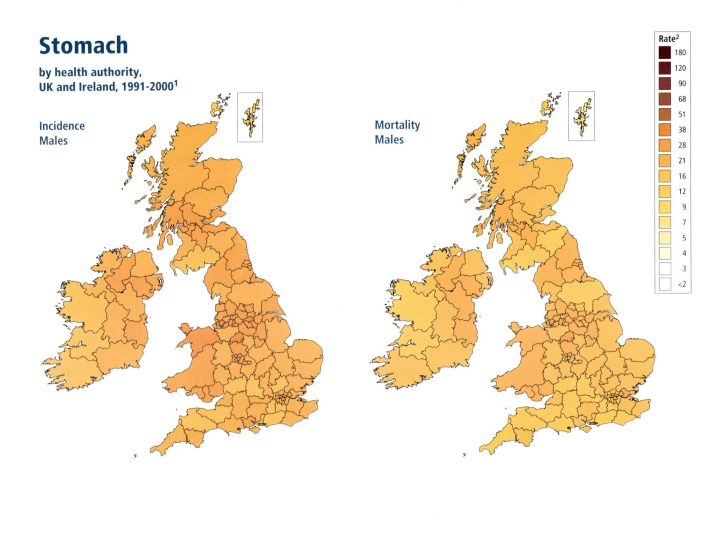

Incidence
Males

Mortality
Males

Rate[2]

	180
	120
	90
	68
	51
	38
	28
	21
	16
	12
	9
	7
	5
	4
	3
	<2

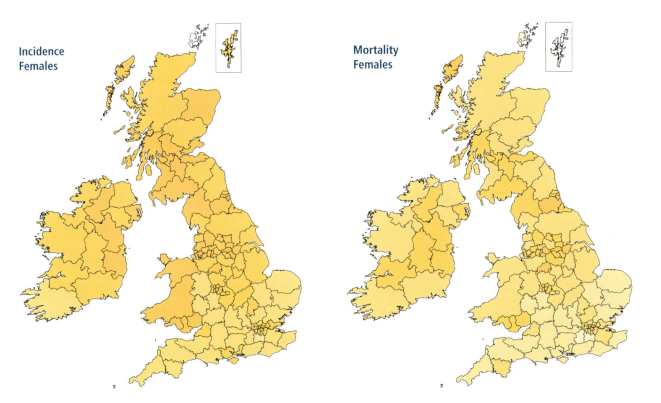

Incidence
Females

Mortality
Females

1 Incidence 1991-99, except Northern Ireland 1993-99, Ireland 1994-99. Mortality 1991-2000, except Scotland 1991-99, Ireland 1994-2000.
2 Rate per 100,000 population directly age standardised using the European standard population.

Testis

**by health authority,
UK and Ireland, 1991-2000[1]**

Incidence

Mortality

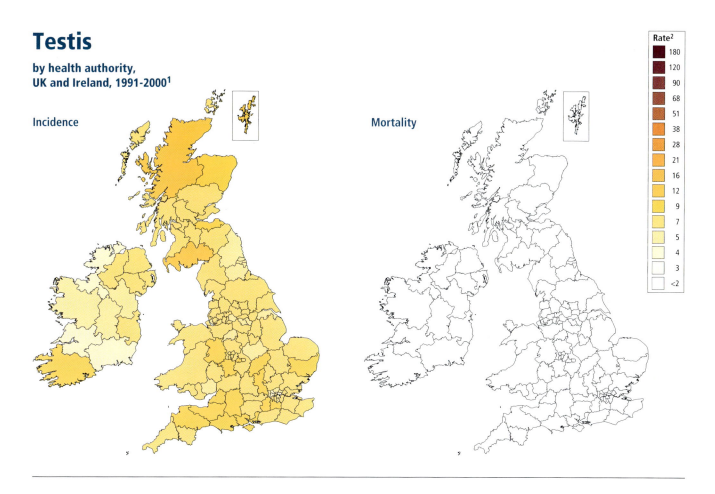

Rate[2]	
	180
	120
	90
	68
	51
	38
	28
	21
	16
	12
	9
	7
	5
	4
	3
	<2

Uterus

**by health authority,
UK and Ireland, 1991-2000[1]**

Incidence

Mortality

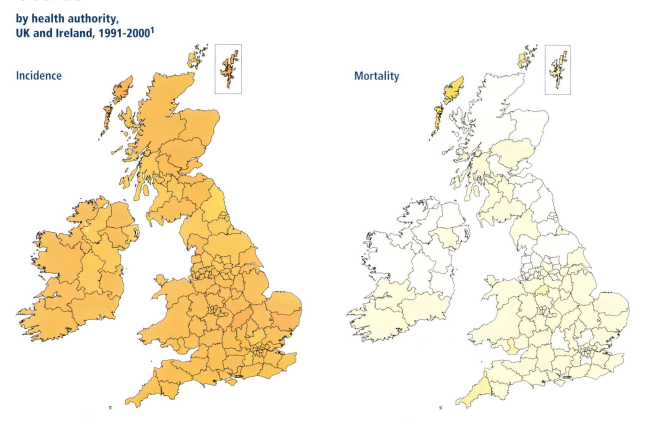

1 Incidence 1991-99, except Northern Ireland 1993-99, Ireland 1994-99. Mortality 1991-2000, except Scotland 1991-99, Ireland 1994-2000.
2 Rate per 100,000 population directly age standardised using the European standard population.

Appendix F

Socio-economic deprivation

Steve Rowan, Mike Quinn

The level of an individual's socio-economic deprivation can be measured using only data for that individual, for example, social class, which is based on occupation.[1] Alternatively, all people living within a particular (usually small) area may be assigned a 'score' from an index such as that of Carstairs and Morris.[2] Such indices are based on data for individuals but usually aggregated to either enumeration districts or wards. The Carstairs index was developed to help explain geographical variations in health data.[3]

The Carstairs index has been calculated for all 1991 census enumeration districts and wards in England and Wales, and 1991 postcode sectors in Scotland, using the following four variables obtained from the census: overcrowding; male unemployment; low social class; and no access to a car.

The four variables are each standardised (by subtracting the average and dividing by the standard deviation of its distribution) and then combined, with each variable having equal weight.

To produce deprivation scores by local authority (LA) in England and Wales, and by health board in Scotland, the following method was used:

- each deprivation score by ward or postcode sector was multiplied by its corresponding population (persons) to give a weighted score;

- the weighted scores for each ward or postcode sector were summed to the appropriate LA or health board (boundaries as at 1991);

- similarly, the populations for each ward or postcode sector were summed to the appropriate LA or health board;

- the resulting weighted LA or health board scores were then divided by their corresponding LA or health board populations to give a deprivation score for each LA and health board.

LAs and health boards have been divided into five deprivation categories, that is, 20 per cent of areas in Great Britain have been classified to each category according to the level of

deprivation. Category 1 is the least deprived (white on the map) and category 5 is the most deprived (dark blue). Since the deprivation scores were normalised using the Great Britain average for each factor, comparisons of the deprivation levels can be made across the whole of Great Britain.

Map F1 shows the distribution of deprivation categories by LA in England and Wales, and by health board within Scotland. However, for consistency with other maps in this atlas, a grid (in red) has been superimposed to show the health authority and health board boundaries as at 2001. Corresponding information is not available for either Northern Ireland or Ireland.

Deprivation maps at ward level for every region of England, and for Wales, and at the postcode sector level for Scotland are given in *Geographic Variations in Health*.[4]

References

1. Rose D, O'Reilly K. *Constructing Classes - Towards a new social classification for the UK*. Swindon: ONS and ESRC, 1997.

2. Carstairs V, Morris R. Deprivation and mortality: an alternative to social class? *Community Medicine* 1989; 11: 213-219.

3. Carstairs V. Deprivation indices: their interpretation and use in relation to health. *Journal of Epidemiology and Community Health* 1995; 49 Suppl 2: S3-S8.

4. Griffiths C, Fitzpatrick J (eds). *Geographic Variations in Health*. Decennial Supplement No. 16. London: The Stationery Office, 2001.

Map F1

Deprivation by local authority (England and Wales) and health board (Scotland) in 1991 showing health authorities in 2001

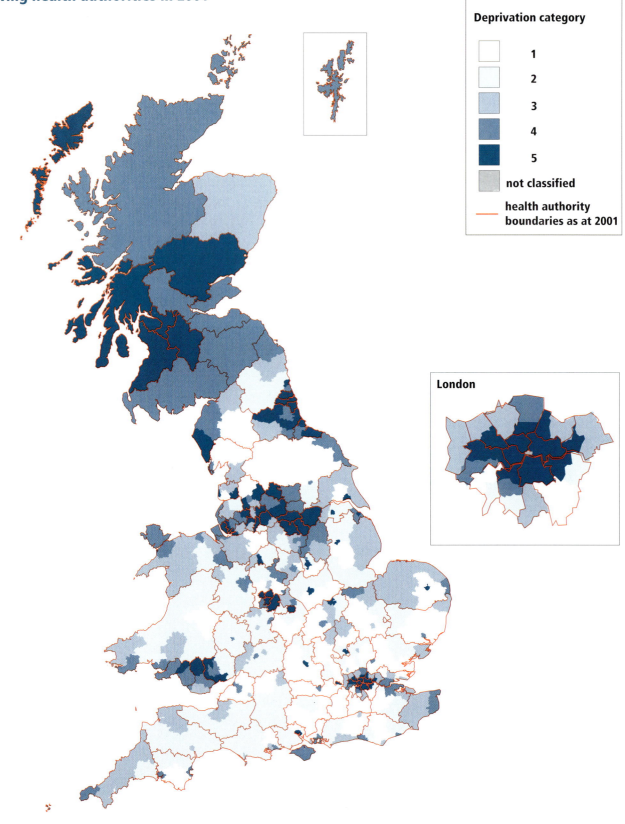

Deprivation category

1
2
3
4
5
not classified
health authority boundaries as at 2001

London

Appendix G

Data – background information

Steve Rowan, Helen Wood, Mike Quinn

Availability of data

At the time the work on this atlas began, cancer incidence and mortality data were available for the following periods:

Table G1

Time periods covered by incidence and mortality data

	Incidence	Mortality
England	1991-1999	1991-2000
Wales	1991-1999	1991-2000
Scotland	1991-1999	1991-1999
Northen Ireland	1993-1999	1991-2000
Ireland	1994-1999	1994-2000

As there may have been varying trends in the incidence of, and mortality from, some cancers in the five countries since 1991, the slightly different time periods covered should be borne in mind when interpreting any apparent differences in cancer rates between countries.

International Classification of Diseases (ICD)

Over the period 1991 to 2000, the ninth and tenth revisions of the International Classification of Diseases (ICD) were in use at various times.[1,2] The constituent countries of the UK and Ireland were asked to supply their data coded to whichever ICD revision they were using at the time of diagnosis or death. Table G2 gives the site codes in both the ICD9 and ICD10 classifications used for each of the main cancer types included in this atlas. All results have been presented in terms of ICD10.

ONS has been advised by both expert epidemiologists and by members of the former Steering Committee on Cancer Registration that non-melanoma skin cancer (ICD10 C44) is greatly under-registered. Registration varies widely depending on a cancer registry's degree of access to outpatient records and general practitioners. It also frequently happens that a person has more than one tumour of this type, and registries had adopted different practices in recording these multiple tumours. Figures in this publication for 'all cancers' therefore exclude non-melanoma skin cancer (nmsc).

Table G2

Cancer site codes in ICD9 and ICD10

Cancer site	ICD9	ICD10
All cancers	140-208 x173	C00-C97 xC44
Lip, mouth and pharynx	140-149	C00-C14
Oesophagus	150	C15
Stomach	151	C16
Colorectal	153-154.1	C18-C20
Pancreas	157	C25
Larynx	161	C32
Lung	162.2-162.9	C34
Melanoma of skin	172	C43
Breast (female)	174	C50
Cervix	180	C53
Uterus	182 (+179*)	C54 (+C55*)
Ovary	183.0	C56
Prostate	185	C61
Testis	186	C62
Kidney	189.0	C64
Bladder	188	C67
Brain	191	C71
Hodgkin's disease	201	C81
Non-Hodgkin's lymphoma	200, 202.0-202.2, 202.8	C82-C85
Multiple myeloma	203.0-203.1	C90
All leukaemias	202.4, 204-208	C91-C95

* Deaths only.

Cancer registrations

More is known about the incidence of, and survival from, cancer than for most other diseases. This is because in the UK and Ireland there are population-based cancer registration systems with 100 per cent geographical coverage and mechanisms in place to follow up cases. However, cancer registration is not statutory and the implications of this are discussed below.

England

Cancer registration in England is conducted by nine regional registries which submit notifications to the National Cancer Intelligence Centre (NCIC) at the Office for National Statistics (ONS, www.statistics.gov.uk, formerly the Office of Population Censuses and Surveys). Most registries get their principal information from hospitals' patient administrations systems (PAS) – usually electronically – and pathology laboratories. Some registries also use hospital records staff to collect data, while others employ peripatetic clerks who visit hospitals, but information is also obtained from coroners, GPs and private nursing homes. In addition, the registries regularly receive from ONS notifications of deaths where cancer is mentioned on the death certificate. Registries match these against their records to indicate possible cases not already known to them, or to update details of existing records. Cancer registrations are recorded in the National Health Service Central Register (NHSCR, part of ONS), as are deaths supplied by the statutory civil registration process (see below). The subsequent linkage of incidence and death records enables calculation of survival (that is, time from diagnosis to death).

Wales

In Wales, cancer registration is carried out by the Welsh Cancer Intelligence and Surveillance Unit (WCISU, a division within the Velindre NHS Trust, www.velindre-tr.wales.nhs.uk/wcisu). They receive their cancer notifications from hospital trusts and the NHS Clearing Service via the computerised Patient Episode Database Wales, as well as from pathology records and other registries. Welsh cancer notifications are – like those for England – submitted to the NCIC at ONS, and recorded in the NHSCR and linked with death records.

Although cancer registration began in 1962, the Welsh Cancer Registry did not cover the whole of Wales until 1974 when an electronic registry was established based on the hospital activity database. It remained essentially unchanged until 1997, when all matters relating to health were devolved to the National Assembly of Wales (NAW), and WCISU took over responsibility for the service, publishing all cancer incidence data for Wales from 1995 onwards.

An innovative information system for clinical oncologists has been in use in the Velindre Trust since 1990 for clinical care of patients. Under the Cancer Information Framework, it has been extended to the whole of Wales to support multi-disciplinary teams across hospital sites. It will form the basis of a new way of capturing clinically rich cancer data items in real time, including better quality for clinical audit and population-based cancer registration data.

Scotland

In Scotland, up to 1997, cancer registration was carried out by five regional cancer registries; they relied mainly on hospital in-patient sources, pathology records and death records. The role of the Scottish Cancer Registry (part of the NHS National Services Scotland Information Services (ISD Scotland: www.isdscotland.org/cancer_information)) was limited to co-ordination, data collection from the regional registries, analysis and publication. From 1997, the national registry has been responsible for all aspects of cancer registration (with the core registration function funded by the Scottish Executive Health

Department) and has established a network of out-posted peripatetic cancer registration officers based in the main hospitals.

Registrations are identified from four main computerised sources: hospital discharge records; oncology records; pathology records; and death records. Information from these and non-computerised sources is linked to create provisional registrations made available to the cancer registration officers. They in turn refer to medical records to validate the provisional registration and abstract additional information not currently collected electronically. Follow-up is achieved by linkage to computerised death records supplied by the Registrar General for Scotland (via the General Register Office for Scotland (GROS)).

The Scottish Cancer Registry has an active programme of quality assurance and the results of many of its studies in this field have been published.

Northern Ireland

The Northern Ireland Cancer Registry (NICR, www.qub.ac.uk/nicr) was established in 1994 and replaced an existing Department of Health and Social Services registry – this had relied on clinicians to complete registration cards and, consequently, ascertainment of cases was incomplete. Complete data are available from 1993.

The registry uses an automated electronic system with the main source being the patient administration system (PAS) used by all hospital trusts. From the PAS, the registry obtains basic tumour information on cancer site and behaviour and this is supplemented by electronic downloads from pathology laboratories which give additional information on the morphology of the tumour. Death certificates are supplied by the Registrar General of Northern Ireland (via the General Register Office for Northern Ireland (GRONI)) and matched against the registry database. Cases notified only from PAS or a death certificate have case records checked to validate the diagnosis.

In addition, there are three disease-specific registries in Northern Ireland, with which the registry data are compared. These were set up independently from the cancer registry and contain information on specific sites – malignant melanomas (incorporated into the NICR), colorectal cancers (historical only) and leukaemia and lymphomas (active).

NICR is part of the Centre for Clinical and Population Sciences at the Queen's University of Belfast, and is involved in a programme of cancer-related research. It also provides a cancer information service for the region and has close connections with the breast and cervical screening services.

Ireland

The National Cancer Registry of Ireland (NCRI, www.ncri.ie/ncri) was founded in 1991, taking over the functions of the Southern Tumour Registry, which had provided population-based registration for about one sixth of the country since 1975. Collection of cancer data for the entire country began in 1994 and is fully funded by the Department of Health and Children.

Most notifications come from pathology departments, with a small number from other hospital sources, death certificates and general practitioners. Death certificates received from the Central Statistics Office are followed up with the hospital of death or the certifying doctor, if the cancer is not already registered. At the time of preparing this atlas, the registry did not accept an unconfirmed death certificate (DCO) as a basis for registration. This position subsequently changed and the implications for the data and their interpretation are discussed below.

Quality of cancer registration data

Level of ascertainment

As noted above, cancer registries differ considerably in their methods of data collection. They probably also differ in terms of the level of ascertainment of their data, that is, the proportion of cancer incidence in the population that is actually recorded by registries. General indications of the level of ascertainment can be obtained by comparing the numbers of registrations by cancer site with deaths from the same cancer in a given period and within the same geographical area. Such mortality-to-incidence ratios (see Appendix H) by sex and site are presented in Appendix B. These ratios have several limitations, but it would be difficult to explain any major differences between countries and regions unless there were similar differences in ascertainment. The ratios also provide a crude indication of survival: cancers with poorer survival rates usually have higher mortality-to-incidence ratios.

A high proportion of cancer registrations made solely on the basis of information from a death certificate (DCO) also implies under-ascertainment. This is because the registry is failing to register some patients who die from cancer while they were alive. Also, some cancer patients will not die of their disease and so the cancer will not be mentioned on their death certificate. Therefore, some of these patients would not be registered either when they were alive, or at death. As noted above, the National Cancer Registry of Ireland initially did not accept DCOs as a basis for registration, but have recently 'added back' DCOs to incidence cases from 1994. Although it was not feasible to re-compute numbers and rates for this

Table G3

Mortality-to-incidence ratios for Ireland: effect of DCO registrations

Cancer site	ICD10	DCOs included[1]		DCOs excluded[2]	
		Male	Female	Male	Female
All cancers	C00-C97 xC44	0.63	0.53	0.64	0.54
Lip, mouth and pharynx	C00-C14	0.48	0.49	0.47	0.50
Oesophagus	C15	1.02	0.94	1.05	0.95
Stomach	C16	0.76	0.80	0.77	0.83
Colon	C18	0.62	0.55	0.63	0.57
Rectum	C19-C20	0.36	0.33	0.35	0.34
Pancreas	C25	1.08	1.00	1.13	1.03
Larynx	C32	0.49	0.52	0.50	0.53
Lung	C34	0.95	0.93	0.98	0.99
Melanoma of skin	C43	0.21	0.13	0.21	0.13
Breast	C50	-	0.36	-	0.37
Cervix	C53	-	0.42	-	0.42
Uterus	C54	-	0.23	-	0.23
Ovary	C56	-	0.65	-	0.67
Prostate	C61	0.42	-	0.43	-
Testis	C62	0.09	-	0.09	-
Kidney	C64	0.51	0.48	0.54	0.49
Bladder	C67	0.34	0.35	0.33	0.36
Brain	C71	0.82	0.85	0.85	0.88
Hodgkin's disease	C81	0.39	0.24	0.37	0.25
Non-Hodgkin's lymphoma	C82-C85	0.53	0.48	0.55	0.51
Multiple myeloma	C90	0.77	0.81	0.80	0.82
All leukaemias	C91-C95	0.62	0.58	0.65	0.60

1 Calculated using 1994-99 mortality and incidence age-standardised rates.
2 Calculated using 1994-99 incidence and 1994-2000 mortality age-standardised rates.

publication, Table G3 shows the differences in the mortality-to-incidence ratios for Ireland when DCOs are included and excluded.

Completeness

Completeness is the extent to which all appropriate data items have been recorded in a registry database. If high proportions of essential data items are missing, this is an indicator of poor quality. For cases that have been registered solely from the information on a death certificate (DCO), the incidence date is unknown and has to be taken as the date of death. Other quality indicators include the proportion of cases where the primary site is unknown, or where important information, such as the age of the patient or their postcode, is missing. The proportions of such cases are extremely low.

Accuracy

As with completeness, the accuracy of the data (that is, the proportion of cases that truly have the recorded characteristic) is only occasionally known directly, usually from special studies. Various indirect measures, however, suggest that there is considerable variation between areas. A report of a project to audit the quality and comparability of cancer registration data in the UK, carried out under the aegis of the United Kingdom Association of Cancer Registries (see Appendix J), was published in 1995.[3] Variations among the registries were found in data quality for diagnostic factors, incidence date, stage of disease, treatment information, and use of death information. The review concluded, however, that cancer registry records were largely complete, accurate and reliable. The review also found that the quality of cancer registry data depended heavily on the competence and experience of staff in the registry, on maintaining good relationships with clinicians, staff in health authorities, and scientists, and on the registry's active involvement in research.

Timeliness

Registration of newly diagnosed cases of cancer is a dynamic process in the sense that the data files are always open. Cancer records may be amended – for example, the site code may be modified should more accurate information become available at a later date. The date of death is added for cases registered when the person was alive. Also, complete new 'late' registrations may be made after publication of what was thought at the time to be virtually complete results for a particular year. These are often prompted by information from a death certificate.

Duplicate registrations

While late registrations result in previously published figures being too low, duplicate registrations can artificially inflate them. Such duplication may arise if a patient is resident in one region but treated in another; this is particularly so for those resident in areas bordering another country for example, patients resident in North Wales and treated in Liverpool. Registrations are therefore carefully examined to distinguish duplicate records from true multiple primary cancers.

Other quality issues

Inaccuracies and incompleteness may arise from diagnostic practice, and changes in it, although such errors and changes come from outside the cancer registration system and are not under its control. Misclassification of cancers is more likely to occur when there is no opportunity to obtain histological confirmation of disease, or if the tumour has a pre-malignant stage which can be confused with invasive carcinoma. Misclassification may also result from mistakes in the collection, abstraction or coding of information, both before and after it reaches a registry. Also, clinical and pathological (and registry) definitions of cancer may change over time and between places, particularly for borderline malignant conditions.

Mortality

For the constituent countries of the UK, there is a statutory requirement to register a death within five days. In Ireland, deaths have to be registered within one year of occurrence.

Most deaths are certified by a medical practitioner. The death certificate is then (normally) taken to a registrar of births and deaths by a person known as an informant – usually a near relative of the deceased. In certain cases, deaths are referred to (and sometimes then investigated by) a coroner who sends information to the registrar of births and deaths which is used instead of that from the medical practitioner.

Details of the system of registration of deaths for the constituent countries of the UK and Ireland have been published elsewhere.[4-7]

Advantages and disadvantages of incidence and mortality data

Cancer incidence data are coded to both cancer site (using ICD) and histological type (using ICD-O) in the countries of Great Britain. Incidence is coded only to ICD in Northern Ireland, and in Ireland to ICD-O with translation to ICD. Mortality data are coded only to ICD in the UK, and in Ireland to ICD with translation to ICD-O.

Around 10 per cent of deaths in England and Wales are coded to 'site unspecified'[4] whereas the corresponding proportion for incidence data is only 3 per cent.[8] Consequently, diagnostic accuracy is less certain for mortality than for incidence.

Mortality data are generally more timely than incidence data because of the statutory requirement to register a death within five days (UK only), and for the large majority of deaths there is only one source document. The data are also virtually 100 per cent complete. Cancer registration is not statutory and collating information from the variety of data sources is time consuming. Final results are only published once it is believed data have been received from all the relevant sources, but this can be difficult to quantify and data may still be incomplete due to late registrations.

In the UK, there is a long time series for deaths data, although this has been affected by coding and classification changes over the years. Details of the effects of these changes during

Table G4

Summary of advantages and disadvantages of incidence and mortality data

Incidence	Mortality
Advantages	**Disadvantages**
- high quality coding	- diagnostic accuracy less certain than for incidence
- both cancer site and histology	- cancer site only, no histology
- very low proportion of 'site unspecified'	- around 10 per cent 'site unspecified'
- date of diagnosis known	- deaths in any one year result from cancers diagnosed over a long time period
Disadvantages	**Advantages**
- may not be complete	- virtually 100 per cent complete
- may not be sufficiently timely	- timely
- data for England and Wales available only from 1971, with evidence of under-ascertainment in the 1970s; data for Scotland, Northern Ireland and Ireland available from 1959, 1993 and 1994, respectively	- long time series (UK only), but affected by ICD changes, and in England and Wales by coding and other changes in 1984 and 1993

the 1990s for England and Wales have been published,[9,10] giving an indication of the likely effect on cancer registrations.

Even if survival rates remain unchanged, trends in mortality give only a delayed indication of trends in new cases because for cancers with moderate or good survival, those dying in any one year may have been diagnosed and treated many years earlier. Cancer mortality trends are therefore a 'fuzzy' indicator of trends in the efficacy of treatment – they reflect earlier trends in incidence and survival, and cannot be interpreted sensibly without them.[11] This has made incidence data increasingly more important for early monitoring of trends, and for assessment of major public health interventions such as breast and cervical screening.

Geography

The United Kingdom (UK) comprises England, Wales, Scotland and Northern Ireland, but does not include the Isle of Man or the Channel Isles.

In this publication, incidence and mortality are presented by country, health regional office (England), health authority (England and Wales), health board (Scotland and Ireland), and health and social services board (Northern Ireland). For simplicity, the term 'health authority' has been used throughout this atlas to refer to the 127 administrative health areas within Scotland, Northern Ireland and Ireland as well as those in England and Wales (see Table G5). For consistency over the period 1991-2000, the health authorities are based on boundaries as at April 2001.

The ordering of the health regional offices and countries in the tables and charts in this atlas is the standard for National

Table G5

Health authorities in the UK and Ireland

Country	Number of health authorities
England	95
Wales	5
Scotland	15
Northern Ireland	4
Ireland	8
Total	127

Statistics. However, for ease of reference, the health authorities appear alphabetically within each country or region of England in the tables in Appendix B.

Population

Population figures have been used in this atlas as denominators to calculate incidence and mortality rates. For the countries of the UK, mid-year population estimates (revised in light of the 2001 UK Census) were used. For Ireland, the population figures used for 1996 are official national census figures, and for 1994-95 and 1997-2000 are official intercensal estimates published by the Central Statistics Office.

Although the UK census population figures for 2001 were overall some one million lower than the previously published population estimates, the differences were concentrated largely in the younger age groups, particularly for males. Cancer is a disease predominantly of the elderly, and checks on data for England and Wales have shown that in general, the effects on previously published cancer incidence rates of using populations for the 1990s that have been revised in the light of the results of the 2001 census, and subsequently, are very small.

Appendix C contains population estimates for 1996 by sex and country, health regional office in England, and health authority.

References

1. World Health Organisation. *International Classification of Diseases, Ninth Revision.* Geneva: WHO, 1977.

2. World Health Organisation. *International Classification of Diseases and Related Health Problems, Tenth Revision* . Geneva: WHO, 1992.

3. Huggett, C. *Review of the Quality and Comparability of Data held by Regional Cancer Registries.* Bristol: Bristol Cancer Epidemiology Unit incorporating the South West Cancer Registry, 1995.

4. ONS. *Mortality statistics 2003: cause, England and Wales.* Series DH2 No.30. London: ONS, 2004.

5. General Register Office for Scotland. *Scotland's Population 2003: The Registrar General's Annual Review of Demographic Trends.* Edinburgh: General Register Office for Scotland, 2004.

6. Northern Ireland Statistics and Research Agency. *Registrar General Annual Report 2003.* Belfast: Northern Ireland Statistics and Research Agency, 2004.

7. Central Statistics Office. *Report on Vital Statistics, 2002.* Dublin: The Stationery Office, 2002.

8. Quinn MJ, Babb PJ, Brock A, Kirby L et al. *Cancer Trends in England and Wales 1950-1999.* Studies on Medical and Population Subjects No. 66. London: The Stationery Office, 2001.

9. Rooney C, Devis T. Mortality trends by cause of death in England and Wales 1980-94: the impact of introducing automated cause coding and related changes in 1993. *Population Trends* 1996: 29-35.

10. Rooney C, Griffiths C, Cook L. The implementation of ICD10 for cause of death - some preliminary results from the bridge coding study. *Health Statistics Quarterly* 2002; 13: 31-41.

11. Coleman MP, Babb PJ, Stockton D, Forman D et al. Trends in breast cancer incidence, survival and mortality in England and Wales. *Lancet* 2000; 356: 590-591.

Appendix H
Methods

Steve Rowan, Mike Quinn

Incidence

Incidence is defined as the number of newly diagnosed cases of cancer registered in a given period, usually a calendar year.

Mortality

Mortality is defined as the number of deaths that occurred in a given period, again usually a calendar year (UK) or were registered in a given year (Ireland).

Numbers of cancers and deaths

In Appendix B, the average number of cancer registrations or deaths each year by sex and by country, health regional office and health authority are given to three meaningful figures, for all cancers and for each of the main cancer sites covered in Chapters 3-23. The average was calculated by dividing the total cancer registrations or deaths for any particular area by the number of years of data available for that area.

Average number of cancer registrations or deaths = N_a/Y_a

where N = number of cancer registrations/deaths
Y = number of years of data available
a = country, health regional office or health authority

At the time of preparing this publication, cancer registration data were available for 1991-99 for England, Wales and Scotland, 1993-99 for Northern Ireland and 1994-99 for Ireland. Mortality data were available for 1991-2000 for England, Wales and Northern Ireland, 1991-99 for Scotland and 1994-2000 for Ireland (see Appendix G).

Crude rate

The crude rate of incidence or mortality for a given period (c) is the total number of cancer registrations or deaths divided by the total corresponding mid-year population, and is usually expressed per 100,000 population:

$c = (\sum N_k / \sum P_k)*100,000$

where N = number of cancer registrations or deaths
P = population
k = 1, … ,18 and the 18 age groups are 0-4, 5-9, … , 80-84, 85 and over.

Crude rates are not suitable for use in making comparisons between different areas or between different periods in time in the same area. This is because the incidence of (and mortality from) cancer varies greatly with age. As the incidence of, and mortality from, most cancers is very low in children, adolescents and young adults, and rates rise rapidly with age from around 50, crude rates will be highest in those areas with the highest proportions of elderly people. Crude rates would also rise over time if the age distribution of the population shifted towards the more elderly, even if there was no change in the rate in any age group.

Age-specific incidence and mortality rates

The age-specific rate of incidence or mortality for a given period (a) is the number of cancer registrations or deaths for a particular sex and age group divided by the corresponding sex- and age-specific mid-year population, and is usually expressed per 100,000 population:

$a_k = (N_k/P_k)*100,000$

where N = number of cancer registrations or deaths
P = population
k = 1, … ,18 and the 18 age groups are 0-4, 5-9, … , 80-84, 85 and over.

Direct age standardisation of rates

Differences in the age structure of populations between geographical areas (or over time) need to be controlled to give unbiased comparisons of incidence or mortality. This can be achieved through direct age standardisation, where the age- and sex-specific rates are weighted by the corresponding number of people in a standard population and then summed to give an overall rate per 100,000 population (d). We have used the European standard population – see Appendix C. The formula for the variance (v) of the directly age-standardised rate is also given below:

$d = \left[\sum_k \{(N_k/P_k)*E_k\} / \sum_k E_k \right]*100,000$

$v = \sum_k \{(E_k^2 *N_k)/P_k^2\}$

where N = number of cancer registrations or deaths
P = population
E = European standard population
k = 1, … ,18 and the 18 age groups are 0-4, 5-9, … , 80-84, 85 and over.

Appendix B, contains tables of directly age-standardised rates by sex and by country, health regional office and health authority, for all cancers, and for each individual cancer site.

Confidence intervals and statistical significance

The 95 per cent confidence intervals (CI) for a directly age-standardised rate (d) are:[1]

CI = d ± (1.96*√(**v**))

where   **d** = the directly age-standardised rate

       v = the variance of **d**

The problem of assessing whether two rates are statistically significantly different from each other, and the subsequent reporting of any such differences, is discussed in detail in Appendix K.

The underlying, crucial, assumption for a significance test of the difference between any two rates is that the rates are independent – and that other, similar, rates on which statistical tests are carried out in the same study are also independent.

For any particular cancer, the rates in an area for males and females, and for incidence and mortality, are clearly not independent. Nor are the rates independent for areas that are adjacent or whose populations are subject to similar levels of any relevant risk factors. And the variations at the health authority level in the rates for related cancers, such as those of the lung and larynx, are also not independent of each other.

In these circumstances, any straightforward statistical test of the difference between two rates would over-estimate the significance level, and so many differences between rates would appear to be significantly different when they were not.

The usual approach to the problem of multiple statistical testing and non-independence is to require a much higher apparent level of statistical significance than 5 per cent.

This can be done by taking into account the number of tests being performed. For example, if 20 such tests were carried out, a significance level of (0.05/20) = 0.0025 or 0.25 per cent could be required. The difficulty with this approach for this atlas is that we do not know how many non-independent tests there would be.

A simple alternative (which also avoids the need to perform hundreds of statistical tests) is to note whether the 95 per cent confidence intervals around the two rates overlap or not. If the two rates were in fact independent, then (assuming roughly equal variances) the non-overlapping of the 95 per cent confidence intervals is roughly equivalent to the rates being significantly different at a significance level of about 0.6 per cent (p=0.006).

In the data tables in Appendix B, the directly age-standardised rates that are above or below the relevant UK and Ireland average rate, and the 95 per cent CIs of the rates do not

overlap, are marked with '*' or '~', respectively. In the tables, health authorities are also ranked (by sex) within each country, or region of England, with 1 indicating the lowest rate.

In each of Chapters 3-23, Figures x.1 and x.2 show directly age-standardised rates by country, or region of England, and Figures x.3 and x.4 show directly age-standardised rates by health authority. In these figures, the 95 per cent CIs around each rate are also given. Figures x.3 and x.4 show the rates for health authorities in ascending order within each country, or region of England.

Mortality-to-incidence ratios

As discussed in Appendix G, mortality-to-incidence ratios (M:I) are useful as an indicator of ascertainment as well as a crude measure of survival by cancer site. M:I ratios are given in Appendix B.

Mortality-to-incidence ratio   = **M/I**
Crude measure of survival using M:I ratios   = (1-(**M/I**))*100

where   **M** and **I** are the directly age-standardised rates for mortality and incidence, respectively.

Maps

In each of Chapters 3-23, Maps x.1 and x.2 show the ratio (r) of the directly age-standardised rate by health authority to the average rate for the UK and Ireland for incidence and mortality, by sex.

r = d$_a$/d$_A$

where   **d** = directly age-standardised rate

       a = health authority

       A = UK and Ireland average

The colouring for the 'ratio' or 'relative' maps is based on seven intervals. The intervals are defined by values of 10 per cent, 33 per cent and 50 per cent or more, above and below the average for the UK and Ireland, that is, ratios of 0.67, 0.75, 0.91, 1.10, 1.33 and 1.50, as shown in Table H1.

In Appendix E, the maps show directly age-standardised rates for both incidence and mortality, by sex. There are 16 intervals in the range 0 to 180 per 100,000 population, spanning the lowest and highest rates (incidence or mortality) for the major cancer sites covered in this atlas.

The colouring for the 'absolute' scale maps is based on intervals calculated using 120 (the lower boundary point of the highest interval) as the starting point, as follows:

n-1 = **n**/1.333

where   **n** is the lower boundary point of an interval, and **n**-1 is the preceding value.

Table **H1**

Colouring and ranges of values used for the 'relative scale' maps in Chapters 3-23

Colour	Minimum value	Maximum value	Legend label
Dark purple	1.50	-	1.50 and over
Mid purple	1.33	1.499	1.33 to 1.50
Light purple	1.10	1.329	1.10 to 1.33
White	0.91	1.099	0.91 to 1.10
Light blue	0.75	0.909	0.75 to 0.91
Mid blue	0.67	0.749	0.67 to 0.75
Dark blue	0.00	0.669	less than 0.67

The resulting values were rounded to the nearest integer, as shown in Table H2. Each boundary point is about one third higher (lower) than the point below (above).

Estimates of preventable cancer cases and deaths

The numbers of preventable cancer cases and deaths presented in Chapter 2 were estimated for ten of the major cancers for which there were marked geographical variations in rates at the health authority level.

Table **H2**

Colouring and ranges of values used for the 'absolute scale' maps in Appendix E

Colour	Minimum value	Maximum value
Darkest red	120	180.00
	90	119.99
	68	89.99
	51	67.99
	38	50.99
	28	37.99
Red	21	27.99
	16	20.99
	12	15.99
Orange	9	11.99
	7	8.99
	5	6.99
Yellow	4	4.99
	3	3.99
	2	2.99
White	0	1.99

It was assumed that it should theoretically be possible to reduce the rates in all health authorities to those in the areas with among the lowest values. With the exception of the incidence of bladder cancer, plausible achievable low rates were chosen by inspection from the data tables in Appendix B and Figures x.3 and x.4 in each of the relevant cancer-specific chapters. The rates chosen were not always the very lowest of all 127 health authorities, particularly for cancers where one or two health authorities had rates that were 'outliers', which were possibly due to random fluctuation in areas with small populations. The rates selected are shown in Table H3.

For bladder cancer, there were recognised variations in incidence due to different coding practices among the cancer registries (see Chapter 3). Estimates were therefore made first of the numbers of preventable deaths. The numbers of preventable cases were then estimated from the numbers of preventable deaths by dividing these by the mortality-to-incidence ratios of 0.45 for males and 0.48 for females, typically found for areas covered by cancer registries that did not register transitional cell papillomas of the bladder as malignant.

Estimates were not made for either the incidence of, or mortality from, cancer of the larynx in females, as the overall numbers of cases and deaths were small relative to those in males. Estimates were not made for the incidence of melanoma of the skin, for the reasons given in Chapter 2.

Table **H3**

'Low' cancer incidence and mortality rates[1] at the health authority level

Cancer	Incidence		Mortality	
	Males	Females	Males	Females
Lung	60.0	22.0	54.0	20.0
Bladder	†	†	8.5	2.5
Larynx	4.0	..	1.4	..
Lip, mouth and pharynx	6.0	3.0	2.5	1.1
Oesophagus	10.2	4.7	10.0	4.1
Pancreas	9.5	6.5	9.1	6.1
Stomach	15.0	5.5	10.8	4.2
Colorectal	42.0	28.0	23.0	15.5
Melanoma of skin	..	..	1.9	1.5
Cervix	:	7.5	:	2.8

1 Directly age standardised using the European standard population; rate per 100,000 population
† Preventable cases estimated from preventable deaths – see text
.. Not included
: Not applicable

The assumption made was that the rates in every health authority could be reduced to the 'low' rate shown in Table H3. This is equivalent to reducing the average rate to the low rate. The numbers of preventable cancer cases and deaths were therefore calculated by subtracting the relevant low rate from the observed average rate for each country and region of England, and multiplying by the relevant estimate of the mid-year population for 1996 (see Appendix C).

For example, the directly age-standardised incidence rate for lung cancer in males in the Northern and Yorkshire region of England was 94.1 per 100,000 population (Table B13.1). The selected achievable 'low' rate for the incidence of lung cancer in males was 60 per 100,000 (Table H3). The difference between these rates is 94.1 − 60 = 34.1 per 100,000. The population of males in the region was 3,028,700. The estimated number of preventable cases of lung cancer in males in the region was therefore (34.1 × 3,028,700)/100,000 = 1,033. The numbers shown in Table 2.6 have been rounded to the nearest 10.

The figures were then summed to give a UK and Ireland total. Estimates are given in Table 2.6 for each country and region of England for lung cancer, and for the UK and Ireland in total for the other nine cancers.

References

1. Esteve J, Benhamou E, Raymond L. *Statistical Methods in Cancer Research, Volume IV - Descriptive Epidemiology.* IARC Scientfic Publications. Lyon: International Agency for Research on Cancer, 1994.

Appendix I
Country profiles

Helen Wood (Editor)

Introduction

This appendix gives brief descriptions of Wales, Scotland, Northern Ireland and Ireland and the Government Office Regions (GORs) of England (North East, North West, Yorkshire and the Humber, East Midlands, West Midlands, East of England, London, South East and South West). The GORs are not the co-terminus with the health regional office areas, but are similar, and as demographic and economic information was not available for the health regional office areas, descriptions of GORs have been included instead.

These profiles give an overview of the geography, demography, economy, and cancer services within each country or region (of England).

England

England is the largest of the five countries covered by this atlas, with a total area of 150,400 square kilometres.

Geography

England is predominantly a lowland country, although there are upland regions in the north and in the south west. The greatest concentrations of population are in London and the South East, South and West Yorkshire, Greater Manchester and Merseyside, the West Midlands conurbation, and the north-east conurbations on the rivers Tyne and Tees. About 35 per cent of the population live in metropolitan areas, with a further 34 per cent in smaller urban districts and 20 per cent in mixed urban/rural areas.

In contrast to Wales, Scotland and Northern Ireland, England has no separate elected national body or department exclusively responsible for its central administration. Instead, there are a number of government departments, whose responsibilities in some cases also cover aspects of affairs in Wales and Scotland. A network of nine Government Office Regions (GOR) is responsible for the implementation of several government programmes in the English regions. The regions are: North East, North West, Yorkshire and the Humber, East Midlands, West Midlands, East of England, London, South East and South West.

Demography

The population of England in 2000 was almost 49 million (51 per cent female). Nineteen per cent of the population were aged under 15 and 18 per cent were of pensionable age or over. In 2003, life expectancy at birth was 76 for boys and 81 for girls. The total fertility rate in 2003 was 1.71 children per woman, which was a slight increase from the record low of 1.63 in 2001. Since the turn of the century there have been more births than deaths in the UK in every year since 1901, except 1976. This means that the population has grown due to natural change. Until the mid-1990s, this natural increase was the main driver of population growth. Since the late 1990s, there has still been natural increase but net international migration into the UK from abroad has been an increasingly important factor in population growth.

England has an average population density of 380 people per square kilometre, regionally, the average ranges from 4,565 in London to 206 in the South West. There are over 16,700 rural towns, villages and hamlets in England with populations of 10,000 or fewer, and over three-quarters of these have populations under 500.

Data from the 1991 census indicated that 94 per cent of the population described themselves as white, 1.7 per cent as Indian, 1.2 per cent as either Pakistani or Bangladeshi, 0.3 per cent as Chinese (with 0.4 per cent in other Asian groups), one per cent as black Caribbean, 0.4 per cent as black African (with 0.4 per cent in other black groups), and 0.6 per cent as belonging to other ethnic groups.

Cancer services

The Department of Health (DH) set out its overall cancer strategy in the NHS Cancer Plan in September 2000. This aimed to reduce cancer deaths, to improve the quality of cancer care and treatment, and to reduce inequalities in health. Cancer care is implemented by Cancer Networks, of which there are currently 34, each covering a population of 1 to 2 million people.

The NHS Breast Screening Programme was set up in 1988 and achieved national coverage by the mid-1990s. Up to 2003, the programme provided free mammographic screening every three years to all women aged from 50 to 64; it is now (2005) being extended to include those aged 65 to 70. Cervical screening began in Britain in the mid-1960s, but was not population based. The NHS Cervical Screening Programme was set up in 1988 and screens all women aged 25 to 64 every three to five years.

Government Office Regions

Regional profiles for the health regional office areas of England, as used in this atlas, were not available. Regional profiles of the Government Office Regions are given below. This information was taken from the overview in the 2001 edition of Regional Trends, which also contains much more detailed information about the GORs within England.[1]

North East

In 1999, the North East had a population of 2.6 million people. The population density within the North East was highest in Middlesbrough unitary authority at over 2,600 people per square kilometre, and lowest in the local authority district of Berwick-upon-Tweed at 27 people per square kilometre.

The standardised mortality ratio for the North East was 10 per cent higher than the UK as a whole in 1999, but the infant mortality rate for 1998-2000 was similar to the rate for the UK (5.7 and 5.8 deaths of infants under one year per 1,000 live births, respectively). The employment rate in the North East was among the lowest in the UK, at 67.4 per cent in 2000, and average gross weekly earnings were lower than the UK average (£365.80 and £409.20, respectively).

Manufacturing industry accounted for a larger proportion of GDP in the North East in 1998 than for the UK as a whole, 27.3 per cent compared with 20.3 per cent, while agriculture, forestry and fishing accounted for 0.7 per cent of GDP, compared with 1.3 per cent for the UK.

North West

In 1999, the North West had a population of 6.9 million people. The population density within the North West was highest in Blackpool unitary authority at over 4,200 people per square kilometre, and lowest in the local authority district of Eden at 23 people per square kilometre.

The standardised mortality ratio for the North West was 8 per cent higher than the UK as a whole in 1999, and the infant mortality rate for 1998-2000 was also higher than the UK rate at 6.3 deaths of infants under one year per 1,000 live births. The employment rate in the North West was among the lowest in the UK, at 72.7 per cent in 2000, and average gross weekly earnings were lower than the UK average at £385.70.

In the North West, manufacturing industry accounted for a larger proportion of GDP than for the UK as a whole in 1998 – 25.1 per cent, while agriculture, forestry and fishing accounted for 0.9 per cent of GDP, lower than for the UK as a whole.

Yorkshire and the Humber

In 1999, Yorkshire and the Humber had a population of 5.0 million people. Within Yorkshire and the Humber, the population density was highest in City of Kingston upon Hull unitary authority at over 3,632 people per square kilometre, and lowest in the local authority district of Ryedale in North Yorkshire at 32 people per square kilometre.

The standardised mortality ratio for Yorkshire and the Humber was the same as the UK as a whole in 1999, while the infant mortality rate for 1998-2000 was higher than the UK rate at 6.8 deaths of infants under one year per 1,000 live births. The employment rate in Yorkshire and the Humber was among the highest in the UK, at 73.5 per cent in 2000, but average gross weekly earnings were lower than the UK average, at £373.70.

In the Yorkshire and the Humber, manufacturing industry accounted for some 26.1 per cent of GDP, a larger proportion than for the UK as a whole in 1998, and agriculture, forestry and fishing also accounted for a higher proportion than the UK as a whole, at 1.6 per cent.

East Midlands

In 1999, the East Midlands had a population of 4.2 million people. The population density within the East Midlands was highest in Leicester unitary authority at 3,985 people per square kilometre, and lowest in the local authority district of West Lindsey in Lincolnshire at 67 people per square kilometre.

The standardised mortality ratio for the East Midlands was the same as the UK as a whole in 1999, and the infant mortality rate for 1998-2000 was similar to the UK rate at 5.7 deaths of infants under one year per 1,000 live births. The employment rate in the East Midlands, at 76.8 per cent, was among the highest in the UK, but average gross weekly earnings were lower than the UK average, at £371.40.

In the East Midlands, manufacturing industry accounted for some 28.8 per cent of GDP in 1998, a larger proportion than for the UK as a whole, and the proportion accounted for by agriculture, forestry and fishing was also higher than that for the UK, at 2.0 per cent.

West Midlands

In 1999, the West Midlands had a population of 5.3 million people. The population density within the West Midlands was highest in Birmingham at over 3,800 people per square kilometre, and lowest in the local authority district of South Shropshire in Shropshire at 41 people per square kilometre.

The standardised mortality ratio for the West Midlands was 2 per cent higher than the UK as a whole in 1999, and the infant mortality rate for 1998-2000 was also higher than the UK rate at 6.8 deaths of infants under one year per 1,000 live births. The employment rate in the West Midlands was among the highest in the UK, at 73.1 per cent, but average gross weekly earnings were lower than the UK average, at £385.90.

In the West Midlands, manufacturing industry accounted for some 28.9 per cent of GDP in 1998, a much larger proportion than for the UK as a whole, and the proportion accounted for by agriculture, forestry and fishing was slightly higher than the UK, at 1.5 per cent.

East of England

In 1999, the East of England had a population of 5.4 million people. The population density was highest in Luton unitary authority at 4,264 people per square kilometre, and lowest in the local authority district of Breckland in Norfolk at 92 people per square kilometre.

The standardised mortality ratio for the East of England was 7 per cent lower than the UK as a whole in 1999, and the infant mortality rate for 1998-2000 was also lower than the UK rate at 4.7 deaths of infants under one year per 1,000 live births. The employment rate in the East of England was among the highest in the UK, at 78.3 per cent, and average gross weekly earnings, at £412.70, were higher than the UK average.

In the East of England, manufacturing industry accounted for some 17 per cent of GDP in 1998, a smaller proportion than for the UK as a whole, but the proportion accounted for by agriculture, forestry and fishing was slightly higher than that for the UK, at 1.7 per cent.

London

In 1999, London had a population of 7.3 million people. Within London the population density was highest in Kensington and Chelsea at 14,930 people per square kilometre, and lowest in Havering at 1,957 people per square kilometre.

Overall, the standardised mortality ratio for London was 5 per cent lower than the UK as a whole in 1999, but the infant mortality rate for 1998-2000 was the same as that for the UK as a whole. The employment rate in London was among the lowest in the UK, at 71.1 per cent, but average gross weekly earnings, at £529.80, were considerably higher than the UK average.

In London, manufacturing industry accounted for 10.9 per cent of GDP in 1998, around half that for the UK as a whole, and agriculture, forestry and fishing accounted for virtually none of London's GDP.

South East

In 1999, the South East had a population of 8.0 million people. Within the South East the population density was highest in Portsmouth unitary authority at 4,720 people per square kilometre, and lowest in the local authority districts of West Oxfordshire and Chichester at 138 people per square kilometre.

The standardised mortality ratio for the South East was 8 per cent lower than the UK as a whole in 1999, and the infant mortality rate for 1998-2000 was also lower than the UK rate at 4.5 deaths of infants under one year per 1,000 live births. The employment rate in the South East, at 80.6 per cent, was the highest in the UK, and average gross weekly earnings, at £434.20, were higher than the UK average.

In the South East, manufacturing industry accounted for some 17 per cent of GDP in 1998, a smaller proportion than for the UK as a whole, but the proportion accounted for by agriculture, forestry and fishing was slightly higher than the UK average, at 1.7 per cent.

South West

In 1999, the South West had a population of 4.9 million people. Within the South West the population density was highest in City of Bristol unitary authority at 3,684 people per square kilometre, and lowest in the local authority districts of West Devon at 41 people per square kilometre.

The standardised mortality ratio for the South West was 10 per cent lower than the UK as a whole in 1999, and the infant mortality rate for 1998-2000 was also lower than the UK rate at 4.7 deaths of infants under one year per 1,000 live births. The employment rate in the South West, at 78.6 per cent, was the second highest in the UK, but average gross weekly earnings, at £379.10, were lower than the UK average.

In the South West, manufacturing industry accounted for some 19.6 per cent of GDP in 1998, close to the proportion for the UK as a whole, but the proportion accounted for by agriculture, forestry and fishing was almost double that for the UK, at 2.5 per cent.

Wales

The principality of Wales occupies the land to the west of the English midlands, with a total land area of just over 20,000 square kilometres. It is largely rural and mountainous with 1,300 kilometres of coastline, 5,000 kilometres of river, and many reservoirs and lakes. A large portion of the land mass is included in the three National Parks of Snowdonia, Brecon Beacons and Pembrokeshire Coast.

Geography

Wales consists of 22 local authority areas known as unitary authorities (UA), each with a local health board (LHB) co-terminous with it. These are aggregated into three Regional Office areas of North Wales, Mid and West Wales, and South East Wales. For planning of provision of health and social services these are regarded as distinct 'health economies', and also correspond to the boundaries of the three cancer networks in Wales.

Much of Wales is hilly and mountainous, making transport and communications difficult. The mountainous area of Snowdonia lies to the north west. The Brecon Beacons extend from Monmouthshire in the south east to Carmarthenshire in the south west. The South Wales coal field lies roughly to the south of the Brecon Beacons. Coal mining was important to the Welsh economy in the nineteenth and twentieth centuries. The former mining communities of South Wales, often referred to as the 'valleys', extend along river valleys from north to south.

The southern ribbon of lower-lying land is often referred to as the M4 corridor. The main cities of Newport, Cardiff and Swansea lie on this route with access to England via two bridges across the River Severn. There is a similar but narrower corridor along the northern coast. Communications between north and south tend to be poor, and many cancer patients in north and mid-Wales are treated in England.

Demography

Wales is a culturally distinct part of the UK and the Welsh language is one of the oldest in Europe, well suited to poetry and song. The Welsh Assembly Government operates a bilingual policy and the language forms part of the curriculum in all Welsh schools. In 2001, 21 per cent of the population spoke at least some Welsh, with 69 per cent of people in Gwynedd speaking the language.

The total population in 2002 was estimated at 2.9 million with a population density of 140 persons per square kilometre. There is a distinct division between the rural and urban parts of Wales. Population density varies from 2,290 persons per square kilometre in the capital city, Cardiff, to 24 persons per square kilometre in rural Powys. Approximately half of the population of Wales live in the South East Wales Regional Office area, which includes the capital, Cardiff.

The age structure of Wales is similar to the UK as a whole with a similar percentage of children aged 0-15 (19 per cent compared with 20 per cent) and a similar proportion of those over retirement age (17 per cent compared with 18 per cent). There are higher percentages of older people in rural areas. The average ethnic mix in Wales is different from the UK, with

only 2.1 per cent of the population reporting themselves as non-white at the 2001 UK Census, compared with 7.9 per cent in the UK; the proportions are, however, higher than elsewhere in Wales in the urban areas of Cardiff and Newport: 8.5 and 4.8 per cent, respectively.

There has been a recent tendency for net immigration into Wales with a net gain of 14,000 during the 1990s, but further analysis reveals that this is greatest in the post-retirement (over 65) age group. In younger age groups, especially males in their 20s, there is net emigration from Wales. Deaths have exceeded births since the mid-1990s, but the trend in the overall projected total population is upwards. Demographic ageing will have particular implications for the future burden on cancer services in Wales.

Wales is relatively poor in comparison with the UK as a whole. The average weekly income in Wales in April 2003 was £414 per week, 12.5 per cent below the English figure of £474 per week. The GDP per head in 1994-96 was rated at 82 per cent of the EU average, compared with 98 percent for the UK as a whole. The rural area of West Wales and the valleys in the south are particularly poor, GDP being less than 75 per cent of the EU average.

Manufacturing is still the most important section of GDP in Wales at 27 per cent in 1998, higher than the UK average. Mining is now less than one per cent of GDP. Wales' considerable natural assets make tourism increasingly important to the Welsh economy, with the hotel and restaurant sector contributing 3.5 per cent of GDP and real estate/rental 15 per cent of GDP in 1998.

Wales has a high percentage of economically inactive people among those of working age: 26 per cent, compared with 21 per cent for the UK as a whole. The percentage of household income from social security benefits is higher in Wales than elsewhere in the UK and there is evidence of a higher proportion of chronic sickness related to this. The unemployment rate in Wales in 2000 was 6.2 per cent, compared with 5.5 per cent in the UK as a whole. The gap was higher for males (6.8 versus 6.0 per cent) than females (5.2 versus 4.9 per cent).

Wales had a slightly higher standardised mortality ratio (SMR) compared with the UK as a whole, of 101 in 1998. The infant mortality rate for Wales in 1997-98, at 6 per 1,000 live births, was similar to that of the UK as a whole.

Cancer Services

Following creation of the Welsh Assembly Government (WAG) in 1999, responsibility for health services passed to the Welsh Health Minister. Divergence from the English NHS has increased. Instead of Primary Care Trusts, there are 22 Local Health Boards with health-commissioning powers which includes most oncology services. There are no Strategic Health Authorities but Health Commission Wales covers certain specialised services. Public health services are provided by a National Public Health Service which includes some aspects of the Health Protection Agency in England.

Prior to devolution, the former Welsh Office convened the Cancer Services Coordinating Group (CSCG) in response to the Calman-Hine report of 1996. It has set down minimum standards for cancer care in Wales as well as convening tumour site steering groups of expert clinicians. The CSCG has also formally defined the information needs for cancer services in Wales in the form of the 'Cancer Information Framework' and the WAG has endorsed this plan.

The principal oncology centre in Wales is the Velindre Trust in Cardiff, but there is also a major oncology centre at Singleton Hospital in Swansea, as well as a smaller oncology centre with radiotherapy services at Ysbyty Glan Clwyd in Rhyl. These three correspond to the three cancer networks in Wales, which are linked together under the CSCG. However, centres in Manchester and Liverpool also continue to provide much of the specialist oncology services to patients living in North Wales.

Scotland

Scotland forms the northern-most part of the United Kingdom and consists of 32 Local Council Areas (local authorities).

Geography

Scotland contains large areas of unspoilt and wild landscape, and many of the UK's mountains, including its highest peak, Ben Nevis. Western Scotland is fringed by the large island chains known as the Inner and Outer Hebrides, and to the north east of the Scottish mainland are the Orkney and Shetland islands. Over an eighth of the total area in Scotland is designated as National Scenic Area and 2 per cent is Green Belt land.

Demography

In 2000, Scotland had a population of just over 5 million (52 per cent females). Scotland has the lowest population density in the UK, with an average of 66 people per square kilometre. Three-quarters of the population live in the central lowlands, and population density is much higher in these areas (3,493 people per square kilometre in Glasgow City). The chief cities are the capital Edinburgh, Glasgow, Aberdeen and Dundee.

In Scotland, the proportion of the population of pension age and over was virtually the same as that for the UK in 1999, at 18 per cent. In Scotland, 19.7 per cent of the population were

aged under 16, compared to 20.4 in the UK as a whole. The majority (90 per cent) of the population were born in Scotland but 7 per cent were born elsewhere in the United Kingdom (England, Wales or Northern Ireland). A relatively small per centage (0.6 per cent) were born elsewhere in Europe, 0.2 per cent in India or Bangladesh, 0.2 per cent in Africa or the Caribbean, and 2.2 per cent in the rest of the world.

In 1999, the Standardised Mortality Ratio for Scotland was 18 per cent higher than that for the UK, at 118 (UK = 100), although this ranged from 94 in Renfrewshire to 144 in Glasgow City. Infant mortality was lower for Scotland than the UK, with a rate of 5.4 deaths of infants under one year per 1,000 live births, compared to 5.8 for the UK as a whole.

The proportion of people of working age qualified to GCE A level or equivalent or higher in Scotland was 56.8 per cent in spring 2001, higher than the UK average of 47.6 per cent. The employment rate for people of working age was among the lowest in the UK, at 71.9 per cent in spring 2000 (74.3 overall for the UK), and average weekly earnings were lower than the UK average, at £380 compared with £409.

In Scotland, the manufacturing industry accounted for some 21.0 per cent of GDP in 1998, compared to 20.3 per cent for the UK as a whole. In 2000, 31 per cent of businesses were in distribution, hotels, catering and repairs industries, compared to a UK average of 28.9 per cent. Nearly 69 per cent of export trade in 2000 was to the EU, higher than the UK average of 58.6 per cent. The distribution of the workforce in Scotland in 1991 showed 34 per cent of the employed population to be working in public administration and other services, 9.8 per cent in manufacturing, 21.2 per cent in distribution, hotels and catering, 7.6 per cent in metal goods, engineering and vehicles industries, 10.4 per cent in banking, finance and insurance, 5.8 per cent in construction, 5.6 per cent in transport and communication, 2.9 per cent in energy and water supplies, 1.9 per cent in metals, minerals and chemicals, and 1.4 per cent in agriculture, forestry and fishing.

Cancer Services

Scotland has had its own Parliament since July 1999, which has taken over responsibility for many functions formerly exercised by the Parliament at Westminster. Since devolution, the Scottish Executive has had responsibility for the National Health Service (NHS) in Scotland. In Scotland, the NHS is funded mainly through taxation and is mostly free at the point of use. Access to hospital care is controlled by a well-developed system of primary care. The private health care sector in Scotland is relatively small, especially in relation to oncology services. The NHS is organised into 15 Health Boards, although the populations covered by each Health Board are quite disparate, ranging from just under 20,000 to over 900,000.

Radiotherapy facilities are provided at five main centres (Inverness, Aberdeen, Dundee, Edinburgh, and Glasgow) but many patients with cancer are diagnosed and receive surgery and chemotherapy at district general hospitals. Since 1998, there has been a determined effort to re-organise the delivery of cancer care in Scotland through formal 'Managed Clinical Networks'.

Cervical screening began in parts of Scotland in the early 1960s, but coverage was uneven until computerised call-recall systems were introduced in the late 1980s. A national breast screening programme was phased in to Scotland in the late 1980s. Until recently, women aged 50 to 64 years were invited for a routine screen once every three years, and women aged over 64 years were screened on request. An expansion of the breast screening programme is now underway: the age range for invitation is being extended to include women up to the age of 70 years (women over 70 years will continue to be screened on request). Although there is not a nationwide screening programme for colorectal cancer, a pilot study designed to test the feasibility of screening with the faecal occult blood test was initiated in two parts of the UK (including three Scottish Health Board areas) in 2000. The prostate-specific antigen (PSA) test was introduced to Scotland in 1989 and use of the test has increased strikingly since then. Although screening with PSA is not currently recommended in Scotland, there is considerable variation in the incidence of prostate cancer across the 15 Scottish health board areas despite much less variation in mortality.

Northern Ireland

Northern Ireland consists of the six counties in the north of the island of Ireland and is part of the United Kingdom.

Geography

Geographically, Northern Ireland is situated on the northwest periphery of the United Kingdom and Europe with an area of 14,160 square kilometres, 584 square kilometres of which is inland water. It is comparable in size to the West Midlands and East Anglia regions of England. Northern Ireland is bordered by the Atlantic Ocean to the north and the Irish Sea to the east. To the south and west a continuous land border with the Republic of Ireland stretches for 500 kilometres. After Scotland, Northern Ireland is the most sparsely populated region of the United Kingdom, with a population density of 125 persons per square kilometre in 2002 (roughly half of the UK average of 245 persons per square kilometre). The population of Northern Ireland was 1.7 million (51 per cent of which were female) at the last census year of 2001, and constitutes 2.9 per cent of the total UK population.

Northern Ireland is predominantly rural, with two-thirds of its population living within a 50 kilometre radius of the capital, Belfast, in the east of the region. The only other sizeable concentration of population is in and around the city of Londonderry (local government district population 107,000) in the north west.

Demography

The age structure of the Northern Ireland population differs from that of the UK as a whole in that it has a slightly higher proportion of children aged 0-15 (24 per cent in NI compared with 20 per cent in the UK) and slightly fewer people over retirement age (16 per cent in NI compared with 18 per cent in the UK). The population of Northern Ireland is relatively racially homogeneous with only a small number of ethnic minorities, mainly Chinese, totalling around 13,000.

In the 1990s, the net effect of migration on the Northern Ireland population has been a loss of about 600 people during the 1990s. In the context of annual migration, flows of around 20,000 immigrants and 20,000 emmigrants mean that migration is broadly in balance. Within the overall picture, migration patterns have varied for different ages and between the sexes. There has been more emmigration than immigration among young people in their twenties, over the decade, with higher numbers among males than females. The net emmigration of younger people has been broadly balanced the immigration of slightly older people mostly in their thirties and forties.

The manufacturing industry, including engineering, metal, food, drink, tobacco and chemical industries, is currently the most important sector (in terms of GDP) in the Northern Ireland economy. Following close behind are financial/business services, education, social work and health services. Most of the heavy industry is located in the two main urban areas.

Northern Ireland has long experienced one of the highest rates of unemployment in the UK. In 2000, the unemployment rate for Northern Ireland (7 per cent) was higher than that for the UK as a whole (5.5 per cent) and equalled rates in London. The distribution of unemployment is, however, not spread evenly across the country. Unemployment is lowest in and around Belfast, but local pockets of high unemployment within the city are concealed due to the sub-division of the country into twelve self-contained labour units for labour market purposes.

Cancer Services

Northern Ireland's health care system is similar to that of the rest of the United Kingdom. There are primary prevention services provided by the Health Promotion Agency for Northern

Ireland, four Health & Social Services Boards, and various voluntary organisations such as the Ulster Cancer Foundation, Action Cancer and the Northern Ireland Chest, Heart and Stroke Association. The Department of Health, Social Services and Public Safety have a strategy called 'Investing for Health'. There is a population-based breast screening service, with all women aged 50-64 invited every three years, and the option for older to women to be screened if they wish. This programme was set up in 1993. Screening for cervical cancer has been offered to all women over 20 at least every five years by invitation since the late 1980s. Before then cervical screening was carried out, but not on an organised population-wide basis. A population-based age-sex register is held by the Central Services Agency (CSA). This is quite up-to-date as the population use medical cards as one form of identification for voting. Various (the Department of Health) government departments have cancer-related strategies, for example for the control of malignant melanoma and other skin cancers, and targeting social need is a major theme. In 1996 the Campbell Report was published, which heralded major reorganisation of cancer services to a more centralised service with disease-specific experts working in teams. There is one cancer centre at the Belfast City Hospital and four cancer units, at the Ulster, Altnagelvin, Antrim and Craigavon Hospitals.

Most cancer patients are treated within the National Health Service system and even most private patients would receive some of their treatment in NHS hospitals. There are 13 separate hospital trusts which treat cancer patients among the four regional health boards. There are five pathology laboratories which also serve the private sector. We have four hospices and active domicilliary hospice care. The vast majority of patients receive all their care within the region. There are some supra-regional services provided elsewhere in the UK; for example, some bone cancers in Birmingham.

Ireland

Ireland (Éire) comprises 83 per cent of the area of the island of Ireland. It is divided for administrative purposes into 26 counties, grouped into four provinces (Leinster, Munster, Connacht and Ulster).

Geography

The country has a total land area of 70,280 square kilometres, with a coastline of 3,170 kilometres (km). It extends for 485 km (301 miles) from north to south and 275 km (171 miles) from east to west. Central lowlands are framed by hillier areas. The River Shannon, which runs from north-east to south-west, is the longest river, and there are a large number of lakes, of which Lough Corrib is the largest. Mean annual temperature is

9°C and mean annual rainfall about 1000 mm, both highest on the Atlantic coasts. The island's lush vegetation earns it the sobriquet 'Emerald Isle'.

Demography

In 2000, Ireland had a population of around 3.78 million (50 per cent female). Around 11 per cent of the population were aged 65 years or over and around 22 per cent were under the age of 15. The Greater Dublin area accounted for 29 per cent of the total population in 2000, while the other four major cities (Cork, Limerick, Galway and Waterford) together accounted for 10 per cent. At the time of the 2002 census of population, 40 per cent of the population lived in rural areas (centres with fewer than 1,500 inhabitants).

The birth rate in Ireland is one of the highest in Europe, with 14.3 births per thousand population in 2000. The total fertility rate in 2000 was 1.9 live births per 1,000 females, and the infant mortality rate was 6.2 per 1,000 live births.

The ethnicity of the population is predominantly white, but information on ethnic background is not collected either at the Census or by the cancer registry. Just under 90 per cent of Irish residents in 2002 were born in Ireland, 6 per cent were born in the UK, one per cent elsewhere in the EU and 3 per cent outside the EU. There was a net inward migration of 154,000 persons (4 per cent of the 1996 population) between 1996 and 2002. Seventy per cent of usual residents lived in their county of birth.

The official languages are Irish (Gaeilge) and English. However, English is by far the predominant language. In 2002, 9 per cent of people reported speaking Irish on a daily basis. About 90,000 people, about 50 per cent of whom reported speaking Irish daily, live in officially designated Irish-speaking communities (Gaeltacht areas) largely on the Western seaboard. All schoolchildren are taught the Irish language as a compulsory part of the school curriculum with a relatively small (though growing) number of schools teaching all subjects through the medium of Irish. Public signs are usually bilingual and there are both a national Irish language television and radio channel.

In 2000, 4.2 per cent of the working-age population were unemployed. Between 1996 and 2002 employment grew by over 25 per cent, largely due to the increasing participation of females in the labour force. Around 8 per cent of the population reported a long-lasting health problem or disability in 2002.

In 2002, 15 per cent of the population aged 15 years and over were in full-time education. Of the remainder, 21 per cent had left the educational system before the age of 16, and 21 per cent had been educated to degree or equivalent standard.

Cancer Services

The National Cancer Strategy was launched in 1996, and the National Cancer Forum was set up to advise on its implementation. Cancer patients in Ireland can use either private or public hospitals, although the majority (about 84 per cent of incident cases) attend public hospitals. All public and private hospitals allow the cancer registry full access to case information. There are two main publicly funded radiotherapy centres in Ireland, located in Dublin and Cork, and three smaller private centres, two in Dublin and one in Galway. Almost all cancer treatment is provided within the country. No population-based cancer screening programmes existed in Ireland in the period 1994 to 1997. A national breast screening programme (BreastCheck) was established in 1998, and screening services commenced in 2000, covering an area in the east of the country with about 50 per cent of the female population. Opportunistic but unorganised cervical screening has existed for many years, but it is not possible to estimate the proportion of the population covered. Phase One of the National Cervical Cancer Screening Programme commenced in 2000, covering the Mid-Western Health Board, and there is a commitment to extend this programme to cover the rest of the country.

References

1. McGinty J, Williams T. *Regional Trends 2001 edition No. 36*. London: The Stationery Office, 2001.

Appendix J

The United Kingdom Association of Cancer Registries

Mike Quinn

In the early 1990s, the cancer registration system in the UK was subject to rapid change. With the development of information technology, the pace of change in registration practice quickened, and increasing demands for accurate and timely information were made on the cancer registration system. Changes in the organisation of the health service and in the methods of health care delivery contributed to an increased interest from various authorities and scientists. There were new uses which could and should be made of registration data, such as medical audit and quality assurance of health care, as well as the routine uses which have been made of these data in the past, such as estimation of incidence and evaluation of survival and mortality.

There was widespread awareness both of the need to improve the quality and completeness of cancer registration data, and of the opportunities to do so through the use of information technology. Together with the increased interest from external bodies in using the data, this led to the creation of several groups bringing together cancer registry staff and personnel from the Office for Population Censuses and Surveys (OPCS, as it then was) to discuss and resolve matters of common interest.

The longest standing of these was the *Cancer Registries' Consultative Group* (CRCG) which concerned itself essentially with issues of data collection, including coding and data quality. It had representation from all cancer registries, and its members were for the most part registry managers and others closely involved in the day-to-day business of data collection. The *Cancer Surveillance Group* (CSG) was set up in 1989 to meet a perceived need for a forum bringing together those with an interest in the use of cancer data. It had a loose, open and informal membership and structure. Its members included epidemiologists and statisticians, as well as other registry staff. The *Cancer Registries' Information Technology Group* (CRITG) brought together technical experts from the various registries. Education and training was another area of activity thought to be of such importance that it could justify the establishment of

another group. There was, however, no forum which brought together registry directors on a regular basis. There was a danger, therefore, with so many different perspectives and forums in which different points of view could be expressed, that the cancer registries might fail to speak with a united voice when, for example, making representations or giving advice to government. With no coherent framework of organisation, there would be a strong possibility of duplication of effort and inadequate communication between the various groups.

It was therefore proposed that a United Kingdom Association of Cancer Registries be established. Following preliminary meetings at which almost all of the UK registries were represented, the Association was brought into being on 2nd April 1992 in Cardiff.

The Association has a federal structure. All affiliated population-based cancer registries in England, the Welsh Cancer Intelligence and Surveillance Unit, ONS, the Information and Statistics Division of the NHS in Scotland and the Northern Ireland Cancer Registry are full members with their representative, usually the director, having a vote on the Executive Committee. Associate (non-voting) members currently (2005) comprise the National Cancer Registry of Ireland, the Childhood Cancer Research Group in Oxford, the CRUK Paediatric and Familial Cancer Research Group in Manchester, the Northern Region Children and Young Persons Malignant Disease Registry in Newcastle, the West Midlands Regional Children's Tumour Registry in Birmingham, the Yorkshire Specialist Register of Cancer in Children and Young People in Leeds, and the charities Cancer Research UK and Marie Curie Cancer Care. Some years after the formation of the UKACR, a Quality Assurance group was set up to standardise the methodology for, and report on, various registry performance indicators included in the national core contract[1,2] such as timeliness and the percentage of registrations made solely from a death certificate. A Training Group and a Coding and Classification Group were established to oversee and co-ordinate the implementation of developments in those particular aspects of cancer registries' work. And a Clinical Effectiveness Group took forward issues relating to the registries' expanding role in clinical audit and performance monitoring on cancer. The Chairs of the various sub-groups, were invited, as appropriate, to attend Executive Committee meetings as observers.

In 2003, the structure of the UKACR's sub-groups was reorganised. Three new sub-groups were established, chaired by a registry director, and with new terms of reference and some decision-making powers delegated from the Executive Committee. The Registration Sub-group, which effectively replaced the CRCG, has the former Coding and Classification

Group and the Quality Assurance Group reporting to it. The other two groups are the Information, Communications and Technology Sub-group; and the Analysis Sub-group.

The current (2005) officers are: Chair – Professor D Forman, Director of Information and Research at the Northern and Yorkshire Cancer Registry and Information Service; Vice Chair – Dr D Brewster, Director of the Scottish Cancer Registry; and Treasurer – Mrs S Reynolds, of the Welsh Cancer Intelligence and Surveillance Unit. It was agreed that ONS was the most appropriate body to provide secretariat facilities; Dr MJ Quinn (Director of the National Cancer Intelligence Centre) was nominated by ONS to be the Association's Executive Secretary.

The UKACR provides:

- a focus for national initiatives in cancer registration;

- a coherent voice for representation of cancer registries in the United Kingdom;

- a channel for liaison between registries and for agreeing policy on matters connected with cancer registration;

- a framework to facilitate the operation of special interest groups and regional registries; and

- a means of stimulating the development of cancer registration, of information procedures and practices, and of research based on cancer registry data.

The UKACR represents the views of its members to government and other bodies operating at national level on issues concerned with data quality, the definition of information requirements, and the development of health information systems where these have implications for cancer registration, in particular where matters of overall policy are concerned. The Association was represented on the former National Advisory Committee on Cancer Registration and is currently represented on the Cancer Registration Advisory Group (CRAG). The establishment of such close links is very important given the intimate ties many regional registries have with NHS information systems, and the potential importance of cancer registration to NHS functions such as medical audit and contracting.

The UKACR has, through consensus, examined and improved coding and classification issues; agreed the complex interface document for transmission of data to and from ONS; developed performance indicators; produced a training manual and cancer-specific training packs for registry staff; developed guidelines for the release of data, including for the rapidly expanding field of genetic counselling; developed guidelines for standardisation of reported results; and established a forum for sharing the latest epidemiological research. Consensus may

be slower to achieve than coercion, but may in practice be stronger and more valuable as there is often a better chance that an agreed procedure will actually be followed. Even near consensus requires those disagreeing to continually justify their minority position.

References

1. NHS Executive. *Core contract for purchasing Cancer Registration.* EL(96)7. London: NHS Executive, 1996.

2. Winyard G. EL(96)7: *Core contract for purchasing cancer registration* (letter). Leeds: NHS Executive, 1998.

Cancer registries in the UK and Ireland: contact details

United Kingdom Association of Cancer Registries website: www.ukacr.org.uk

a) England

East Anglian

Dr J Rashbass, General Director
jem@cbcu.cam.ac.uk
Dr C H Brown, Medical Director
Eastern Cancer Registration and Information Centre
Box 193, Level 5 Oncology
Addenbrooke's Hospital
Hills Road
CAMBRIDGE, CB2 2QQ
www.srl.cam.ac.uk/ciu
eacr@medschl.cam.ac.uk
Tel: 01223 216644
Fax: 01223 245636

Merseyside and Cheshire

Post vacant, Director
Merseyside and Cheshire Cancer Registry
2nd Floor
Muspratt Building
The University of Liverpool
LIVERPOOL, L69 3GB
www.mccr.nhs.uk
Tel: 0151 794 5691
Fax: 0151 794 5700

North Western

Dr A Moran, Medical Director
tony.moran@cce.man.ac.uk
North Western Cancer Registry
Centre for Cancer Epidemiology
Christie Hospital NHS Trust
Kinnaird Road
Withington
MANCHESTER, M20 9QL
Tel: 0161 446 3579
Fax: 0161 446 3590

Northern and Yorkshire

Professor R Haward, Medical Director
bob.haward@nycris.leedsth.nhs.uk
Professor D Forman, Director of Information and Research
david.forman@nycris.leedsth.nhs.uk
Northern and Yorkshire Cancer Registry
  and Information Service
Arthington House
Cookridge Hospital
LEEDS, LS16 6QB
www.nycris.org.uk
Tel: 0113 392 4309
Fax: 0113 392 4178

Oxford

Dr M Roche, Medical Director
monica.roche@ociu.nhs.uk
Oxford Cancer Intelligence Unit
4150 Chancellor Court
Oxford Business Park South
OXFORD, OX4 2JY
www.ociu.nhs.uk
ociu.staff@ociu.nhs.uk
Tel: 01865 334770
Fax: 01865 334794

South and West

Dr J Verne, Director
jverne.gosw@go-regions.gsi.gov.uk
South and West Cancer Intelligence Service
Grosvenor House
149 Whiteladies Road
BRISTOL, BS8 2RA
www.theswcis.nhs.uk
Tel: 0117 970 6474
Fax: 0117 970 6481

Mr T Malik, Deputy Director
tariq.malik@swcis.nhs.uk
South and West Cancer Intelligence Service
Highcroft
Romsey Road
WINCHESTER, SO22 5DH
Tel: 01962 863511
Fax: 01962 878360

Thames

Professor H Møller, Director and Professor of Cancer Epidemiology
henrik.moller@kcl.ac.uk
Thames Cancer Registry
1st Floor, Capital House
42 Weston Street
LONDON, SE1 3QD
www.thames-cancer-reg.org.uk
tcr@kcl.ac.uk
Tel: 020 7378 7688
Fax: 020 7378 9510

Trent

Mr D Meechan, Director
Trent Cancer Registry
5 Old Fulwood Road
SHEFFIELD, S10 3TG
www.trentcancer.nhs.uk
Tel: 0114 226 3560
Fax: 0114 226 3561

West Midlands

Dr G Lawrence, Director
gill.lawrence@wmciu.nhs.uk
West Midlands Cancer Intelligence Unit
Public Health Building
The University of Birmingham
BIRMINGHAM, B15 2TT
wmciu@wmciu.nhs.uk
Tel: 0121 414 7711
Fax: 0121 414 7712

b) Wales

Dr J Steward, Director
john.steward@velindre-tr.wales.nhs.uk
Welsh Cancer Intelligence and Surveillance Unit
14 Cathedral Road
CARDIFF, CF11 9LJ
www.velindre-tr.wales.nhs.uk
Tel: 029 20 373500
Fax: 029 20 373511

c) Scotland

Dr D Brewster, Director
david.brewster@isd.csa.scot.nhs.uk
Scottish Cancer Registry
NHS National Services Scotland
Information Services
Gyle Square
1 South Gyle Crescent
EDINBURGH, EH12 9EB
www.show.scot.nhs.uk/isd/cancer/cancer/htm
cancerstats@isd.csa.scot.nhs.uk
Tel: 0131 275 6092
Fax: 0131 275 7511

d) Northern Ireland

Dr A Gavin, Director
a.gavin@qub.ac.uk
Northern Ireland Cancer Registry
Queen's University of Belfast
Dept of Epidemiology and Public Health
Mulhouse Building
Grosvenor Road
BELFAST, BT12 6BJ
www.qub.ac.uk/nicr
NICR@qub.ac.uk
Tel: 028 9063 2573
Fax: 028 9024 8017

e) Ireland*

Dr H Comber, Director
h.comber@ncri.ie
National Cancer Registry of Ireland
Elm Court
Boreenmanna Road
CORK
Ireland
Tel: +353 21 4318014
Fax: +353 21 4318016

* Associate member of the UKACR

Appendix K

Methodological issues considered

Mike Quinn

This appendix contains a discussion of the range of methodological issues involved in producing a cancer atlas, and explains the reasons behind our choice of methodologies.

Cancer incidence and mortality

For investigations of the aetiology (causes and risk factors) of cancer and for health care planning, incidence is the measure of primary interest, but mortality data are useful in planning resources for palliative care and hospices. Cancer incidence and mortality data each have their own advantages and disadvantages. The diagnostic accuracy is generally better for incidence than for mortality, and a much lower proportion of cases than deaths are 'unspecified' as to the type of cancer. Incidence data, however, may not be as complete or as timely as mortality data. The main problems with mortality data are that the trends and the geographical distribution reflect a combination of the incidence and the survival rates in each area; and for cancers with moderate or good survival, deaths in any one year result from cases diagnosed and treated many years earlier. Mortality data are therefore imperfect and 'fuzzy' indicators of trends and patterns related to the causes of cancer.[1] These and other aspects of data quality are discussed further below and in more detail in Appendix G.

There is strong evidence that survival rates for almost all cancers vary among the countries of Europe.[2-4] But the figures from the EUROCARE studies show that for the major cancers there was little variation between Scotland, Wales and the participating regions of England. There is also evidence that there is relatively little variation for any of the major cancers among all the regions of England[5] or among the health authorities of England.[6,7] To date (2005) all the cancer survival figures published separately for England and Wales, Scotland, Northern Ireland and Ireland have not been comparable owing to differences in the time periods covered, the exclusion criteria for the individual records (for example, relating to multiple tumours), the methodology used, the age groups used, and the weights for age standardisation (if any). A project is currently in progress to produce comparable survival figures for the five countries and the regions of England for 20 or so major

cancers.[8] The inclusion of cancer survival figures at the health authority level would have required a substantial amount of further work.

We therefore chose to produce maps of both cancer incidence and mortality, but not of survival.

Time period covered by this atlas

We decided to use cancer incidence and mortality data that related mostly to the 1990s. The most recently available nine or ten years of data provided totals of around 2,400,000 cancer cases (all malignancies excluding non-melanoma skin cancer) and 1,600,000 deaths from cancer in the UK and Ireland. The inclusion of even more (earlier) years of data would have given further reliability to the rates for whichever type of small geographical area was chosen. There have, however, been noticeable long-term trends in both incidence and mortality for many cancers, and the inclusion of data from the 1980s might have obscured the more recent geographical patterns.

Types of cancer included in this atlas

This atlas covers in detail the 16 most common cancers in males and the 17 most common in females (20 separate cancers in total*) for which the average number of newly diagnosed cases each year in one or the other sex in England during the 1990s was at least 1,000. Hodgkin's disease was also included, despite the smaller number of cases (about 800 in males and 600 in females each year in the UK and Ireland), to give coverage of all the lymphomas and leukaemias. These cancers together constitute almost 90 per cent of all malignant cancers (excluding non-melanoma skin cancer – see Appendix G).

Mesothelioma was not included, despite an annual average incidence of about 1,000 cases in England during the 1990s, and rising numbers throughout the period.[9] This cancer is caused by exposure to asbestos.[10,11] The vast majority of such exposure occurred in an industrial setting in places where asbestos was processed, asbestos products such as brake linings for vehicles were manufactured, or asbestos was used, such as shipbuilding yards and railway carriage works.[12] Trends and the unique geographical patterns in this cancer have been described in many scientific papers.[13-16]

Data collection and quality

Incidence

The cancer incidence data included in this atlas were collected by the cancer registries in the UK and Ireland. There are nine population-based regional cancer registries in England; their

---

* *Counting as one type: cancers of the lip, mouth and pharynx; cancers of the colon and rectum (colorectal); and all leukaemias.*

data are collated by the National Cancer Intelligence Centre (NCIC) at ONS. There are population-based cancer registries in each of Wales, Scotland, Northern Ireland and Ireland. See Appendix G for descriptions of the operation of each national registry.

Several aspects of the cancer registration systems in the UK and Ireland that affect data quality and therefore the interpretation of cancer incidence (and survival) data have been discussed in detail by Swerdlow,[17] Swerdlow and dos Santos Silva[18] and Quinn et al.[19] These include geographic coverage; methods of data collection; ascertainment (or completeness of registration); completeness of recording of data items; validity; accuracy; late registrations, deletions and amendments; duplicate and multiple registrations; registrations made solely on the information from death certificates; clinical and pathological definitions and diagnoses; changes in coding systems; changes in the definition of resident population; and error. Brief details of these are given in Appendix G.

A report of a project to audit the quality and comparability of cancer registration data in the UK,[20] carried out under the aegis of the United Kingdom Association of Cancer Registries, found some variation among the registries in data quality for diagnostic factors, incidence date, stage of disease, treatment information and use of death information. It is known that there is some variation among the UK and Irish registries in, for example, the proportions of cases registered solely on the basis of information from death certificates (and these proportions vary within each registry across the different cancers, and over time).[21] There will be corresponding variations in the ascertainment of cases. A study in one English registry found that data quality varied by the age of the patient, the type of cancer, and area of residence.[22] However, a substantial audit of Scottish cancer registry data in the early 1990s,[23] in which information was re-abstracted from the available records, found that severe discrepancies had occurred in under 3 per cent of cases. During the 1990s, almost all of the registries were engaged in some type of ongoing audit of data quality, mostly involving re-abstraction of information from case notes. In addition, all registries receive continuous feedback on data quality from clinicians and other staff in health authorities, and from their own research and collaborative studies with other scientists. The review by Huggett[20] concluded that although comparisons between the various published studies of data quality was difficult, cancer registry records were largely complete, accurate and reliable.

Regional differences in the classification and registration of some cancers can, however, contribute substantially to the apparent geographical patterns. This is particularly a problem for bladder cancer (Chapter 3), for which a particular sub-type

of the cancer was considered to be malignant by some registries but non-invasive (benign) by others. For cancer of the ovary (Chapter 18), cases judged to be of 'borderline' malignant potential were not classified as malignant under the Ninth Revision of the WHO International Classification of Diseases (ICD9)[24] but were under the Tenth Revision (ICD10).[25] Coding using ICD10 was introduced at different times in each of the national registries in the mid-1990s. For cancer of the uterus (Chapter 23), cases should not normally be registered without sufficient information being available to classify them to the cervix or body of the uterus. The proportion of cases assigned to the 'non-specific' code for uterus was only about 10 per cent overall, but this varied among the five countries and among the registries in England. The apparent geographical patterns in incidence may have been affected both by this and by the geographic variation in the prevalence of women who have had a hysterectomy.[26]

Mortality

Information on cause of death is obtained through the statutory process of registration of death.[27] In England and Wales, this is carried out by the Local Registration Service in partnership with the General Register Office (GRO), which is part of ONS. There are General Register Offices in Scotland, Northern Ireland and Ireland which operate in a similar way.[28-30] Cancer mortality figures are generally not affected by differences among the cancer registries, although there may sometimes be 'attribution bias' – an increased probability of a (registered) cancer being mentioned on a death certificate. But mortality data never were free from bias or criticism.[31] The diagnostic accuracy of cancer on death certificates is much less certain than for cancer incidence, and (in England and Wales) for about 10 per cent of deaths from cancer, a specific cancer site is not given.[32] Many studies have shown wide variability in death certification and coding, particularly between countries,[33-45] and there have been several changes in coding and other procedures in England and Wales during the 1990s[46-49] (see Appendix G).

The time periods covered by the cancer incidence and mortality data analyses in this atlas are described in detail in Appendix G. In brief, for incidence, data were included from 1991 onwards for England, Wales and Scotland. For Northern Ireland, data were available from the start of that registry's operation in 1993; and similarly for Ireland from 1994. The latest available complete year of data from all of the national registries when the work on the analyses began was 1999. Mortality data for England, Wales and Northern Ireland cover 1991 to 2000 (after which the basis of coding the cause of death changed), for Scotland 1991 to 1999 (after which coding changed there) and from 1994 to 2000 in Ireland.

Populations

The calculation of area-specific rates requires accurate estimates of the population at risk.[50] In general, potential problems arising from inaccuracies in population data receive far less attention than those in the incidence or mortality data.[51] As the areas of interest become smaller, the problems become more acute, as small changes in the populations can have large effects on the resulting incidence and mortality rates.[52]

Scandinavian and some other European countries have population registers that are updated continuously, and so accurate population counts for small areas are always available.[50,53] In the UK and Ireland, the primary source of population data is the census. It is the only attempt at a complete count of the population and in theory gives true population counts at very small levels of aggregation. There are two main problems. The first is undercount: people are missed by the census, and the proportions missed are not uniform across geographical areas or socio-economic groups. For the 2001 census in the UK, the methodology was designed to integrate adjustments for undercount into the census database using information from a very large follow-up survey.[54] Census under-enumeration is generally highest for men and women in the 20-24 age group, particularly in inner city areas. This will only marginally affect the age-standardised rates for most cancers, as the age-specific rates in young people are very low; the principal exception is testicular cancer (Chapter 22).

The second problem is that the census is conducted only every ten years in the UK and every five years in Ireland, and estimates of the population therefore need to be made for the years between the censuses. This is done for England and Wales at the local and health authority levels by ONS.[55] Between censuses, quite substantial changes may occur in the size and composition of the population in a given area. The births and deaths in an area are readily accounted for, but the estimation of migration into and out of an area is usually imprecise. As much migration occurs within short distances, the smaller the areas being considered, the greater the migration problem.

The population denominators used in this atlas to calculate age-specific cancer incidence and mortality rates were the sums of the mid-year population estimates for each year of data included; as noted above the time periods covered by the data differed between the countries and between the incidence and mortality data. The mid-year population estimates for 1996, the approximate mid-point of the time period spanned by the vast majority of the data, are given in

Appendix C as a guideline. For the UK, the original population estimates made in the 1990s were adjusted using information from the 2001 census (these were slightly revised subsequently for a few areas[56]). For Ireland, the 1996 populations are official national census figures published by the Central Statistics Office.

Measures of incidence and mortality for comparison of areas

It is clearly not valid simply to compare the numbers of cases of (or deaths from) cancer in one area with those in another, as the areas may have widely different total populations (for example, the population of England is about ten times that of Scotland). The *crude rate*, the total number of cases (or deaths) divided by the total population, takes the population sizes of different areas into account. But as cancer incidence and mortality rates generally increase with age, and populations in different areas usually have different age structures, it is essential to adjust, or standardise, any rates for age.[57,58]

The method of *indirect* standardisation takes the ratio of the observed number of cases (or deaths) in a given small area to the number that would be expected if the rates in a reference (usually the national) population applied in the small area. This ratio is usually multiplied by 100 and is called the standardised incidence (or mortality) ratio (SIR or SMR). A value of 132 means that 32 per cent more cases were observed in that particular small area than if the age-specific incidence rates had been the same as in the country as a whole. The SIR or SMR generally provides a more precise statistical estimate than the directly standardised rate[57] (see below). But the main problem with indirect standardisation is that comparisons of SIRs or SMRs between areas are not valid, and may be misleading. It is possible for all the age-specific rates in one area to be lower than those in a second area, yet the SIR or SMR in the first area is higher than that for the second because of widely different population structures in the two areas. Given that the invalidity of comparing SIRs or SMRs was recognised over 80 years ago[59-61] it is surprising how often it has been done in recent times.

The method of *direct* standardisation uses the age-specific rates in each small area concerned to determine the incidence (or mortality) rate that would be observed in that area if it had the same population structure as a given or standard (usually theoretical) population. The two most commonly used standard populations are the World and European standards[62] (see also Appendix C). As the World standard population is much more heavily weighted than the European towards younger ages, and cancer is mainly a disease of elderly people, for most European countries the directly age-standardised cancer rates using the World standard tend to be about half the crude rates, whereas rates using the European standard are broadly comparable with the crude rates.

A third measure is the *cumulative rate*,[57] which gives an approximation of the risk of being diagnosed with the cancer (in the absence of mortality) or of dying from the cancer, either before a given age or between two ages. It is calculated by summing the age-specific incidence or mortality rates over the desired age range. There are problems with doing this calculation over a whole lifetime because the last age group is open-ended; cumulative incidence rates are rarely calculated above 75 years, as competing causes of death then play a major role. The cumulative rate is proportional to the simple arithmetic average of the age-specific rates; in other words, it is effectively a standardised rate using a standard population in which every age group contains the same number of people. The interpretation of the cumulative rate assumes that the age-specific rates from the cross-sectional incidence or mortality age curve for a given time period correctly represent the risk for any one individual. But the risk estimate is actually for a fictitious person who synthesises the risks for various birth cohorts:[57] for example, using data for (say) 1996, the incidence rate for the 30-34 age group relates to those born in 1962-66, while the rate for the 65-69 age group relates to those born in 1927-31. The cumulative rate is the indicator that is the most readily interpretable on a probability scale: for example, the risk of being diagnosed with cancer of the colon up to age 75 in a given area may be 2.73 per cent. An alternative, but more complicated, calculation of the lifetime risk using the life table method[63] does not have a problem with the upper-most age group, and allows for competing mortality – but does require appropriate life tables. The widespread use of either the cumulative rate or lifetime risk would greatly increase the comparability of cancer maps and atlases.

In this atlas, all the incidence and mortality rates have been directly age-standardised using the European standard population. Further details are given in Appendix H.

Choice of 'small' geographical area

The potential types of sub-national area (below the very large regional level in England) that could be used for mapping cancer incidence and mortality include wards (either 'frozen' census wards, or boundaries at a particular (other) point in time), local authorities, counties, health authorities, primary care trusts (PCT), strategic health authorities (SHA), and cancer networks. Although cancer is a major cause of morbidity and mortality, in small geographical areas the numbers of cases of

any particular type of cancer occurring each year can be quite low, and of course the corresponding numbers of cancer deaths are even lower.

There are about 10,000 wards in England and Wales with an average population of about 5,000 people, and so on average in each ward there would be just two cases of lung cancer diagnosed in males, and three cases of, and one death from, breast cancer in females each year. Even if ten years of data were aggregated, the numbers of cases of even the most common cancers would be small, and the corresponding numbers of deaths even smaller; for the less common cancers, with about 1,000 cases each year in England, there would be an average of only one case in each ward over the whole period. And many wards are much smaller than the average, with populations of only about 2,000 people. Even if reliable population estimates were available for wards for the years between censuses, all incidence and mortality rates would be highly unreliable, with very wide confidence intervals (see Appendix H).

Local authorities (LA), of which there are about 350 in England, have an average population of about 140,000 people, and much more reliable rates can be calculated for them than for wards. LAs were the geographical unit chosen as the basis for analysis in the book *Geographic Variations in Health*,[64] but this included only the three most common cancers in each sex. Uncertainty in the rates, as indicated by wide confidence intervals, would have been a problem for all of the cancers less common than these. As noted above, counties were used by Swerdlow and dos Santos Silva in their atlas of cancer incidence in England and Wales over the period 1962-85;[18] at that time there were about 60 counties in England and Wales (counting Greater London as one; and the three parts of Lincolnshire and Yorkshire and the two parts of Suffolk and Sussex separately), with an average population of just under 1 million people. But the boundaries of several counties were affected by the formation of unitary authorities during the 1990s. Also, most counties are quite large in terms of land area, and generally consist of a mixture of densely populated cities or large towns with relatively high levels of socio-economic deprivation, more affluent suburban areas, and sparsely populated rural areas. Real and important differences in cancer rates *within* counties could be obscured if this level were used for analysis. In addition, a major disadvantage of local government boundaries (both LAs and counties) is that they are generally not contiguous with the boundaries of the administrative units in the NHS which are responsible for delivering health care.

In England and Wales the latest of several major re-organisations in the NHS, to the current (2005) PCT and SHA boundaries, took place in 2001, after the period to which all of the incidence and mortality data relate. And there have been several boundary changes in PCTs and SHAs since 2001. There are about the same number of PCTs (about 300) as LAs in England, and slightly fewer SHAs (28) than counties, so these units suffer from the same problems as LAs and counties of being too small and too large, respectively. Before the re-organisation of the NHS organisations and boundaries in 2001, there had been about 100 health authorities with an average population of about half a million people. With ten years of aggregated data, these areas would have sufficient numbers of cases and deaths, even for the less common cancers, to enable the calculation of reliable directly age-standardised rates. And they were the areas that were in existence for most of the period covered by the data. In addition, most of the current SHAs consist of aggregations of the former health authorities, and most of the current PCTs aggregate to the former health authorities. We therefore decided to present the data at the health authority level for England.

In consultation with our colleagues in Wales, Scotland, Northern Ireland and Ireland, and taking into account the implications of the various possible geographical sub-units, we decided to use the 'small' areas based on health administrative boundaries in this atlas as summarised in the table below. For simplicity and convenience, these slightly different types of areas are referred to as 'health authorities' throughout this atlas. A 'key' map to the 127 health authorities is given in Appendix A.

Tables of cancer incidence and mortality, and populations, by health authority

Mid-year population estimates for 1996 for the 127 health authorities, by sex and age group, are given in Appendix C. The age-standardised rates adjust for differences in the age structure of the population between countries, regions and health authorities, and hence they allow unbiased comparisons of the rates across areas (see Appendix H).

We had estimated that nine or ten years of data would give an average of over 100 newly diagnosed cases of all the different types of cancer at the health authority level, and hence reasonably reliable directly age-standardised rates, with quite narrow confidence intervals – except for Hodgkin's disease, for which the average numbers would be only about 55 in males and 40 in females. The numbers of cases are also small for several cancers for some of the health authorities with small populations, especially three in Scotland: the Western Isles, Orkney and Shetland. We decided not to omit either incidence or mortality rates for these three areas from the charts and maps; but it must be borne in mind that for the less common cancers, and particularly for mortality, the rates for these areas are based on relatively small numbers (of cases or deaths) and therefore have very wide confidence intervals.

Tables giving the numbers of cases and deaths, and the corresponding directly age-standardised rates, together with the ratio of the mortality to incidence rates, by cancer, by sex where appropriate, for all 127 health authorities, the eight regions of England, and the five countries, are given in Appendix B. These tables also indicate where the 95 per cent confidence interval about a country, regional or health authority rate does not overlap the 95 per cent confidence interval about the corresponding average rate for the UK and Ireland. This is *not* equivalent to the area rate being statistically significantly different from the average at the 5 per cent level – see Appendix H for further details.

Table **K1**

Health authority areas and their equivalents by country, UK and Ireland

Country	Area	Number of areas	Average population
England	Health authority	95	510,000
Wales	Health authority	5	577,000
Scotland	Health board	15	340,000
Northern Ireland	Health and social service board	4	415,000
Ireland	Regional health board[1]	8	453,000
Total		127	

1 *Except for the Eastern Regional Health Authority, which includes Dublin, and is divided into three area health boards – this subdivision was not used.*

Divergence from the methodological guidelines of Walter and Birnie[65]

Much thought has been given by many authors to the methodological issues involved in disease mapping.[66-70] This atlas meets most, but not all, of the methodological guidelines for health atlases set out by Walter and Birnie.[65] For five reasons, we have not indicated on the maps in each of the cancer-specific chapters where the incidence and mortality rates are statistically significantly different from the relevant overall UK and Ireland average. First, the choice of significance level is arbitrary. Second, allowance has to be made for the very large numbers of comparisons of rates in any one area for the different cancers, some of which will be related to each other, for example, where there is a common aetiological factor or a similar relationship with socio-economic deprivation. Also, the rates for incidence and mortality in any one area will often be closely related to each other. These factors would tend to cause statistical tests to give a falsely high level of significance. Third, statistically significant deviations in risk (from the average) are more likely to occur in areas with large populations, even if the actual deviations are small.[71] Fourth, it is clear from the charts of rates by health authority for most cancers that there is a relatively narrow range of values, and most rates do not differ widely from those in the other health authorities in their own countries and regions or elsewhere. Any rates which are not quite statistically significantly different from the average (or from each other) would probably become so if two or three more years of data were included. But fifth, and most important, we were interested in the broad geographical patterns in each cancer, how the patterns in incidence generally related to those in mortality, and how the patterns in both were related to known aetiological factors and to socio-economic deprivation. For these purposes, the statistical significance (or otherwise) – even if this could be accurately assessed – of a rate of either incidence or mortality for a particular cancer in a given small area is not highly relevant.

We have included descriptions of any broad time trends in incidence and mortality in each of the cancer-specific chapters – but we have not investigated temporal changes in geographic patterns over what is a relatively short period (mostly ten years) covered by the data.

We have not formally analysed the spatial structure in the data. There are several statistical measures available,[72-81] but, as mentioned above, we were principally interested in relating the broad geographical patterns in incidence to those in mortality and to aetiological (risk) factors and socio-economic deprivation. With about ten years of both incidence and mortality data, the health authority areas we used as the basic unit for analysis were individually generally large enough to

give reliable age-standardised rates. Health authorities were also large enough that real differences between them could exist both in the underlying levels of the relevant risk factors and in the levels of the disease. Any artificial measure of 'association' in rates for adjacent areas was therefore not of great interest. For similar reasons, we have not geographically 'smoothed' the data. As noted above, rates for small areas, especially wards, would be highly unreliable, even with the large amount of available data, and would need to be smoothed.[82-87] Using health authorities as the lowest level of geography, with an average population of about half a million people (in England), and ten years of data, effectively smoothes the results to a large extent. There would be a danger in smoothing the data further, by Bayesian or other statistical techniques including 'head banging',[88-91] of obscuring real differences in rates between adjacent areas, especially, for example, where one area was a highly industrialised and socio-economically deprived city with high population density, and the other a large, relatively affluent and sparsely populated rural area.

We have not included maps of potentially relevant aetiological factors. A comprehensive (although now slightly out of date) guide to information on geographical variations in a very large number of risk factors for cancer was given by Swerdlow and dos Santos Silva,[18] including abortions; age at first birth; air pollution; alcohol consumption; asbestos; blood group; diet; education; ethnicity; fallout from nuclear explosions; geochemistry; height, weight and obesity; hepatitis B; immigration; income; indoor radiation; medical conditions; nuclear installations; occupation and industry; operated conditions (surgical interventions); overcrowding; parity; population mobility; post-neonatal mortality (four weeks to one year); rainfall; screening; smoking (cigarettes); social class; sunshine; venereal diseases; and Welsh language speaking (which it was thought 'might correlate with other cultural behaviours'). The analyses presented in Chapters 3 to 23 show that for many cancers the known risk factors could explain only a small proportion of the incident cases, and so the geographical distribution of these factors could not greatly influence the observed patterns. For those cancers related to smoking and/or drinking alcohol and related 'lifestyle' factors, overall measures of socio-economic deprivation are often a highly accurate marker. For example, the strength of the relationship of the incidence of lung cancer with deprivation[19] measured on a small-area basis by the Carstairs index[92] is closely similar to that with social class measured on an individual basis in the Longitudinal Study;[93,94] and the prevalence of smoking is much higher in social classes IV and V than in classes I and II.[95] Details of the Carstairs index are given in Appendix F, along with a map showing its distribution at the LA level in Great Britain.

Finally, we have not included tables of age-specific numerators. Such tables for all of the 21 cancers by sex where appropriate for all of the 127 health authorities, the eight regions of England, and the five countries, would require considerably more pages in what is already a thick book. We expect that not many readers would be interested in such detail, or would want to calculate age-standardised rates using a different standard population. These tables are available, however, on request from the NCIC at ONS.

References

1. Coleman MP, Babb PJ, Stockton D, Forman D et al. Trends in breast cancer incidence, survival and mortality in England and Wales. *Lancet* 2000; 356: 590-591 (letter).

2. Berrino F, Sant M, Verdecchia A, Capocaccia R et al. *Survival of cancer patients in Europe: The EUROCARE Study* . IARC Scientific Publications No.132. Lyon: International Agency for Research on Cancer, 1995.

3. Berrino F, Capocaccia R, Esteve J, Gatta G et al. Survival of *Cancer Patients in Europe: the EUROCARE-2 Study.* IARC Scientific Publications No.151. Lyon: International Agency for Research on Cancer, 1999.

4. Berrino F, Capocaccia R, Coleman MP, Esteve J et al. Survival of Cancer Patients in Europe: the EUROCARE-3 Study. *Annals of Oncology* 2003; 14, Suppl 5.

5. Coleman MP, Babb P, Damiecki P, Grosclaude P et al. *Cancer Survival Trends in England and Wales, 1971-1995: Deprivation and NHS Region.* Studies on Medical and Population Subjects No. 61. London: The Stationery Office, 1999.

6. Romanengo M, Cooper N, Robinson C, Malalagoda M et al. Cancer survival in the health authorities of England, 1993-2000. *Health Statistics Quaterly* 2002;13:95-103.

7. ONS. Cancer Survival in England by Strategic Health Authority. April 2005. Available at *http://www.statistics.gov.uk/StatBase/Product.asp?vlnk=11991&Pos=1&ColRank=1&Rank=272.*

8. Rachet B, Quinn M, Brewster D, Gavin A, Steward J, Comber H. Relative survival for twenty common adult cancers in the UK and Republic of Ireland *Proceedings of the 13th Annual Scientific Conference of the UK Association of Cancer Registries.* London: UKACR, 2004.

9. McElvenny DM, Darnton AJ, Price MJ, Hodgson JT. Mesothelioma mortality in Great Britain from 1968 to 2001. *Occupational Medicine* 2005; 55: 79-87.

10. Greenberg M, Lloyd Davies TA. Mesothelioma register 1967-68. *British Journal of Industrial Medicine* 1974; 31: 91-104.

11. Doll R, Peto J. *Effects on Health of Exposure to Asbestos.* London: HMSO, 1985.

12. Gardner MJ, Acheson ED, Winter PD. Mortality from mesothelioma of the pleura during 1968-78 in England and Wales. *British Journal of Cancer* 1982; 46: 81-88.

13. Peto J, Hodgson JT, Matthews FE, Jones JR. Continuing increase in mesothelioma mortality in Britain. *Lancet* 1995; 345: 535-539.

14. Hodgson JT, McElvenny DM, Darnton AJ, Price MJ et al. The expected burden of mesothelioma mortality in Great Britain from 2002 to 2050. *British Journal of Cancer* 2005; 92: 587-593.

15. Hodgson JT, Peto J, Jones R, Matthews FE. Mesothelioma mortality in Britain: patterns by birth cohort and occupation. *Annals of Occupational Hygiene* 1997; 41 (Suppl 1): 129-133.

16. Health and Safety Executive. *Occupational Health Statistics Bulletin 2002/03*, 2003.

17. Swerdlow AJ. Cancer regstration in England and Wales: some aspects relevant to interpretation of the data. *The Journal of the Royal Statistical Society Series A (Statistics in Society)* 1986; 149: 146-160.

18. Swerdlow AJ, dos Santos Silva I. *Atlas of Cancer Incidence in England and Wales 1962-85.* Oxford: Oxford University Press, 1993.

19. Quinn MJ, Babb PJ, Brock A, Kirby L, Jones J. *Cancer Trends in England and Wales 1950-1999.* Studies on Medical and Population Subjects No. 66. London: The Stationery Office, 2001.

20. Huggett, C. *Review of the Quality and Comparability of Data held by Regional Cancer Registries.* Bristol: Bristol Cancer Epidemiology Unit incorporating the South West Cancer Registry, 1995.

21. UK Association of Cancer Registries QA Group. UKACR Quality Performance Indicators 2004 *Proceedings of the 13th Annual Scientific Conference of the UK Association of Cancer Registries.* London: UKACR, September 2004.

22. Seddon DJ, Williams EMI. Data quality in population-based cancer registration: an assessment of the Merseyside and Cheshire Cancer Registry. *British Journal of Cancer* 1997; 76: 667-674.

23. Brewster D, Crichton J, Muir C. How accurate are Scottish cancer registration data? *British Journal of Cancer* 1994; 70: 954-959.

24. World Health Organisation. *International Classification of Diseases, Ninth Revision.* Geneva: WHO, 1977.

25. World Health Organisation. *International Classification of Diseases and Related Health Problems, Tenth Revision.* Geneva: WHO, 1992.

26. Redburn JC, Murphy MFG. Hysterectomy prevalence and adjusted cervical and uterine cancer rates in England and Wales. *British Journal of Obstetrics and Gynaecology* 2001; 108: 388-395.

27. ONS. *Mortality statistics 2003: cause, England and Wales.* Series DH2 No.30. London: ONS, 2004.

28. General Register Office for Scotland. *Scotland's Population 2003: The Registrar General's Annual Review of Demographic Trends.* Edinburgh: General Register Office for Scotland, 2004.

29. Northern Ireland Statistics and Research Agency. *Registrar General Annual Report 2003.* Belfast: Northern Ireland Statistics and Research Agency, 2004.

30. Central Statistics Office. *Report on Vital Statistics, 2002.* Dublin: The Stationery Office, 2002.

31. Defoe D. *A journal of the plague year.* London, 1722.

32. ONS. *Mortality statistics 1995: cause, England and Wales.* Series DH2 No.25. London: The Stationery Office, 1996.

33. Heasman MA, Lipworth L. *Accuracy of certification of cause of death.* Studies on Medical and Population Subjects No.20. London: HMSO, 1966.

34. Alderson MR, Meade TW. Accuracy of diagnosis on death certificates compared with that in hospital records. *British Journal of Preventative and Social Medicine* 1967; 21: 22-29.

35. Cameron HM, McGoogan E. A prospective study of 1152 hospital autopsies: II. Analysis of inaccuracies in clinical diagnoses and their significance. *Journal of Pathology* 1981; 133: 285-300.

36. Grulich AE, Swerdlow AJ, dos SS, I, Beral V. Is the apparent rise in cancer mortality in the elderly real? Analysis of changes in certification and coding of cause of death in England and Wales, 1970-1990. *International Journal of Cancer* 1995; 63: 164-168.

37. Percy C, Muir C. The international comparability of cancer mortality data. Results of an international death certificate study. *American Journal of Epidemiology* 1989; 129: 934-946.

38. Ashworth TG. Inadequacy of death certification: proposal for change. *Journal of Clinical Pathology* 1991; 44: 265-268.

39. Percy CL, Dolman AB. Comparison of the coding of death certificates related to cancer in seven countries. *Public Health Reports* 1978; 93: 335-350.

40. Percy C, Stanek E, III, Gloeckler L. Accuracy of cancer death certificates and its effect on cancer mortality statistics. *American Journal of Public Health* 1981; 71: 242-250.

41. Percy CL, Miller BA, Gloeckler Ries LA. Effect of changes in cancer classification and the accuracy of cancer death certificates on trends in cancer mortality. *Annals of the New York Academy of Sciences* 1990; 609: 87-97.

42. Hoel DG, Ron E, Carter R, Mabuchi K. Influence of death certificate errors on cancer mortality trends. *Journal of the National Cancer Institute* 1993; 85: 1063-1068.

43. Lindahl BI, Johansson LA. Multiple cause-of-death data as a tool for detecting artificial trends in the underlying cause statistics: a methodological study. *Scandinavian Journal of Social Medicine* 1994; 22: 145-158.

44. Garne JP, Aspegren K, Balldin G. Breast cancer as cause of death--a study over the validity of the officially registered cause of death in 2631 breast cancer patients dying in Malmo, Sweden 1964-1992. *Acta Oncologica* 1996; 35: 671-675.

45. Coleman MP, Aylin P. *Death certification and mortality statistics: an international perspective.* Studies on Medical and Population Subjects No.64. London: The Stationery Office, 2000.

46. Rooney C, Devis T. Mortality trends by cause of death in England and Wales 1980-94: the impact of introducing automated cause coding and related changes in 1993. *Population Trends* 1996: 29-35.

47. Rooney C, Griffiths C, Cook L. The implementation of ICD10 for cause of death - some preliminary results from the bridge coding study. *Health Statistics Quarterly* 2002; 13: 31-41.

48. Report. Results of the ICD-10 bridge coding study, England and Wales, 1999. *Health Statistics Quaterly* 2002; 14: 75-83.

49. Brock A, Griffiths C, Rooney C. The effect of the introduction of ICD-10 on cancer mortality trends in England and Wales. *Health Statistics Quarterly* 2004; 23: 7-17.

50. Arnold RA, Diamond ID, Wakefield JC. The use of population data in spatial epidemiology. In: Elliott P, Wakefield JC, Best NG, Briggs DJ (eds) *Spatial Epidemiology - Methods and Applications.* Oxford: Oxford University Press, 2000.

51. Smans M, Esteve J. Practical approaches to disease mapping. In: Elliott P, Cuzick J, English D, Stern R (eds) *Geographical and Environmental Epidemiology: Methods for Small-Area Studies.* Oxford: Oxford University Press, 1992.

52. Diamond ID. Population counts in small areas. In: Elliott P, Cuzick J, English D, Stern R (eds) *Geographical and Environmental Epidemiology: Methods for Small-Area Studies.* Oxford: Oxford University Press, 1992.

53. Redfern P. Population registers: some administrative and statistical pros and cons. *Journal of the Royal Statistical Society Series A* 1989; 152: 1-41.

54. ONS. *One Number Census Consultation Paper.* Titchfield: ONS, 1998.

55. Chappell R. ONS precedures for compiling population estimates at sub-national levels. In: Arnold R, Elliott P, Wakefield J, Quinn MJ (eds) *Population Counts in Small Areas: Implications for Studies of Environment and Health.* Studies on Medical and Population Subjects No.62. London: The Stationery Office, 1999.

56. ONS. *2001 Census - Local Authority Population Studies: Full report.* London: ONS, 2004.

57. Esteve J, Benhamou E, Raymond L. *Statistical Methods in Cancer Research, Volume IV - Descriptive Epidemiology.* IARC Scientfic Publications. Lyon: International Agency for Research on Cancer, 1994.

58. dos Santos Silva I. *Cancer Epidemiology: Principles and Methods.* Lyon: International Agency for Research on Cancer, 1999.

59. Wolfenden HH. On the methods of comparing the mortalities of two or more communities, and the standardisation of death rates. *Journal of the Royal Statistical Society* 1923; 86: 399-411.

60. Yule GU. On some points relating to vital statistics, more especially statistics of occupational mortality. *Journal of the Royal Statistical Society* 1933; 97: 1-84.

61. Rothman KJ. *Modern Epidemiology.* Boston/Toronto: Little, Brown & Co., 1986.

62. Breslow NE, Day NE. *Statistical Methods in Cancer Research, Volume II - The Design and Analysis of Cohort Studies.* IARC Scientific Publications No.82. Lyon: International Agency for Research on Cancer, 1987.

63. Schouten LJ, Straatman H, Kiemeney LALM, Verbeek ALM. Cancer incidence: Life table risk versus cumulative risk. *Journal of Epidemiology and Community Health* 1994; 48: 596-600.

64. Griffiths C, Fitzpatrick J. *Geographic Variations in Health.* Decennial Supplements No. 16. London: The Stationery Office, 2001.

65. Walter SD, Birnie SE. Mapping mortality and morbidity patterns: an international comparison. *International Journal of Epidemiology* 1991; 20: 678-689.

66. Howe GM. Historical evolution of disease mapping in general and specifically of cancer mapping. In: Boyle P, Muir CS, Grundmann E (eds) *Cancer Mapping.* Berlin: Springer-Verlag, 1989.

67. Cleveland W, McGill R. Graphical perception: theory, experimentation and application to the development of graphical methods. *Journal of the American Medical Association* 1984; 79: 531-554.

68. Cliff AD, Haggett P. *Atlas of Disease Distributions: Analytic Approaches to Epidemiologic Data.* Oxford: Blackwell, 1988.

69. Pickle LW, White AA. Effects of the choice of age-adjustment method on maps of death rates. *Statistics in Medicine* 1995; 14: 615-627.

70. Diggle PJ. Overview of statistical methods for disease mapping and its relationship to cluster detection. In: Elliott P, Wakefield JC, Best NG, Briggs DJ (eds) *Spatial Epidemiology - Methods and Applications.* Oxford: Oxford University Press, 2000.

71. Walter SD. Disease mapping: a historical perspective. In: Elliott P, Wakefield JC, Best NG, Briggs DJ (eds) *Spatial Epidemiology - Methods and Applications.* Oxford: Oxford University Press, 2000.

72. Kemp I, Boyle P, Smans M, Muir C. *Atlas of Cancer in Scotland 1975-1980. Incidence and epidemiological perspective.* IARC Scientific Publications No. 72. Lyon: International Agency for Research on Cancer, 1985.

73. Cliff AD, Ord JK. *Spatial Process: Models in Applications.* London: Pion, 1982.

74. Moran PAP. The interpretation of statistical maps. *Journal of the Royal Statistical Society Series B* 1948; 10: 243-251.

75. Moran PAP. Notes on continuous stochastic phenomena. *Biometrika* 1950; 37: 17-23.

76. Oden N. Adjusting Moran's I for population density. *Statistics in Medicine* 1995; 14: 17-26.

77. Geary RC. The contiguity ratio and statistical mapping. *The Incorporated Statistician* 1954; 5: 115-145.

78. Walter SD. Assessing spatial patterns in disease rates. *Statistics in Medicine* 1993; 12: 1885-1894.

79. Mantel N. The detection of disease clustering and a generalized regression approach. *Cancer Research* 1967; 27: 209-220.

80. Ohno Y, Aoki K. Cancer deaths by city and county in Japan (1969-1971): a test of significance for geographic clusters of disease. *Social Science and Medicine* 1981; 15D: 251-258.

81. Smans M. Analysis of spatial aggregation. In: Boyle P, Muir CS, Grundmann E (eds) *Recent Results in Cancer Research - Cancer Mapping.* Berlin: Springer-Verlag, 1989.

82. Clayton DG, Kaldor J. Empirical Bayes estimates of age-standardised relative risks for use in disease mapping. *Biometrics* 1987; 43: 671-682.

83. Manton KG, Woodbury MA, Stallard E, Riggan WB et al. Empirical Bayes procedures for stabilizing maps of U.S. cancer mortality rates. *Journal of the American Statistical Association* 1989; 84: 637-650.

84. Besag J, York J, Mollie A. Bayesian image restoration, with two applications in social statistics. *Annals of the Institute of Statistical Mathematics* 1991; 43: 1-21.

85. Clayton D, Bernardinelli L. Bayesian methods for mapping disease risk. In: Elliott P, Cuzick J, English D, Stern R (eds) *Geographical and Environmental Epidemiology: Methids for Small-Area Studies.* Oxford: Oxford University Press, 1992.

86. Wakefield JC, Best NG, Waller L. Bayesian approaches to disease mapping. In: Elliott P, Wakefield JC, Best NG, Briggs DJ (eds) *Spatial Epidemiology - Methods and Applications.* Oxford: Oxford University Press, 2000.

87. Wakefield JC, Kelsall JE, Morris SE. Clustering, cluster detection, and spatial variation in risk. In: Elliott P, Wakefield JC, Best NG, Briggs DJ (eds) *Spatial Epidemiology - Methods and Applications.* Oxford: Oxford University Press, 2000.

88. Pickle LW, Mungiole M, Jones GK, White AA. Analysis of mapped mortality data by mixed effects model. *Proceedings of the Biometrics Section, American Statistical Association 1996 Meeting, Chicago* 1996: 227-232.

89. Tukey PA, Tukey JW. Graphical display of data sets in 3 or more dimensions. In: Barnett V (ed) *Interpreting Multivariate Data.* New York: Wiley, 1981.

90. Hansen KM. Headbanging: Robust smoothing in the plane. *IEEE Transactions on Geoscience and Remote Sensing* 1991; 29: 369-378.

91. Mungiole M, Pickle LW, Simonson KH. Application of a weighted head-banging algorithm to mortality data maps. *Statistics in Medicine* 1999; 18: 3201-3209.

92. Carstairs V, Morris R. Deprivation and mortality: an alternative to social class? *Community Medicine* 1989; 11: 213-219.

93. Davey-Smith G, Leon D, Shipley MJ, Rose G. Socioeconomic differentials in cancer among men. *International Journal of Epidemiology* 1991; 20: 339-345.

94. Pugh H, Power C, Goldblatt P, Arber S. Women's lung cancer mortality, socio-economic status and changing smoking patterns. *Social Science and Medicine* 1991; 32: 1105-1110.

95. Rickards L, Fox K, Roberts C, Fletcher L et al. *Living in Britain: Results from the 2002 General Household Survey.* London: The Stationery Office, 2004.

Appendix L
Glossary

Adenocarcinoma – a malignant epithelial tumour derived from glandular tissue (tissue which produces a secretion).

Adenoma – a benign epithelial tumour derived from glandular tissue.

Adjuvant – in terms of treating cancer, a second form of treatment that is given in addition to the main form. For example, adjuvant radiotherapy before or after surgery, adjuvant chemotherapy after surgery or radiotherapy.

Aetiology – the cause(s) of a disease.

Age-specific rate – the number of cancer registrations or deaths for a particular sex and age group divided by the corresponding sex- and age-specific mid-year population; usually expressed per 100,000 population (see Appendix H for fuller explanation).

Age standardisation – a way of controlling for differences in the age structure of populations between geographical areas or over time, to allow unbiased comparison of incidence or mortality rates (see Appendix H for fuller explanation).

Age-standardised rate – an incidence or mortality rate which has been weighted using a standard population (in this book the European standard population) to control for differences in populations between geographical areas or over time, to allow unbiased comparison; usually expressed per 100,000 population (see Appendix H for fuller explanation).

Ascertainment (level) – the proportion of all newly diagnosed cases of cancer that are registered by a cancer registry.

Astrocytoma – a type of brain tumour arising from supporting cells in the brain called astrocytes.

Benign – tumours which are usually slow growing, in which the cells resemble those of their tissue of origin, which do not invade surrounding tissue or metastasise to distant sites, and which are not usually fatal.

Carcinogen – a chemical that can modify the structure of a cell's DNA to initiate or promote malignant transformation (cancer development).

Carcinogenesis – the development of cancer.

Carcinoma – a malignant tumour derived from epithelial tissue (tissue covering the internal organs and other internal surfaces of the body; also forms glands).

Carstairs deprivation index – small area level (enumeration district or ward) index of socio-economic deprivation based on data for male unemployment, car ownership, house overcrowding and social class from a census.

Case-control study – an epidemiological study in which exposure to a putative risk factor is compared between a group of people who have a disease or condition (cases) and a group who do not, and are representative of the population from which the cases originated (controls). For example, examining past exposure to chemicals in people with and without bladder cancer to determine whether people with bladder cancer were more likely to have been exposed.

Cervical intraepithelial neoplasia (CIN) – the pre-invasive stage of cervical cancer, when the tumour is confined to the surface epithelium. Cervical screening is designed to detect CIN.

Chemotherapy – the use of drugs to treat cancer by killing tumour cells (chemotherapeutic drugs).

Cohort – a defined group of people. A birth cohort is a group of people, selected by their year of birth, whose characteristics can be followed as they enter successive age and time periods.

Cohort study – an epidemiological study in which rates of disease are compared in groups with different exposures. For example, a group of smokers (exposed) and a group of non-smokers (not exposed) are followed up over time to see which ones develop lung cancer. The rates of lung cancer in the two groups are then compared to determine whether smoking (the exposure) increases the risk of developing lung cancer (is a risk factor for lung cancer, in other words).

Confidence interval – a range of values for a variable (for example, a rate) constructed so that this range has a specified probability of including the true value of the variable (see Appendix **H** for fuller explanation).

Confounding factor – a factor which may appear to be a cause or risk factor for a disease when it is not, because it is related to a real cause or risk factor for the disease. For example, people who smoke cigarettes and drink alcohol are more likely to develop lung cancer. It could be concluded (falsely) that alcohol consumption causes lung cancer, because smoking (which does cause it) is associated with drinking – in this case alcohol consumption is a confounding factor.

Crude survival – the proportion of a cohort of subjects alive at the end of a specified time interval since diagnosis, irrespective of the cause of death.

Cryptorchidism – failure of one or both testicles to descend into the scrotum.

Death certificate only – cases of cancer registered solely from information provided on the death certificate. These patients necessarily appear to have zero survival time (as the date of diagnosis has to be taken to be the date of death).

Deprivation – usually refers to socio-economic deprivation indicated by poor housing conditions and low income. Defined here using the Carstairs index, which combines several variables, taken as indicative of socio-economic status, into a score for each enumeration district or ward. All individuals living in a particular enumeration district or ward are assigned the same deprivation score (see Appendix F for further details).

Dysplasia – disordered cell growth.

Endometrium – the lining of the uterus.

Germ cell – cell specialised to produce gametes (spermatozoa in males, oocytes in females).

Glioma – a tumour of the central nervous system arising from supporting cells, called glial cells, of which there are several types including astrocytes.

Grade – an estimate of the degree of malignancy of a tumour, based on the proportion of its cells which resemble the cells of origin. Grade I has the least degree of malignancy and grade IV has the greatest.

Great Britain – England including the Scilly Isles, Wales and Scotland including Orkney and Shetland (excludes the Isle of Man and the Channel Islands, which are Crown Dependencies).

Heritable – capable of being passed from one generation to the next; for example, the tendency to develop a disease can be inherited.

Histology – the study of cells and tissues at the microscopic level; in terms of cancer, the type of cell from which the tumour arises.

Histologically verified – a tumour from which a sample has been examined microscopically (also known as microscopically verified).

Incidence – the number or rate (per head of population) of new cases of a disease diagnosed in a given population during a specified time period (usually a calendar year). The crude rate is the total number of cases divided by the mid-year population, usually expressed per 100,000 population (see also age-standardised rate).

In situ – localised tumour which has not invaded surrounding tissues or spread to other parts of the body.

Invasive – tumour which has spread to surrounding tissues.

Ireland – the country of the Republic of Ireland, occupying most of the island of Ireland.

Latent period – the interval between disease initiation to onset of clinical symptoms and signs (also known as latency period).

Lead-time bias – if an individual participates in a screening programme which detects a disease earlier than it would have been detected in the absence of screening, the amount of time by which diagnosis is advanced as a result of screening is called the lead time. Since the point of diagnosis is brought forward in time, survival as measured from diagnosis is lengthened, even if total length of life is not increased.

Leukaemia – a group of cancers of the white blood cells in the bone marrow and/or the lymph nodes, classified according to whether they arise from lymphocytes (lymphocytic and lymphoblastic leukaemias) or from granulocytes (myeloid leukaemias), and according to whether they progress rapidly (acute) or slowly and intermittently (chronic).

Leukoplakia – an abnormal condition characterised by white spots or patches on mucous membranes, especially of the mouth and vulva. Also called leukoplasia.

Life tables – tables giving statistics on life expectancy of a population, based on mortality rates. Used to calculate lifetime risk and relative survival.

Lifetime risk – the chance of an individual being diagnosed with a specific type of cancer during their lifetime; usually estimated using the 'life table' method, and expressed as a percentage.

Lymphocyte – an agranulocytic leukocyte (white blood cell) that normally makes up a quarter of the white blood cell count but increases in the presence of infection.

Lymphoma – a tumour of the lymphatic system. The two main types are Hodgkin's disease (which mainly affects lymph nodes) and non-Hodgkin's lymphoma (which can affect diffuse lymphatic tissues throughout the body, as well as lymph nodes).

Male-to-female ratio – the number of cases or deaths (or the age-standardised rate) in males divided by that in females.

Malignant – tumours which grow by invasion into surrounding tissues and have the ability to metastasise to distant sites.

Menarche – the first menstrual period.

Meta-analysis – technique of synthesising research results by using various statistical methods to retrieve, select, and combine results from previous separate but related studies.

Metastases – secondary cancers, formed by the process of metastasis.

Metastasis – the spread or transfer of cancer cells from the site of the original tumour to another place in the body where a new tumour starts to form. This usually occurs by way of the bloodstream or the lymphatic system.

Misclassification – an error in the process of cancer registration whereby a primary tumour could be classified as a secondary or vice versa, or a primary tumour could be classified to the wrong ICD site code.

Morphology – in terms of cancer, the type of cell from which the tumour arises and the behaviour of the tumour (benign, *in situ*, malignant, borderline, uncertain, microinvasive).

Mortality – the number or rate (per head of population) of deaths in a given population during a specified time period (usually a calendar year). The crude rate is the total number of deaths divided by the mid-year population, usually expressed per 100,000 population (see also age-standardised rate).

Mortality-to-incidence ratio – the number of deaths (or age-standardised mortality rate) in a particular period (usually a calendar year) divided by the number of new cases (or age-standardised incidence rate) in the same period (see Appendix H for fuller explanation).

Mutagenic – capable of inducing mutation in DNA (refers mainly to extracellular factors such as X-rays or chemical pollution).

Neoplasm – a growth of abnormal tissue (also known as a tumour).

Pap smear – cells are taken from the cervix using a spatula or brush, fixed on a slide, stained with a special stain (Papanicolao stain), and examined under a microscope for abnormalities (also known as smear test, cervical smear).

Papilloma – a small benign epithelial tumour.

Parity – the number of live-born children a woman has delivered.

Prevalence – a measure of the commonality of disease, or an activity, such as smoking; for example, the number of cases of a disease present in a population at a specified time, or the proportion of the population who undertake a certain activity at a specified time.

Primary tumour – the tumour which forms where the cancer originally begins (at the primary site).

Radiotherapy – the use of high-energy radiation to treat cancer, either by directing an external beam of radiation at the affected area of the body (teletherapy) or by placing small radioactive sources inside the body, close to the location of a tumour (brachytherapy).

Relative survival – the ratio of the observed survival in the cohort being studied and the survival that would have been expected had they been subject only to the mortality rates of the general population. In other words, survival from cancer in the absence of other causes of death. The expected survival is derived from a life table.

Risk – the proportion of people in a population who develop a disease within a specified time period.

Risk factor – a variable associated with an increased risk of disease, not necessarily causal. Risk factors are evaluated by comparing the risk of disease in those exposed to the potential risk factor with the risk of disease in those not exposed (relative risk).

Sarcoma – a malignant tumour of connective tissue: bone, muscle, blood vessel, cartilage, fat, or fibrous tissue.

Screening – the routine examination of apparently healthy individuals in order to detect a particular disease at an early stage, before it becomes clinically symptomatic.

Secondary tumours – tumours formed at sites distant from the site of the original tumour (also known as metastases).

Site – the anatomical location of a tumour, as specified by the ICD code.

Socio-economic gradient – measure of the relationship between incidence, mortality or survival rates and levels of socio-economic deprivation. For cancer there is commonly a positive gradient (incidence rates are higher in more deprived groups for the smoking-related cancers, for example), although for some cancers there is a negative gradient (incidence rates are higher in more affluent groups for breast and prostate cancers). It is usually a marker for some risk factor associated with deprivation.

Squamous cell – type of epithelial cell found in many parts of the body, for example lungs, kidneys, mouth, oesophagus, and skin.

Stage – a measure of the size and extent of a tumour at the time of diagnosis. The TNM system is the principal staging system for most types of cancer, and is based on the extent of the primary tumour (T), the extent of lymph node involvement (N), and the presence or absence of metastases (M).

Sub-site – more specific anatomical location of a tumour, as specified by the fourth digit of the ICD code.

Transitional cell – type of epithelial cell which forms the lining of the bladder and urinary tract.

Tumour – a mass of abnormal tissue, the growth of which exceeds and is uncoordinated with the normal tissue from which it originates, and which persists in the same excessive manner after the stimuli which evoked the change have ceased (also known as a neoplasm).

United Kingdom (UK) – England including the Scilly Isles, Wales, Scotland including Orkney and Shetland, and Northern Ireland (excludes the Isle of Man and the Channel Islands, which are Crown Dependencies).